DATE DUE

Spectra of Atoms and Molecules

Topics in Physical Chemistry

Series Editor, Donald G. Truhlar

Spectra of Atoms and Molecules
Peter F. Bernath

Electronic Structure Calculations on Fullerenes and Their Derivatives
Jerzy Cioslowski

Algebraic Theory of Molecules
F. Iachello and R. D. Levine

Quantum Mechanics in Chemistry
Jack Simons and Jeff Nichols

R

Spectra of Atoms and Molecules

Peter F. Bernath

New York Oxford
OXFORD UNIVERSITY PRESS
1995

Oxford University Press

Oxford New York Toronto
Delhi Bombay Calcutta Madras Karachi
Kuala Lumpur Singapore Hong Kong Tokyo
Nairobi Dar es Salaam Cape Town
Melbourne Auckland Madrid

and associated companies in

Berlin Ibadan

Published by Oxford University Press, Inc.,
198 Madison Avenue, New York, New York 10016-4314

Library of Congress Cataloging-in-Publication Data
Bernath, Peter F.
Spectra of atoms and molecules/Peter F. Bernath.
p. cm.—(Topics in physical chemistry)
Includes bibliographical references and index.
ISBN 0-19-507598-6
1. Atomic spectroscopy.
2. Molecular spectroscopy.
3. Spectrum analysis.
I. Title. II. Series: Topics in physical chemistry series.
QC454.A8B47 1995
543′.0858—dc20 93—48550

9 8 7 6 5 4 3 2

Printed in the United States of America
on acid-free paper

For Robin, Elizabeth, and Victoria

Preface

This book is designed as a textbook to introduce advanced undergraduates and, particularly, new graduate students to the vast field of spectroscopy. It presumes that the student is familiar with the material in an undergraduate course in quantum mechanics. I have taken great care to review the relevant mathematics and quantum mechanics as needed throughout the book. Considerable detail is provided on the origin of spectroscopic principles. My goal is to demystify spectroscopy by showing the necessary steps in a derivation, as appropriate in a textbook.

The digital computer has permeated all of science including spectroscopy. The application of simple analytical formulas and the nonstatistical graphical treatment of data are long dead. Modern spectroscopy is based on the matrix approach to quantum mechanics. Real spectroscopic problems can be solved on the computer more easily if they are formulated in terms of matrix operations rather than differential equations. I have tried to convey the spirit of modern spectroscopy, through the extensive use of the language of matrices.

The infrared and electronic spectroscopy of polyatomic molecules makes extensive use of group theory. Rather than assume a previous exposure or try to summarize group theory in a short chapter, I have chosen to provide a more thorough introduction. My favorite book on group theory is the text by Bishop, *Group Theory and Chemistry,* and I largely follow his approach to the subject.

This book is not a monograph on spectroscopy, but it can be profitably read by physicists, chemists, astronomers, and engineers who need to become acquainted with the subject. Some topics in this book, such as parity, are not discussed well in any of the textbooks or monographs that I have encountered. I have tried to take particular care to address the elementary aspects of spectroscopy that students have found to be most confusing.

To the uninitiated, the subject of spectroscopy seems enshrouded in layers of bewildering and arbitrary notation. Spectroscopy has a long tradition so many of the symbols are rooted in history and are not likely to change. Ultimately all notation is arbitrary, although some notations are more helpful than others. One of the goals of this book is to introduce the language of spectroscopy to the new student of the subject. Although the student may not be happy with some aspects of spectroscopic notation, it is easier to adopt the notation than to try to change long-standing spectroscopic habits.

The principles of spectroscopy are timeless, but spectroscopic techniques are more transient. Rather than focus on the latest methods of recording spectra (which will be out of fashion tomorrow), I concentrate on the interpretation of the spectra themselves. This book attempts to answer the question: What information is encoded in the spectra of atoms and molecules?

A scientific subject cannot be mastered without solving problems. I have therefore provided many spectroscopic problems at the end of each chapter. These problems have been acquired over the years from many people, including M. Barfield, S. Kukolich, R. W. Field, and F. McCourt. In addition I have "borrowed" many problems either directly or with only small changes from many of the books listed as general references at the end of each chapter and from the books listed in Appendix D. I thank these people and apologize for not giving them more credit!

Spectroscopy needs spectra and diagrams to help interpret the spectra. Although the ultimate analysis of a spectrum may involve the fitting of line positions and intensities with a computer program, there is much qualitative information to be gained by the inspection of a spectrum. I have therefore provided many spectra and diagrams in this book. In addition to the specific figure acknowledgments at the end of the appendices, I would like to thank a very talented group of undergraduates for their efforts. J. Ogilvie, K. Walker, R. LeBlanc, A. Billyard, and J. Dietrich are responsible for the creation of most of the figures in this book.

I also would like to thank the many people who read drafts of the entire book or of various chapters. They include F. McCourt, M. Dulick, D. Klapstein, R. Le Roy, N. Isenor, D. Irish, M. Morse, C. Jarman, J. Ogilvie, P. Colarusso, R. Bartholomew, and C. Zhao. Their comments and corrections were very helpful. Please contact me about other errors in the book and with any comments you would like to make. I thank Heather Hergott for an outstanding job typing the manuscript.

Finally, I thank my wife Robin for her encouragement and understanding. Without her this book would never have been written.

Ontario P. F. B.
January 1994

Contents

Spectra of Atoms and Molecules

Chapter 1
Introduction

Waves, Particles, and Units

Spectroscopy is the study of the interaction of light with matter. To begin, a few words about light, matter, and the effect of light on matter are in order.

Light is an electromagnetic wave represented (for the purposes of this book) by the plane waves

$$\mathbf{E}(\mathbf{r}, t) = \mathbf{E}_0 \cos(\mathbf{k} \cdot \mathbf{r} - \omega t + \phi_0) \tag{1.1}$$

or

$$\mathbf{E}(\mathbf{r}, t) = \mathbf{E}_0 e^{i(\mathbf{k} \cdot \mathbf{r} - \omega t + \phi_0)}. \tag{1.2}$$

Note that the real form, equation (1.1), and the complex form, equation (1.2), are different mathematical representations of the same physical plane wave. Either the real or complex form can be chosen for convenience, although implicitly only the real part of equation (1.2) has a physical interpretation. In this book vectors and matrices are written in bold type. There is an electric field $\mathbf{E}$ (in volts per meter) perpendicular to $\mathbf{k}$ that propagates in the direction $\mathbf{k}$ and has an angular frequency $\omega = 2\pi\nu = 2\pi/T$. The frequency ν (in hertz) is the reciprocal of the period T (in seconds), that is, $\nu = 1/T$. The period T and the wavelength λ are defined in Figure 1.1 with $\mathbf{k}$ in the z direction. The wavevector $\mathbf{k}$ has a magnitude $|\mathbf{k}| = k = 2\pi/\lambda$ and a direction given by the normal to the plane of constant phase. $|\mathbf{E}_0|$ is the amplitude of the electric field, while $\mathbf{k} \cdot \mathbf{r} - \omega t + \phi_0$ is the phase (ϕ_0 is an initial phase angle of arbitrary value).

The presence of a magnetic field, also oscillating at angular frequency ω and orthogonal to both $\mathbf{E}$ and $\mathbf{k}$, is ignored in this book. Other "complications" such as Maxwell's equations, Gaussian laser beams, birefringence, and vector potentials are also not considered. These subjects, although part of spectroscopy in general, are discussed in books on optics, quantum optics, lasers, or electricity and magnetism.

Wavelength and frequency are related by the equation $\lambda\nu = c$, in which c is the speed of the electromagnetic wave. In vacuum $c = c_0$, but in general $c = c_0/n$ with n as the index of refraction of the propagation medium. Since ν has the same value in any medium, the wavelength also depends on the index of refraction of the medium. Thus since

$$\lambda\nu = c \tag{1.3}$$

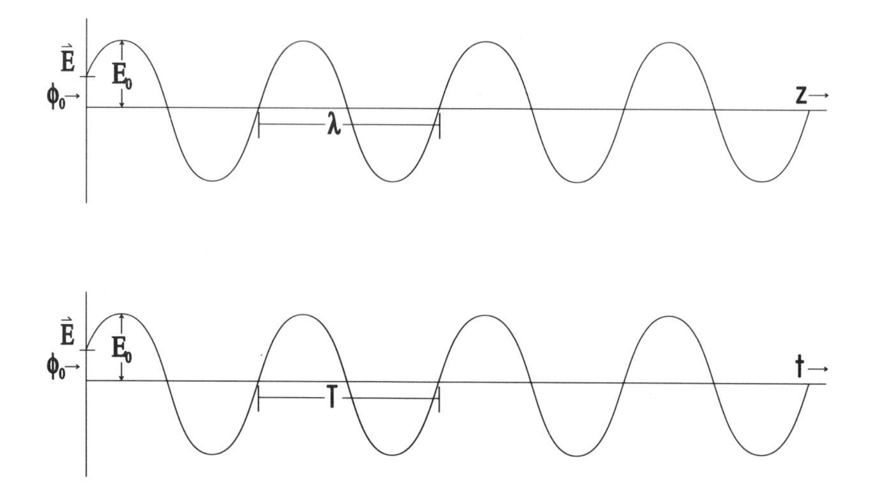

Figure 1.1. The electric field at $t = 0$ as a function of z is plotted in the upper panel, while the lower panel is the corresponding plot at $z = 0$ as a function of time.

we must have

$$\frac{\lambda_0}{n}\nu = \frac{c_0}{n}. \qquad (1.4)$$

Historically, direct frequency measurements were not possible in the infrared and visible regions of the spectrum. It was therefore convenient to measure (and report) λ in air, correct for the refractive index of air to give λ_0, and then define $\tilde{\nu} = 1/\lambda_0$. Before SI units were adopted, the centimeter was more widely used than the meter so that $\tilde{\nu}$ represents the number of wavelengths in one centimeter in vacuum and as a consequence $\tilde{\nu}$ is called the wavenumber. The units of the wavenumber are cm^{-1} (reciprocal centimeters) but common usage also calls the "cm^{-1}" the "wavenumber." Fortunately, the SI unit for $\tilde{\nu}$, the m^{-1}, is almost never used so that this sloppy, but standard, practice causes no confusion.

The oscillating electric field is a function of both spatial (**r**) and temporal (t) variables. If the direction of propagation of the electromagnetic wave is along the z-axis and the wave is examined at one instant of time, say $t = 0$, then for $\phi_0 = 0$,

$$\mathbf{E} = \mathbf{E}_0 \cos(kz) = \mathbf{E}_0 \cos\frac{2\pi z}{\lambda}. \qquad (1.5)$$

Alternatively, the wave can be observed at a single point, say $z = 0$, as a function of time

$$\mathbf{E} = \mathbf{E}_0 \cos(\omega t) = \mathbf{E}_0 \cos(2\pi\nu t). \qquad (1.6)$$

Both equations (1.5) and (1.6) are plotted in Figure 1.1 with arbitrary initial phases.

In contrast to longitudinal waves, such as sound waves, electromagnetic waves are transverse waves. If the wave propagates in the z direction, then there are two possible independent transverse directions, x and y. This leads to the polarization of light, since $\mathbf{E}$ could lie either along x or along y, or more generally, it could lie anywhere in the xy plane. Therefore we may write

$$\mathbf{E} = E_x\hat{\mathbf{i}} + E_y\hat{\mathbf{j}}, \tag{1.7}$$

with $\hat{\mathbf{i}}$ and $\hat{\mathbf{j}}$ representing unit vectors lying along the x and y axes.

The wave nature of light became firmly established in the nineteenth century, but by the beginning of the twentieth century, light was also found to have a particle aspect. The wave–particle duality of electromagnetic radiation is difficult to visualize since there are no classical, macroscopic analogs. In the microscopic world, electromagnetic waves seem to guide photons (particles) of a definite energy E and momentum p with

$$E = h\nu = \hbar\omega = \frac{hc}{\lambda} = hc\tilde{\nu} \tag{1.8}$$

and

$$p = \frac{h}{\lambda} = \hbar k. \tag{1.9}$$

In 1924 it occurred to de Broglie that if electromagnetic waves could display properties associated with particles, then perhaps particles could also display wavelike properties. Using equation (1.9), he postulated that a particle should have a wavelength,

$$\lambda = \frac{h}{p} = \frac{h}{mv}. \tag{1.10}$$

This prediction of de Broglie was verified in 1927 by Davisson and Germer's observation of an electron beam diffracted by a nickel crystal.

In this book SI units and expressions are used as much as possible, with the traditional spectroscopic exceptions of the angstrom (Å) and the wavenumber (cm^{-1}). The symbols and units used will largely follow the International Union of Pure and Applied Chemistry (IUPAC) recommendations of the "Green Book" by I. Mills et al.[1] The fundamental physical constants, as supplied in Appendix A, are the 1986 Cohen and Taylor[2] values. The value of h changed substantially in this revision of the constants, and the speed of light in vacuum (c_0) was fixed exactly at 299,792,458 m/s. The atomic masses used are taken from the Green Book, and they correspond to the 1983 Wapstra and Audi[3] values.

The Electromagnetic Spectrum

There are traditional names associated with the various regions of the electromagnetic spectrum. The radio frequency region (3 MHz–3 GHz) has photons of sufficient energy to flip nuclear spins [nuclear magnetic resonance (NMR)] in magnetic fields of a few tesla ($1\,T = 10^4$ gauss). In the microwave region (3 GHz–3000 GHz) energies correspond to rotational transitions in molecules and to electron spin flips [electron spin resonance (ESR)]. Unlike all the spectra discussed in this book, NMR and ESR transitions are induced by the oscillating magnetic field of the electromagnetic radiation. Infrared quanta ($100\,cm^{-1}$–$13{,}000\,cm^{-1}$) excite the vibrational motion in matter. Visible and ultraviolet (uv) transitions (10,000 Å–100 Å) involve valence electron rearrangements in molecules (1 nm = 10 Å). Core electronic transitions are promoted at x-ray wavelengths (100 Å–0.1 Å). Finally, below 0.1 Å in wavelength, γ-rays are associated with nuclear processes. Chemists customarily use the units of MHz or GHz for radio and microwave radiation, cm^{-1} for infrared radiation, and Å or nm for visible, uv, and x-ray radiation (Figure 1.2). These customary units are units of frequency (MHz), inverse wavelength (cm^{-1}), and wavelength (Å).

It is worth noting that the different regions of the spectrum do not possess sharp borders and that the type of molecular motion associated with spectroscopy in each region is only approximate. For example, overtone vibrational spectra can be found in the visible region of the spectrum (causing the blue color of the oceans). Infrared electronic transitions are also not rare, for example, the A–X electronic transitions of CN and C_2.

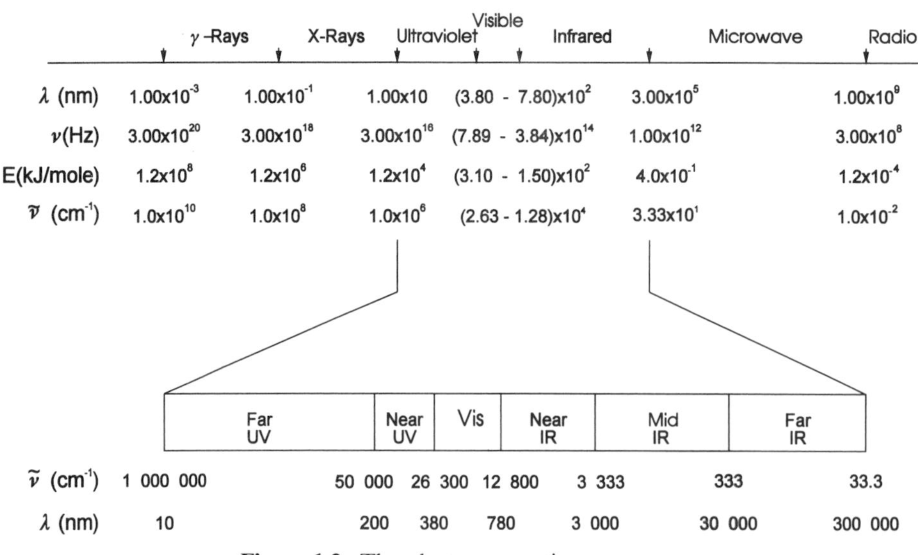

Figure 1.2. The electromagnetic spectrum.

A further subdivision of the infrared, visible and ultraviolet regions of the spectrum is customary. The infrared region is divided into the far-infrared (33–333 cm^{-1}), mid-infrared (333–3333 cm^{-1}), and near-infrared (3333–13,000 cm^{-1}) regions. In the far-infrared region are found rotational transitions of light molecules, phonons of solids, and metal-ligand vibrations, as well as ring-puckering and torsional motions of many organic molecules. The mid-infrared is the traditional infrared region in which the fundamental vibrations of most molecules lie. The near-infrared region is associated with overtone vibrations and a few electronic transitions. The visible region is divided into the colors of the rainbow from the red limit at about 7800 Å to the violet at 4000 Å. The near-ultraviolet region covers 4000 Å–2000 Å, while the vacuum ultraviolet region is 2000 Å–100 Å. This latter region is so named because air is opaque below 2000 Å, so that only evacuated instruments can be used when spectra are taken in this region.

It is a spectroscopic custom to report all infrared, visible and near-ultraviolet wavelengths as air wavelengths (λ), rather than as vacuum wavelengths (λ_0). Of course, below 2000 Å all wavelengths are vacuum wavelengths since measurements in air are not possible. The wavenumber is related to energy, $E = hc\tilde{\nu}$, and is the reciprocal of the vacuum wavelength in centimeters, $\tilde{\nu} = 1/\lambda_0(\text{cm}) \neq 1/\lambda$. For accurate work, it is often necessary to correct for the refractive index of air. This can be seen, for example, by considering dry air at 15°C and 760 Torr for which $n = 1.0002781$ at 5000 Å.[4] Thus $\lambda = 5000.000$ Å in air corresponds to $\lambda_0 = 5001.391$ Å in vacuum and $\tilde{\nu} =$ 19,994.44 cm^{-1} rather than 20,000 cm^{-1}!

Interaction of Radiation with Matter

Blackbody Radiation

The spectrum of the radiation emitted by a blackbody is important both for historical reasons and for practical applications. Consider a cavity (Figure 1.3) in a material that is maintained at constant temperature T. The emission of radiation from the cavity walls is in equilibrium with the radiation that is absorbed by the walls. It is convenient to define a radiation density ρ (with units of joules/m^3) inside the cavity. The

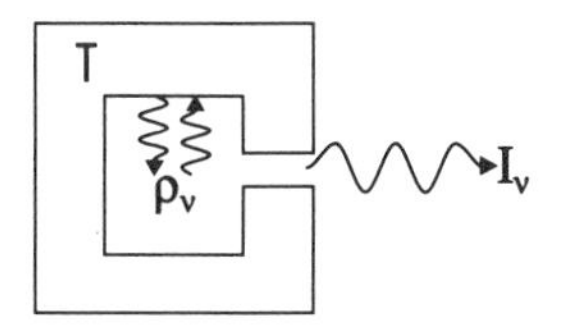

Figure 1.3. Cross section of a blackbody cavity at a temperature T with a radiation density ρ_ν emitting radiation with intensity I_ν from a small hole.

frequency distribution of this radiation is represented by the function ρ_ν, which is the radiation density in the frequency interval between ν and $\nu + d\nu$ defined such that

$$\rho = \int_0^\infty \rho_\nu \, d\nu. \tag{1.11}$$

Therefore, the energy density function ρ_ν has units of joule-seconds per cubic meter ($\mathrm{J\,s\,m^{-3}}$). The distribution function characterizing the intensity of the radiation emitted from the hole is labeled I_ν (units of watt-seconds per square meter). The functions ρ_ν and I_ν are universal functions depending only on the temperature and frequency, and are independent of the shape or size of the cavity and of the material of construction as long as the hole is small.

Planck derived the universal function,

$$\rho_\nu(T) = \frac{8\pi h\nu^3}{c^3} \frac{1}{e^{h\nu/kT} - 1}, \tag{1.12}$$

named in his honor. Geometrical considerations give the relationship between I_ν and ρ_ν as

$$I_\nu = \rho_\nu \frac{c}{4}. \tag{1.13}$$

Figure 1.4 shows ρ_ν as a function of ν and the dependence on the temperature T.

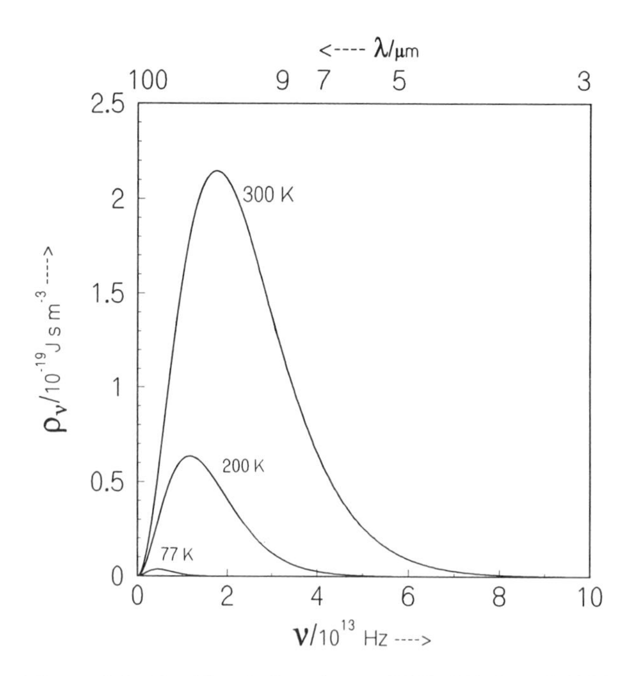

Figure 1.4. The Planck function at 77 K, 200 K, and 300 K.

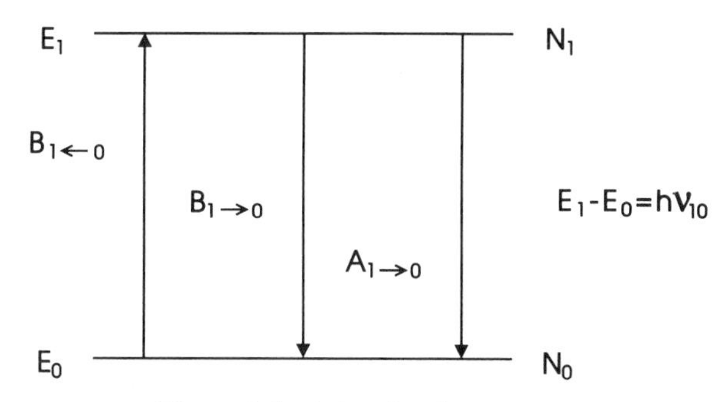

Figure 1.5. A two-level system.

Einstein A and B Coefficients

Consider a collection of N two-level systems (Figure 1.5) in a volume of $1\ \mathrm{m}^3$ with upper energy E_1 and lower energy E_0, all at a constant temperature T and bathed by the radiation density $\rho_\nu(T)$. Since the entire collection is in thermal equilibrium, if the number of systems with energy E_1 is N_1 and the number of systems with energy E_0 is N_0, then the populations N_1 and N_0 ($N = N_1 + N_0$) are necessarily related by

$$\frac{N_1}{N_0} = e^{-h\nu_{10}/kT}, \tag{1.14}$$

in which $h\nu_{10} = E_1 - E_0$. This is the well-known Boltzmann expression for thermal equilibrium.

There are three possible processes that can change the state of the system from E_0 to E_1 or from E_1 to E_0: absorption, spontaneous emission, and stimulated emission (Figure 1.6). Absorption results from the presence of a radiation density $\rho_\nu(\nu_{10})$ of the precise frequency needed to drive a transition from the ground state to the excited state at the rate

$$\frac{dN_1}{dt} = B_{1\leftarrow 0}\rho_\nu(\nu_{10})N_0. \tag{1.15}$$

The coefficient $B_{1\leftarrow 0}$ is thus a "rate constant" known as the Einstein absorption coefficient or Einstein B coefficient. Similarly if the system is already in an excited state, then a photon of energy $h\nu_{10}$ (provided by ρ_ν) can induce the system to make the transition to the ground state. The rate for stimulated emission is given by

$$\frac{dN_1}{dt} = -B_{1\rightarrow 0}\rho_\nu(\nu_{10})N_1, \tag{1.16}$$

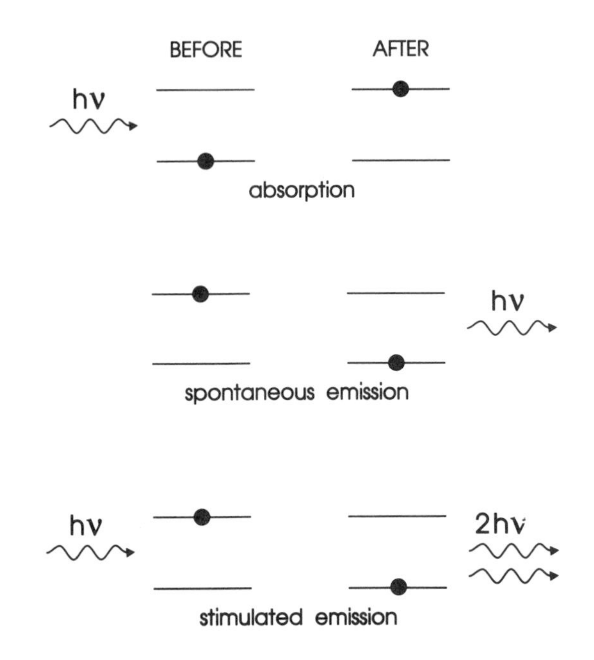

Figure 1.6. Schematic representations of absorption (top), spontaneous emission (middle), and stimulated emission (bottom) processes in a two-level system.

in which $B_{1\to 0}$ is the stimulated emission coefficient. Finally the system in the excited state can spontaneously emit a photon at a rate

$$\frac{dN_1}{dt} = -A_{1\to 0}N_1. \tag{1.17}$$

Since the system is at equilibrium, the rate of population of the excited state by absorption must balance the rate of depopulation by stimulated and spontaneous emission, so that

$$N_0 B_{1\leftarrow 0}\rho_\nu = A_{1\to 0}N_1 + B_{1\to 0}\rho_\nu N_1 \tag{1.18}$$

and hence

$$\frac{N_1}{N_0} = \frac{B_{1\leftarrow 0}\rho_\nu}{A_{1\to 0} + B_{1\to 0}\rho_\nu} = e^{-h\nu_{10}/kT}. \tag{1.19}$$

Solving for ρ_ν in equation (1.19) then yields

$$\rho_\nu(\nu_{10}) = \frac{A_{1\to 0}}{B_{1\leftarrow 0}e^{h\nu_{10}/kT} - B_{1\to 0}}. \tag{1.20}$$

However, $\rho_\nu(\nu_{10})$ is also given by the Planck function (1.12)

$$\rho_\nu(\nu_{10}) = \frac{8\pi h\nu_{10}^3}{c^3}\frac{1}{e^{h\nu_{10}/kT}-1}.$$

For expressions (1.12) and (1.20) both to be valid, it is necessary that

$$B_{1\leftarrow 0} = B_{1\rightarrow 0} \tag{1.21}$$

and that

$$A_{1\rightarrow 0} = \frac{8\pi h\nu_{10}^3}{c^3}B_{1\leftarrow 0}. \tag{1.22}$$

Remarkably, the rate constants for absorption and stimulated emission—two apparently different physical processes—are identical. Moreover, the spontaneous emission rate (lifetime) can be determined from the absorption coefficient (1.22). Note, however, the ν_{10}^3 factor (1.22), which plays an important role in the competition between the induced and spontaneous emission processes.

Absorption and Emission of Radiation

The interaction of electromagnetic radiation with matter can be described by a simple semiclassical model. In the semiclassical treatment the energy levels of molecules are obtained by solution of the time-independent Schrödinger equation

$$\hat{H}\psi_n = E_n\psi_n,$$

while the electromagnetic radiation is treated classically. Consider a two-level system described by lower and upper state wavefunctions, ψ_0 and ψ_1 (Figure 1.7), respectively.

Electromagnetic radiation that fulfills the Bohr condition, $E_1 - E_0 = h\nu = \hbar\omega$, is applied to the system in order to induce a transition from the lower energy state at E_0

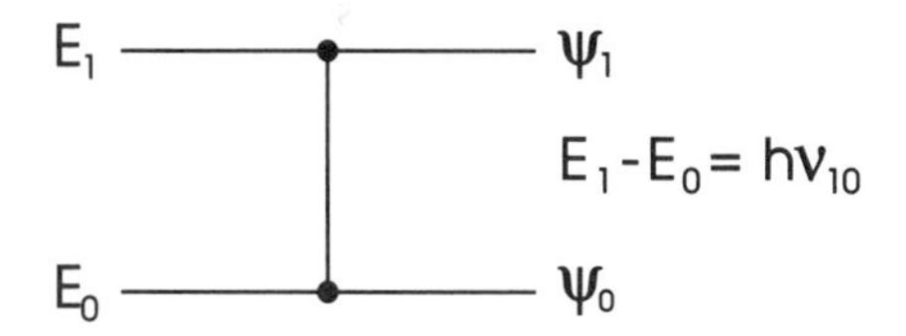

Figure 1.7. Two-level system.

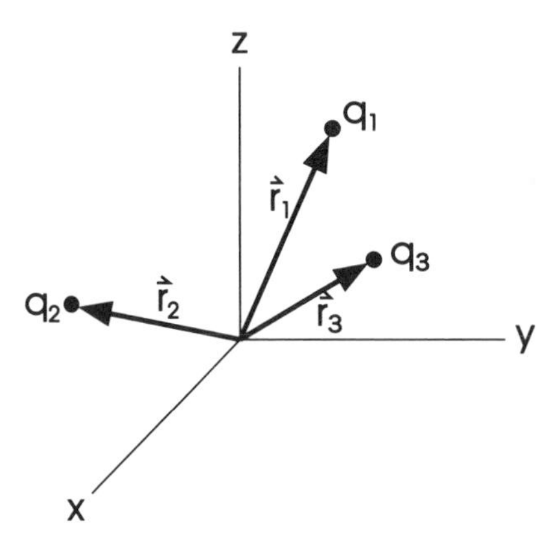

Figure 1.8. Three particles with charges q_1, q_2, and q_3 located at positions $\mathbf{r}_1$, $\mathbf{r}_2$, and $\mathbf{r}_3$.

to the upper energy state at E_1. The molecule consists of nuclei and electrons at positions $\mathbf{r}_i$ possessing charges q_i. The system as a whole thus has a net dipole moment $\boldsymbol{\mu}$ with Cartesian components

$$\mu_x = \sum x_i q_i, \tag{1.23}$$

$$\mu_y = \sum y_i q_i, \tag{1.24}$$

$$\mu_z = \sum z_i q_i, \tag{1.25}$$

where x, y, and z are the coordinates of the particles relative to the center of mass of the molecule (Figure 1.8).

The interaction of the radiation with the material system is taken into account by the addition of the time-dependent perturbation (see the first section of this chapter and of Chapter 4),

$$\begin{aligned} \hat{H}' &= -\boldsymbol{\mu} \cdot \mathbf{E}(t) \\ &= -\boldsymbol{\mu} \cdot \mathbf{E}_0 \cos(\mathbf{k} \cdot \mathbf{r} - \omega t). \end{aligned} \tag{1.26}$$

If the oscillating electric field is in the z direction and the system is at the origin, $\mathbf{r} = 0$ ($\lambda \gg$ dimensions of the system to avoid having different electric field strengths at different parts of the molecule), then

$$\hat{H}' = -\mu_z E_{0z} \cos(\omega t) \tag{1.27}$$

$$= -\mu E \cos(\omega t). \tag{1.28}$$

The transition probability is obtained by solving the time-dependent Schrödinger equation

$$i\hbar \frac{\partial \Psi}{\partial t} = [\hat{H} + \hat{H}'(t)]\Psi. \tag{1.29}$$

In the absence of $\hat{H}'$, the two time-dependent solutions of equation (1.29) are

$$\Psi_0 = \psi_0 e^{-iE_0 t/\hbar} = \psi_0 e^{-i\omega_0 t}$$

$$\Psi_1 = \psi_1 e^{-iE_1 t/\hbar} = \psi_1 e^{-i\omega_1 t},$$

with $\omega_i = E_i/\hbar$.

The wavefunction for the perturbed system is given by the linear combination of the complete set of functions Ψ_0 and Ψ_1:

$$\Psi(t) = a_0(t)\psi_0 e^{-iE_0 t/\hbar} + a_1(t)\psi_1 e^{-iE_1 t/\hbar} = a_0\psi_0 e^{-i\omega_0 t} + a_1\psi_1 e^{-i\omega_1 t}, \tag{1.30}$$

where a_0 and a_1 are time-dependent coefficients. Substitution of the solution (1.30) into the time-dependent Schrödinger equation (1.29) leads to the equation

$$i\hbar(\dot{a}_0\psi_0 e^{-i\omega_0 t} + \dot{a}_1\psi_1 e^{-i\omega_1 t}) = \hat{H}'a_0\psi_0 e^{-i\omega_0 t} + \hat{H}'a_1\psi_1 e^{-i\omega_1 t}, \tag{1.31}$$

where the dot notation $\dot{a}_0 = da_0/dt$ is used to indicate derivatives with respect to time.

Multiplication by $\psi_0^* e^{i\omega_0 t}$, or $\psi_1^* e^{i\omega_1 t}$, followed by integration over all space then gives two coupled differential equations

$$i\hbar\dot{a}_0 = a_0\langle\psi_0|\,\hat{H}'\,|\psi_0\rangle + a_1\langle\psi_0|\,\hat{H}'\,|\psi_1\rangle e^{-i\omega_{10}t} \tag{1.32a}$$

$$i\hbar\dot{a}_1 = a_0\langle\psi_1|\,\hat{H}'\,|\psi_0\rangle e^{i\omega_{10}t} + a_1\langle\psi_1|\,\hat{H}'\,|\psi_1\rangle \tag{1.32b}$$

using the Dirac bracket notation $\langle f_1|\,\hat{A}\,|f_3\rangle = \int f_1^*\hat{A}f_3\,d\tau$. At this stage no approximations have been made (other than the restriction to the two states ψ_1 and ψ_0) and the two equations (1.32) are entirely equivalent to the original Schrödinger equation. Now if $\hat{H}'$ is taken as $-\mu E\cos(\omega t)$ in the electric-dipole approximation, then the time-dependent perturbation $\hat{H}'$ has odd parity (see the fifth section of Chapter 5 and the fourth section of Chapter 9). In other words, H' is an odd function since $\mu = ez$, while the products $|\psi_1|^2$ or $|\psi_0|^2$ are even functions so that the integrands $\psi_1^*\hat{H}'\psi_1$ and $\psi_0^*\hat{H}'\psi_0$ are also odd functions. All atomic and molecular states also have a definite parity (either even or odd) with respect to inversion in the space-fixed coordinate system (see the fifth section of Chapter 5 and the fourth section of Chapter 9) so that $\langle\psi_0|\,\hat{H}'\,|\psi_0\rangle = \langle\psi_1|\,\hat{H}'\,|\psi_1\rangle = 0$ and equations (1.32) reduce to

$$i\hbar\dot{a}_0 = -a_1 M_{01} E e^{-i\omega_{10}t}\cos\omega t, \tag{1.33a}$$

$$i\hbar\dot{a}_1 = -a_0 M_{01} E e^{i\omega_{10}t}\cos\omega t. \tag{1.33b}$$

The integral $M_{01} = M_{10} = \langle\psi_1|\,\mu\,|\psi_0\rangle$ is the transition dipole moment and is the most critical factor in determining selection rules and line intensities. In general $\mathbf{M}_{10}$ is a vector quantity and the symbol $\boldsymbol{\mu}_{10}$ ($\equiv \mathbf{M}_{10}$) is often used. It is convenient to define

$$\omega_R = \frac{M_{10}E}{\hbar} \tag{1.34}$$

as the Rabi frequency and use the identity

$$\cos(\omega t) = \frac{e^{i\omega t} + e^{-i\omega t}}{2}$$

to rewrite equations (1.33) as

$$\dot{a}_0 = \frac{ia_1\omega_R(e^{-i(\omega_{10}-\omega)t} + e^{-i(\omega_{10}+\omega)t})}{2} \tag{1.35a}$$

$$\dot{a}_1 = \frac{ia_0\omega_R(e^{i(\omega_{10}-\omega)t} + e^{i(\omega_{10}+\omega)t})}{2}. \tag{1.35b}$$

The physical meaning of the Rabi frequency becomes clear later in this section.

At this stage an approximation can be made by noting that $\omega_{10} \approx \omega$ since the system with Bohr frequency $(E_1 - E_0)/\hbar = \omega_{10}$ is resonant or nearly resonant with the optical angular frequency $\omega = 2\pi\nu$. The terms $e^{i(\omega_{10}-\omega)t}$ and $e^{-i(\omega_{10}-\omega)t}$ thus represent slowly varying functions of time compared to the rapidly oscillating nonresonant terms $e^{i(\omega_{10}+\omega)t}$ and $e^{-i(\omega_{10}+\omega)t}$.

In what is known as the rotating wave approximation the nonresonant high-frequency terms can be neglected because their effects essentially average to zero since they are rapidly oscillating functions of time. Defining $\Delta = \omega - \omega_{10}$, equations (1.35a) and (1.35b) become

$$\dot{a}_0 = \frac{i\omega_R e^{i\Delta t} a_1}{2} \tag{1.36a}$$

$$\dot{a}_1 = \frac{i\omega_R e^{-i\Delta t} a_0}{2}. \tag{1.36b}$$

The equations (1.36a) and (1.36b) can be solved analytically. The difference $\Delta = \omega - \omega_{10}$ is often referred to as the *detuning frequency* since it measures how far the electromagnetic radiation of frequency ω is tuned away from the resonance frequency ω_{10}. The solution (see Problem 14) to these two simultaneous first-order differential equations with initial conditions $a_0(0) = 1$ and $a_1(0) = 0$ for the system initially in the ground state at $t = 0$ is

$$a_0(t) = \left[\cos\left(\frac{\Omega t}{2}\right) - i\left(\frac{\Delta}{\Omega}\right)\sin\left(\frac{\Omega t}{2}\right)\right]e^{i\Delta t/2} \tag{1.37}$$

and

$$a_1(t) = i\left(\frac{\omega_R}{\Omega}\right)\sin\left(\frac{\Omega t}{2}\right)e^{-i\Delta t/2}, \tag{1.38}$$

where $\Omega = [(\omega_R)^2 + \Delta^2]^{1/2}$. These solutions can be checked by substitution into equations (1.36).

The time-dependent probability that the system will be found in the excited state is given by

$$|a_1(t)|^2 = \frac{\omega_R^2}{\Omega^2}\sin^2\left(\frac{\Omega t}{2}\right), \tag{1.39}$$

while the corresponding time-dependent probability that the system will be found in the ground state is

$$|a_0|^2 = 1 - |a_1|^2 = 1 - \frac{\omega_R^2}{\Omega^2}\sin^2\left(\frac{\Omega t}{2}\right). \tag{1.40}$$

At resonance $\Delta = 0$ and $\Omega = \omega_R$ so that in this case

$$|a_1|^2 = \sin^2\left(\frac{\omega_R t}{2}\right) \tag{1.41}$$

$$|a_0|^2 = 1 - \sin^2\left(\frac{\omega_R t}{2}\right) = \cos^2\left(\frac{\omega_R t}{2}\right). \tag{1.42}$$

The transition probability $|a_1|^2$ is plotted in Figure 1.9 for various detuning frequencies.

The meaning of the Rabi frequency becomes clear from equations (1.34), (1.41), and Figure 1.9. The system is coherently cycled (i.e., no abrupt changes in the phases or amplitudes of the wavefunctions) between the ground and the excited states by the

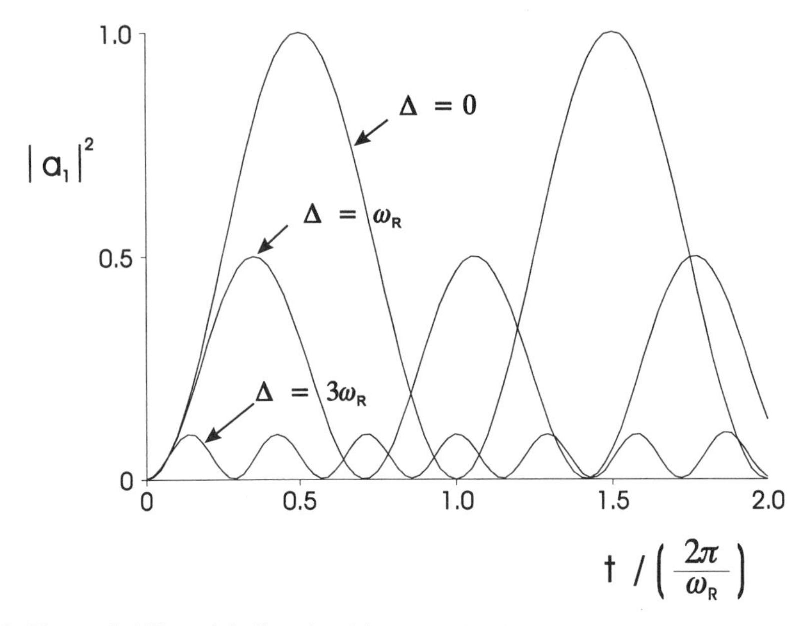

Figure 1.9. The probability of finding the driven two-level system in the excited state for three detunings: $\Delta = 0$, $\Delta = \omega_R$, and $\Delta = 3\omega_R$.

electromagnetic radiation. At resonance the system is completely inverted after a time $t_\pi = \pi/\omega_R$, while off-resonance there is a reduced probability for finding the system in the excited state.

This simple picture of a coherently driven system has ignored all decay processes such as spontaneous emission from the excited state. Spontaneous emission of a photon would break the coherence of the excitation and reset the system to the ground state (this is referred to as a T_1 process). Similarly, collisions can also cause relaxation in the system. In fact, collisions can reset the phase of the atomic or molecular wavefunction (only the relative phases of ψ_1 and ψ_0 are important) without changing any of the populations (this is referred to as a T_2 process). These phase-changing collisions also interrupt the coherent cycling of the system. Such processes were first studied in NMR (which is the source of the names T_1 and T_2 processes) and are now extensively studied in the field of quantum optics.

The effect of collisions and other relaxation phenomena is to damp out the coherent cycling of the excited system (called *Rabi oscillations*). However, Rabi oscillations can be observed in any quantum system simply by increasing the intensity of the radiation. This increases the applied electric field **E** so that at some point the Rabi cycling frequency exceeds the relaxation frequency, $\omega_R \gg \omega_{\text{relaxation}}$, and coherent behavior will be observed. This is easily achieved in NMR where spin relaxation processes are slow and many watts of radio frequency power can be applied to the system. In the infrared and visible region of the spectrum relaxation processes are much faster and Rabi oscillations are normally damped. For example, a real system would oscillate briefly when a strong field is applied suddenly to it, but it soon loses coherence and saturates

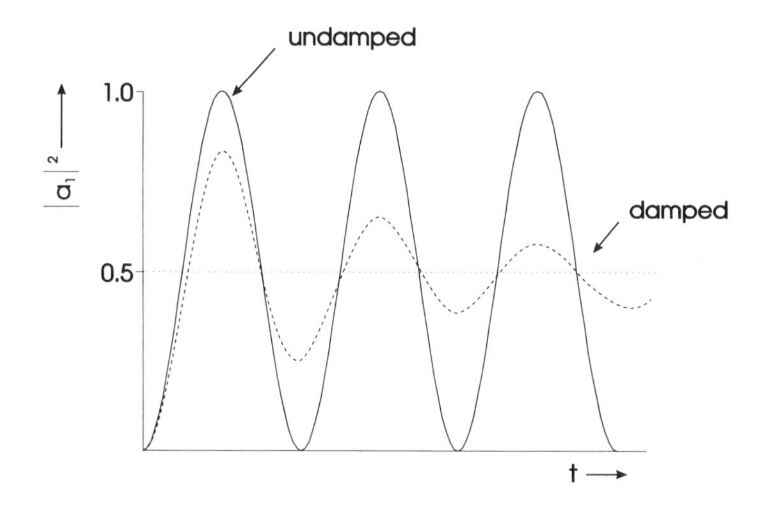

Figure 1.10. The driven two-level system saturates when relaxation processes (damping) are included.

(Figure 1.10). When the system is saturated, half of the molecules in the system are in the lower state and half are in the upper state. The rate of stimulated emission (down) matches the rate of absorption (up).

For example, consider a 1-W laser beam, 1 mm in diameter interacting with a two-level system that has a transition dipole moment of 1 debye ($1\,\mathrm{D} = 3.33564 \times 10^{-30}\,\mathrm{C\,m}$). What is the Rabi frequency? The intensity of the laser beam is $1.3 \times 10^6\,\mathrm{W/m^2}$ and the electric field is calculated from

$$I = \left(\frac{1}{2}\varepsilon_0 E^2\right)c \tag{1.43}$$

so that

$$E = 27.4\sqrt{I} = 3.1 \times 10^4\,\mathrm{V/m}$$

and

$$\omega_R = \frac{\mu E}{\hbar} = 9.8 \times 10^8\,\mathrm{rad/s} \quad \text{or} \quad \nu_R = 156\,\mathrm{MHz}, \quad \frac{1}{\nu_R} = 6.4\,\mathrm{ns}.$$

In equation (1.43), $\varepsilon_0 = 8.854187 \times 10^{-12}\,\mathrm{C^2\,N^{-1}\,m^{-2}}$ (Appendix A) is the permittivity of vacuum that appears in Coulomb's law for the magnitude of the force of interaction between two electrical charges,

$$F = \frac{q_1 q_2}{4\pi\varepsilon_0 r^2}.$$

Since a typical electronic transition may have a natural lifetime of 10 ns ($\omega_{\mathrm{natural}} = 6.3 \times 10^8\,\mathrm{rad/s}$), the effects of Rabi cycling are already present at 1 W. At the megawatt or higher power levels of typical pulsed lasers, the coherent effects of strong radiation are even more pronounced, provided electrical breakdown is avoided. At these high electric field strengths, however, the simple two-level model is not a good description of an atomic or molecular system.

The case of weak electromagnetic radiation interacting with the system is also common. In fact before the development of the laser in 1960 the weak-field case was applied to all regions of the spectrum except in the r.f. and microwave regions, where powerful coherent sources were available. In the weak-field case there is a negligible buildup of population in the excited state, so that $a_1 \approx 0$, $a_0 \approx 1$, so

$$\dot{a}_1 = \frac{i\omega_R}{2} e^{-i\Delta t}. \tag{1.44}$$

Equation (1.44) is readily integrated to give

$$\begin{aligned} a_1 &= \frac{i\omega_R}{2} \int_0^t e^{-i\Delta t} \, dt \\ &= \frac{\omega_R}{-2\Delta} (e^{-i\Delta t} - 1). \end{aligned} \tag{1.45}$$

The probability for finding the system in the excited state after a time t is given in equation (1.45) by

$$P_{1\leftarrow 0} = |a_1|^2 = \frac{\omega_R^2}{\Delta^2} \sin^2\left(\frac{\Delta t}{2}\right) = \frac{\mu_{10}^2 E^2}{\hbar^2} \frac{\sin^2[(\omega - \omega_{10})t/2]}{(\omega - \omega_{10})^2}. \tag{1.46}$$

This formula is very deceptive because it assumes monochromatic radiation and short interaction times. These two requirements are inconsistent with one another because the Heisenberg uncertainty principle

$$\Delta E \, \Delta t \geq \hbar \qquad \text{or} \qquad \Delta \nu \, \Delta t \geq \frac{1}{2\pi} \tag{1.47}$$

must always be satisfied. If monochromatic radiation is applied to the system for a time Δt, then the system sees radiation of width $\Delta \nu = 1/(2\pi \Delta t)$ in frequency space (this is certainly not monochromatic!). For example, a pulse of radiation 10 ns long has an intrinsic width of at least 160 MHz in frequency space.

Before equation (1.46) can be used, the effects of the finite frequency spread of the radiation must be included. Consider the radiation applied to the system to be broad band rather than monochromatic and to have a radiation density $\rho = \varepsilon_0 E^2/2$. The total transition probability is given by integrating over all frequencies, that is by

$$
\begin{aligned}
P_{1\leftarrow 0} &= \frac{2\mu_{10}^2}{\varepsilon_0 \hbar^2} \int \rho_v(\omega) \frac{\sin^2(\omega - \omega_{10})t/2}{(\omega - \omega_{10})^2} d\omega \\
&= \frac{2\mu_{10}^2}{\varepsilon_0 \hbar^2} \rho_v(\omega_{10}) \int \frac{\sin^2(\omega - \omega_{10})t/2}{(\omega - \omega_{10})^2} d\omega \\
&= \frac{\mu_{10}^2}{\varepsilon_0 \hbar^2} \rho_v(\omega_{10}) \pi t
\end{aligned}
\tag{1.48}
$$

in which $\rho(\omega)$ is assumed to be slowly varying near ω_{10} so that it can be removed from the integration. This is the case because $[\sin^2(\omega - \omega_{10})t/2]/(\omega_{10} - \omega)^2$ is sharply peaked at $\omega = \omega_{10}$ (see Figure 1.25). The absorption rate per molecule is thus given by

$$
\frac{dP_{1\leftarrow 0}}{dt} = \frac{\pi \mu_{10}^2}{\varepsilon_0 \hbar^2} \rho_v(\omega_{10}). \tag{1.49}
$$

In order to derive an expression for the absorption coefficient in terms of the transition dipole moment, the preceding equation (1.49) needs to be compared with equation (1.15), in which $N_0 \approx N$ for the weak-field case, so dividing by N gives the transition probability per molecule

$$
\frac{d(N_1/N)}{dt} = B_{1\leftarrow 0} \rho_v(v_{10})
$$

or

$$
\frac{dP_{1\leftarrow 0}}{dt} = B_{1\leftarrow 0} \rho_v(v_{10}). \tag{1.50}
$$

A factor of 3 is missing from equation (1.49) because equation (1.50) has been derived using isotropic radiation traveling in the x, y and z directions, while equation (1.49) has been derived using a plane wave traveling in the z direction. Since only the z component of the isotropic radiation is effective in inducing a transition, and since $\rho(v) = 2\pi\rho(\omega)$, we have

$$
B_{1\leftarrow 0} = \frac{1}{6\varepsilon_0 \hbar^2} \mu_{10}^2 = \frac{2\pi^2}{3\varepsilon_0 h^2} \mu_{10}^2 \tag{1.51}
$$

and

$$
A_{1\rightarrow 0} = \frac{16\pi^3 v^3}{3\varepsilon_0 h c^3} \mu_{10}^2 . \tag{1.52}
$$

These equations, (1.51) and (1.52), are key results because they relate the observed macroscopic transition rates to the microscopic transition dipole moment of an atom or molecule. Upon substitution of the values of the constants we obtain $A_{1\to 0} = 3.136 \times 10^{-7}(\tilde{\nu})^3\mu_{10}^2$, with $\tilde{\nu}$ expressed in cm^{-1} and μ_{10} in debye. Although these equations are essentially correct, one factor that has been ignored is the possibility of relaxation.

Collisions or the spontaneous radiative lifetime of the upper state have all been ignored so far. When these losses are considered, the molecular absorption line shape changes from a Dirac delta function $\delta(\nu - \nu_{10})$ that is infinitely sharp and infinitely narrow, but with unit area, to a real molecular line shape. As described below, the line shape function $g(\nu - \nu_{10})$ is typically either a Lorentzian or a Gaussian function with unit area but finite width and height, and now equations (1.51) and (1.52) are replaced by

$$(B_{1\leftarrow 0})_\nu = \frac{2\pi^2}{3\varepsilon_0 h^2}\mu_{10}^2 g(\nu - \nu_{10}) \tag{1.53}$$

and

$$(A_{1\to 0})_\nu = \frac{16\pi^3\nu^3}{3\varepsilon_0 hc^3}\mu_{10}^2 g(\nu - \nu_{10}), \tag{1.54}$$

where $\int (B_{1\leftarrow 0})_\nu \, d\nu = B_{1\leftarrow 0}$ and $\int (A_{1\to 0})_\nu \, d\nu = A_{1\to 0}$. In practice the ν subscripts in equations (1.53) and (1.54) are suppressed and the same symbols $A_{1\to 0}$ and $B_{1\leftarrow 0}$ are used with and without, equations (1.51) and (1.52), line shape functions.

Beer's Law

Consider a system (Figure 1.11) with N_0 molecules per cubic meter in the ground state and N_1 in the excited state. A flux of photons $F = I_0/h\nu$ (units of photons/m^2 s) is incident upon the system from the left. As these photons travel through the system they can be absorbed or they can induce stimulated emission. What is the intensity of radiation after a distance l?

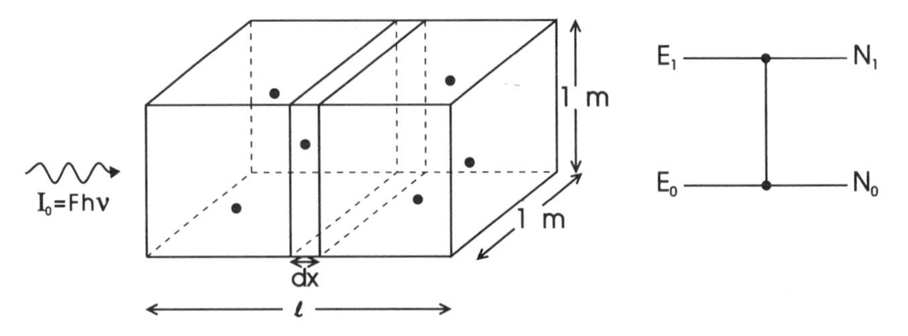

Figure 1.11. A system with dimensions 1 m × 1 m × l m that contains molecules.

If only absorption and stimulated emission are considered, then we can write

$$\begin{aligned}\frac{dN_1}{dt} &= -B_{1\to 0}\rho N_1 + B_{1\leftarrow 0}\rho N_0 \\ &= \frac{2\pi^2\mu_{10}^2}{3\varepsilon_0 h^2}(N_0 - N_1)g(\nu - \nu_{10})\rho \\ &= \frac{2\pi^2\mu_{10}^2\nu}{3\varepsilon_0 hc}(N_0 - N_1)g(\nu - \nu_{10})F \\ &= \sigma F(N_0 - N_1), \end{aligned} \tag{1.55}$$

in which $\rho = I/c = h\nu F/c$ has been used. The absorption cross section is defined in this way as

$$\sigma = \frac{2\pi^2\mu_{10}^2}{3\varepsilon_0 hc}\nu g(\nu - \nu_{10}), \tag{1.56}$$

with dimensions of m^2. The physical interpretation of σ is as the "effective area" that a molecule presents to the stream of photons of flux F. Notice that equations (1.56) and (1.54) can be combined to give the convenient equation

$$\sigma = \frac{A\lambda^2 g(\nu - \nu_{10})}{8\pi} = \frac{\lambda^2 g(\nu - \nu_{10})}{8\pi\tau_{sp}} \tag{1.57}$$

which relates the cross section to the lifetime of a transition.

If a flux F is incident to the left of a small element of thickness dx with cross-sectional area of $1\ m^2$, then the change in flux caused by passing through the element is

$$dF = -\sigma F(N_0 - N_1)\,dx. \tag{1.58}$$

Upon integrating over the absorption path, this becomes

$$\int_{F_0}^{F} \frac{dF}{F} = -\sigma(N_0 - N_1)\int_0^l dx$$

or

$$\ln\left(\frac{F}{F_0}\right) = \ln\left(\frac{I}{I_0}\right) = -\sigma(N_0 - N_1)l. \tag{1.59}$$

Expression (1.59) can also be rewritten in the form

$$I = I_0 e^{-\sigma(N_0 - N_1)l}, \tag{1.60}$$

which is equivalent to Beer's law,

$$I = I_0 e^{-\varepsilon c l}. \tag{1.61}$$

It is common to report σ in cm^2, N in molecules per cm^3, and l in cm rather than the corresponding SI units. The units used in Beer's law are customarily moles per liter for c, cm for l, and liter mole^{-1} cm^{-1} for the molar absorption coefficient, ε. Sometimes the cross section and concentration are combined to define an absorption coefficient $\alpha = \sigma(N_0 - N_1)$ for a system, in which case we write

$$I = I_0 e^{-\alpha l}. \tag{1.62}$$

Line-Shape Functions: *Homogeneous and Inhomogeneous Line Shapes*

A real spectrum of a molecule, such as that for CO_2 (Figure 1.12), contains many absorption features called *lines* organized into a band associated with a particular mode of vibration. For the spectrum illustrated in Figure 1.12 the lines are associated with the antisymmetric stretching mode of CO_2, ν_3. At high resolution the spectrum seems to consist of very narrow features, but if the scale is expanded the lines are observed to have definite widths and characteristic shapes. What are the possible line-shape functions $g(\nu - \nu_{10})$ and what physical processes are responsible for the shapes?

Line-shape functions fall into one of two general categories: homogeneous and inhomogeneous. A homogeneous line shape occurs when all molecules in the system have identical line-shape functions. For example, if an atomic or molecular absorber in the gas phase is subject to a high pressure, then all molecules in the system are found to have an identical pressure-broadened line shape for a particular transition. Pressure broadening of a transition is said, therefore, to be a homogeneous broadening.

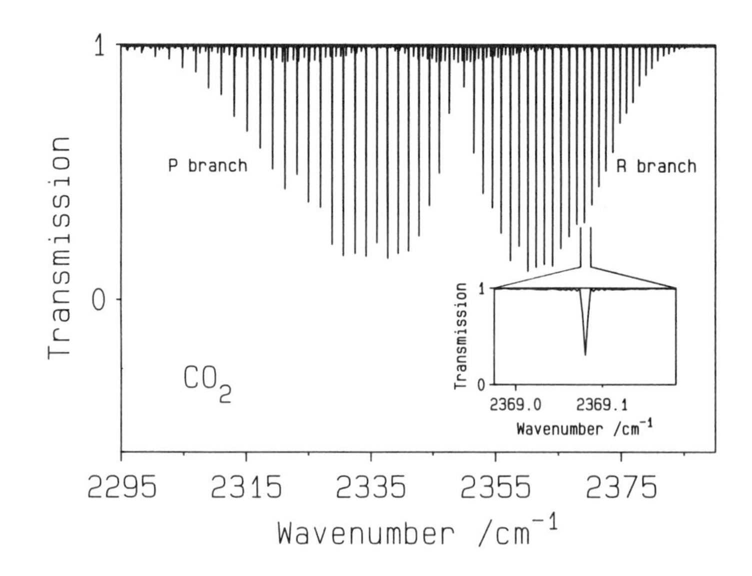

Figure 1.12. A typical molecular spectrum, the antisymmetric stretching mode of carbon dioxide.

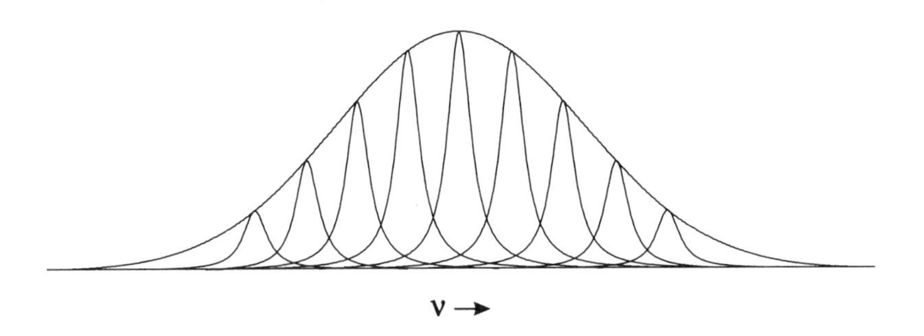

Figure 1.13. An inhomogeneously broadened line made up of many homogeneously broadened components.

In contrast, if a molecule is dissolved in a liquid, then the disorder inherent in the structure of the liquid provides numerous different solvent environments for the solute. Each solute molecule experiences a slightly different solvent environment and therefore has a slightly different absorption spectrum. The observed absorption spectrum (Figure 1.13) is made up of all of the different spectra for the different molecular environments; it is said to be inhomogeneously broadened.

The most important example of gas phase inhomogeneous broadening occurs because of the Maxwell–Boltzmann distribution of molecular velocities and is called *Doppler broadening.* The different molecular velocities give the incident radiation a frequency shift of $\nu = (1 \pm v/c)\nu_0$ in the molecular frame of reference. This results in slightly different spectra for the molecules moving at different velocities and results in an inhomogeneous line shape.

Natural Lifetime Broadening

Consider a two-level system with an intrinsic lifetime of τ_{sp} seconds for the level at E_1 for the spontaneous emission of radiation (Figure 1.14). The wavefunction that describes the state of the system is given as

$$\begin{aligned} \Psi &= a_0\Psi_0 + a_1\Psi_1 \\ &= a_0\psi_0 e^{-iE_0t/\hbar} + a_1\psi_1 e^{-iE_1t/\hbar}, \end{aligned} \tag{1.63}$$

where a_0 and a_1 are simply constants because no electromagnetic radiation is present

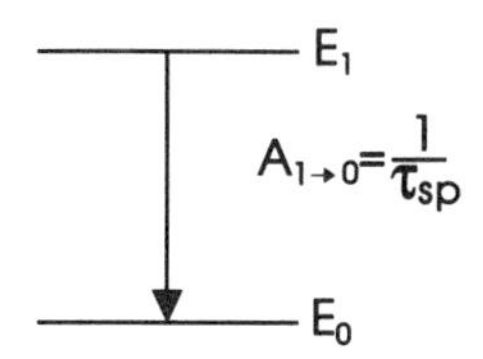

Figure 1.14. Spontaneous emission in a two-level system.

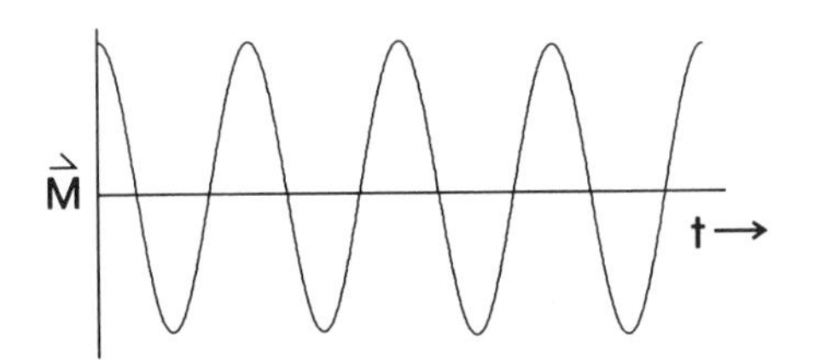

Figure 1.15. Oscillating dipole moment of a system in a superposition state.

in the system. When the system is excited into this superposition state (for example, by a pulse of electromagnetic radiation), the dipole moment of the system can be computed as

$$\begin{aligned}\mathbf{M} &= \langle\Psi|\,\boldsymbol{\mu}\,|\Psi\rangle \\ &= \boldsymbol{\mu}_{10}(a_0^* a_1 e^{-i\omega_{10}t} + a_0 a_1^* e^{i\omega_{10}t}) \\ &= \mathrm{Re}(2\boldsymbol{\mu}_{10} a_0 a_1^* e^{i\omega_{10}t}),\end{aligned} \tag{1.64}$$

assuming space-fixed dipole moments $\langle\psi_0|\,\boldsymbol{\mu}\,|\psi_0\rangle = \langle\psi_1|\,\boldsymbol{\mu}\,|\psi_1\rangle = 0$. (NB: Dipole moments are still possible in the molecular frame.) The dipole moment of the system oscillates at the Bohr frequency with

$$\mathbf{M} = 2a_0 a_1 \boldsymbol{\mu}_{10} \cos(\omega_{10}t) \tag{1.65}$$

if a_0 and a_1 are chosen as real numbers. A system in such a superposition state has a macroscopic oscillating dipole in the laboratory frame (Figure 1.15).

Now if the amount of population in the excited state decreases slowly in time (relative to the Bohr frequency) due to the spontaneous emission, then the amplitude of the oscillation will also decrease. This corresponds to a slow decrease in a_1 (equation (1.65)) at a rate of $\gamma/2$ where $\gamma = 1/\tau_{sp} = A_{1\to 0}$. Thus the oscillating dipole moment is now

$$\mathbf{M} = \mathbf{M}_0 e^{-\gamma t/2} \cos(\omega_{10}t) \tag{1.66}$$

as shown in Figure 1.16.

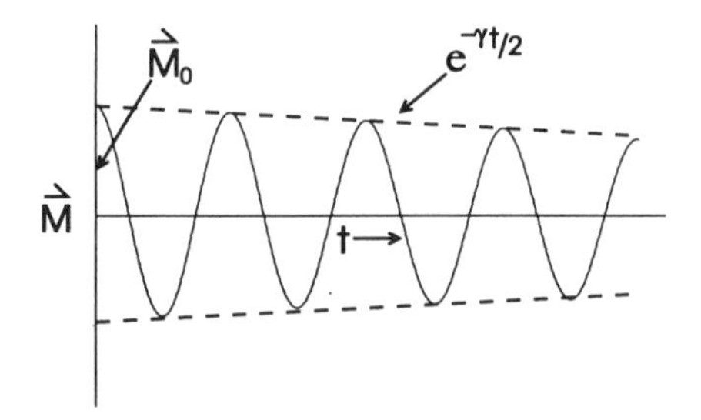

Figure 1.16. Slowly damped oscillating dipole moment.

What frequencies are associated with a damped cosine wave? Clearly the undamped wave oscillates infinitely at exactly the Bohr frequency ω_{10}. The distribution of frequencies $F(\omega)$ present in a waveform $f(t)$ can be determined by taking a Fourier transform

$$F(\omega) = \int_{-\infty}^{\infty} f(t)e^{-i\omega t}\, d\omega. \tag{1.67}$$

Note also that an arbitrary waveform $f(t)$ can be written as a sum (integral) over plane waves $e^{i\omega t}$, each with amplitude $F(\omega)$, as

$$f(t) = \frac{1}{2\pi}\int_{-\infty}^{\infty} F(\omega)e^{i\omega t}\, d\omega. \tag{1.68}$$

Thus $F(\omega)$ measures the "amount" of each "frequency" required to synthesize $f(t)$ out of sine and cosine functions ($e^{i\omega t} = \cos \omega t + i \sin \omega t$). Taking the Fourier transform of the time-dependent part of $\mathbf{M}(t)$ gives

$$\begin{aligned} F(\omega) &= \int_{-\infty}^{\infty} e^{-\gamma t/2} \cos(\omega_{10} t) e^{-i\omega t}\, dt \\ &= \frac{1}{2}\int_{0}^{\infty} e^{-\gamma t/2}(e^{-i(\omega-\omega_{10})t} + e^{-i(\omega+\omega_{10})t})\, dt \\ &= \frac{1}{2}\left[\frac{1}{\gamma/2 + i(\omega - \omega_{10})} + \frac{1}{\gamma/2 + i(\omega + \omega_{10})}\right] \end{aligned} \tag{1.69}$$

for the decay process beginning at $t = 0$. The nonresonant term containing $\omega + \omega_{10}$ is dropped because $\omega \approx \omega_{10}$ and $\omega_{10} \gg \gamma$ so that it is negligible in comparison to the resonant term containing $\omega - \omega_{10}$ (c.f. rotating wave approximation) and

$$F(\omega) \approx \frac{1}{2}\frac{1}{\gamma/2 + i(\omega - \omega_{10})}. \tag{1.70}$$

In the semiclassical picture an oscillating dipole moment radiates power at a rate proportional to $|\mu_{10}|^2$ (i.e., $A_{1\to 0} \propto |\mu_{10}|^2$) and the line-shape function

$$|F(\omega)|^2 = \frac{1}{4}\frac{1}{(\gamma/2)^2 + (\omega - \omega_{10})^2} \tag{1.71}$$

is an unnormalized Lorentzian. Normalization requires that

$$\int_{-\infty}^{\infty} g(\omega - \omega_{10})\, d\omega = \int_{-\infty}^{\infty} g(\nu - \nu_{10})\, d\nu = 1. \tag{1.72}$$

Note that

$$2\pi g(\omega - \omega_{10}) = g(\nu - \nu_{10}).$$

The final normalized Lorentzian line-shape functions are

$$g(\omega - \omega_{10}) = \frac{\gamma/(2\pi)}{(\gamma/2)^2 + (\omega - \omega_{10})^2}, \tag{1.73}$$

and

$$g(\nu - \nu_{10}) = \frac{\gamma}{(\gamma/2)^2 + (2\pi)^2(\nu - \nu_{10})^2}. \tag{1.74}$$

Without spontaneous emission the line-shape function would be $\delta(\nu - \nu_{10})$ since the infinite cosine wave oscillates at a frequency of exactly ν_{10}. The decaying cosine wave caused by spontaneous emission gives a Lorentzian function of finite width (Figure 1.17). At the peak $g(\nu - \nu_{10}) = 4/\gamma$, and the function drops to half this value when

$$(2\pi)^2(\nu_{1/2} - \nu_{10})^2 = \left(\frac{\gamma}{2}\right)^2 \quad \text{or} \quad \pm(\nu_{1/2} - \nu_0) = \frac{\gamma}{4\pi}.$$

The full width at half maximum (FWHM) $\Delta\nu_{1/2}$ is given by

$$\Delta\nu_{1/2} = \frac{\gamma}{2\pi} = \frac{1}{2\pi\tau_{sp}}, \tag{1.75}$$

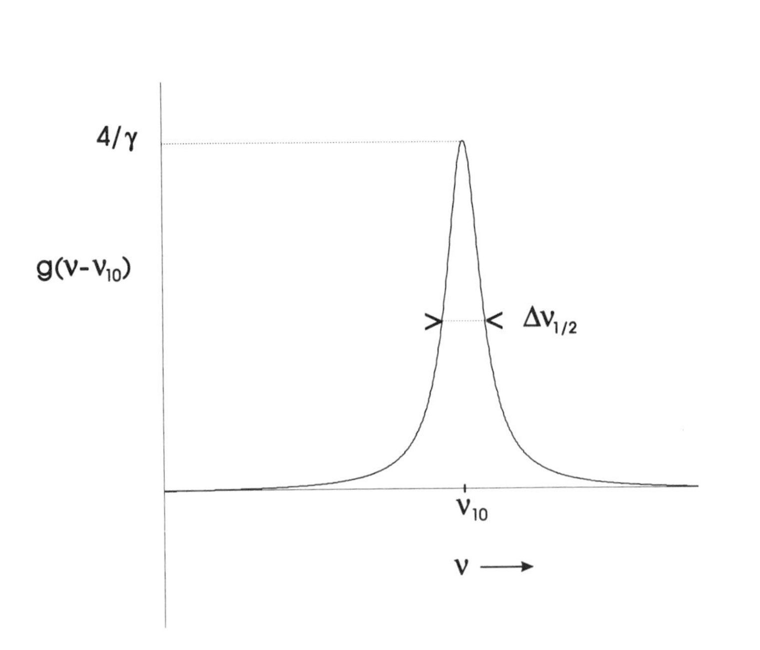

Figure 1.17. A normalized Lorentzian function.

since $\gamma = 1/\tau_{sp}$. The Lorentzian line-shape function ($\nu_0 = \nu_{10}$) can thus be expressed as

$$g(\nu - \nu_0) = \frac{\Delta\nu_{1/2}/(2\pi)}{(\Delta\nu_{1/2}/2)^2 + (\nu - \nu_0)^2} \tag{1.76}$$

in terms of the full width at half maximum.

The important result $\Delta\nu_{1/2} = 1/(2\pi\tau_{sp})$ agrees with the uncertainty principle $\Delta E\, \Delta t \geq \hbar$, or

$$\frac{\Delta E}{h}\tau_{sp} \sim \frac{1}{2\pi},$$

$$\Delta\nu \sim \frac{1}{2\pi\tau_{sp}}. \tag{1.77}$$

The spontaneous lifetime of the excited state means that the atom or molecule cannot be found at E_1 for more than τ_{sp} on average. This provides a fundamental limit on the linewidth arising from the transition between the two states (Figure 1.18). Formula (1.77) has been checked experimentally such as in the case of the sodium $3^2P_{3/2} \rightarrow 3^2S_{1/2}$ transition (D line) at 5890 Å. The experimentally measured lifetime of $\tau_{sp} = 16$ ns and the observed homogeneous linewidth $\Delta\nu_{1/2} = 10$ MHz are consistent with equation (1.77). In terms of the uncertainty principle if an excited state exists for only τ_{sp} seconds on average, then the energy level E_1 cannot be measured relative to E_0 with an accuracy greater than $\Delta\nu_{1/2}$ Hz.

The expression $\Delta\nu_{1/2} = 1/(2\pi\tau)$ has widespread use in chemical physics. For example, if H_2O is excited by vacuum ultraviolet light, it can dissociate very rapidly:

$$H_2O \xrightarrow{h\nu} HO + H. \tag{1.78}$$

If the H_2O molecule exists in its excited electronic state for only one vibrational period ($\tilde{\nu} = 3600\ \text{cm}^{-1}$ corresponding to an OH stretch), then $\tau = 9.3 \times 10^{-15}$ s $= 9.3$ fem-

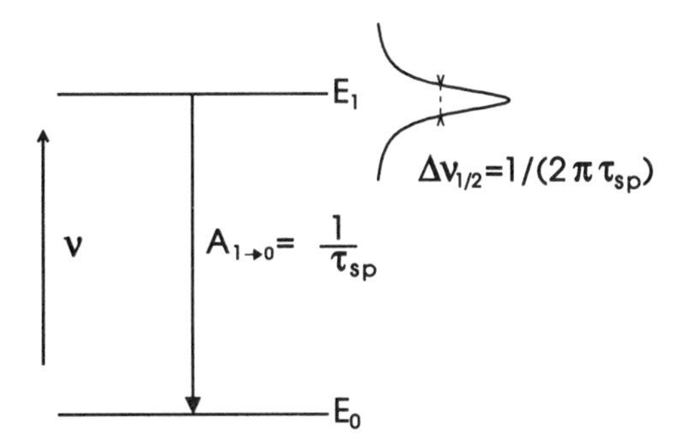

Figure 1.18. The lifetime τ_{sp} gives the transition $E_1 \rightarrow E_0$ a finite linewidth.

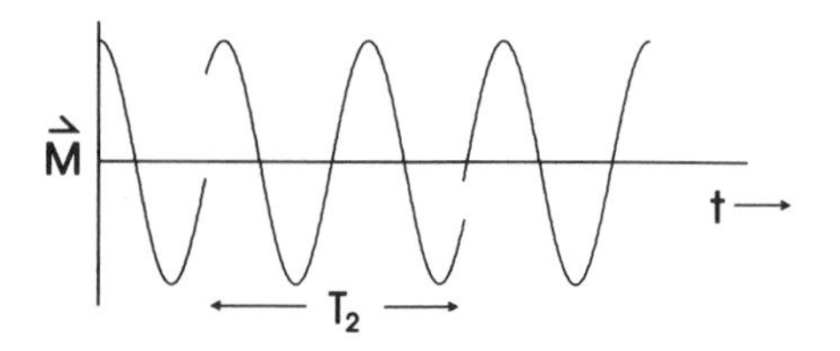

Figure 1.19. The phase of an oscillating dipole moment randomly interrupted by collisions.

toseconds (fs). Thus the width (FWHM) of a line in the spectrum will be $\Delta\nu_{1/2} = 1.7 \times 10^{13}$ Hz or $\Delta\tilde{\nu}_{1/2} = 570\ \text{cm}^{-1}$. A measurement of the homogeneous width of the spectral line can thus provide an estimate of the lifetime of the excited state.

Pressure Broadening

The derivation of the pressure-broadening line shape is a difficult problem because it depends on the intermolecular potentials between the colliding molecules. However, a simplified model within the semiclassical picture gives some estimation of the effect.

Consider the two-level system discussed in the previous section with the wavefunction written as a superposition state. The dipole moment oscillates at the Bohr frequency except during a collision. If the collision is sufficiently strong, then the phase of the oscillating dipole moment is altered in a random manner by the encounter. Let the average time between collisions be T_2 (Figure 1.19). The infinite cosine wave is broken into pieces of average length T_2. The effect of collisions will be to convert the infinitely narrow line shape associated with an infinitely long cosine wave to a line-shape function of finite width. The application of Fourier transform arguments (using autocorrelation functions[5]) to decompose the broken waveform into frequency components results in a Lorentzian line shape with a width (FWHM) given by

$$\Delta\nu_{1/2} = \frac{1}{\pi T_2}. \tag{1.79}$$

Since the average time between collisions is proportional to the reciprocal of the pressure, p, then

$$\Delta\nu_{1/2} = bp, \tag{1.80}$$

with b referred to as the pressure-broadening coefficient. The quantitative calculation of b without recourse to experiment poses a difficult theoretical problem. Experimentally, typical values for b are about 10 MHz per Torr of the pressure-broadening gas.

In general not only are the lines broadened by increasing pressure but they are also shifted in frequency. These shifts are generally small, less than 1 MHz/Torr, but they are important when very accurate spectroscopic measurements are to be made.

Doppler Broadening

Doppler broadening results in an inhomogeneous line-shape function. If the transition has an intrinsic homogeneous line shape $g_H(\nu - \nu_0')$ centered at ν_0', then the inhomogeneous distribution function $g_I(\nu_0' - \nu_0)$, centered at ν_0, is required to describe the total line-shape function

$$g(\nu - \nu_0) = \int_{-\infty}^{\infty} g_I(\nu_0' - \nu_0) g_H(\nu - \nu_0')\, d\nu_0'. \tag{1.81}$$

The distribution function $g_I(\nu_0' - \nu_0)$ gives the probability density that a system has a resonance frequency in the interval ν_0' to $\nu_0' + d\nu_0'$

$$dp = g_I(\nu_0' - \nu_0)\, d\nu_0'. \tag{1.82}$$

The line-shape integral (1.81) is a convolution of the two functions g_I and g_H, as can be made more apparent by making the substitution $x = \nu_0' - \nu_0$,

$$g(\nu - \nu_0) = \int_{-\infty}^{\infty} g_I(x) g_H((\nu - \nu_0) - x)\, dx. \tag{1.83}$$

Commonly the homogeneous line-shape function g_H is Lorentzian, while the inhomogeneous function g_I is a Gaussian, and the convolution of the two is called a Voigt line-shape function (Figure 1.20).

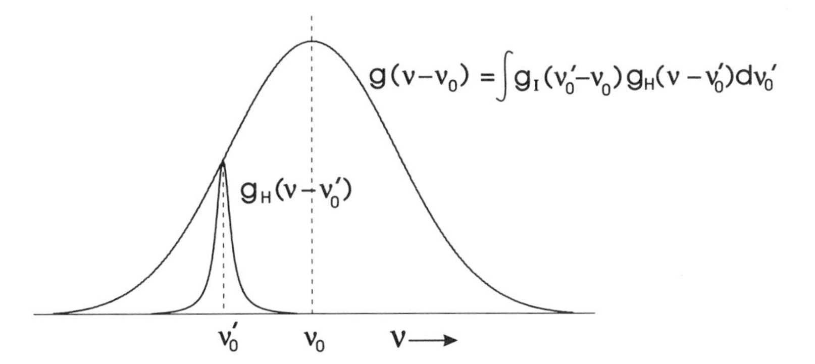

Figure 1.20. The Voigt line shape is a convolution of an inhomogeneous line-shape function with a homogeneous line-shape function.

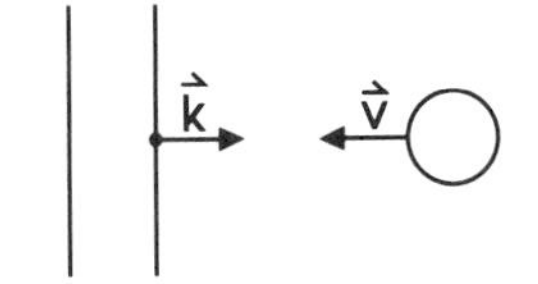

Figure 1.21. Interaction of a plane electromagnetic wave with a moving atom.

The Voigt line-shape function is a general form that can include purely homogeneous or purely inhomogeneous line shapes as limiting cases. If the width of the inhomogeneous part is much greater than the homogeneous part, that is, if $\Delta\nu_I \gg \Delta\nu_H$, then $g_H(\nu - \nu_0') \approx \delta(\nu - \nu_0')$ and

$$\begin{aligned} g(\nu - \nu_0) &= \int_{-\infty}^{\infty} g_I(\nu_0' - \nu_0)\delta(\nu - \nu_0')\, d\nu_0' \\ &= g_I(\nu - \nu_0). \end{aligned} \tag{1.84}$$

Conversely, if $\Delta\nu_I \ll \Delta\nu_H$, then $g_I(\nu_0' - \nu_0) \approx \delta(\nu_0' - \nu_0)$ and $g(\nu - \nu_0) = g_H(\nu - \nu_0)$.

Consider an atom with velocity $\mathbf{v}$ interacting with a plane wave with a wave vector $\mathbf{k}$. If $\mathbf{k}$ is parallel to $\mathbf{v}$, then the atom sees a Doppler shifted frequency $\nu' = \nu(1 \pm v/c)$, depending upon whether the atom is moving in a direction that is the same as $(-)$ or opposite to $(+)$ that of the electromagnetic radiation (Figure 1.21). In general, it is only the component of $\mathbf{v}$ along $\mathbf{k}$ (i.e., $v \cos\theta$) that matters, so that

$$\nu' = \nu\left(1 - \frac{\mathbf{v}\cdot\mathbf{k}}{c\,|\mathbf{k}|}\right), \tag{1.85}$$

neglecting a small relativistic correction ("second-order Doppler effect").

The Doppler effect can be viewed in two equivalent ways. In the frame of the atom it is the frequency of the electromagnetic wave that has been shifted, with the atom at rest at the origin of the atomic coordinate system. Alternatively, in the fixed laboratory frame the electromagnetic wave is unshifted at ν, but the atomic resonance frequency ν_0 (of the atom moving at velocity v) has been shifted to the new value of

$$\nu_0' = \frac{\nu_0}{1 \pm v/c}. \tag{1.86}$$

All that is required to obtain a line-shape function is the distribution of velocities.

The distribution of molecular velocity components along a given axis (such as $\mathbf{k}$) in a gaseous system is given by the Maxwell–Boltzmann distribution function

$$p_v\, dv = \left(\frac{m}{2\pi kT}\right)^{1/2} e^{(-mv^2)/(2kT)}\, dv \tag{1.87}$$

for particles of mass m at a temperature T. Using equation (1.87) and $dv = (c/\nu_0)\,d\nu_0'$ gives the normalized inhomogeneous line-shape function

$$g_D(\nu - \nu_0) = \frac{1}{\nu_0}\left(\frac{mc^2}{2\pi kT}\right)^{1/2} e^{(-mc^2(\nu-\nu_0)^2)/(2kT\nu_0^2)}. \tag{1.88}$$

The FWHM $\Delta\nu_D$ is easily shown to be

$$\Delta\nu_D = 2\nu_0\sqrt{\frac{2kT\ln(2)}{mc^2}}, \tag{1.89}$$

or

$$\Delta\tilde{\nu}_D = 7.1\times10^{-7}\tilde{\nu}_0\sqrt{\frac{T}{M}}, \tag{1.90}$$

in which T is in K, M in amu, $\tilde{\nu}_0$ in cm^{-1}, and $\Delta\tilde{\nu}_D$ in cm^{-1}.

The Doppler line-shape function is thus given by

$$g_D(\nu - \nu_0) = \frac{2}{\Delta\nu_D}\sqrt{\frac{\ln(2)}{\pi}}e^{-4\ln 2[(\nu-\nu_0)/\Delta\nu_D]^2} \tag{1.91}$$

in terms of the Doppler FWHM.

The Gaussian function is the bell-shaped curve well known in statistics. The Gaussian line-shape function is more sharply peaked around $\nu - \nu_0$ than the corresponding Lorentzian line-shape function (Figure 1.22). Notice the much more extensive "wings" on the Lorentzian function in comparision to the Gaussian function.

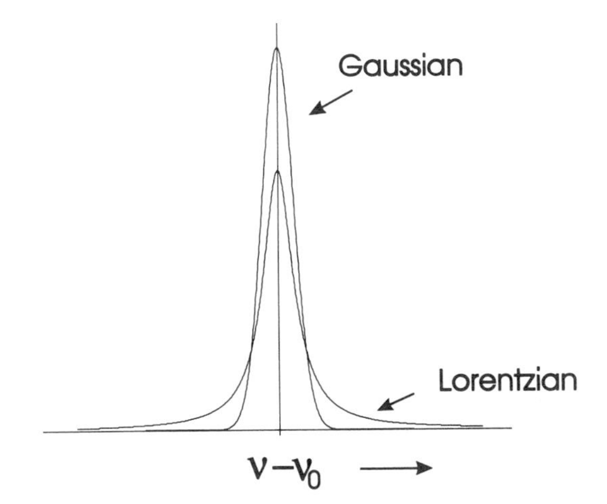

Figure 1.22. Normalized Gaussian and Lorentzian line-shape functions.

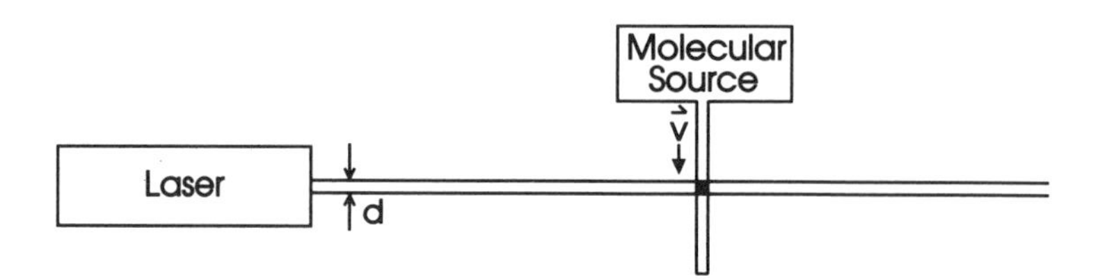

Figure 1.23. A laser beam interacts with a molecular beam inside a vacuum chamber.

For example, the Doppler width of the $3^2P_{3/2} - 3^2S_{1/2}$ transition of the Na atom at 300 K is already $\Delta\nu_D = 0.044\ \text{cm}^{-1} = 1317$ MHz. The Doppler width is much larger than the natural linewidth of 10 MHz. In addition, if the Na atom is surrounded by Ar atoms at 1 Torr total pressure, then pressure broadening contributes 27 MHz to the total homogeneous linewidth of 37 MHz.[6] Visible and ultraviolet transitions of gas phase atoms and molecules typically display Doppler broadening because the inhomogeneous linewidth greatly exceeds the total homogeneous linewidth.

Transit-Time Broadening

Consider the experimental arrangement of Figure 1.23 in which a laser beam of width d is crossed at right angles with a beam of molecules moving at a speed of v in a vacuum chamber. The molecules can only interact with the radiation for a finite time, $\tau = d/v$, called the *transit time.* The time τ corresponds to the time required by a molecule in the molecular beam to cross through the laser beam. If the laser is considered to be perfectly monochromatic with a frequency ν, then a molecule experiences an electric field as shown in Figure 1.24, assuming a constant weak light intensity from one side of the laser beam to the other. If an intense laser beam is used in these experiments, Rabi oscillations are observed.

Suppose an infinitely long (in time) oscillating electric field with an infinitely narrow frequency distribution has been chopped into a finite length with a finite frequency width. The finite time allowed for the laser–molecule interaction has resulted in a

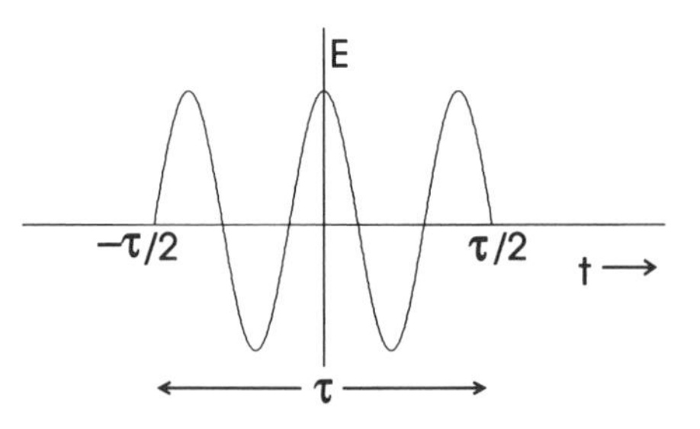

Figure 1.24. A molecule experiences an electromagnetic wave of finite length τ.

broadening of the transition. As far as the observed line shape is concerned it does not matter whether the molecular resonance is broadened or the frequency distribution of the applied radiation is increased. The result is the same: a broader line.

The frequency distribution associated with the electric field of Figure 1.24 is determined, as for lifetime broadening, by taking the Fourier transform. Thus we write

$$\begin{aligned} F(\omega) &= \int_{-\infty}^{\infty} E_0 \cos(\omega_0 t) e^{-i\omega t}\, d\omega \\ &= \frac{E_0}{2} \int_{-\tau/2}^{\tau/2} (e^{i\omega_0 t} + e^{-i\omega_0 t}) e^{-i\omega t}\, d\omega \\ &= \frac{E_0}{2} \left[\frac{e^{i(\omega_0 - \omega)t}}{i(\omega_0 - \omega)} + \frac{e^{-i(\omega_0 + \omega)t}}{-i(\omega_0 + \omega)} \right]_{-\tau/2}^{\tau/2} \\ &\approx E_0 \frac{\sin(\omega_0 - \omega)\tau/2}{(\omega_0 - \omega)} \end{aligned} \tag{1.92}$$

in which the nonresonant term in $\omega_0 + \omega$ has, as usual, been discarded. Since the intensity of the light is proportional to $|E|^2$, the unnormalized line shape is proportional to

$$|F(\omega)|^2 = \frac{\sin^2(\omega_0 - \omega)\tau/2}{(\omega_0 - \omega)^2}. \tag{1.93}$$

The normalized line-shape function

$$g(\omega - \omega_0) = \frac{2}{\pi\tau} \frac{\sin^2(\omega - \omega_0)\tau/2}{(\omega - \omega_0)^2}, \tag{1.94}$$

or

$$g(\nu - \nu_0) = \frac{1}{\pi^2\tau} \frac{\sin^2[2\pi(\nu - \nu_0)\tau/2]}{(\nu - \nu_0)^2}, \tag{1.95}$$

is plotted in Figure 1.25. The FWHM of this function is about $\Delta\nu_{1/2} = 5.6/\tau$. For an atom traveling at 500 m/s through a laser beam of 1 mm width, $\tau = 2 \times 10^{-6}$ s and $\Delta\nu_{1/2} = 2.8$ MHz. Although transit-time broadening is relatively small, it is not negligible for very precise Doppler-free measurements or for microwave-molecular beam measurements.

Power Broadening

The high-power laser has become an ubiquitous tool of modern spectroscopy. The application of intense electromagnetic radiation to a system will cause the spectral

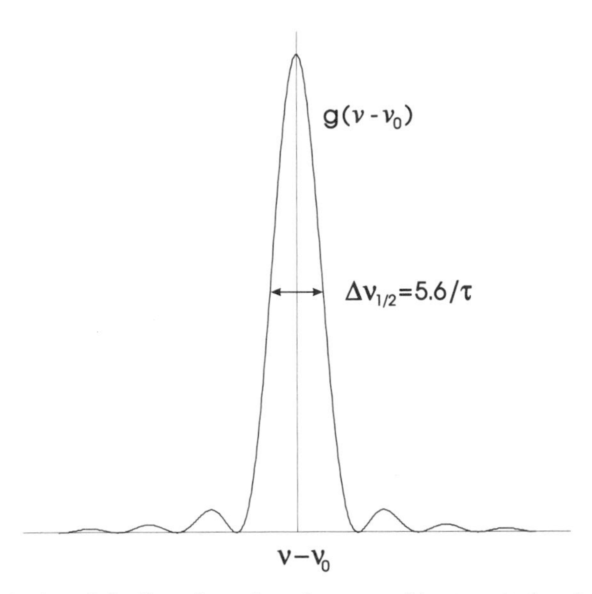

Figure 1.25. A plot of the line-shape function caused by transit-time broadening.

lines to broaden and even split. The detailed calculation of a molecular line shape at high power is complicated, but a simple estimate of the linewidth is available from the uncertainty principle, equation (1.47),

$$\Delta E \, \Delta t \geq \hbar.$$

At high powers the molecular system undergoes Rabi oscillations at an angular frequency $\omega_R = \mu_{10}E/\hbar$. The system is thus in the excited state E_1 only for a period of about $h/\mu_{10}E$, which is therefore an estimate for Δt,

$$\Delta E \frac{h}{\mu_{10}E} \sim \hbar$$

or

$$\Delta \nu \sim \frac{\mu_{10}E}{2\pi h} = \frac{\omega_R}{4\pi^2}. \tag{1.96}$$

For example, a 1-W laser beam of 1-mm diameter interacting with a two-level system with a transition dipole moment of 1 D, $\omega_R = 9.8 \times 10^8$ rad/s and the value of $\Delta\nu$ is already 25 MHz. Pulsed lasers can easily achieve peak powers of 1 MW (10 mJ in 10 ns), which will increase E, equation (1.43), by a factor of 1000, 3.1×10^7 V/m for the preceding example. The power-broadened linewidth is then of the order 25,000 MHz or 0.83 cm^{-1}, which is larger than a typical Doppler width for a visible or uv electronic transition.

In this chapter the interaction of light with matter has been discussed and key equations derived. For spectroscopy perhaps the most important equations are (1.53), (1.54), (1.56), and (1.59) because they relate the microscopic molecular world to macroscopic absorption and emission rates. The transition dipole moment

$$\mathbf{M}_{10} = \boldsymbol{\mu}_{10} = \int \psi_1^* \boldsymbol{\mu} \psi_0 \, d\tau$$

will be used numerous times to derive selection rules. The absorption strength, equation (1.56), of a transition and the emission rate, equation (1.54), are proportional to the square of the transition dipole moment. The amount of absorption (I/I_0) of a molecular transition depends on three factors: the cross section σ, the population difference between the two levels $N_0 - N_1$, and the optical path length l:

$$\frac{I}{I_0} = e^{-\sigma(N_0 - N_1)l}.$$

Spectroscopists use these relationships constantly.

Problems

1. The helium–neon laser operates at a wavelength of 6328.165 Å in air. The refractive index of air is 1.0002759 at this wavelength.
   (a) What is the speed of light in air and the vacuum wavelength?
   (b) Convert the wavelength to wavenumbers (cm^{-1}) and calculate the magnitude of the wave vector in air.
   (c) What are the frequency and period of oscillation?
   (d) Calculate the energy and momentum of a photon with an air wavelength of 6328.165 Å.
   (e) What will be the wavelength and frequency in water if the refractive index is 1.3333?
2. The refractive index of dry air at 15°C and 760 Torr pressure is given by the Cauchy formula

$$(n - 1) \times 10^7 = 2726.43 + 12.288 \times \frac{10^8}{\lambda^2} + 0.3555 \times \frac{10^{16}}{\lambda^4}$$

   where λ is in Å (*CRC Handbook of Chemistry and Physics,* CRC Press). [The formula due to Edlén (*Metrologia* **2,** 71 (1966)) is slightly more accurate but less convenient to use.]
   (a) Calculate the refractive index of air at 4000 Å, 6000 Å, and 8000 Å.
   (b) Convert the air wavelengths into vacuum and calculate the corresponding wavenumbers (cm^{-1}).

3. (a) What is the momentum and the de Broglie wavelength associated with a human weighing 150 lb and walking at 4 mi/hr?
   (b) What is the momentum and the de Broglie wavelength of an electron accelerated through a voltage of 100 V?
4. A crystal lattice has a typical spacing of 2 Å.
   (a) What velocity, momentum, and kinetic energy should be used for the electrons in an electron diffraction experiment? (*Hint*: $\lambda \approx d$ (lattice spacing) for a good diffraction experiment.)
   (b) What voltage needs to be applied to the electron gun for the diffraction experiment?
   (c) Answer part (a) if neutrons were used instead of electrons.
   (d) If the diffraction experiment were carried out with photons, then what wavelength, energy, and momentum would be required?
5. (a) Make the necessary conversions in order to fill in the table:

Wavelength (Å)	420			
Wavenumber (cm^{-1})		100		
Energy (J)				
Energy (kJ/mole)			490	
Frequency (Hz)				8.21×10^{13}

   (b) Name the spectral region associated with each of the last four columns of the table.
6. There are two limiting cases associated with the Planck function

$$\rho_\nu(T) = \frac{8\pi h\nu^3}{c^3} \frac{1}{e^{h\nu/kT} - 1}.$$

   (a) Calculate kT at room temperature (20°C) in J, kJ/mole, eV, and cm^{-1}.
   (b) For long wavelengths (microwave frequencies at room temperature) $h\nu \ll kT$. In this case, derive a simpler, approximate equation called Rayleigh–Jeans law for $\rho_\nu(T)$. This is the formula derived using classical arguments prior to Planck's quantized oscillator approach.
   (c) For high-energy photons (near-infrared wavelengths at room temperature) $h\nu \gg kT$. In this case, derive a simpler, approximate expression for $\rho_\nu(T)$ called Wien's formula.
7. (a) Differentiate the Planck function to determine the frequency at which ρ_ν is a maximum (Figure 1.4).
   (b) Derive the Wien displacement law for blackbody radiation

$$\lambda_{\max} T = 2.898 \times 10^{-3} \text{ m K}$$

   using $\rho_\lambda \, d\lambda$.

(c) What wavelengths correspond to the maximum of the Planck function in interstellar space at 3 K, at room temperature (20°C), in a flame (2000°C), and in the photosphere of the sun (6000 K)?

8. The total power at all frequencies emitted from a small hole in the wall of a blackbody cavity is given by the Stefan–Boltzmann law

$$I = \sigma T^4,$$

where σ is the Stefan–Boltzmann constant that has a value of $5.6705 \times 10^{-8}\ \mathrm{W\,m^{-2}\,K^{-4}}$.

(a) Derive the Stefan–Boltzmann law.

(b) Determine an expression for σ in terms of fundamental physical constants and obtain a numerical value. (*Hint*: $\int_0^\infty x^3/(e^x - 1)\,dx = \pi^4/15$.)

9. Convert the Planck law from a function of frequency to a function of wavelength, that is, derive $\rho_\lambda\,d\lambda$ from $\rho_\nu\,d\nu$.

10. Consider the two-level system (Figure 1.5) at room temperature, 20°C, and in the photosphere of the sun at 6000 K. What are the relative populations N_1/N_0 corresponding to transitions that would occur at 6000 Å, 1000 $\mathrm{cm^{-1}}$, 100 GHz, and 1 GHz?

11. A 100-W tungsten filament lamp operates at 2000 K. Assuming that the filament emits like a blackbody, what is the total power emitted between 6000 Å and 6001 Å? How many photons per second are emitted in this wavelength interval?

12. (a) What is the magnitude of the electric field for the beam of a 1-mW helium–neon laser with a diameter of 1 mm?

(b) How many photons per second are emitted at 6328 Å?

(c) If the laser linewidth is 1 kHz, what temperature would a blackbody have to be to emit the same number of photons from an equal area over the same frequency interval as the laser?

13. Derive the relationship (1.13) between the energy density ρ_ν and the intensity I_ν for a blackbody

$$I_\nu = \rho_\nu \frac{c}{4}.$$

14. Solve equations (1.36) for the interaction of light with a two-level system.

(a) First convert the two first-order simultaneous differential equations into a single second-order equation by substituting one equation (1.36a) into the other (1.36b).

(b) The general solution for a second-order differential equation with constant coefficients is

$$a_0(t) = Ae^{i\alpha t} + Be^{i\beta t}.$$

Show this implies

$$a_1(t) = \frac{2}{i\omega_R} e^{-i\Delta t}[Ai\alpha e^{i\alpha t} + Bi\beta e^{i\beta t}].$$

(c) To obtain α and β substitute the general solution into the second-order equation and obtain the characteristic equation

$$\alpha^2 - \Delta\alpha - \frac{\omega_R^2}{4} = 0.$$

The two solutions of this equation are α and β. Find α and β and simplify the answer using the definition

$$\Omega = (\omega_R^2 + \Delta^2)^{1/2}.$$

(d) Determine A and B from the initial conditions at $t = 0$, $a_0(0) = 1$, and $a_1(0) = 0$, and derive equations (1.37) and (1.38).

(e) Verify that the final answer (1.37) and (1.38) satisfies the differential equations (1.36a) and (1.36b).

15. The $3^2P_{3/2} - 3^2S_{1/2}$ transition of Na (actually the $F = 3$–$F = 2$ hyperfine transition) has a Rabi frequency of 4.15×10^8 rad/s with a laser intensity of 560 mW/cm^2. What is the transition dipole moment in debye?

16. The lifetime of the $3^2P_{3/2} \rightarrow 3^2S_{1/2}$ transition of the Na atom at 5890 Å is measured to be 16 ns.
    (a) What are the Einstein A and B coefficients for the transition?
    (b) What is the transition dipole moment in debye?
    (c) What is the peak absorption cross section for the transition in Å^2, assuming that the linewidth is determined by lifetime broadening?

17. What are the Doppler linewidths for the pure rotational transition of CO at 115 GHz, the infrared transition of CO_2 at 667 cm^{-1}, and the ultraviolet transition of the Hg atom at 2537 Å, all at room temperature (20°C)?

18. Calculate the transit-time broadening for hydrogen atoms traversing a 1-mm-diameter laser beam. For the velocity of the hydrogen atoms use the rms speed ($v = (3kT/m)^{1/2}$) at room temperature (20°C).

19. At what pressure will the Doppler broadening (FWHM) equal the pressure broadening (FWHM) for a room temperature (20°C) sample of CO gas for a pure rotational transition at 115 GHz, a vibration-rotation transition at 2140 cm^{-1}, and an electronic transition at 1537 Å? Use a "typical" pressure-broadening coefficient of 10 MHz/Torr in all three cases.

20. What are the minimum spectral linewidths of pulsed lasers with pulse durations of 10 fs, 1 ps, 10 ns, and 1 μs?

21. (a) For Na atoms in a flame at 2000 K and 760-Torr pressure calculate the peak absorption cross section (at line center) for the $3^2P_{3/2}$–$3^2S_{1/2}$ transition at 5890 Å. Use 30 MHz/Torr as the pressure-broadening coefficient and the data in Problem 16.
    (b) If the path length in the flame is 10 cm, what concentration of Na atoms will produce an absorption (I/I_0) of $1/e$ at line center?

(c) Is the transition primarily Doppler or pressure broadened?

(d) Convert the peak absorption cross section in cm^2 to the peak molar absorption coefficient ε.

22. For Ar atoms at room temperature (20°C) and 1-Torr pressure, estimate a collision frequency for an atom from the van der Waals radius of 1.5 Å. What is the corresponding pressure-broadening coefficient in MHz/Torr?

23. A stationary atom of mass m emits a photon of energy $h\nu$ and momentum $\hbar k$.

(a) Using the laws of conservation of energy and momentum, show that the shift in frequency of the emitted photon due to recoil of the atom is

$$\frac{\nu - \nu_0}{\nu_0} \approx \frac{-h\nu_0}{2mc^2}.$$

(b) What is the shift in frequency due to recoil of the atom for the Na D line at 5890 Å?

(c) What is the shift in frequency for a γ-ray of energy 1369 keV emitted from ^{24}Na?

24. Consider the introduction of a spontaneous decay at a rate γ from state 1 to 0 for the two-level system.

(a) Explain why the differential equations

$$\dot{a}_0 = +\frac{\gamma}{2} a_1 + i\omega_R e^{i\Delta t} \frac{a_1}{2}$$

and

$$\dot{a}_1 = -\frac{\gamma}{2} a_1 + i\omega_R e^{-i\Delta t} \frac{a_0}{2}$$

are reasonable equations, provided $a_1 \approx 0$ and $a_0 \approx 1$.

(b) Using first-order perturbation theory, show that

$$|a_1(t)|^2 = \frac{(\omega_R/2)^2}{(\omega_{10} - \omega)^2 + (\gamma/2)^2} [1 - 2e^{-\gamma t/2} \cos(\omega_{10} - \omega)t + e^{-\gamma t}].$$

Notice the Lorentzian line-shape function.

References

1. Mills, I. et al. *Quantities, Units and Symbols in Physical Chemistry,* Blackwell, Oxford, 1988.
2. Cohen, E. R. and Taylor, B. N. The 1986 Adjustment of the Fundamental Constants, CODATA Bull. **63,** 1 (1986).

3. Wapstra, A. H. and Audi, G. The 1983 Atomic Mass Evaluation. I. Atomic Mass Table, Nucl. Phys. A. **432,** 1 (1985).
4. Lide, D. R., Ed. *CRC Handbook of Chemistry and Physics,* 73rd ed., CRC Press, Boca Raton, Fla., 1992.
5. Svelto, O. *Principles of Lasers,* 3rd ed., Plenum, New York, 1989, pp. 30–36.
6. Demtröder, W. *Laser Spectroscopy,* Springer-Verlag, Berlin, 1982, p. 98.

General References

Bracewell, R. N. *The Fourier Transform and Its Applications,* 2nd ed., McGraw-Hill, New York, 1986.

Corney, A. *Atomic and Laser Spectroscopy,* Oxford, 1977.

Demtröder, W. *Laser Spectroscopy,* Springer-Verlag, Berlin, 1982.

Fowles, G. R. *Introduction to Modern Optics,* 2nd ed., Dover, New York, 1989.

Letokhov, V. S. and Chebotayev, V. P., *Nonlinear Laser Spectroscopy,* Springer-Verlag, Berlin, 1977.

Levenson, M. D. and Kano, S. S. *Introduction to Nonlinear Laser Spectroscopy,* 2nd ed., Academic Press, San Diego, 1988.

Milonni, P. W. and Eberly, J. H. *Lasers,* Wiley, New York, 1988.

Steinfeld, J. I. *Molecules and Radiation,* 2nd ed., MIT Press, Cambridge, 1985.

Svelto, O. *Principles of Lasers,* 3rd ed., Plenum, New York, 1989.

Yariv, A. *Quantum Electronics,* 3rd ed., Wiley, New York, 1989.

Chapter 2
Molecular Symmetry

The language of group theory has become the language of spectroscopy. The concept of molecular symmetry and its application to the study of spectra of atoms and molecules (in the form of group theory) has proved to be of great value. Group theory is used to label and classify the energy levels of molecules. Group theory also provides qualitative information about the possibility of transitions between these energy levels. For example, the vibrational energy levels of a molecule can be labeled quickly by symmetry type and transitions between energy levels sorted into electric-dipole allowed and electric-dipole forbidden categories.

The concept of molecular symmetry is more subtle than expected because of the continuous motion of the atoms. As the molecule vibrates and rotates, which positions of the nuclei should be chosen as representative? In this book only the symmetry of a molecule at its equilibrium geometry is considered in detail. Only in a few isolated examples, such as in the inversion of ammonia or in bent-linear correlation diagrams, is the possibility of fluxional behavior considered.

In some areas of spectroscopy, such as the study of hydrogen bonded and van der Waals complexes (for example, $(H_2O)_2$), fluxional behavior is the norm rather than the exception.[1] The weak intermolecular bonds between the monomeric units in these systems allow many different geometrical isomers to interconvert rapidly. In this case group theory based on the permutations and inversions of nuclei rather than on the customary symmetry operations is more useful.

Symmetry Operations

The idea of molecular symmetry can be quantified by the introduction of symmetry operations. A symmetry operation is a geometrical action (such as a reflection) that leaves the nuclei of a molecule in indistinguishable positions. These geometrical operations can be classified into four types: reflections ($\hat{\sigma}$), rotations ($\hat{C}_n$), rotation-reflections ($\hat{S}_n$), and inversions ($\hat{i}$). For mathematical reasons a fifth operation, the "do nothing" operation of identity ($\hat{E}$) needs to be added.

Associated with each symmetry operation (except the identity) is a symmetry element. For example, associated with a particular reflection symmetry operation ($\hat{\sigma}$)

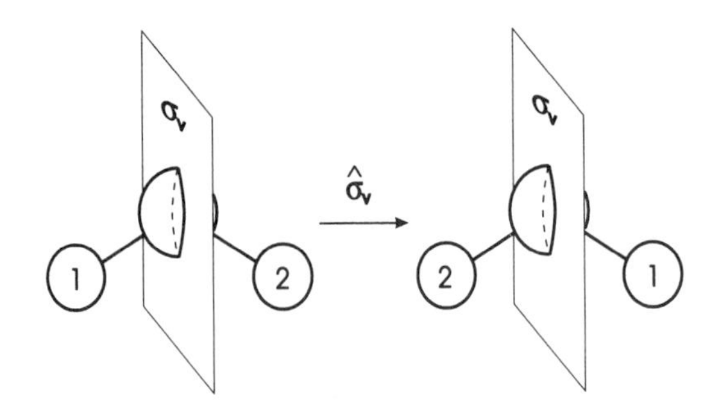

Figure 2.1. For the H_2O molecule, the $\hat{\sigma}_v$ operation reflects hydrogen atom 1 into hydrogen atom 2 (and 2 into 1) through the associated plane of symmetry, σ_v.

is a plane of symmetry (σ) (see Figure 2.1). The distinction between symmetry operators and symmetry elements is quite important and is a source of some confusion. The symmetry operation is the actual action, while the symmetry element is the point, line, or plane about which the action occurs. In this book operators are marked with carets (e.g., $\hat{\sigma}$) in order to distinguish between symmetry operators and symmetry elements.

Operator Algebra

An operator is a mathematical prescription for transforming one function into another. For example, if $\hat{A}$ represents an operator and f and g are two functions that are related by

$$\hat{A}f \rightarrow g, \tag{2.1}$$

then $\hat{A}$ is said to transform f into g. For example, for symmetry operations a reflection might move an atom (represented by Cartesian coordinates x, y, z) to a new location (x', y', z'); this operation is represented by the expression

$$\hat{\sigma}(x, y, z) \rightarrow (x', y', z'). \tag{2.2}$$

Other examples of operators include the differential operator, $\hat{D} \equiv d/dx$, and the exponential operator, $\exp \equiv e^{(\,)}$. Linear operators such as the symmetry operators or the differential operator obey the following rule, namely

$$\hat{A}k(f_1 + f_2) = k\hat{A}f_1 + k\hat{A}f_2, \tag{2.3}$$

where k is a constant and f_1 and f_2 are functions.

Operators can be combined together by addition, namely

$$\hat{C}f = (\hat{A} + \hat{B})f \equiv \hat{A}f + \hat{B}f \qquad \text{or} \qquad \hat{C} = (\hat{A} + \hat{B}) = \hat{A} + \hat{B} \tag{2.4}$$

and multiplication:

$$\hat{Z}f = (\hat{X}\hat{Y})f = \hat{X}(\hat{Y}f) \quad \text{or} \quad \hat{Z} = (\hat{X}\hat{Y}) = \hat{X}\hat{Y}. \tag{2.5}$$

For multiplication of operators (2.5) the operator on the right operates first. Multiplication of operators is simply defined as repeated operations, for example, $\hat{X} \cdot \hat{X} \cdot \hat{Y} \equiv \hat{X}^2\hat{Y}$. The inverse of an operator simply undoes the operation, that is

$$\hat{A}\hat{A}^{-1} = \hat{A}^{-1}\hat{A} = \hat{E} = \hat{1}, \tag{2.6}$$

in which $\hat{E}$ and $\hat{1}$ are the identity or unit operators. For example, ln and exp are an operator–inverse operator pair since $\ln e^{(\,)} = \hat{1}$.

Although operators always have an implied function f to the right, this function is usually suppressed. This leads to a compact notation, but can also lead to confusion or fallacious conclusions, for example, for $\hat{x} = x$ and $\hat{D} = d/dx$,

$$\hat{D}\hat{x} = \hat{1} + \hat{x}\hat{D} \neq \hat{x}\hat{D}.$$

The addition and multiplication of operators strongly resembles ordinary algebra, with the exception that $\hat{A}\hat{B}$ is not necessarily equal to $\hat{B}\hat{A}$. Operators, in general, do not commute, so that

$$\begin{aligned}(\hat{A} + \hat{B})^2 &= \hat{A}^2 + \hat{A}\hat{B} + \hat{B}\hat{A} + \hat{B}^2 \\ &\neq \hat{A}^2 + 2\hat{A}\hat{B} + \hat{B}^2.\end{aligned} \tag{2.7}$$

The similarity between ordinary algebra and operator algebra occurs because addition and multiplication are defined, and the associative law,

$$\hat{A}(\hat{B}\hat{C}) = (\hat{A}\hat{B})\hat{C}, \tag{2.8}$$

and distributive law,

$$\hat{A}(\hat{B} + \hat{C}) = \hat{A}\hat{B} + \hat{A}\hat{C}, \tag{2.9}$$

hold.

$\hat{E}$ Operator

The identity operator leaves a molecule unchanged. The symbol $\hat{E}$ comes from the German word Einheit, which means unity.

$\hat{C}_n$ Operator

The rotation operator rotates a molecule by an angle of $2\pi/n$ radians in a clockwise direction about a C_n-axis (Figure 2.2). A molecule is said to possess an n-fold axis of symmetry if a rotation of $2\pi/n$ radians leaves the nuclei in indistinguishable positions. Although we are able (as a matter of convenience) to label the nuclei in our drawings of molecules, real nuclei carry no labels. When a molecule has several rotational axes of symmetry the one with the largest value of n is called the principal axis.

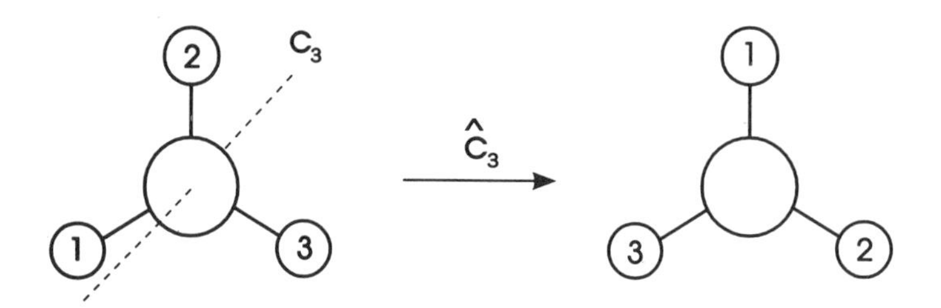

Figure 2.2. For the BF_3 molecule, the $\hat{C}_3$ operator rotates the F atoms by $2\pi/3$ radians = 120° in a clockwise direction about the C_3-axis out of the molecular plane.

Rotations can be repeated

$$\hat{C}_n^k = \hat{C}_n \cdot \hat{C}_n \cdots \hat{C}_n \qquad (k \text{ times}) \tag{2.10}$$

and $\hat{C}_n^n = \hat{E}$ since a rotation by $n(2\pi/n) = 2\pi$ radians corresponds to no rotation at all. A rotation in the counterclockwise direction $\hat{C}_n^{-1}$ (that is, by $-2\pi/n$ radians), has the effect of undoing the $\hat{C}_n$ operation, so

$$\hat{C}_n \hat{C}_n^{-1} = \hat{C}_n^{-1} \hat{C}_n = \hat{E}. \tag{2.11}$$

$\hat{\sigma}$ Operator

The reflection operator reflects a molecule through a plane. If the nuclei are in indistinguishable positions after a reflection operation, then the molecule has a plane of symmetry. The Greek letter σ (sigma) originates from the German word Spiegel for mirror. Since a second reflection undoes the effect of the first reflection, $\hat{\sigma}^2 = \hat{E}$, and the $\hat{\sigma}$ operation is its own inverse. There are three types of mirror planes. They are labeled by subscripts v, h, and d (standing for vertical, horizontal, and dihedral). The vertical σ_v and the horizontal σ_h mirror planes are easy to spot because they either contain the principal axis of the molecule (σ_v) or are perpendicular to it (σ_h). Dihedral planes are more difficult to find and are special cases of vertical planes. A σ_d plane is a vertical plane that also bisects the angle between two adjacent twofold axes (C_2), that are perpendicular to the principal axis. For example, the two types of vertical planes can be seen in Figure 2.3 for benzene; the σ_v planes bisect atoms, while the σ_d planes bisect bonds.

$\hat{S}_n$ Operator

The rotation–reflection operator is made up of a clockwise rotation by $2\pi/n$ radians followed by a reflection in a plane perpendicular to that axis,

$$\hat{S}_n = \hat{\sigma}_h \hat{C}_n \qquad \text{or} \qquad \hat{S}_n = \hat{C}_n \hat{\sigma}_h \tag{2.12}$$

since, in this case, the operations commute (Figure 2.4). If the atomic framework is

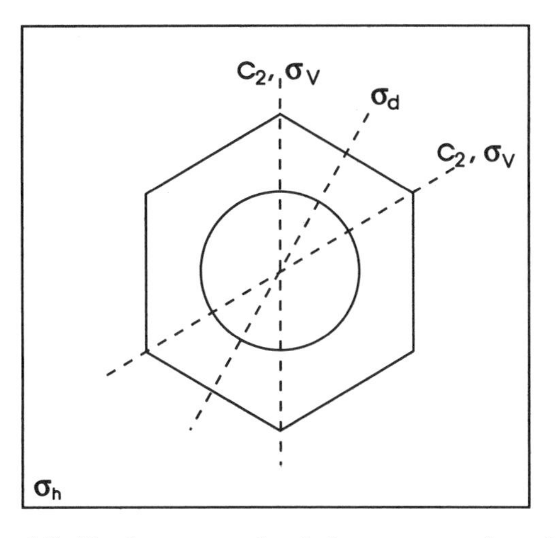

Figure 2.3. The benzene molecule has σ_h, σ_v, and σ_d planes.

unchanged by this operation, then the molecule is said to possess a rotation–reflection axis of symmetry or an improper axis of symmetry. All molecules that have a C_n-axis of symmetry and a σ_h plane of symmetry must also have an S_n-axis along the C_n-axis.

$\hat{\imath}$ Operator

A special case of an improper rotation operation is the inversion operation $\hat{\imath}$. The inversion operator $\hat{\imath}$ inverts all of the atoms of the molecule through a point. If the molecule is coincident with itself after inversion, then it is said to possess a center (point) of symmetry.

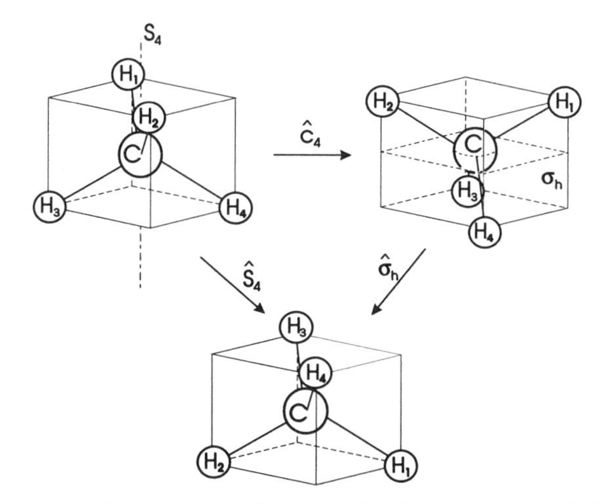

Figure 2.4. The CH_4 molecule has an S_4-axis that coincides with a C_2-axis but does not have a C_4-axis. The $\hat{C}_4$ operator is not a symmetry operator, but $\hat{S}_4 = \hat{\sigma}_h \hat{C}_4$ is one.

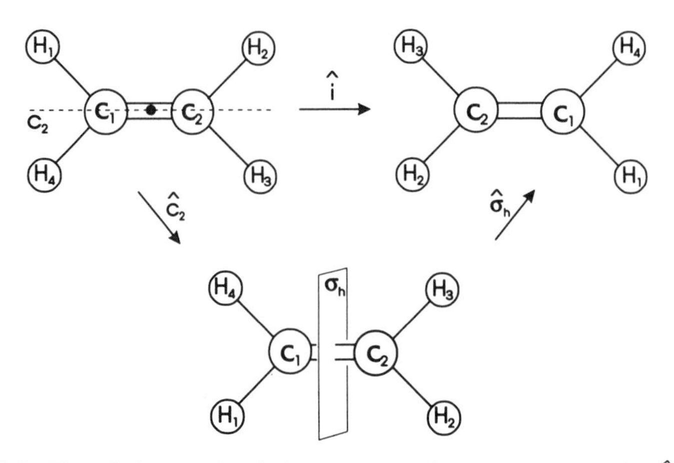

Figure 2.5. The ethylene molecule has a center of symmetry. Note that $\hat{i} = \hat{\sigma}_h \hat{C}_2$.

If the origin of the molecular coordinate system coincides with the center of symmetry (as is customary) (Figure 2.5), then the inversion operation changes the signs of the coordinates of an atom, that is,

$$\hat{i}(x, y, z) \rightarrow (-x, -y, -z). \tag{2.13}$$

The $\hat{S}_2 = \hat{\sigma}_h \hat{C}_2 = \hat{C}_2 \hat{\sigma}_h$ operation is equivalent to inversion because

$$\hat{\sigma}_h \hat{C}_2(x, y, z) \rightarrow \hat{\sigma}_h(-x, -y, z) \rightarrow (-x, -y, -z), \tag{2.14}$$

where the z-axis lies along the C_2-axis. The inversion operator is also its own inverse since

$$(\hat{i})^2 = (\hat{C}_2 \hat{\sigma})^2 = \hat{C}_2^2 \cdot \hat{\sigma}_h^2 = \hat{E}. \tag{2.15}$$

Symmetry Operator Algebra

Symmetry operators can be applied successively to a molecule to produce new operators. For example, consider the ammonia molecule and the rotation and reflection operators. Figure 2.6 shows that the successive application of a rotation operator and a reflection operator generates a new reflection operator. Notice that the operator on the right operates first and that, in general, the operators do not commute, for example,

$$\hat{C}_3 \hat{\sigma}_v'' \neq \hat{\sigma}_v'' \hat{C}_3, \tag{2.16}$$

(Figures 2.6 and 2.7).

Although "division" of symmetry operators is not defined, there is always an

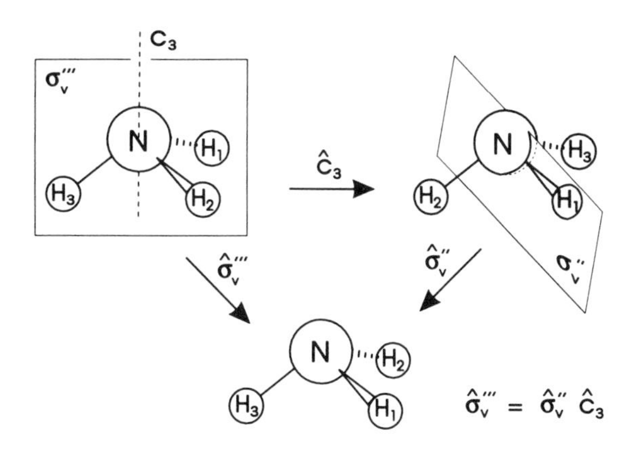

Figure 2.6. The application of $\hat{C}_3$ and then $\hat{\sigma}''$ to the ammonia molecule is equivalent to $\hat{\sigma}'''_v$ ($\hat{\sigma}''_v\hat{C}_3 = \hat{\sigma}'''_v$). Notice that the $\hat{\sigma}'_v$, $\hat{\sigma}''_v$, $\hat{\sigma}'''_v$ planes contain the original H_1, H_2 and H_3 atoms, respectively. The application of symmetry operators interchanges atoms, but the symmetry planes (and their labels) are unaffected.

inverse operator that serves the same function. The inverse operator is useful in algebraic manipulations. For example, consider the product

$$\hat{C}_3\hat{\sigma}''_v = \hat{\sigma}'_v. \tag{2.17}$$

If we operate on the left by the inverse operator $\hat{C}_3^{-1}$, we obtain

$$\hat{C}_3^{-1}\hat{C}_3\hat{\sigma}''_v = \hat{C}_3^{-1}\hat{\sigma}'_v, \tag{2.18}$$

or

$$\hat{\sigma}''_v = \hat{C}_3^{-1}\hat{\sigma}'_v. \tag{2.19}$$

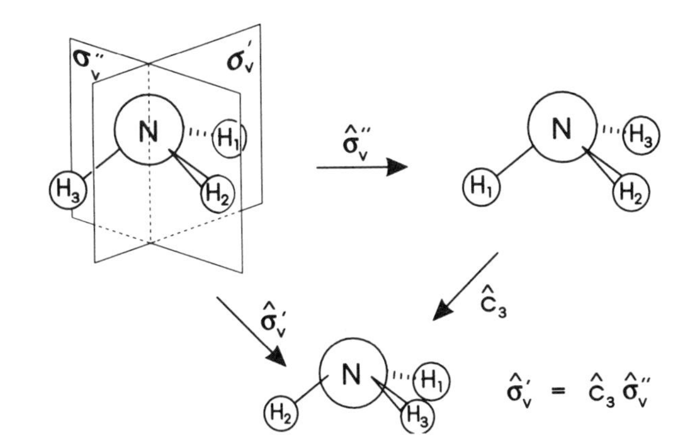

Figure 2.7. The application of $\hat{\sigma}''_v$ and then $\hat{C}_3$ to the ammonia molecule is equivalent to $\hat{\sigma}'_v$ ($\hat{C}_3\hat{\sigma}''_v = \hat{\sigma}'_v$).

However, if we operate on the right by the inverse operator $(\hat{\sigma}''_v)^{-1} = \hat{\sigma}''_v$, we obtain

$$\hat{C}_3(\hat{\sigma}''_v)^2 = \hat{\sigma}'_v\hat{\sigma}''_v \tag{2.20}$$

or

$$\hat{C}_3 = \hat{\sigma}'_v\hat{\sigma}''_v. \tag{2.21}$$

In the preceding examples multiplication from the left by $\hat{C}_3^{-1}$ (equation (2.18)) or from the right by $\hat{\sigma}''_v = (\hat{\sigma}''_v)^{-1}$ (equation (2.20)) produces two new equations, (2.19) and (2.21).

Care must be taken to preserve the order of operators. Taking the inverse of a product operator reverses the order of the component operators, that is

$$(\hat{A}\hat{B}\hat{C})^{-1} = \hat{C}^{-1}\hat{B}^{-1}\hat{A}^{-1}. \tag{2.22}$$

To illustrate $(\hat{A}\hat{B})^{-1} = \hat{B}^{-1}\hat{A}^{-1}$ we can use the symmetry operators associated with the NH_3 molecule. For instance, from expression (2.17) we obtain

$$(\hat{C}_3\hat{\sigma}''_v)^{-1} = (\hat{\sigma}'_v)^{-1} = \hat{\sigma}'_v. \tag{2.23}$$

Multiplying equation (2.23) by $\hat{\sigma}''_v$ from the right

$$(\hat{C}_3\hat{\sigma}''_v)^{-1}(\hat{\sigma}''_v) = \hat{\sigma}'_v\hat{\sigma}''_v, \tag{2.24}$$

and then $\hat{C}_3^{-1}$ from the right, we obtain, using equations (2.17) and (2.21),

$$(\hat{C}_3\hat{\sigma}''_v)^{-1}(\hat{\sigma}''_v)\hat{C}_3^{-1} = \hat{\sigma}'_v\hat{\sigma}''_v\hat{C}_3^{-1} = \hat{C}_3\hat{C}_3^{-1} = \hat{E}. \tag{2.25}$$

This expression (2.25) is equivalent to

$$(\hat{C}_3\hat{\sigma}''_v)^{-1}((\hat{\sigma}''_v)^{-1}\hat{C}_3^{-1}) = \hat{E} \tag{2.26}$$

or

$$(\hat{C}_3\hat{\sigma}''_v)^{-1} = (\hat{\sigma}''_v)^{-1}\hat{C}_3^{-1} \tag{2.27}$$

as required.

Groups

A group is a set of elements along with a combining operation ("multiplication") which obey the following four rules:

1. *Closure.* The product of any two elements must also be in the group. If P and Q are members of the group and $PQ = R$, then R is also a member of the group.

2. *Associative Law.* As long as the elements are not interchanged the order of multiplication is immaterial, $(PQ)R = P(QR)$.

3. *Identity element.* There is an identity element E in the group, $RE = ER = R$.

4. *Inverse.* Every element R has an inverse R^{-1} in the group, $RR^{-1} = R^{-1}R = E$.

This definition of a group is very general. The elements could be, for example, numbers, matrices, or symmetry operators and the combining operators could be addition, multiplication, or matrix multiplication. Note also that the elements of a group do not necessarily commute, that is $PQ \neq QP$. If the group elements *do* commute with one another, then the group is called an *Abelian group.*

An example of a group (with an infinite number of members) is the set of positive and negative integers, including zero, under the operation of addition. The numbers 1, -1, $i = \sqrt{-1}$, and $-i$ form a group if the combining operation is multiplication. The number of elements in the group is called the *order*, so that $\{1, -1, i, -i\}$ forms a group of order 4 under multiplication.

Point Groups

Point symmetry groups are groups whose elements are the symmetry operations of molecules. They are called *point groups* because the center of mass of the molecule remains unchanged under all symmetry operations and all of the symmetry elements meet at this point.

For example, the point group associated with the ammonia molecule has six members, $\{\hat{E}, \hat{C}_3, \hat{C}_3^{-1}, \hat{\sigma}'_v, \hat{\sigma}''_v, \hat{\sigma}'''_v\}$, associated with the three vertical planes of symmetry and the C_3 axis of symmetry (Figure 2.8). The six symmetry operations can be

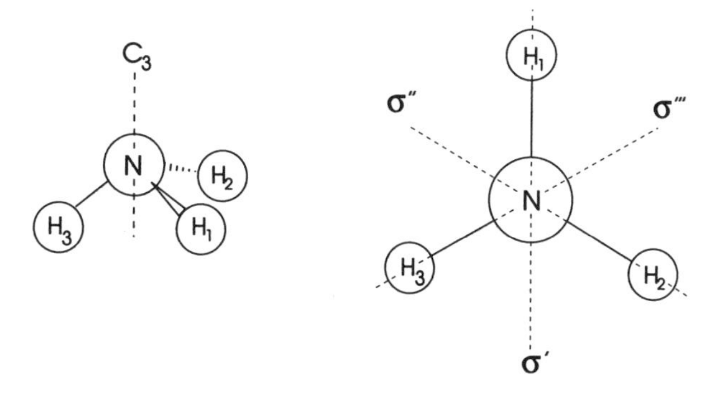

Figure 2.8. The elements of symmetry for the ammonia molecule.

Table 2.1. Group Multiplication Table for the Symmetry Operations Associated with NH_3

C_{3v}	$\hat{E}$	$\hat{\sigma}'_v$	$\hat{\sigma}''_v$	$\hat{\sigma}'''_v$	$\hat{C}_3$	$\hat{C}_3^{-1}$
$\hat{E}$	$\hat{E}$	$\hat{\sigma}'_v$	$\hat{\sigma}''_v$	$\hat{\sigma}'''_v$	$\hat{C}_3$	$\hat{C}_3^{-1}$
$\hat{\sigma}'_v$	$\hat{\sigma}'_v$	$\hat{E}$	$\hat{C}_3$	$\hat{C}_3^{-1}$	$\hat{\sigma}''_v$	$\hat{\sigma}'''_v$
$\hat{\sigma}''_v$	$\hat{\sigma}''_v$	$\hat{C}_3^{-1}$	$\hat{E}$	$\hat{C}_3$	$\hat{\sigma}'''_v$	$\hat{\sigma}'_v$
$\hat{\sigma}'''_v$	$\hat{\sigma}'''_v$	$\hat{C}_3$	$\hat{C}_3^{-1}$	$\hat{E}$	$\hat{\sigma}'_v$	$\hat{\sigma}''_v$
$\hat{C}_3$	$\hat{C}_3$	$\hat{\sigma}'''_v$	$\hat{\sigma}'_v$	$\hat{\sigma}''_v$	$\hat{C}_3^{-1}$	$\hat{E}$
$\hat{C}_3^{-1}$	$\hat{C}_3^{-1}$	$\hat{\sigma}''_v$	$\hat{\sigma}'''_v$	$\hat{\sigma}'_v$	$\hat{E}$	$\hat{C}_3$

combined, and the results of all possible products are summarized in the group multiplication table (Table 2.1). The multiplication table is read by picking the column headed by the operator applied first (e.g., $\hat{\sigma}'_v$) and the row with the second operator (e.g., $\hat{\sigma}''_v$) and finding the symmetry operator $\hat{C}_3^{-1}$ at their intersection; thus, $\hat{\sigma}''\hat{\sigma}'_v = \hat{C}_3^{-1}$. Notice that each operator appears in a given row or column of the multiplication table just once but in a different position. This result is known as the *Rearrangement Theorem.*

Two important terms in group theory are isomorphic and homomorphic. Two groups are isomorphic if there is a one-to-one correspondence between the elements of the two groups such that $AB = C$ implies that $A'B' = C'$ (Figure 2.9). The two groups, therefore, have the same multiplication table except for a change in symbols or in the meaning of the operators. For example, the two groups G_1 and G_2

$$G_1 = \{1, i, -1, -i\}$$

$$G_2 = \left\{ \begin{pmatrix} 1 & 0 \\ 0 & 1 \end{pmatrix} \begin{pmatrix} 0 & 1 \\ -1 & 0 \end{pmatrix} \begin{pmatrix} -1 & 0 \\ 0 & -1 \end{pmatrix} \begin{pmatrix} 0 & -1 \\ 1 & 0 \end{pmatrix} \right\}$$

are isomorphic with the combining operations for G_1 and G_2 being multiplication and matrix multiplication, respectively. Two groups are homomorphic if there is a many-to-one relationship between some of the elements of the two groups. The structure of the two homomorphic groups is no longer identical in form, but the multiplication rules are preserved since $AB = C$ and $A'B' = C'$. A homomorphic

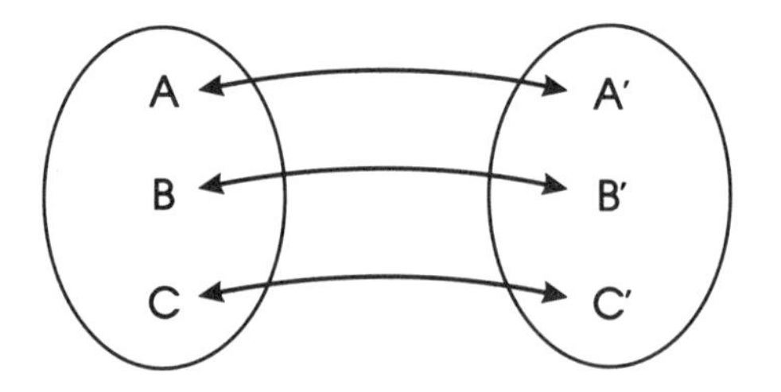

Figure 2.9. An isomorphic mapping.

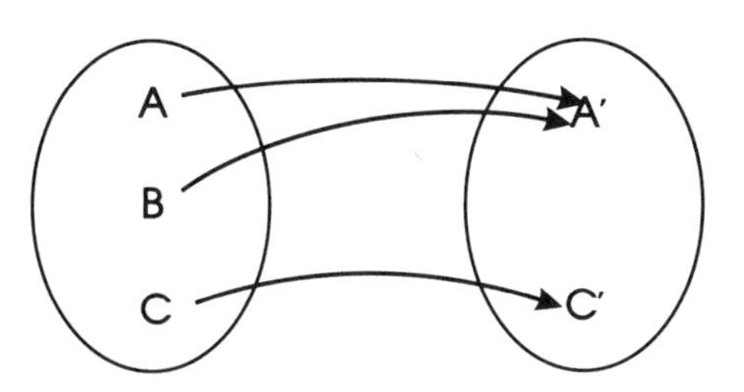

Figure 2.10. A homomorphic mapping.

mapping allows a many-to-one correspondence between the elements of two groups, Figure 2.10. For example, there is the trivial homomorphic relationship between any group (e.g., G_1) and the group G_3, containing the number 1 as the only element,

$$G_1 = \{1, i, -1, -i\}$$

$$G_3 = \{1\}.$$

Classes

The members of a group can be divided into classes. Two members of a group, P and R, belong to the same class if another member Z can be found such that $P = Z^{-1}RZ$: P and R are said to be conjugate to each other and they form a class.

For example, consider all possible classes associated with the symmetry group of ammonia. Clearly $\hat{E}$ is in a class by itself since $\hat{A}^{-1}\hat{E}\hat{A} = \hat{A}^{-1}\hat{A} = \hat{E}$ for any $\hat{A}$. For $\hat{C}_3$ all possible other members of the class are

$$\hat{\sigma}'_v\hat{C}_3\hat{\sigma}'_v = \hat{\sigma}'_v\hat{\sigma}'''_v = \hat{C}_3^{-1} \tag{2.28}$$

$$\hat{\sigma}''_v\hat{C}_3\hat{\sigma}''_v = \hat{\sigma}''_v\hat{\sigma}'_v = \hat{C}_3^{-1} \tag{2.29}$$

$$\hat{\sigma}'''_v\hat{C}_3\hat{\sigma}'''_v = \hat{\sigma}'''_v\hat{\sigma}''_v = \hat{C}_3^{-1} \tag{2.30}$$

$$\hat{C}_3^{-1}\hat{C}_3\hat{C}_3 = \hat{E}\hat{C}_3 = \hat{C}_3 \tag{2.31}$$

$$\hat{C}_3\hat{C}_3\hat{C}_3^{-1} = \hat{C}_3\hat{E} = \hat{C}_3 \tag{2.32}$$

so that $\hat{C}_3$ and $\hat{C}_3^{-1}$ are in the same class. Similarly for $\hat{\sigma}'_v$ all possible other members of the class are:

$$\hat{\sigma}'_v\hat{\sigma}'_v\hat{\sigma}'_v = \hat{\sigma}'_v\hat{E} = \hat{\sigma}'_v \tag{2.33}$$

$$\hat{\sigma}''_v\hat{\sigma}'_v\hat{\sigma}''_v = \hat{\sigma}''_v\hat{C}_3 = \hat{\sigma}'''_v \tag{2.34}$$

$$\hat{\sigma}'''_v\hat{\sigma}'_v\hat{\sigma}'''_v = \hat{\sigma}'''_v\hat{C}_3^{-1} = \sigma''_v \tag{2.35}$$

$$\hat{C}_3^{-1}\hat{\sigma}'_v\hat{C}_3 = \hat{C}_3^{-1}\hat{\sigma}''_v = \hat{\sigma}'''_v \tag{2.36}$$

$$\hat{C}_3\hat{\sigma}'_v\hat{C}_3^{-1} = \hat{C}_3^{-1}\hat{\sigma}'''_v = \hat{\sigma}''_v \tag{2.37}$$

so that $\hat{\sigma}'_v$, $\hat{\sigma}''_v$, and $\hat{\sigma}'''_v$ are in the same class. The symmetry group of order six for ammonia thus has three classes: $\hat{E}$; $\hat{C}_3$, and $\hat{C}_3^{-1}$; $\hat{\sigma}'_v$, $\hat{\sigma}''_v$, and $\hat{\sigma}'''_v$. Although it is not obvious from the mathematical definition of a class, the members of a class of a point group have a geometrical relationship to each other, for example, all involve reflections or rotations of a certain type.

Subgroups

A subgroup is a subset of the group operations of the full group that also is a group. For example, the rotational subgroup associated with the NH_3 point group is

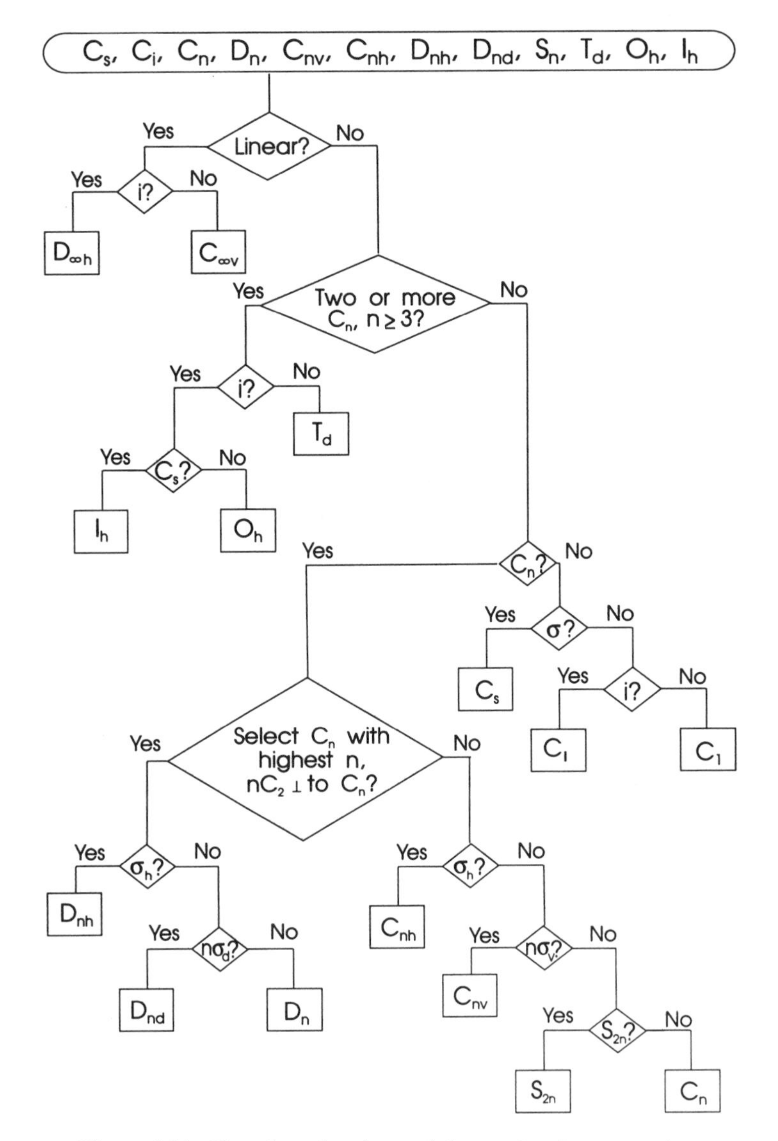

Figure 2.11. Flowchart for determining molecular symmetry.

$\{\hat{E}, \hat{C}_3, \hat{C}_3^{-1}\}$. The order of a subgroup is always a factor of the order of the full group, that is, if g is the order of the full group and g_1 is the order of the subgroup, then g is exactly divisible by g_1.

Notation for Point Groups

In this book the Schoenflies notation is used to label the possible point groups. In the list below only the essential symmetry elements are specified for the various groups, except for T_d, O_h, and K_h, which can be recognized by inspection. Notice that the structures are assumed to be rigid; the possibility of rotation around carbon-carbon single bonds is ignored. A flowchart is provided (Figure 2.11) as an aid for determining molecular symmetry.

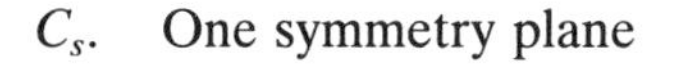

C_s. One symmetry plane

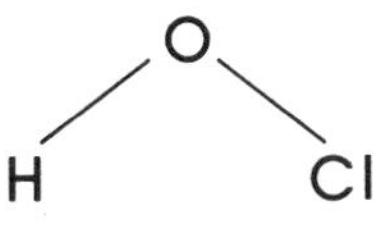

C_i. One center of symmetry

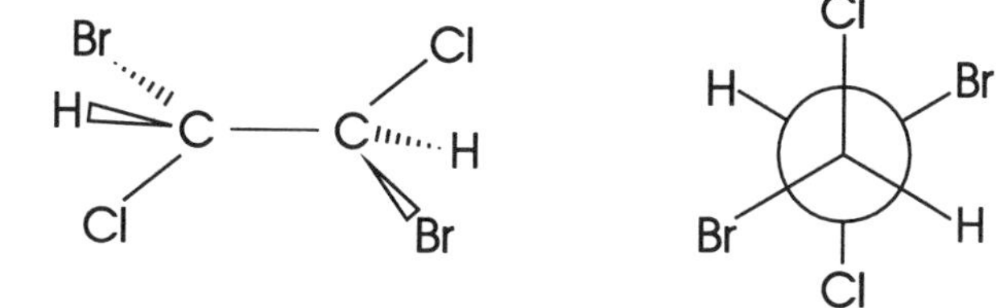

C_n. A simple C_n-axis.

C_1. No elements of symmetry.

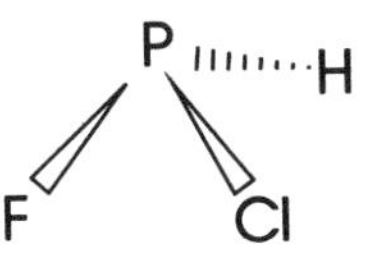

C_3. CCl_3—CH_3. Neither eclipsed nor staggered.

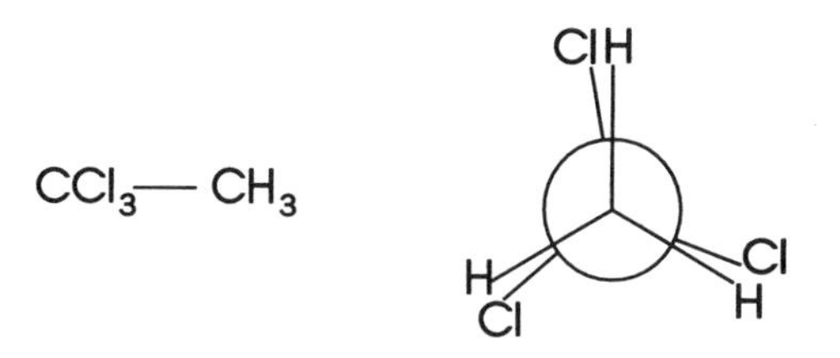

D_n. One C_n-axis and nC_2-axes perpendicular to it.

D_3. CH_3—CH_3. Neither eclipsed nor staggered.

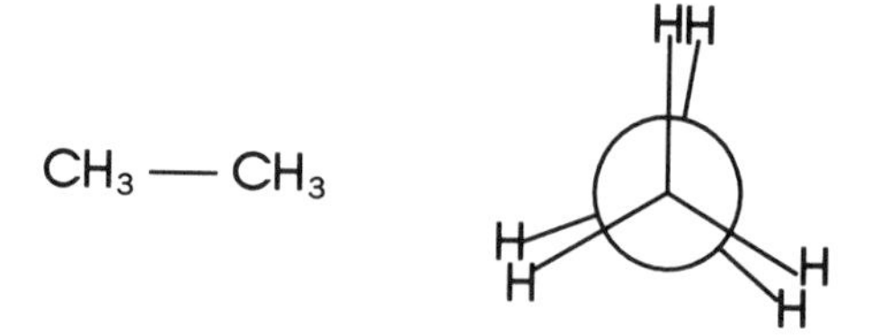

C_{nv}. One C_n-axis with $n\sigma_v$ planes.

C_{3v}

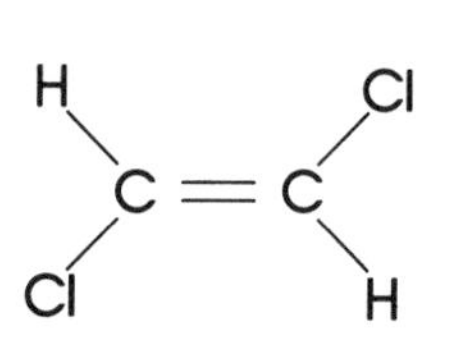

C_{nh}. One C_n-axis with a σ_h plane.

C_{2h}. *trans*-Dichloroethylene.

D_{nh}. One C_n-axis, nC_2-axes perpendicular to the C_n-axis, and one σ_h plane.

D_{6h}. Benzene $D_{\infty h}$ Cl—Cl

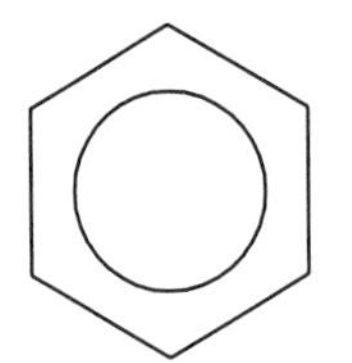

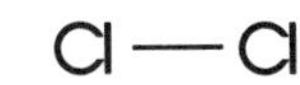

D_{nd}. One C_n-axis, nC_2-axes perpendicular to the C_n-axis, and $n\sigma_d$ planes.

D_{2d}. Allene

S_n. One S_n-axis.

S_4. Puckered octagon

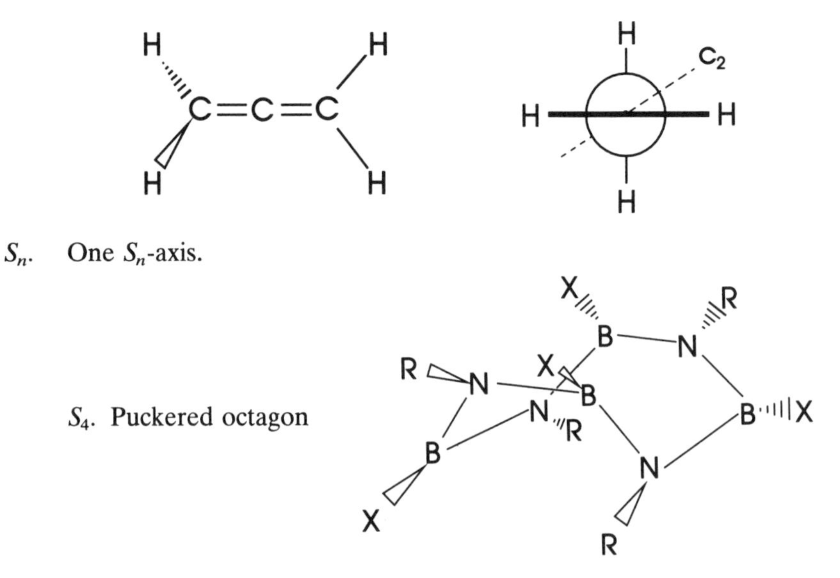

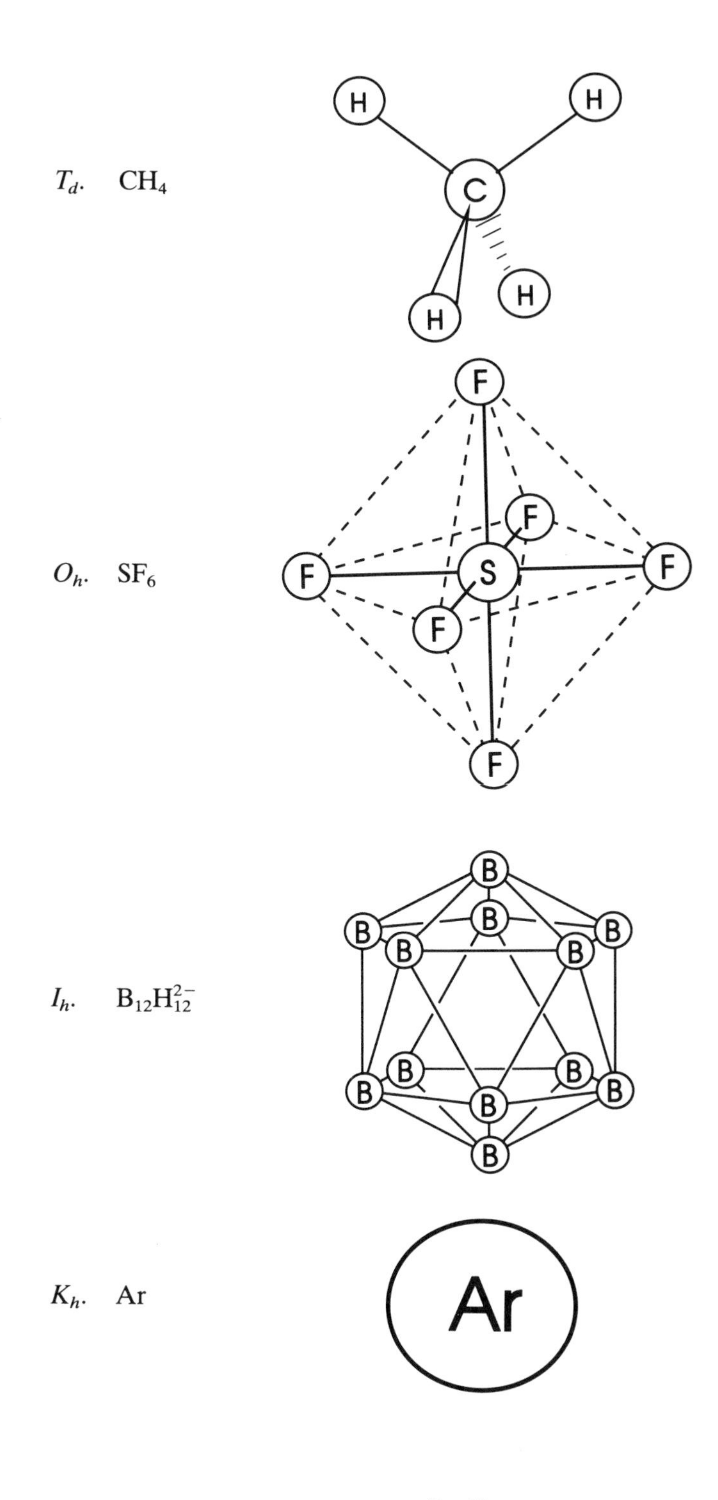

Problems

1. Construct the group multiplication table for the symmetry operations of the H_2CO molecule.

2. Construct the group multiplication table for the symmetry operations of the BF_3 molecule.
3. What are the point groups for the following molecules:
   (a) CF_4
   (b) CH_2ClBr
   (c) CH_3Br
   (d) PH_3
   (e) BCl_3
   (f) Cyclohexane, chair
   (g) Cyclohexane, boat
   (h) Ferrocene, staggered
   (i) Acetylene
   (j) Ethylene
4. List all the symmetry operations for the following molecules:
   (a) *trans*-Dichloroethylene
   (b) CH_3—CH_3, neither eclipsed nor staggered
   (c) SF_6
   (d) Allene
   (e) NC—CN
5. What are the point groups of the following square planar complexes:
   (a) $[Pt(Cl)_4]^{2-}$
   (b) $[Pt(Cl)_3CN]^{2-}$
   (c) *cis*-$[Pt(Cl)_2(CN)_2]^{2-}$
   (d) *trans*-$[Pt(Cl)_2(CN)_2]^{2-}$
6. What are the point groups of the following octahedral molecules:
   (a) MA_6
   (b) MA_5B
   (c) cis-MA_4B_2
   (d) *trans*-MA_4B_2
   (e) *fac*-MA_3B_3
   (f) *mer*-MA_3B_3
7. It is generally stated that a molecular species is optically active if its mirror image is not superimposable upon the original structure. A more universally applicable test is for the presence of an improper rotation axis. When an improper axis is present, the structure is optically inactive. Using this criterion, determine whether or not the following structures are optically active:
   (a) *trans*-1,2-Dichlorocyclopropane
   (b) Ethane (neither staggered nor eclipsed)
   (c) CHFClBr
   (d) $Co(en)_3^{3+}$
   (e) *cis*-$[Co(en)_2Cl_2]^+$
   (f) *trans*-$[Co(en)_2Cl_2]^+$

Reference

1. Cohen, R. C. and Saykally, R. J. J. Phys. Chem. **96,** 1024 (1992).

General References

Bishop, D. M. *Group Theory and Chemistry,* Dover, New York, 1993.

Boardman, A. D., O'Connor, D. E. and Young, P. A. *Symmetry and Its Applications in Science,* McGraw-Hill, Maidenhead, England, 1973.

Bunker, P. R. *Molecular Symmetry and Spectroscopy,* Academic Press, New York, 1979.

Cotton, F. A. *Chemical Applications of Group Theory,* 3rd ed., Wiley, New York, 1990.

Douglas, B. E. and Hollingsworth, C. A. *Symmetry in Bonding and Spectra,* Academic Press, San Diego, 1985.

Hamermesh, M. *Group Theory and Its Application to Physical Problems,* Dover, New York, 1989.

Hochstrasser, R. M. *Molecular Aspects of Symmetry,* Benjamin, New York, 1966.

Chapter 3
Matrix Representation of Groups

So far the symmetry operations associated with a molecule have been defined in geometric terms. If the atoms are represented by their Cartesian coordinates, then the symmetry operators can be represented by matrices. In general a set of matrices homomorphic with the point group operations can be found by the methods described in this chapter. These matrices form a representation of the group. A brief summary of some properties of matrices is made first. Proofs of these properties are left as exercises or can be found in various standard references.

Vectors and Matrices

A point in space can be described by the vector (Figure 3.1)

$$\mathbf{r} = x\hat{\mathbf{i}} + y\hat{\mathbf{j}} + z\hat{\mathbf{k}} \tag{3.1}$$

where in this context the circumflexes are used to denote unit vectors rather than operators. The magnitude of the vector is given by

$$|\mathbf{r}| = r = (x^2 + y^2 + z^2)^{1/2} \tag{3.2}$$

and its direction is specified by the angles α, β, and γ.

If $\mathbf{A} = A_x\hat{\mathbf{i}} + A_y\hat{\mathbf{j}} + A_z\hat{\mathbf{k}}$ and $\mathbf{B} = B_x\hat{\mathbf{i}} + B_y\hat{\mathbf{j}} + B_z\hat{\mathbf{k}}$, then the dot product of two vectors $\mathbf{A}$ and $\mathbf{B}$ is the scalar

$$\mathbf{A} \cdot \mathbf{B} = A_xB_x + A_yB_y + A_zB_z = |\mathbf{A}|\,|\mathbf{B}| \cos\theta, \tag{3.3}$$

where θ is the angle between the two vectors. The cross product of two vectors $\mathbf{A}$ and $\mathbf{B}$ gives a third vector, $\mathbf{C} = \mathbf{A} \times \mathbf{B}$. The magnitude of $\mathbf{C}$ is given by

$$|\mathbf{C}| = |\mathbf{A}|\,|\mathbf{B}| \sin\theta \tag{3.4}$$

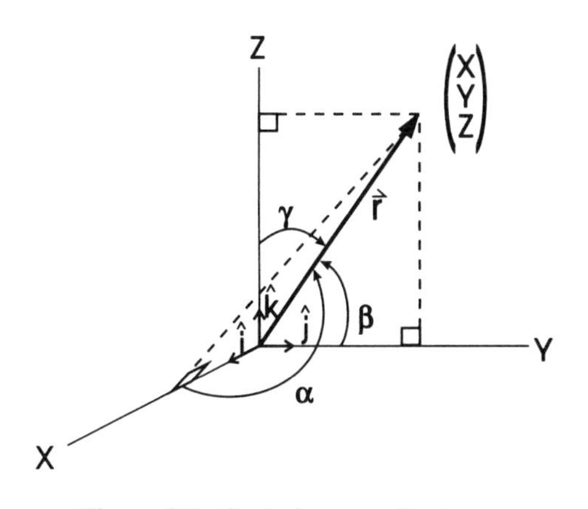

Figure 3.1. Cartesian coordinates.

and the direction of **C** is perpendicular to the plane defined by **A** and **B**. In terms of components

$$\mathbf{C} = (A_yB_z - A_zB_y)\hat{\mathbf{i}} + (A_zB_x - A_xB_z)\hat{\mathbf{j}} + (A_xB_y - A_yB_x)\hat{\mathbf{k}} \tag{3.5}$$

or in terms of a determinant (defined later)

$$\mathbf{C} = \begin{vmatrix} \hat{\mathbf{i}} & \hat{\mathbf{j}} & \hat{\mathbf{k}} \\ A_x & A_y & A_z \\ B_x & B_y & B_z \end{vmatrix}. \tag{3.6}$$

Polar coordinates r, θ, ϕ are defined by the following diagram (Figure 3.2) so that **r** is represented by

$$\mathbf{r} = r(\sin\theta\cos\phi\hat{\mathbf{i}} + \sin\theta\sin\phi\hat{\mathbf{j}} + \cos\theta\hat{\mathbf{k}}). \tag{3.7}$$

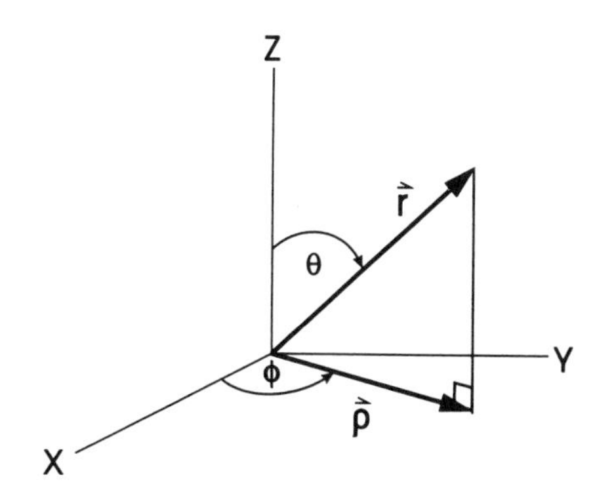

Figure 3.2. Polar coordinates.

A matrix is in general an $m \times n$ rectangular array of numbers

$$\mathbf{A} = \begin{pmatrix} A_{11} & A_{12} & A_{13} & \cdots & A_{1n} \\ A_{21} & A_{22} & A_{23} & \cdots & A_{2n} \\ \vdots & \vdots & \vdots & \cdots & \vdots \\ A_{m1} & A_{m2} & A_{m3} & \cdots & A_{mn} \end{pmatrix} \tag{3.8}$$

with a typical element A_{ij}. A vector is, in general, a one-dimensional matrix—either an $n \times 1$ matrix (column vector) or a $1 \times n$ matrix (row vector). In this book only square $n \times n$ matrices are used and all vectors are assumed to be column vectors, for example

$$\mathbf{x} = \begin{pmatrix} x_1 \\ x_2 \\ \vdots \\ x_n \end{pmatrix}. \tag{3.9}$$

Row vectors $\mathbf{x}' = (x_1 x_2 \cdots x_n)$ are the transpose of the corresponding column vectors.

A number of matrix operators are employed, including:

The complex conjugate $\mathbf{A}^*$, where $(A^*)_{ij} = (A_{ij})^*$; (3.10)

The transpose $\mathbf{A}'$, where $(A')_{ij} = A_{ji}$; (3.11)

The Hermitian conjugate $\mathbf{A}^\dagger = (\mathbf{A}')^* = (\mathbf{A}^*)'$, where $(A^\dagger)_{ij} = (A_{ji})^*$; (3.12)

The inverse $\mathbf{A}^{-1}$, where $\mathbf{A}\mathbf{A}^{-1} = \mathbf{A}^{-1}\mathbf{A} = \mathbf{1}$; (3.13)

The trace $\mathrm{tr}(\mathbf{A})$, where $\mathrm{tr}(\mathbf{A}) = \sum_{i=1}^{n} A_{ii}$; (3.14)

The determinant $|\mathbf{A}|$, where $|\mathbf{A}| = \sum_{j=1}^{n} A_{ij} M_{ij}$. (3.15)

The M_{ij} terms are called *cofactors* and they are obtained by striking out the ith row and the jth column of the original determinant and multiplying by $(-1)^{i+j}$:

$$M_{ij} = (-1)^{i+j} \begin{vmatrix} A_{11} & \cdots & A_{1j} & \cdots & A_{1n} \\ \vdots & \cdots & & \cdots & \vdots \\ A_{i1} & \cdots & A_{ij} & \cdots & A_{in} \\ \vdots & \cdots & & \cdots & \vdots \\ A_{n1} & \cdots & A_{nj} & \cdots & A_{nn} \end{vmatrix}. \tag{3.16}$$

The smaller determinants ($(n-1) \times (n-1)$ in size) in the cofactors can be expanded in a similar fashion to create in the end after repeated expansion, the sum of $n!$ terms.

For example,

$$\begin{vmatrix} 1 & 2 & 3 \\ 4 & 5 & 6 \\ 3 & 2 & 1 \end{vmatrix} = 1\begin{vmatrix} 5 & 6 \\ 2 & 1 \end{vmatrix} - 2\begin{vmatrix} 4 & 6 \\ 3 & 1 \end{vmatrix} + 3\begin{vmatrix} 4 & 5 \\ 3 & 2 \end{vmatrix}$$
$$= 1(5-12) - 2(4-18) + 3(8-15)$$
$$= 0.$$

Matrix addition

$$\mathbf{A} + \mathbf{B} = \mathbf{C}$$

is defined only if **A** and **B** have the same number of rows and columns, so that

$$A_{ij} + B_{ij} = C_{ij}. \tag{3.17}$$

The matrix product **C** of two matrices **A** and **B** is written as

$$\mathbf{A} \cdot \mathbf{B} = \mathbf{AB} = \mathbf{C},$$

with the matrix element C_{ij} of **C** given by

$$C_{ij} = \sum_{k=1}^{n} A_{ik} B_{kj}. \tag{3.18}$$

As an example

$$\begin{pmatrix} 1 & 2 \\ 3 & 4 \end{pmatrix} \begin{pmatrix} 5 & 6 \\ 7 & 8 \end{pmatrix} = \begin{pmatrix} 19 & 22 \\ 43 & 50 \end{pmatrix}.$$

Matrix multiplication, however, is not commutative since

$$(\mathbf{BA})_{ij} = \sum_{k=1}^{n} B_{ik} A_{kj} \qquad \text{while} \qquad (\mathbf{AB})_{ij} = \sum_{k=1}^{n} A_{ik} B_{kj}. \tag{3.19}$$

This can be illustrated, for example, by interchanging the order of multiplication in the preceding example, resulting in a different matrix:

$$\begin{pmatrix} 5 & 6 \\ 7 & 8 \end{pmatrix} \begin{pmatrix} 1 & 2 \\ 3 & 4 \end{pmatrix} = \begin{pmatrix} 23 & 34 \\ 31 & 46 \end{pmatrix}.$$

Special matrices include:

The identity matrix **1**, where $(\mathbf{1})_{ij} = \delta_{ij}$; (3.20)

(the Kronecker delta δ_{ij} has a value of 1 for $i = j$ or 0 for $i \neq j$)

A symmetric matrix, where $\mathbf{A}' = \mathbf{A}$; (3.21)

A Hermitian matrix, where$\mathbf{A}^{\dagger} = \mathbf{A}$; (3.22)

An orthogonal matrix, where $\mathbf{A}' = \mathbf{A}^{-1}$; (3.23)

A unitary matrix, where $\mathbf{A}^{\dagger} = \mathbf{A}^{-1}$. (3.24)

Matrices are useful in transforming the magnitude and direction of a vector. Thus if $\mathbf{y}$ is an $n \times 1$ vector and $\mathbf{A}$ is an $n \times n$ matrix, then a new $n \times 1$ vector $\mathbf{y}'$ is related to $\mathbf{y}$ by the matrix equation $\mathbf{y}' = \mathbf{A}\mathbf{y}$, or

$$\begin{pmatrix} y_1' \\ y_2' \\ \vdots \\ y_n' \end{pmatrix} = \begin{pmatrix} A_{11} \cdots A_{1n} \\ \\ A_{n1} \cdots A_{nn} \end{pmatrix} \begin{pmatrix} y_1 \\ \vdots \\ y_n \end{pmatrix}. \tag{3.25}$$

Unitary (or orthogonal) transformations of vectors do not alter the magnitude of a vector, but do change its direction. Matrices can also be used in inverting a set of linear equations. Thus, if a set of linear equations is represented by

$$\mathbf{A}\mathbf{x} = \mathbf{c}, \tag{3.26}$$

then multiplication from the left by $\mathbf{A}^{-1}$ gives $\mathbf{x}$ in terms of $\mathbf{c}$ via

$$\begin{aligned} \mathbf{A}^{-1}\mathbf{A}\mathbf{x} &= \mathbf{A}^{-1}\mathbf{c} \\ \mathbf{x} &= \mathbf{A}^{-1}\mathbf{c}, \end{aligned} \tag{3.27}$$

where $\mathbf{c}$ is a column vector of constants.

The inverse matrix can be calculated in a number of ways, such as

$$(A^{-1})_{ij} = \frac{M_{ji}}{|\mathbf{A}|} \tag{3.28}$$

or

$$\mathbf{A}^{-1} = \frac{1}{|\mathbf{A}|} \begin{pmatrix} M_{11} & \cdots & M_{n1} \\ \vdots & & \vdots \\ M_{1n} & \cdots & M_{nn} \end{pmatrix}, \tag{3.29}$$

where M_{ij} is the cofactor of $\mathbf{A}$ defined in equation (3.16).

Matrix Eigenvalue Problem

The $n \times n$ matrix $\mathbf{A}$ has n eigenvalues and n eigenvectors defined by the equation

$$\mathbf{A}\mathbf{x}_i = \lambda_i \mathbf{x}_i, \qquad i = 1, \ldots, n.$$

The eigenvectors are special vectors whose directions are unchanged (but are stretched or shrunk by the factor λ) when transformed by **A**. These n eigenvectors are determined by solving a set of n linear homogeneous equations

$$\mathbf{Ax} = \lambda \mathbf{1x}, \tag{3.30}$$

or

$$(\mathbf{A} - \lambda \mathbf{1})\mathbf{x} = 0. \tag{3.31}$$

These equations have a trivial solution, $\mathbf{x} = \mathbf{0}$, and a set of n nontrivial solutions obtained from the nth-order polynomial in λ (also referred to as the secular determinant or secular equation),

$$|\mathbf{A} - \lambda \mathbf{1}| = 0. \tag{3.32}$$

Each eigenvector $\mathbf{x}_i$ associated with λ_i can be determined by substituting λ_i in equation (3.31) and solving. Usually the eigenvectors are normalized to unity:

$$|\mathbf{x}| = (\mathbf{x}^\dagger \mathbf{x})^{1/2} = \sqrt{\sum_{i=1}^{n} x_i^* x_i} = \sqrt{\sum_{i=1}^{n} |x_i|^2}. \tag{3.33}$$

These n eigenvectors, expressed as column vectors, can be arranged in a matrix. $\mathbf{X} = (\mathbf{x}_1, \mathbf{x}_2 \cdots \mathbf{x}_n)$ and the n eigenvalue equations written as

$$\mathbf{AX} = \mathbf{X\Lambda}, \tag{3.34}$$

in which $\mathbf{\Lambda}$ is the diagonal matrix

$$\mathbf{\Lambda} = \begin{pmatrix} \lambda_1 & 0 & \cdots & 0 \\ 0 & \lambda_2 & \cdots & 0 \\ \vdots & \vdots & \cdots & \vdots \\ 0 & 0 & \cdots & \lambda_n \end{pmatrix}$$

made up of the eigenvalues λ_i. From this relation (3.34), we see that $\mathbf{\Lambda}$ can be determined from **A** as

$$\mathbf{\Lambda} = \mathbf{X}^{-1}\mathbf{AX}. \tag{3.35}$$

For example, if the matrix **A** is

$$\mathbf{A} = \begin{pmatrix} 4 & 1 \\ 2 & 3 \end{pmatrix}$$

then from equation (3.32) the secular determinant is

$$\begin{vmatrix} 4 - \lambda & 1 \\ 2 & 3 - \lambda \end{vmatrix} = 0,$$

from which the second-degree eigenvalue equation is

$$\lambda^2 - 7\lambda + 10 = 0,$$

with solutions

$$\lambda_1 = 2, \qquad \lambda_2 = 5.$$

Substitution of the eigenvalues λ_1 and λ_2 in turn into equation (3.31) gives

$$\mathbf{x}_1 = \begin{pmatrix} \frac{-1}{\sqrt{5}} \\ \frac{2}{\sqrt{5}} \end{pmatrix}, \qquad \mathbf{x}_2 = \begin{pmatrix} \frac{1}{\sqrt{2}} \\ \frac{1}{\sqrt{2}} \end{pmatrix} \qquad \text{and} \qquad \mathbf{X} = \begin{pmatrix} \frac{-1}{\sqrt{5}} & \frac{1}{\sqrt{2}} \\ \frac{2}{\sqrt{5}} & \frac{1}{\sqrt{2}} \end{pmatrix}.$$

If $\mathbf{A}$ is Hermitian, $\mathbf{A}^\dagger = \mathbf{A}$, then the eigenvalues are real and the eigenvectors can always be made to be orthogonal to each other. In this case $\mathbf{X}$ is a unitary matrix since, in general, the elements of the $\mathbf{x}_i$ will be complex numbers. If $\mathbf{A}$ is real and symmetric, then the $\mathbf{x}_i$ will also be real and $\mathbf{X}$ is then an orthogonal matrix. The determination of eigenvectors is equivalent to finding a matrix $\mathbf{X}$ that transforms $\mathbf{A}$ into the diagonal form (3.35). Surprisingly, this is a simpler problem (for a computer!) than finding the zeros of the nth-order polynomial generated by the secular equation (3.32).

Similarity Transformations

The matrix $\mathbf{A}$ is transformed into the matrix $\mathbf{B}$ via a matrix $\mathbf{Z}$ by the relationship

$$\mathbf{Z}^{-1}\mathbf{A}\mathbf{Z} = \mathbf{B}.$$

This is a similarity transformation (Figure 3.3). If $\mathbf{Z}$ is a unitary (or orthogonal) matrix, then $\mathbf{A}$ and $\mathbf{B}$ are related via a unitary (or orthogonal) transformation. Notice that if $\mathbf{A}$, $\mathbf{B}$, and $\mathbf{Z}$ were matrix representations of symmetry operators, then $\mathbf{A}$ and $\mathbf{B}$ would be in the same class. If $\mathbf{A}$ and $\mathbf{B}$ are similar, then the eigenvalues of $\mathbf{A}$ are the same as $\mathbf{B}$ (but the eigenvectors are different), and $|\mathbf{A}| = |\mathbf{B}|$, $\text{tr}(\mathbf{A}) = \text{tr}(\mathbf{B})$. If $\mathbf{Z}$ transforms $\mathbf{A}$ into

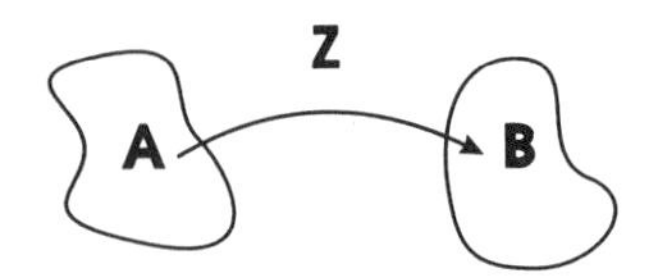

Figure 3.3. A matrix transformation.

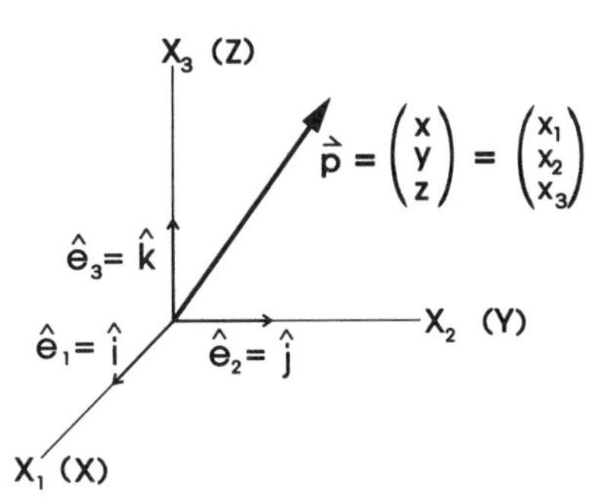

Figure 3.4. Cartesian coordinate system.

the diagonal matrix **B**, then the matrix **Z** diagonalizes **A** and the eigenvalues are found in **B**. A Hermitian (or symmetric) matrix is diagonalized by a unitary (or orthogonal) transformation.

Symmetry Operations and Position Vectors

Perhaps the simplest way to generate a matrix representation of a group is to consider the effect of a symmetry operation on a point

$$\mathbf{p} = \begin{pmatrix} x \\ y \\ z \end{pmatrix}$$

in space. The point is located in real three-dimensional space (Figure 3.4).

It is convenient to replace the familiar X, Y, and Z axes by X_1, X_2, and X_3, respectively, as well as the usual Cartesian unit vectors $\hat{\mathbf{i}}$, $\hat{\mathbf{j}}$, and $\hat{\mathbf{k}}$ by $\hat{\mathbf{e}}_1$, $\hat{\mathbf{e}}_2$, and $\hat{\mathbf{e}}_3$ (Figure 3.4). In this case,

$$\mathbf{p} = x_1\hat{\mathbf{e}}_1 + x_2\hat{\mathbf{e}}_2 + x_3\hat{\mathbf{e}}_3 = \sum_{i=1}^{3} x_i\hat{\mathbf{e}}_i. \tag{3.36}$$

Reflection

A reflection in the X_1–X_3 (X–Z) plane changes the x_2 component of **p** to $-x_2$, so that

$$\hat{\boldsymbol{\sigma}}_{13}\begin{pmatrix} x_1 \\ x_2 \\ x_3 \end{pmatrix} = \begin{pmatrix} x_1 \\ -x_2 \\ x_3 \end{pmatrix}$$

and $\hat{\boldsymbol{\sigma}}_{13}$ is represented by the matrix

$$\mathbf{D}(\hat{\sigma}_{13}) = \begin{pmatrix} 1 & 0 & 0 \\ 0 & -1 & 0 \\ 0 & 0 & 1 \end{pmatrix}.$$

Similarly, a matrix representation for $\hat{\boldsymbol{\sigma}}_{12}$ and $\hat{\boldsymbol{\sigma}}_{23}$ can be found to be

$$\mathbf{D}(\hat{\sigma}_{12}) = \begin{pmatrix} 1 & 0 & 0 \\ 0 & 1 & 0 \\ 0 & 0 & -1 \end{pmatrix} \quad \text{and} \quad \mathbf{D}(\hat{\sigma}_{23}) = \begin{pmatrix} -1 & 0 & 0 \\ 0 & 1 & 0 \\ 0 & 0 & 1 \end{pmatrix}.$$

Rotation

The clockwise rotation of θ radians about the $X_3(Z)$-axis produces a new set of coordinates, Figure 3.5:

$$\hat{C}_\theta^{(Z)}\mathbf{p} = \mathbf{p}' \tag{3.37}$$

or

$$\hat{C}_\theta \begin{pmatrix} x_1 \\ x_2 \\ x_3 \end{pmatrix} = \begin{pmatrix} x_1' \\ x_2' \\ x_3' \end{pmatrix}, \tag{3.38}$$

where

$$\begin{aligned} x_1' = d\cos\phi' &= d\cos(\phi - \theta) \\ &= d\cos\phi\cos\theta + d\sin\phi\sin\theta \\ &= x_1\cos\theta + x_2\sin\theta \end{aligned} \tag{3.39}$$

$$\begin{aligned} x_2' = d\sin\phi' &= d\sin(\phi - \theta) \\ &= d\sin\phi\cos\theta - d\cos\phi\sin\theta \\ &= x_2\cos\theta - x_1\sin\theta \\ &= -x_1\sin\theta + x_2\cos\theta, \end{aligned} \tag{3.40}$$

and

$$x_3' = x_3. \tag{3.41}$$

These equations can be expressed in the compact matrix form

$$\begin{pmatrix} \cos\theta & \sin\theta & 0 \\ -\sin\theta & \cos\theta & 0 \\ 0 & 0 & 1 \end{pmatrix} \begin{pmatrix} x_1 \\ x_2 \\ x_3 \end{pmatrix} = \begin{pmatrix} x_1' \\ x_2' \\ x_3' \end{pmatrix} \tag{3.42}$$

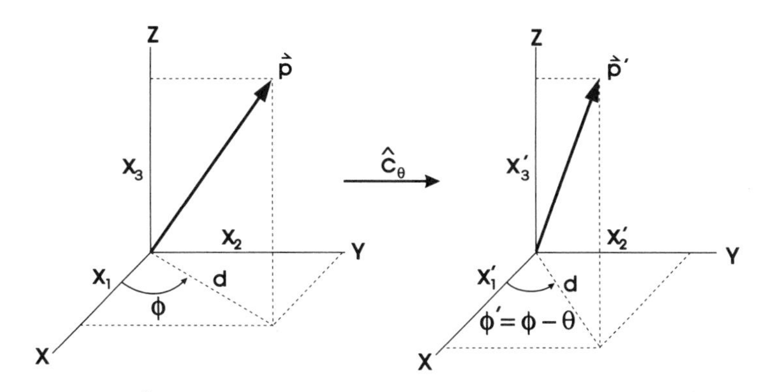

Figure 3.5. The $\hat{C}_\theta^{(Z)}$ operation rotates the point **p** by θ radians about the X_3 axis.

with the matrix representation $\mathbf{D}(\hat{C}_\theta)$ of a rotation by an angle θ about the X_3-axis given by

$$\mathbf{D}(\hat{C}_\theta) = \begin{pmatrix} \cos\theta & \sin\theta & 0 \\ -\sin\theta & \cos\theta & 0 \\ 0 & 0 & 1 \end{pmatrix}. \tag{3.43}$$

The matrix for $\hat{C}_\theta^{-1} = \hat{C}_{-\theta}$ is given by

$$\mathbf{D}(\hat{C}_\theta^{-1}) = \begin{pmatrix} \cos(-\theta) & \sin(-\theta) & 0 \\ -\sin(-\theta) & \cos(-\theta) & 0 \\ 0 & 0 & 1 \end{pmatrix}$$

$$= \begin{pmatrix} \cos\theta & -\sin\theta & 0 \\ \sin\theta & \cos\theta & 0 \\ 0 & 0 & 1 \end{pmatrix}. \tag{3.44}$$

$\mathbf{D}(\hat{C}_\theta)$ is an orthogonal matrix since $\mathbf{D}(\hat{C}_\theta^{-1}) = [\mathbf{D}(\hat{C}_\theta)]^t$ as required.

Inversion

The operation of inversion $\hat{i}$ changes the signs of all coordinates

$$\hat{i}\begin{pmatrix} x_1 \\ x_2 \\ x_3 \end{pmatrix} = \begin{pmatrix} -x_1 \\ -x_2 \\ -x_3 \end{pmatrix} \tag{3.45}$$

so that its matrix representation is

$$\mathbf{D}(\hat{i}) = \begin{pmatrix} -1 & 0 & 0 \\ 0 & -1 & 0 \\ 0 & 0 & -1 \end{pmatrix}. \tag{3.46}$$

Rotation-Reflection

The improper rotation operation, $\hat{S}_\theta$, corresponds to a $\hat{C}_\theta$ operation about the X_3-axis followed by a horizontal reflection $\hat{\sigma}_{12}$ in the X_1-X_2 (X-Y) plane. Since the matrix representations of $\hat{C}_\theta$ and $\hat{\sigma}_{12}$ have already been derived, the matrix for $\hat{S}_\theta$ can be determined by multiplication. Thus

$$\mathbf{D}(\hat{S}_\theta) = \mathbf{D}(\hat{\sigma}_{12})\mathbf{D}(\hat{C}_\theta) = \begin{pmatrix} 1 & 0 & 0 \\ 0 & 1 & 0 \\ 0 & 0 & -1 \end{pmatrix}\begin{pmatrix} \cos\theta & \sin\theta & 0 \\ -\sin\theta & \cos\theta & 0 \\ 0 & 0 & 1 \end{pmatrix} \tag{3.47}$$

$$= \begin{pmatrix} \cos\theta & \sin\theta & 0 \\ -\sin\theta & \cos\theta & 0 \\ 0 & 0 & -1 \end{pmatrix}. \tag{3.48}$$

Identity

The "do-nothing" operation $\hat{E}$ is represented by the unit matrix **1** or

$$\mathbf{D}(\hat{E}) = \begin{pmatrix} 1 & 0 & 0 \\ 0 & 1 & 0 \\ 0 & 0 & 1 \end{pmatrix}. \tag{3.49}$$

For example, the C_{2v} point group of water (Figure 3.6) has four symmetry operators $\{\hat{E}, \hat{C}_2, \hat{\sigma}_{13}, \hat{\sigma}_{23}\}$, so that the corresponding matrix representations of the symmetry operations are

$$\mathbf{D}(\hat{E}) = \begin{pmatrix} 1 & 0 & 0 \\ 0 & 1 & 0 \\ 0 & 0 & 1 \end{pmatrix} \qquad \mathbf{D}(\hat{C}_2) = \begin{pmatrix} -1 & 0 & 0 \\ 0 & -1 & 0 \\ 0 & 0 & 1 \end{pmatrix}$$

$$\mathbf{D}(\hat{\sigma}_{13}) = \begin{pmatrix} 1 & 0 & 0 \\ 0 & -1 & 0 \\ 0 & 0 & 1 \end{pmatrix} \qquad \mathbf{D}(\hat{\sigma}_{23}) = \begin{pmatrix} -1 & 0 & 0 \\ 0 & 1 & 0 \\ 0 & 0 & 1 \end{pmatrix}.$$

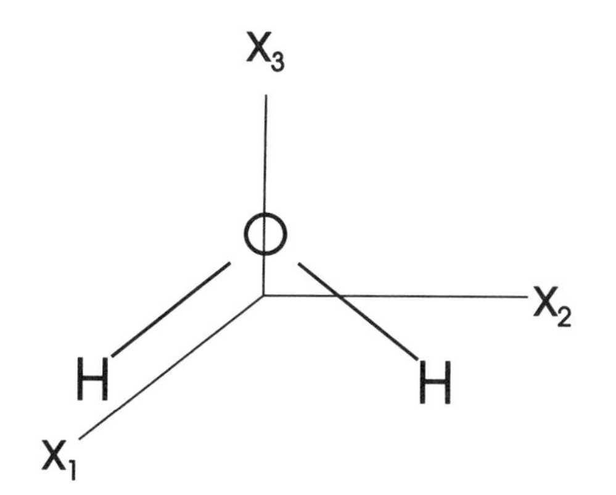

Figure 3.6. The coordinate system for the water molecule.

A multiplication table for the C_{2v} point group (Table 3.1) is obtained by calculating all possible products between the four **D** matrices just given. For example, $\mathbf{D}(\hat{C}_2)\mathbf{D}(\hat{\sigma}_{13}) = \mathbf{D}(\hat{\sigma}_{23})$.

Symmetry Operators and Basis Vectors

A matrix representation of a group can also be generated by considering the effect of symmetry operations on basis vectors. For example, $\hat{C}_\theta$, the rotation operation about the X_3-axis, can occur either by rotating the point **p** in a clockwise direction or by leaving **p** fixed but rotating the coordinate system in the opposite direction (Figure 3.7). For example, the matrix representation for the $\hat{C}_3$ operation generated from the rotation of the point **p** is

$$\mathbf{D}(\hat{C}_3) = \begin{pmatrix} \cos\frac{2\pi}{3} & \sin\frac{2\pi}{3} & 0 \\ -\sin\frac{2\pi}{3} & \cos\frac{2\pi}{3} & 0 \\ 0 & 0 & 1 \end{pmatrix} = \begin{pmatrix} -\frac{1}{2} & \frac{\sqrt{3}}{2} & 0 \\ -\frac{\sqrt{3}}{2} & -\frac{1}{2} & 0 \\ 0 & 0 & 1 \end{pmatrix}$$

so that $\mathbf{x}' = \mathbf{D}(\hat{C}_3)\mathbf{x}$ has components x_i' given in terms of the components x_i by

$$x_1' = -\frac{x_1}{2} + \frac{\sqrt{3}}{2}x_2$$

$$x_2' = -\frac{\sqrt{3}}{2}x_1 - \frac{x_2}{2}$$

$$x_3' = x_3.$$

Table 3.1. Multiplication Table for the C_{2v} Point Group

C_{2v}	$\hat{E}$	$\hat{C}_2$	$\hat{\sigma}_{13}$	$\hat{\sigma}_{23}$
$\hat{E}$	$\hat{E}$	$\hat{C}_2$	$\hat{\sigma}_{13}$	$\hat{\sigma}_{23}$
$\hat{C}_2$	$\hat{C}_2$	$\hat{E}$	$\hat{\sigma}_{23}$	$\hat{\sigma}_{13}$
$\hat{\sigma}_{13}$	$\hat{\sigma}_{13}$	$\hat{\sigma}_{23}$	$\hat{E}$	$\hat{C}_2$
$\hat{\sigma}_{23}$	$\hat{\sigma}_{23}$	$\hat{\sigma}_{13}$	$\hat{C}_2$	$\hat{E}$

However, the rotation of the basis vectors in the opposite direction requires that

$$\hat{\mathbf{e}}_1' = -\frac{\hat{\mathbf{e}}_1}{2} - \frac{\sqrt{3}}{2}\hat{\mathbf{e}}_2$$

$$\hat{\mathbf{e}}_2' = \frac{\sqrt{3}}{2}\hat{\mathbf{e}}_1 - \frac{\hat{\mathbf{e}}_2}{2}$$

$$\hat{\mathbf{e}}_3' = \hat{\mathbf{e}}_3$$

or

$$\begin{pmatrix} \hat{\mathbf{e}}_1' \\ \hat{\mathbf{e}}_2' \\ \hat{\mathbf{e}}_3' \end{pmatrix} = \begin{pmatrix} -\frac{1}{2} & -\frac{\sqrt{3}}{2} & 0 \\ \frac{\sqrt{3}}{2} & -\frac{1}{2} & 0 \\ 0 & 0 & 1 \end{pmatrix} \begin{pmatrix} \hat{\mathbf{e}}_1 \\ \hat{\mathbf{e}}_2 \\ \hat{\mathbf{e}}_3 \end{pmatrix}.$$

For matrix manipulations it is convenient to introduce the basis vectors $\hat{\mathbf{e}}_1 = \hat{\mathbf{i}}$, $\hat{\mathbf{e}}_2 = \hat{\mathbf{j}}$, and $\hat{\mathbf{e}}_3 = \hat{\mathbf{k}}$. The matrix representation of $\hat{C}_3$ cannot depend on whether the representation is generated using the coordinates x_i or the basis functions $\hat{\mathbf{e}}_i$ since the two are

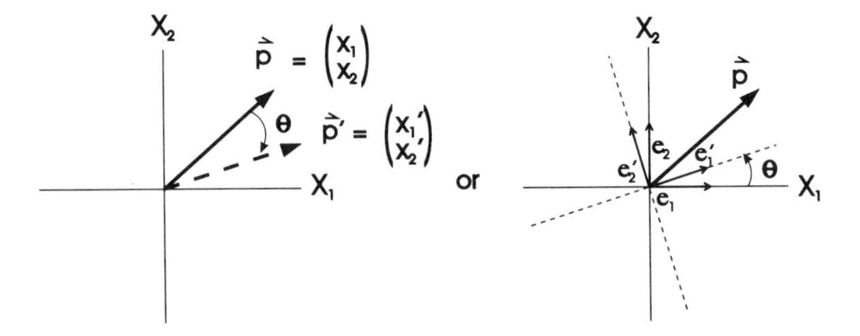

Figure 3.7. The clockwise rotation by θ radians about the Z-axis, $\hat{C}_\theta$, can be accomplished by rotating the point by $+\theta$ or by rotating the coordinate system by $-\theta$.

equivalent. This means that the $\hat{\mathbf{e}}_i$ must be represented by a row vector

$$(\hat{\mathbf{e}}_1' \; \hat{\mathbf{e}}_2' \; \hat{\mathbf{e}}_3') = (\hat{\mathbf{e}}_1 \; \hat{\mathbf{e}}_2 \; \hat{\mathbf{e}}_3) \begin{pmatrix} -\frac{1}{2} & \frac{\sqrt{3}}{2} & 0 \\ -\frac{\sqrt{3}}{2} & -\frac{1}{2} & 0 \\ 0 & 0 & 1 \end{pmatrix}$$

since $(\mathbf{e}')^t = [\mathbf{D}(\hat{C}_3^{-1})\mathbf{e}]^t = \mathbf{e}^t\mathbf{D}(\hat{C}_3^{-1})^t = (\mathbf{e})^t\mathbf{D}(\hat{C}_3)$.

The general operation $\hat{R}$ can operate on the point or on the basis functions:

$$\mathbf{p} = \sum \hat{\mathbf{e}}_i x_i$$

$$\mathbf{p}' = \hat{R}\mathbf{p} = \sum \hat{\mathbf{e}}_i(\hat{R}x_i) = \sum \hat{\mathbf{e}}_i x_i' \tag{3.50}$$

$$= \sum (\hat{R}\hat{\mathbf{e}}_i)x_i = \sum \hat{\mathbf{e}}_i' x_i \tag{3.51}$$

or in matrix form

$$\mathbf{p} = \mathbf{e}^t\mathbf{x} \tag{3.52}$$

$$\begin{aligned} \mathbf{p}' = \hat{R}\mathbf{p} = \mathbf{e}^t\mathbf{D}(\hat{R})\mathbf{x} &= \mathbf{e}^t(\mathbf{D}(\hat{R})\mathbf{x}) = \mathbf{e}^t\mathbf{x}' \\ &= (\mathbf{e}^t\mathbf{D}(\hat{R}))\mathbf{x} = (\mathbf{e}')^t\mathbf{x}. \end{aligned} \tag{3.53}$$

The use of row vectors for basis functions seems surprising, but it is necessary in order to obtain a consistent matrix representation and to maintain the homomorphism.

The use of basis functions rather than a single point is very useful for generating matrix representations larger than 3×3. For example, if each atom of NH_3 is given a set of three Cartesian basis functions, then a 12×12 matrix representation is generated. Consider the effect of $\hat{\sigma}_v'$ on the twelve basis functions shown in Figure 3.8. Thus

$$\mathbf{D}(\hat{\sigma}_v') = \left(\begin{array}{ccc:ccc:ccc:ccc}
-1 & 0 & 0 & 0 & 0 & 0 & 0 & 0 & 0 & 0 & 0 & 0 \\
0 & 1 & 0 & 0 & 0 & 0 & 0 & 0 & 0 & 0 & 0 & 0 \\
0 & 0 & 1 & 0 & 0 & 0 & 0 & 0 & 0 & 0 & 0 & 0 \\
\hdashline
0 & 0 & 0 & -1 & 0 & 0 & 0 & 0 & 0 & 0 & 0 & 0 \\
0 & 0 & 0 & 0 & 1 & 0 & 0 & 0 & 0 & 0 & 0 & 0 \\
0 & 0 & 0 & 0 & 0 & 1 & 0 & 0 & 0 & 0 & 0 & 0 \\
\hdashline
0 & 0 & 0 & 0 & 0 & 0 & 0 & 0 & 0 & -1 & 0 & 0 \\
0 & 0 & 0 & 0 & 0 & 0 & 0 & 0 & 0 & 0 & 1 & 0 \\
0 & 0 & 0 & 0 & 0 & 0 & 0 & 0 & 0 & 0 & 0 & 1 \\
\hdashline
0 & 0 & 0 & 0 & 0 & 0 & -1 & 0 & 0 & 0 & 0 & 0 \\
0 & 0 & 0 & 0 & 0 & 0 & 0 & 1 & 0 & 0 & 0 & 0 \\
0 & 0 & 0 & 0 & 0 & 0 & 0 & 0 & 1 & 0 & 0 & 0
\end{array}\right).$$

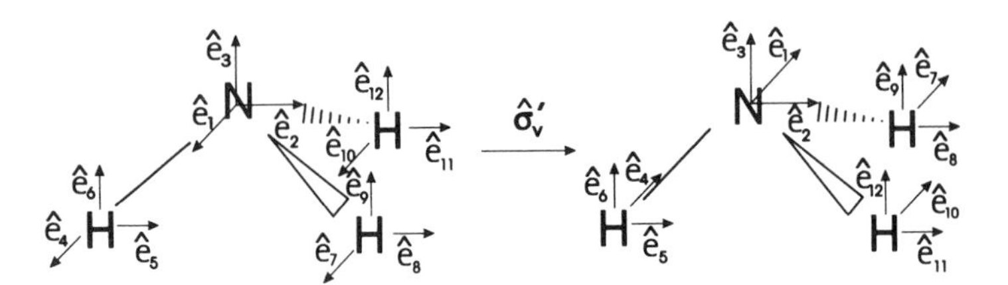

Figure 3.8. The $\hat{\sigma}_v'$ operator reflects the twelve basis functions through the plane of the paper (the plane containing $\hat{e}_5$ and $\hat{e}_6$).

This matrix representation will be useful later in deriving the symmetry of the vibrational modes of NH_3.

Symmetry Operators and Basis Functions

The ordinary 3-dimensional Cartesian vector space can be generalized to an n-dimensional vector space. In the previous section a 12-dimensional vector space was used with the 12 orthogonal basis functions associated with the four atoms of the NH_3 molecule. It is possible to generalize even further by allowing the basis vectors to be functions. In this case the vector space is usually called a *function space.* The properties of a function space parallel those of ordinary 3-dimensional vector space.

Function Spaces

A function space is a set of functions $\{f_1, f_2, \ldots, f_n\}$ (n may be infinite) with the following properties:

1. The addition of any two functions f_i and f_j produces a third function f_k that is also a member of the function space. Thus, $f_i + f_j = f_k$ is analogous to the addition of two vectors $\mathbf{a} + \mathbf{b} = \mathbf{c}$.

2. Multiplication of any function f_i by a constant k produces a new function kf_i that is also a member of the function space. Again, this is analogous to changing the length of a vector $\mathbf{a}$ to $k\mathbf{a}$ by multiplication by a constant k.

3. The scalar or inner product of two functions is given by

$$\langle f_i \mid f_j \rangle = \int f_i^* f_j \, d\tau. \tag{3.54}$$

This parallels the definition of the dot product of two vectors **a** and **b**:

$$\mathbf{a} \cdot \mathbf{b} = |\mathbf{a}|\,|\mathbf{b}| \cos\theta = a_1 b_1 + a_2 b_2 + a_3 b_3. \tag{3.55}$$

4. If there are n linearly independent functions in the function space, $f_1, f_2, \ldots, f_n$, then

$$a_1 f_1 + a_2 f_2 + \cdots + a_n f_n = 0 \qquad \text{only if} \quad a_1 = a_2 = \cdots a_n = 0$$

and any function in the space can be represented by a linear combination of these n linearly independent functions, that is,

$$f = \sum_{i=1}^{n} a_i f_i. \tag{3.56}$$

The n linearly independent functions are said to span the function space of dimension n. It is always possible to find a set of n orthogonal basis functions, that is, functions f_i, f_j such that

$$\langle f_i \,|\, f_j \rangle = \delta_{ij},$$

which span the space. These n basis functions are like the three orthogonal Cartesian basis functions $\hat{\mathbf{e}}_1(=\hat{\mathbf{i}})$, $\hat{\mathbf{e}}_2(=\hat{\mathbf{j}})$, and $\hat{\mathbf{e}}_3(=\hat{\mathbf{k}})$. Any vector in 3-dimensional space can be expressed in terms of these basis functions,

$$\mathbf{a} = a_x \hat{\mathbf{i}} + a_y \hat{\mathbf{j}} + a_z \hat{\mathbf{k}}. \tag{3.57}$$

The integrals of the type $\langle f_i \,|\, f_j \rangle$ are often called overlap integrals because they measure how much the two functions overlap in space.

Function spaces are commonly used in solving the Schrödinger equation, $\hat{H}\psi = E\psi$. A set of degenerate wavefunctions associated with a single energy forms a function space. For example, consider the five degenerate d-orbitals associated with an atom that form a function space. Any d-orbital

$$\psi_d = \sum_{i=1}^{5} a_i d_i \tag{3.58}$$

can be expressed in terms of the five orthogonal d_i functions.

Gram-Schmidt Procedure

Given a set of n linearly independent (but not orthogonal) basis functions it is often desirable to find a set of n orthogonal basis functions. This can always be accomplished by the Gram-Schmidt procedure. Let $\{f_1, f_2, \ldots, f_n\}$ be a set of linearly independent functions, then

$$\phi_1 = \frac{f_1}{(\langle f_1 \mid f_1 \rangle)^{1/2}}, \tag{3.59}$$

$$\phi_2 = (f_2 - \phi_1\langle\phi_1 \mid f_2\rangle)/(\langle f_2 - \phi_1\langle\phi_1 \mid f_2\rangle \mid f_2 - \phi_1\langle\phi_1 \mid f_2\rangle\rangle)^{1/2}, \tag{3.60}$$

$$\begin{aligned}\phi_3 = (f_3 - \phi_1\langle\phi_1 \mid f_3\rangle - \phi_2\langle\phi_2 \mid f_3\rangle)/(\langle f_3 - \phi_1\langle\phi_1 \mid f_3\rangle \\ - \phi_2\langle\phi_2 \mid f_3\rangle \mid f_3 - \phi_1\langle\phi_1 \mid f_3\rangle - \phi_2\langle\phi_2 \mid f_3\rangle\rangle)^{1/2},\end{aligned} \tag{3.61}$$

$$\vdots$$

$$\begin{aligned}\phi_n = (f_n - \phi_1\langle\phi_1 \mid f_n\rangle \cdots - \phi_{n-1}\langle\phi_{n-1} \mid f_n\rangle)/ \\ (\langle f_n - \phi_1\langle\phi_1 \mid f_n\rangle \cdots \mid f_n - \phi_1\langle\phi_1 \mid f_n\rangle \cdots\rangle)^{1/2}.\end{aligned} \tag{3.62}$$

The $\{\phi_n\}$ are now orthogonal and normalized,

$$\langle\phi_i \mid \phi_j\rangle = \delta_{ij}. \tag{3.63}$$

This procedure works because at each step the new ϕ_i function is made orthogonal to the preceding $i-1$ functions by subtracting the overlap integrals and then normalizing the new function. The Gram-Schmidt procedure is useful in vector analysis and in quantum mechanics, where a set of orthogonal functions makes calculations easier because there are no overlap integrals between the basis functions.

Transformation Operators

A set of functions in a function space, like a set of basis vectors, can also be used to generate a matrix representation of a group. For this purpose it is necessary to define a set of linear operators $\{\hat{O}_R\}$ isomorphic with the group of symmetry operators $\{\hat{R}\}$. These operators operate on functions and are defined by the equation

$$(\hat{O}_R)f_i(x_1', x_2', x_3') = f_i(x_1, x_2, x_3), \tag{3.64}$$

where f_i is a member of the set of n linearly independent basis functions and x_i' results from the operation of $\hat{R}$ on x_i. Since the basis functions span the function space, this operation must produce a linear combination of the basis functions:

$$\hat{O}_R f_j = \sum_i f_i[\mathbf{D}(\hat{R})]_{ij}. \tag{3.65}$$

This "backwards" definition of $\hat{O}_R$ is again necessary in order to obtain a consistent $\mathbf{D}(\hat{R})$ matrix when the $\hat{O}_R$ operates on functions that are like basis vectors in a vector space. The $\hat{O}_R$ operators are unitary operators so the scalar product of two functions is unchanged by a unitary transformation. The $\hat{O}_R$ operators can therefore always be represented by unitary matrices. Consider a unitary transformation of two vectors,

$$\mathbf{y}' = \mathbf{U}\mathbf{y} \tag{3.66}$$

$$\mathbf{x}' = \mathbf{U}\mathbf{x} \tag{3.67}$$

and the dot product between the vectors:

$$\begin{aligned} \mathbf{x}' \cdot \mathbf{y}' &= \sum (x_i')^* y_i' = (\mathbf{x}')^\dagger \mathbf{y}' = (\mathbf{U}\mathbf{x})^\dagger (\mathbf{U}\mathbf{y}) \\ &= \mathbf{x}^\dagger \mathbf{U}^\dagger \mathbf{U}\mathbf{y} = \mathbf{x}^\dagger \mathbf{y} \end{aligned} \tag{3.68}$$

since $\mathbf{U}^\dagger = \mathbf{U}^{-1}$ by definition. Thus a unitary transformation leaves the dot product of two vectors unchanged. Also the length (norm) of a vector $|\mathbf{x}| = (\mathbf{x}^\dagger \mathbf{x})^{1/2}$ is unchanged by a unitary transformation.

Similarly, the $\hat{O}_R$ operators work in a function space and they do not change the scalar product of two functions:

$$\langle f \mid g \rangle = \langle \hat{O}_R f \mid \hat{O}_R g \rangle.$$

The proof is based on the definition of the scalar product

$$\langle f \mid g \rangle = \int f^*(x_1, x_2, x_3) g(x_1, x_2, x_3)\, dx_1\, dx_2\, dx_3. \tag{3.69}$$

Now the $\hat{O}_R$ operator will move the element $dx_1\, dx_2\, dx_3$ to $dx_1'\, dx_2'\, dx_3'$ and $f(x_1, x_2, x_3) = \hat{O}_R f(x_1', x_2', x_3')$ and $g(x_1, x_2, x_3) = \hat{O}_R g(x_1', x_2', x_3')$ by definition. Therefore

$$\langle f \mid g \rangle = \int \hat{O}_R f^*(x_1', x_2', x_3') \hat{O}_R g(x_1', x_2', x_3')\, dx_1'\, dx_2'\, dx_3' = \langle \hat{O}_R f \mid \hat{O}_R g \rangle \tag{3.70}$$

since the integration variables are dummy variables. Thus the $\hat{O}_R$ operators are unitary operators that can be represented by unitary matrices.

As an example of using the $\hat{O}_R$ operators consider the effect of the $\hat{C}_3^{(z)}$ operation on the function $f = xyz = x_1 x_2 x_3$.

$$\hat{O}_{C_3} f' = f \tag{3.71}$$

$$\hat{O}_{C_3} f(x_1', x_2', x_3') = x_1 x_2 x_3 \tag{3.72}$$

but

$$x_1' = -\frac{x_1}{2} + \sqrt{3}\frac{x_2}{2} \tag{3.73a}$$

$$x_2' = -\sqrt{3}\frac{x_1}{2} - \frac{x_2}{2} \tag{3.73b}$$

$$x_3' = x_3 \tag{3.73c}$$

or

$$\begin{pmatrix} x_1' \\ x_2' \\ x_3' \end{pmatrix} = \begin{pmatrix} -\frac{1}{2} & \frac{\sqrt{3}}{2} & 0 \\ -\frac{\sqrt{3}}{2} & -\frac{1}{2} & 0 \\ 0 & 0 & 1 \end{pmatrix} \begin{pmatrix} x_1 \\ x_2 \\ x_3 \end{pmatrix}.$$

Inverting this matrix gives

$$\begin{pmatrix} x_1 \\ x_2 \\ x_3 \end{pmatrix} = \begin{pmatrix} -\frac{1}{2} & \frac{\sqrt{3}}{2} & 0 \\ -\frac{\sqrt{3}}{2} & -\frac{1}{2} & 0 \\ 0 & 0 & 1 \end{pmatrix}^{-1} \begin{pmatrix} x_1' \\ x_2' \\ x_3' \end{pmatrix} = \begin{pmatrix} -\frac{1}{2} & -\frac{\sqrt{3}}{2} & 0 \\ \frac{\sqrt{3}}{2} & -\frac{1}{2} & 0 \\ 0 & 0 & 1 \end{pmatrix} \begin{pmatrix} x_1' \\ x_2' \\ x_3' \end{pmatrix}$$

or

$$x_1 = -\frac{x_1'}{2} - \sqrt{3}\frac{x_2'}{2} \tag{3.74a}$$

$$x_2 = \sqrt{3}\frac{x_1'}{2} - \frac{x_2'}{2} \tag{3.74b}$$

$$x_3 = x_3'. \tag{3.74c}$$

Substituting equations (3.74) into equation (3.72) gives

$$\begin{aligned} (\hat{O}_{C_3})f(x_1', x_2', x_3') &= \left(-\frac{x_1'}{2} - \sqrt{3}\frac{x_2'}{2}\right)\left(\sqrt{3}\frac{x_1'}{2} - \frac{x_2'}{2}\right)x_3 \\ &= \left[-\frac{\sqrt{3}}{4}(x_1')^2 + \frac{\sqrt{3}}{4}(x_2')^2 - \frac{3}{4}(x_1'x_2') + \frac{1}{4}x_1'x_2'\right]x_3 \\ &= x_3'\frac{\left(-\frac{\sqrt{3}}{2}(x_1')^2 + \frac{\sqrt{3}}{2}(x_2')^2 - x_1'x_2'\right)}{2} \end{aligned} \tag{3.75}$$

$$\therefore \hat{O}_{C_3}f = \frac{\left(-\frac{\sqrt{3}}{2}x_1^2 + \frac{\sqrt{3}}{2}x_2^2 - x_1x_2\right)x_3}{2}. \tag{3.76}$$

In this way the effect of any symmetry operator on any function can be determined.

Equivalent, Reducible, and Irreducible Matrix Representations

Evidently an infinite number of matrix representations of a point group are possible. The time has come to limit the possibilities.

Equivalent Representations

If two sets of linearly independent basis functions exist and are related by a linear transformation $\mathbf{g} = \mathbf{Af}$, then the matrix representation produced by $\mathbf{f}$ is said to be equivalent to the matrix representation generated by $\mathbf{g}$. In fact, it is found that the two representations are related by a similarity transformation. Consider the effect of a symmetry operator $\hat{R}$ on the basis functions $\mathbf{g}$ where

$$(\mathbf{g}')^t = (\mathbf{g})^t \mathbf{D}^g(\hat{R}). \tag{3.77}$$

Since

$$\mathbf{g} = \mathbf{Af}$$

then

$$(\mathbf{g})^t = (\mathbf{Af})^t = \mathbf{f}^t \mathbf{A}^t \tag{3.78}$$

and

$$(\mathbf{g}')^t = (\mathbf{A}(\mathbf{f}'))^t = (\mathbf{f}')^t \mathbf{A}^t. \tag{3.79}$$

Comparing equation (3.77) and equation (3.79) gives

$$(\mathbf{f}')^t \mathbf{A}^t = \mathbf{f}^t \mathbf{A}^t \mathbf{D}^g(\hat{R}) \tag{3.80}$$

or

$$(\mathbf{f}')^t = \mathbf{f}^t [\mathbf{A}^t \mathbf{D}^g(\hat{R})(\mathbf{A}^t)^{-1}]. \tag{3.81}$$

Let

$$\mathbf{A}^t = \mathbf{B}^{-1}$$

then

$$\mathbf{B}^{-1}\mathbf{D}^g\mathbf{B} = \mathbf{D}^f, \tag{3.82}$$

and so the matrix representations generated by $\mathbf{f}$ and $\mathbf{g}$ are related by a similarity transformation. Equivalent representations have the same eigenvalues, traces, and determinants so that they are not considered to be "different" from the point of view of representing symmetry operators.

Unitary Representations

In general, many different types of matrices could be used to represent symmetry operations. However, if an orthonormal set of basis functions is used in an n-dimensional function space, then the matrices generated will be unitary. Since it is always possible to find a set of n orthogonal functions that span the function space, it is always possible to construct a unitary representation of a group. The properties of unitary matrices are so convenient they will be used exclusively to represent symmetry operators.

Reducible and Irreducible Representations

So far the matrix representations of symmetry operators have been nonequivalent and unitary. However, the dimension of the matrices in a given representation could be very large. Fortunately, it is always possible to find a similarity transformation that reduces the representation to block diagonal form where the nonzero elements occur in blocks along the principal diagonal. For example, the 3-dimensional representation of the group C_{3v} generated by using a point is

$$\mathbf{D}(\hat{E}) = \left(\begin{array}{cc:c} 1 & 0 & 0 \\ 0 & 1 & 0 \\ \hdashline 0 & 0 & 1 \end{array}\right) \qquad \mathbf{D}(\hat{\sigma}_v') = \left(\begin{array}{cc:c} -1 & 0 & 0 \\ 0 & 1 & 0 \\ \hdashline 0 & 0 & 1 \end{array}\right)$$

$$\mathbf{D}(\hat{C}_3) = \left(\begin{array}{cc:c} -\frac{1}{2} & \frac{\sqrt{3}}{2} & 0 \\ -\frac{\sqrt{3}}{2} & -\frac{1}{2} & 0 \\ \hdashline 0 & 0 & 1 \end{array}\right) \qquad \mathbf{D}(\hat{C}_3^{-1}) = \left(\begin{array}{cc:c} -\frac{1}{2} & -\frac{\sqrt{3}}{2} & 0 \\ \frac{\sqrt{3}}{2} & -\frac{1}{2} & 0 \\ \hdashline 0 & 0 & 1 \end{array}\right)$$

$$\mathbf{D}(\hat{\sigma}_v'') = \left(\begin{array}{cc:c} \frac{1}{2} & -\frac{\sqrt{3}}{2} & 0 \\ -\frac{\sqrt{3}}{2} & -\frac{1}{2} & 0 \\ \hdashline 0 & 0 & 1 \end{array}\right) \qquad \mathbf{D}(\hat{\sigma}_v''') = \left(\begin{array}{cc:c} \frac{1}{2} & \frac{\sqrt{3}}{2} & 0 \\ \frac{\sqrt{3}}{2} & -\frac{1}{2} & 0 \\ \hdashline 0 & 0 & 1 \end{array}\right).$$

Notice that all of the matrices have the same form and that the subblocks (highlighted in the equations with dashed lines) also form representations of the group since multiplication does not change the block structure. In this C_{3v} example, the six matrices cannot be reduced further (i.e., no extra zero can be introduced) by the application of a single similarity transformation to all matrices. These simple 2×2 and 1×1 blocks are therefore called *irreducible representations,* while the 3-dimensional representation is termed *reducible.*

The conventional symbol for a representation is Γ, which stands for all members of a representation. A superscript is used to label different representations

$$\begin{aligned}
\Gamma^{\text{red}} &= \{(\mathbf{D}(\hat{R})), (\mathbf{D}(\hat{S})), \ldots\} \\
\Gamma^{\text{red}} &= \Gamma^1 \oplus \Gamma^2 \oplus \cdots \Gamma^n \\
&= \oplus \sum a_\nu \Gamma^\nu.
\end{aligned}$$

The decomposition of a reducible representation into irreducible representations is symbolized with a $\oplus$. The circle indicates that this is a "direct" sum of the representations rather than ordinary addition, such as for numbers or matrices. A given irreducible representation Γ^ν may occur several (a_ν) times in the reducible representation. Note that the superscript on Γ always serves as a label and never indicates repeated multiplication. It turns out that while the number of reducible representations of a group is infinite, the number of irreducible representations is small. In particular the number of irreducible representations of a point group turns out to be equal to the number of classes in that group.

Great Orthogonality Theorem

Since all matrix representations can be reduced to the direct sum of a small number of irreducible representations, these irreducible representations must be very important. The central theorem about irreducible representations of point groups is appropriately called the Great Orthogonality Theorem. The Great Orthogonality Theorem requires that

$$\sum_{\hat{R}} D^{\mu}_{ik}(\hat{R}) D^{\nu}_{mj}(\hat{R}^{-1}) = \frac{g}{n_\nu} \delta_{\mu\nu} \delta_{ij} \delta_{km} \tag{3.83}$$

in which $\mathbf{D}^\mu(\hat{R})$ and $\mathbf{D}^\nu(\hat{R})$ are matrix representations of two nonequivalent irreducible representations Γ^μ and Γ^ν, g is the order of the group and n_ν is the dimension of the νth irreducible representation. The sum is over all $\hat{R}$, the operations in the group. The proof of this theorem is rather involved and will not be reproduced here.

If unitary matrices are used to represent the group, then

$$\mathbf{D}^\nu(\hat{R}^{-1}) = [\mathbf{D}^\nu(\hat{R})]^{-1} = [\mathbf{D}^\nu(\hat{R})]^\dagger \tag{3.84}$$

and

$$D^{\nu}_{mj}(\hat{R}^{-1}) = D^{\nu*}_{jm}(\hat{R}). \tag{3.85}$$

The Great Orthogonality Theorem can therefore be restated in the form

$$\sum_{\hat{R}} D^{\mu}_{ik}(\hat{R})[D^{\nu}_{jm}(\hat{R})]^* = \frac{g}{n_\mu}\,\delta_{\mu\nu}\,\delta_{ij}\,\delta_{km}. \tag{3.86}$$

This theorem states that the corresponding matrix elements in the various irreducible representations can be formed into vectors that are orthogonal to one another. This may be best illustrated with a specific example.

Consider the six 3×3 matrices that represent the C_{3v} point group generated earlier. Since the number of symmetry operators is six, each vector will be of dimension six. If the matrix element in the upper left corner is used, then

$$\mathbf{v_1} = \begin{pmatrix} 1 \\ -1 \\ -\frac{1}{2} \\ -\frac{1}{2} \\ \frac{1}{2} \\ \frac{1}{2} \end{pmatrix}$$

and four other vectors can be generated

$$\mathbf{v_2} = \begin{pmatrix} 0 \\ 0 \\ \frac{\sqrt{3}}{2} \\ -\frac{\sqrt{3}}{2} \\ -\frac{\sqrt{3}}{2} \\ \frac{\sqrt{3}}{2} \end{pmatrix} \quad \mathbf{v_3} = \begin{pmatrix} 0 \\ 0 \\ -\frac{\sqrt{3}}{2} \\ \frac{\sqrt{3}}{2} \\ -\frac{\sqrt{3}}{2} \\ \frac{\sqrt{3}}{2} \end{pmatrix} \quad \mathbf{v_4} = \begin{pmatrix} 1 \\ 1 \\ -\frac{1}{2} \\ -\frac{1}{2} \\ -\frac{1}{2} \\ -\frac{1}{2} \end{pmatrix} \quad \mathbf{v_5} = \begin{pmatrix} 1 \\ 1 \\ 1 \\ 1 \\ 1 \\ 1 \end{pmatrix}.$$

The vectors constructed in this fashion are all orthogonal: if $\mu \neq \nu$ (i.e., the vectors originate from different irreducible representations), then

$$\mathbf{v}_i \cdot \mathbf{v}_5 = 0, \qquad i = 1, 2, 3, 4$$

as required. If $\mu = \nu$ (i.e., the vectors originate from the same representation), then the vectors arising from different rows and columns are orthogonal:

$$\mathbf{v}_i \cdot \mathbf{v}_j = 0 \qquad \text{for} \quad i \neq j.$$

Finally the vectors are normalized to the value g/n_μ (i.e., the order of the group divided by the dimension of the irreducible representation).

In this example, when $\mu = \nu$, $i = j$, $k = m$:

$$\mathbf{v}_5 \cdot \mathbf{v}_5 = 6$$

while

$$\frac{g}{n_\mu} = \frac{6}{1} = 6$$

as required and

$$\mathbf{v}_1 \cdot \mathbf{v}_1 = 3$$

while

$$\frac{g}{n_\mu} = \frac{6}{2} = 3$$

as required.

If a group is of order g, then each vector will be g-dimensional. The maximum number of linearly independent vectors in a g-dimensional vector space is also g. If an irreducible representation is an $n_\mu \times n_\mu$ matrix, it must contribute n_μ^2 orthogonal vectors and the total number of orthogonal vectors in the vector space cannot exceed g, therefore

$$\sum_{\mu} n_\mu^2 \leq g, \tag{3.87}$$

where the sum is over all irreducible representations. In fact it can be proved that the equality holds and

$$\sum_{\nu=1}^{r} n_\nu^2 = g, \tag{3.88}$$

where r is the number of irreducible representations.

Characters

A character is the trace of a matrix that serves as a representation of a symmetry operation,

$$\chi(\hat{R}) = \mathrm{tr}(\mathbf{D}(\hat{R})). \tag{3.89}$$

Characters are represented by the symbol χ and they serve to represent (characterize) a matrix. They are convenient because a single number, rather than the entire matrix, can be used in most applications in spectroscopy.

For example, the characters of the 3-dimensional representation of the C_{3v} point group used previously in the sixth section are

	$\hat{E}$	$\hat{C}_3$	$\hat{C}_3^{-1}$	$\hat{\sigma}'_v$	$\hat{\sigma}''_v$	$\hat{\sigma}'''_v$
$\chi^{\mathrm{red}}(\hat{R})$	3	0	0	1	1	1

Notice that although the matrices that represent the $\hat{E}$ operator, the three reflection planes, and the two rotation operators are all different, the characters of the symmetry operators in the same class are the same. This is because the members of a class are related by a similarity transformation, for example,

$$\mathbf{D}(\hat{\sigma}'''_v) = \mathbf{D}(\hat{C}_3^{-1})\mathbf{D}(\hat{\sigma}'_v)\mathbf{D}(\hat{C}_3)$$

and the traces (characters) of the matrix representations are unchanged by a similarity transformation. This is also convenient because the character of just a single member of each class needs to be worked out.

The Great Orthogonality Theorem can be used for characters as well as matrices since

$$\sum_{\hat{R}} D^{\mu}_{ii}(\hat{R}) D^{\nu *}_{jj}(\hat{R}) = \frac{g}{n_\mu}\,\delta_{\mu\nu}\,\delta_{ij}\,\delta_{ij} = \frac{g}{n_\mu}\,\delta_{\mu\nu}\,\delta_{ij} \tag{3.90}$$

and

$$\begin{aligned}\sum_{\hat{R}} \sum_{i=1}^{n_\mu} D^{\mu}_{ii}(\hat{R}) \sum_{j=1}^{n_\nu} D^{*\nu}_{jj}(\hat{R}) &= \frac{g}{n_\mu}\,\delta_{\mu\nu} \sum_{i=1}^{n_\mu}\sum_{j=1}^{n_\nu} \delta_{ij} \\ &= \frac{g}{n_\mu}\,\delta_{\mu\nu} \min(n_\mu, n_\nu) \\ &= \frac{g}{n_\mu}\,\delta_{\mu\nu} n_\mu \;(n_\mu < n_\nu)\end{aligned} \tag{3.91}$$

therefore

$$\sum_{\hat{R}} \chi^{\mu}(\hat{R})(\chi^{\nu}(\hat{R}))^* = g\delta_{\mu\nu} \tag{3.92}$$

using the definition of the trace of a matrix. Thus the characters of the irreducible representations are also orthogonal in the same sense that the matrix elements of a matrix representation are orthogonal.

For example, the characters of the two irreducible representations generated from the 3-dimensional representation of C_{3v} are

	$\hat{E}$	$\hat{C}_3$	$\hat{C}_3^{-1}$	$\hat{\sigma}_v'$	$\hat{\sigma}_v''$	$\hat{\sigma}_v'''$
$\chi^{\text{red}}(\hat{R})$	3	0	0	1	1	1
$\chi^1(\hat{R})$	1	1	1	1	1	1
$\chi^2(\hat{R})$	2	−1	−1	0	0	0

where the $\chi(\hat{R})$ of the 3-dimensional representation is also listed. Notice the way the characters add

$$\chi^{\text{red}}(\hat{R}) = \chi^1(\hat{R}) + \chi^2(\hat{R})$$

and for the reducible representation

$$\Gamma^{\text{red}} = \Gamma^1 \oplus \Gamma^2.$$

Upon checking, we note that

$$\sum_R \chi^1(\hat{R})\chi^2(\hat{R})^* = 1(2) + 1(-1) + 1(-1) + 1(0) + 1(0) + 1(0) = 0,$$

$$\sum_R \chi^1(\hat{R})\chi^1(\hat{R})^* = 1 + 1 + 1 + 1 + 1 + 1 = 6,$$

$$\sum_R \chi^2(\hat{R})\chi^2(\hat{R})^* = 2(2) + (-1)(-1) + (-1)(-1) = 6,$$

as required by the orthogonality relationship (3.92).

Let the number of classes in a group be k and let the number of members of each class be g_i. For the preceding example, $k = 3$ and $g_1 = 1$, $g_2 = 2$, and $g_3 = 3$ referring to the classes $\{\hat{E}\}$, $\{\hat{C}_3, \hat{C}_3^{-1}\}$ and $\{\hat{\sigma}_v', \hat{\sigma}_v'', \hat{\sigma}_v'''\}$ with

$$\sum_{i=1}^{k} g_i = g. \tag{3.93}$$

The sum over group operations in the orthogonality theorem can be replaced by a sum over classes:

$$\sum_{i=1}^{k} g_i \chi^{\mu}(\hat{R}_i)\chi^{\nu}(\hat{R}_i)^* = \sum_{i=1}^{k} [\sqrt{g_i}\,\chi^{\mu}(\hat{R}_i)][\sqrt{g_i}\,\chi^{\nu}(\hat{R}_i)]^* = g\,\delta_{\mu\nu}. \tag{3.94}$$

This is now an orthogonality relationship in a k-dimensional (k = number of classes) vector space. The maximum number of independent vectors that can be found in a

k-dimensional space is also k, so that the number of irreducible representations r has to be less than or equal to the number of classes, $r \leq k$. In fact, it can be proved that $r = k$ so that the number of classes is identical to the number of irreducible representations.

A reducible matrix representation can be written as a direct sum of irreducible matrix representations, that is,

$$\Gamma^{\text{red}} = a_1\Gamma^1 \oplus a_2\Gamma^2 \oplus \cdots \oplus a_n\Gamma^n. \tag{3.95}$$

There is a parallel equation for characters, namely

$$\chi^{\text{red}}(\hat{R}) = \sum_{\nu} a_\nu \chi^\nu(\hat{R}), \tag{3.96}$$

for all $\hat{R}$ where this is now an arithmetic sum. This equation for characters holds true because the sum of the diagonal elements of the reducible matrix must equal the sum of the diagonal elements of the submatrices of irreducible representations of the block diagonal form.

The orthogonality theorem for characters can be used to quickly determine the number of each type of irreducible representation, provided that the characters are all known:

$$\sum_{\hat{R}} \chi^{\text{red}}(\hat{R})\chi^\mu(\hat{R})^* = \sum_{\hat{R}}\sum_{\nu} a_\nu \chi^\nu(\hat{R})\chi^\mu(\hat{R})^* = a_\nu \sum_{\nu}\sum_{\hat{R}} \chi^\nu(\hat{R})\chi^\mu(\hat{R})^*$$

$$= \sum_{\nu} a_\nu g\, \delta_{\mu\nu} = a_\mu g;$$

therefore

$$a_\mu = \frac{1}{g}\sum_{\hat{R}} \chi^{\text{red}}(\hat{R}_i)\chi^\mu(\hat{R}_i)^* = \frac{1}{g}\sum_{i=1}^{k} g_i \chi^{\text{red}}(\hat{R}_i)\chi^\mu(\hat{R}_i)^*. \tag{3.97}$$

Character Tables

The characters of the irreducible representations of the point groups are used in the applications of group theory. It is therefore very helpful to have tables of the characters available (Appendix B). The character table for the C_{3v} point group is

C_{3v}	$\hat{E}$	$2\hat{C}_3$	$3\hat{\sigma}_v$
A_1	1	1	1
A_2	1	1	−1
E	2	−1	0

Since the number of classes equals the number of irreducible representations, this table is square. The symmetry operations are listed along the top row with only one character provided for each class. The number of members g_i in each class is also provided. Along the leftmost column the names of each of the irreducible representations are provided using Mulliken notation (discussed below).

Character tables can be constructed using various properties of characters without finding the actual irreducible matrices. The properties include the following:

1. The number of irreducible representations r is equal to the number of classes k, making a square table.

2. The sum of the squares of the dimensions of the irreducible representations is equal to the order of the group,

$$\sum n_\mu^2 = g. \tag{3.98}$$

   For every group there exists the totally symmetric representation consisting of all ones. These characters form the first entries along the second row of the table.

3. The rows are orthogonal to each other and normalized according to the equation

$$\sum_{\substack{i=1 \\ \text{classes}}} g_i \chi^\mu(\hat{R}_i)\chi^\nu(\hat{R}_i)^* = g\delta_{\mu\nu}. \tag{3.99}$$

4. The columns are also orthogonal and normalized according to the equation

$$\sum_{\nu=1}^{r(=k)} \chi^\nu(\hat{R}_i)\chi^\nu(\hat{R}_j)^* = \frac{g}{g_j}\delta_{ij}. \tag{3.100}$$

   This equation (3.100) has not been derived here but is very useful in constructing character tables.

Mulliken Notation

Each of the irreducible representations could be numbered in order, for example,

C_{3v}	$\hat{E}$	$2\hat{C}_3$	$3\hat{\sigma}_v$
$\Gamma^1(=A_1)$	1	1	1
$\Gamma^2(=A_2)$	1	1	−1
$\Gamma^3(=E)$	2	−1	0

but this labeling scheme is not very informative. Mulliken[1] proposed a labeling scheme that provides some additional information about the symmetry properties of the irreducible representation. All one-dimensional representations are labeled A or B, depending on whether the irreducible representation is symmetric ($\chi(\hat{C}_n$ or $\hat{S}_n) = +1$) or antisymmetric $\chi(\hat{C}_n$ or $\hat{S}_n) = -1$ with respect to rotation (or improper rotation) about the highest order symmetry axis in the molecule. If there is no rotational axis of symmetry, then the one-dimensional irreducible representations are labeled *A*. All 2-dimensional irreducible representations are labeled *E* (unrelated to the operator $\hat{E}$). Three-dimensional irreducible representations are labeled *T* by most workers, except for some infrared spectroscopists who use *F* to label triply degenerate vibrations. Finally the four- and fivefold degenerate irreducible representations found in I_h are labeled *G* and *H*, respectively.

If a center of symmetry is present in a molecule, then *g* or *u* is used as a subscript to identify even (*g*) and odd (*u*) irreducible representations. The *g* and *u* stand for gerade and ungerade, the German words for even and odd. The irreducible representations are of *g* symmetry if $\chi(\hat{i}) > 0$ and *u* symmetry if $\chi(\hat{i}) < 0$. The point groups that contain $\hat{i}$ [C_{nh}(n even), D_{nh} (n even), D_{nd} (n odd), O_h, $D_{\infty h}$, and I_h] can be written as "direct product" groups $G \otimes C_i$. The direct product operation is discussed in more detail in the section on direct product representations in Chapter 4 and below. Each direct product group $G \otimes C_i$ has a character table twice as large as *G*. There are twice the number of irreducible representations (now labeled by *g* and *u*) and twice the number of symmetry operations of *G*: $G = \{\hat{R}\}$, $G \otimes C_i = \{\hat{R}, \hat{i}\hat{R}\}$. If the character tables are considered to be square matrices, then the direct product groups $G \otimes C_i$ have character tables that are direct products (see below) of the character tables for *G* and C_i.

A similar situation arises for point groups with a $\hat{\sigma}_h$ operation but no $\hat{i}$ operation (C_{2h} and D_{nh} with *n* odd). In this case, the group can be written as $G \otimes C_s$ and all of the characters are either single prime or double prime depending on whether $\chi(\hat{\sigma}_h) > 0$ (single prime) or $\chi(\hat{\sigma}_n) < 0$ (double prime). It is useful to recognize direct product groups because the amount of work can be greatly decreased in most applications. For example, one trick is to use the appropriate subgroup for a problem, such as *O* rather than the full group O_h, and then add *g* and *u* at the end by inspection. Consider the character table for the point group of the octahedron $O_h = O \otimes C_i$. The subgroup *O*, made up of only the rotations of O_h, has the character table

O	$\hat{E}$	$8\hat{C}_3$	$3\hat{C}_2$	$6\hat{C}_4$	$6C_2'$
A_1	1	1	1	1	1
A_2	1	1	1	−1	−1
E	2	−1	2	0	0
T_1	3	0	−1	1	−1
T_2	3	0	−1	−1	1

while the character table for C_i is

C_i	$\hat{E}$	$\hat{i}$
A_g	1	1
A_u	1	−1

and therefore,

O_h	$\hat{E}$	$8\hat{C}_3$	$3\hat{C}_2$	$6\hat{C}_4$	$6\hat{C}_2'$	$\hat{i}$	$8\hat{S}_6$	$3\hat{\sigma}_h$	$6\hat{S}_4$	$6\hat{\sigma}_d$
A_{1g}	1	1	1	1	1	1	1	1	1	1
A_{2g}	1	1	1	−1	−1	1	1	1	−1	−1
E_g	2	−1	2	0	0	2	−1	2	0	0
T_{1g}	3	0	−1	1	−1	3	0	−1	1	−1
T_{2g}	3	0	−1	−1	1	3	0	−1	−1	1
A_{1u}	1	1	1	1	1	−1	−1	−1	−1	−1
A_{2u}	1	1	1	−1	−1	−1	−1	−1	1	1
E_u	2	−1	2	0	0	−2	1	−2	0	0
T_{1u}	3	0	−1	1	−1	−3	0	1	−1	1
T_{2u}	3	0	−1	−1	1	−3	0	1	1	−1

O is a subgroup of O_h made up of the rotational symmetry operators of an octahedron.

The definition of a direct product of two matrices $\mathbf{A}^{(n\times n)}$ and $\mathbf{B}^{(m\times m)}$ is

$$\mathbf{A}\otimes\mathbf{B}=\begin{pmatrix} A_{11}\mathbf{B} & \cdots & A_{1n}\mathbf{B} \\ \vdots & \cdots & \vdots \\ A_{n1}\mathbf{B} & \cdots & A_{nn}\mathbf{B} \end{pmatrix}, \tag{3.101}$$

so $\mathbf{A}\otimes\mathbf{B}$ is a new super matrix of dimension $(n\times m)\times(n\times m)$.

If none of the rules for labeling irreducible representations is sufficient to provide a unique label, then numeric subscripts are added to distinguish among the irreducible representations. As an example, we could take A_1 and A_2 in the C_{3v} point group.

Some of the character tables contain characters that are complex numbers, such as the cyclic group C_n. The cyclic group of order n is made up of the rotation operators $\{\hat{C}_n, \hat{C}_n^2, \ldots, \hat{C}_n^n=\hat{E}\}$. Clearly these groups are all Abelian since any operator commutes with itself and each symmetry operator is in its own class. The number of classes and the number of irreducible representations are therefore equal to g, the order of the group. All irreducible representations must be one dimensional since $\Sigma n_\mu^2 = g$ is satisfied only for $n_\nu = 1$, $\nu = 1, \ldots, g$. The characters that are complex must be paired with their complex conjugates, thereby giving rise to a double degeneracy.

For example, the C_5 point group has the character table.

C_5	$\hat{E}$	$\hat{C}_5$	$\hat{C}_5^2$	$\hat{C}_5^3$	$\hat{C}_5^4$
A_1	1	1	1	1	1
E_1	$\begin{cases}1\\1\end{cases}$	$\begin{matrix}\varepsilon\\ \varepsilon^*\end{matrix}$	$\begin{matrix}\varepsilon^2\\ \varepsilon^{2*}\end{matrix}$	$\begin{matrix}\varepsilon^{2*}\\ \varepsilon^2\end{matrix}$	$\begin{matrix}\varepsilon^*\\ \varepsilon\end{matrix}$
E_2	$\begin{cases}1\\1\end{cases}$	$\begin{matrix}\varepsilon^2\\ \varepsilon^{2*}\end{matrix}$	$\begin{matrix}\varepsilon^*\\ \varepsilon\end{matrix}$	$\begin{matrix}\varepsilon\\ \varepsilon^*\end{matrix}$	$\begin{matrix}\varepsilon^{2*}\\ \varepsilon^2\end{matrix}$

where $\varepsilon = e^{2\pi i/5}$.

The characters for the irreducible representations that are complex pairs are labeled as E. The sums of the complex conjugate pairs are real numbers and can be used in most applications rather than the individual complex components.

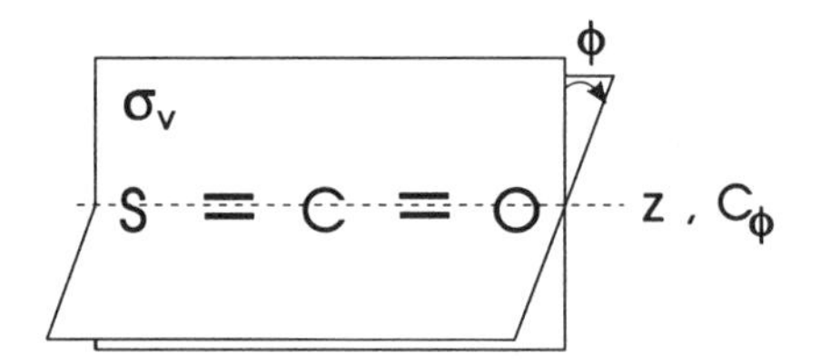

Figure 3.9. The symmetry elements for a linear molecule of $C_{\infty v}$ symmetry.

The point groups $C_{\infty v}$ and $D_{\infty h}$ $(= C_{\infty n} \otimes C_i)$ for linear molecules are of infinite order so the methods discussed so far cannot be used to derive their character tables. The group $C_{\infty v}$ has the rotational symmetry operators $\hat{C}(\phi)$ and their inverses $\hat{C}(-\phi)$ in the same class (Figure 3.9). There are an infinite number of pairs, each with a different ϕ and each pair in a different class. There are also an infinite number of reflection operators, but they all belong to the same class as shown in the character table

$C_{\infty v}$	E	$2\hat{C}(\phi)$	$\infty\hat{\sigma}_v$
$A_1 = \Sigma^+$	1	1	1
$A_2 = \Sigma^-$	1	1	−1
$E_1 = \Pi$	2	$2\cos\phi$	0
$E_2 = \Delta$	2	$2\cos 2\phi$	0
$E_3 = \Phi$	2	$2\cos 3\phi$	0

Mulliken's notation for the characters is on the left, but this notation is not normally used. Instead spectroscopists use the traditional set of Greek labels.

Problems

1. For the vectors

$$\mathbf{a} = \begin{pmatrix} 1 \\ 2 \\ 3 \end{pmatrix} \quad \text{and} \quad \mathbf{b} = \begin{pmatrix} 2 \\ 2 \\ 2 \end{pmatrix},$$

   (a) Calculate $|\mathbf{a}|$, $|\mathbf{b}|$, $\mathbf{a} \cdot \mathbf{b}$, $\mathbf{a} \times \mathbf{b}$.
   (b) Convert **a** and **b** to spherical polar coordinates.

2. For the matrix

$$\mathbf{A} = \begin{pmatrix} 2 & 1 & 2 \\ 3 & 5 & 7 \\ 1 & 1 & 1 \end{pmatrix}$$

   calculate:
   (a) $\mathbf{A}^t$
   (b) $|\mathbf{A}|$
   (c) $\mathbf{A}^{-1}$

3. For the Pauli matrices

$$\boldsymbol{\sigma}_x = \begin{pmatrix} 0 & 1 \\ 1 & 0 \end{pmatrix} \qquad \boldsymbol{\sigma}_y = \begin{pmatrix} 0 & -i \\ i & 0 \end{pmatrix} \qquad \boldsymbol{\sigma}_z = \begin{pmatrix} 1 & 0 \\ 0 & -1 \end{pmatrix}$$

   (a) Verify that $\boldsymbol{\sigma}_x\boldsymbol{\sigma}_y = i\boldsymbol{\sigma}_z$.
   (b) Calculate $\boldsymbol{\sigma}_y\boldsymbol{\sigma}_x$ and evaluate the commutator $[\boldsymbol{\sigma}_x, \boldsymbol{\sigma}_y]$.
4. If

$$\mathbf{A} = \begin{pmatrix} 1 & 2 & 3 \\ 2 & 1 & 2 \\ 4 & 3 & 2 \end{pmatrix} \quad \text{and} \quad \mathbf{B} = \begin{pmatrix} 1 & 0 & 1 \\ 3 & -1 & -2 \\ 2 & 3 & 2 \end{pmatrix},$$

   then calculate $\mathbf{A} + \mathbf{B}$, $\mathbf{A} - \mathbf{B}$, $\mathbf{AB}$, and $\mathbf{BA}$.
5. Prove that
   (a) $(\mathbf{AB})^t = \mathbf{B}^t\mathbf{A}^t$
   (b) $(\mathbf{AB})^\dagger = \mathbf{B}^\dagger\mathbf{A}^\dagger$
6. Prove that $\mathbf{A}^\dagger\mathbf{A}$ and $\mathbf{AA}^\dagger$ are Hermitian for any matrix $\mathbf{A}$.
7. Verify that

$$\begin{pmatrix} \cos\theta & \sin\theta \\ -\sin\theta & \cos\theta \end{pmatrix}$$

   is an orthogonal matrix.
8. Find $\mathbf{A}^{-1}$ if

$$\mathbf{A} = \begin{pmatrix} 2 & 3 & 1 \\ 3 & 5 & 1 \\ 0 & 0 & 2 \end{pmatrix}$$

   and verify that $\mathbf{AA}^{-1} = \mathbf{A}^{-1}\mathbf{A} = \mathbf{1}$.
9. Solve the following set of linear equations using matrix methods

$$\begin{aligned} 4x - 3y + z &= 11 \\ 2x + y - 4z &= -1 \\ x + 2y - 2z &= 1. \end{aligned}$$

10. From the characteristic polynomial prove that:
    (a) The product of the eigenvalues of a matrix equals the determinant, $|\mathbf{A}| = \lambda_1\lambda_2 \cdots \lambda_n$.
    (b) The sum of eigenvalues equals the trace of a matrix, $\text{tr}(\mathbf{A}) = \lambda_1 + \lambda_2 + \cdots + \lambda_n$.
11. Prove that the eigenvalues of a Hermitian matrix are real and that the eigenvectors can be made orthogonal to each other.
12. Find the eigenvalues and normalized eigenvectors of the matrix

$$\begin{pmatrix} 1 & -8 \\ 2 & 11 \end{pmatrix}.$$

13. (a) Find the eigenvalues and normalized eigenvectors of the matrix

$$\mathbf{A} = \begin{pmatrix} 2 & 4-i \\ 4+i & -14 \end{pmatrix}.$$

    (b) Construct the matrix **X** that diagonalizes **A** and verify that it works.
14. (a) Construct the 9-dimensional matrix representation generated by the unit Cartesian vectors associated with each atom of the H_2O molecule. (Pick x_i out of the plane.)
    (b) What are the characters of the reducible representation?
    (c) How many times does each irreducible representation occur in this representation?
15. Consider the four out-of-plane p_z orbitals of cyclobutadiene.
    (a) Assuming D_{4h} symmetry, construct a 4-dimensional matrix representation of the D_4 subgroup.
    (b) What are the characters of this reducible representation?
    (c) How many times does each irreducible representation occur in this representation?
16. Given the set of polynomials $\{1, x, x^2, x^3, \ldots\}$, construct the first three members of a new set of orthonormal polynomials on the interval $-1 \leq x \leq 1$ using the Gram-Schmidt procedure. They are proportional to the Legendre polynomials.
17. (a) For the point group D_{2h} construct a 3-dimensional matrix representation using the set of three real p orbitals.
    (b) To what irreducible representations do these orbitals belong?
18. Construct the character table for the C_{4v} point group, without consulting tables.
19. For the D_{3h} point group verify equations (3.98), (3.99), and (3.100).
20. Construct the D_{2h} character table by taking a direct product of the C_i and D_2 character tables.
21. One matrix representation that can easily be constructed for any group is called the *regular representation.* This is obtained by writing the group multiplication table in such a form that the identity element E lies along the main diagonal. Then the matrix representative for a particular group element R is obtained by replacing that element everywhere in the multiplication table by unity and all other group elements by zero.
    (a) Do so for the group C_{3v} and obtain the corresponding reducible matrix representation group.
    (b) What are the characters of this representation?
    (c) What irreducible representations make up the regular representation of C_{3v}?

Reference

1. Mulliken, R. S. *J. Chem. Phys.* **23,** 1997 (1955).

General References

Arfken, G. *Mathematical Methods for Physicists,* 3rd ed., Academic Press, Orlando, Fla., 1985.

Bishop, D. M. *Group Theory and Chemistry,* Dover, New York, 1993.

Hamermesh, M. *Group Theory and Its Application to Physical Problems,* Dover, New York, 1989.

Mortimer, R. G. *Mathematics for Physical Chemistry,* Macmillan, New York, 1981.

Stephenson, G. *An Introduction to Matrices, Sets and Groups for Science Students,* Dover, New York, 1986.

Tinkham, M. *Group Theory and Quantum Mechanics,* McGraw-Hill, New York, 1964.

Chapter 4
Quantum Mechanics and Group Theory

Matrix Representation of the Schrödinger Equation

The application of quantum mechanics to spectroscopic problems involves solving the appropriate time-independent Schrödinger equation, $\hat{H}\psi = E\psi$. The solutions of this eigenvalue problem are a set of wavefunctions $\{\psi_i\}$ and a corresponding set of eigenvalues $\{E_i\}$. Although solving the Schrödinger equation is, in general, a difficult mathematical problem, steady progress has been made over the years. Spectroscopists, however, do not suffer from such mathematical difficulties—they simply measure the difference between two eigenvalues, $E_i - E_j = h\nu$, and the intensity of the transition.

The most appropriate formulation of quantum mechanics for spectroscopy is based upon the Heisenberg matrix mechanics approach. Although simple spectroscopic models, such as the rigid rotor and the harmonic oscillator, are customarily solved using differential equations, any application of quantum mechanics to real systems is usually best handled by matrix mechanics. The general spectroscopic problem is handled by selection of an appropriate Hamiltonian and selection of a basis set, followed by diagonalization of the Hamiltonian matrix to obtain the wavefunctions and energy levels.

The solution of $\hat{H}\psi = E\psi$ (after selection of $\hat{H}$) proceeds by expanding the wavefunction in terms of a set of appropriate basis functions, that is,

$$\psi = \sum_i c_i f_i, \tag{4.1}$$

or using Dirac notation,

$$|\psi\rangle = \sum c_i \, |f_i\rangle. \tag{4.2}$$

For example, if $\hat{H}$ represents an anharmonic molecular oscillator, then $\{|f_i\rangle\}$ might be the harmonic oscillator wavefunctions. In this basis set the arbitrary wavefunctions $|\psi\rangle$ and $|\phi\rangle$ are represented by column vectors of expansion coefficients, namely

$$|\psi\rangle = \begin{pmatrix} c_1 \\ c_2 \\ c_3 \\ \vdots \end{pmatrix} \quad \text{and} \quad |\phi\rangle = \begin{pmatrix} d_1 \\ d_2 \\ d_3 \\ \vdots \end{pmatrix}.$$

In the vector notation the basis functions can be represented by

$$|f_1\rangle = \begin{pmatrix} 1 \\ 0 \\ 0 \\ 0 \\ \vdots \end{pmatrix} \quad |f_2\rangle = \begin{pmatrix} 0 \\ 1 \\ 0 \\ 0 \\ \vdots \end{pmatrix}$$

and so on. The scalar product of two wavefunctions is expressed in the form

$$\langle\psi \mid \phi\rangle = \int \psi^* \phi \, d\tau = \sum c_i^* d_i = (c_1^* c_2^* \cdots) \begin{pmatrix} d_1 \\ d_2 \\ \vdots \end{pmatrix}. \tag{4.3}$$

The operation of Hermitian conjugation converts the ket vectors ($|\psi\rangle$) to bra vectors ($\langle\psi|$), that is,

$$|\psi\rangle^\dagger = \langle\psi| = (c_1^* c_2^* \cdots), \tag{4.4}$$

so that the scalar product can be interpreted as a matrix product, as in equation (4.3).

The Dirac notation is particularly useful for algebraic manipulations. Notice, for example, $\langle f_i \mid f_j\rangle = \delta_{ij}$ is a number, but $\hat{P}_{ij} = |f_i\rangle\langle f_j|$ is a matrix operator. If $i = j$ then, for example, in a 5-dimensional space

$$\hat{P}_{33} = \begin{pmatrix} 0 & 0 & 0 & 0 & 0 \\ 0 & 0 & 0 & 0 & 0 \\ 0 & 0 & 1 & 0 & 0 \\ 0 & 0 & 0 & 0 & 0 \\ 0 & 0 & 0 & 0 & 0 \end{pmatrix}$$

for $i = j = 3$, and so $\hat{P}_{ii} \equiv \hat{P}_i$ is represented by a matrix with a 1 in the ith position on the diagonal and zeros elsewhere. This is an example of a projection operator, since

$$\hat{P}_i |b\rangle = |f_i\rangle\langle f_i \mid b\rangle, \tag{4.5}$$

and $\hat{P}_i$ projects out of an arbitrary vector $|b\rangle$ the ith component. A useful identity is that of completeness of the basis set, namely that

$$\sum_i \hat{P}_i = \mathbf{1} = \sum_i |f_i\rangle\langle f_i|, \tag{4.6}$$

which can be used to derive the expansion coefficients $\langle f_i \mid \psi\rangle$ of a wavefunction. Thus, we can write

$$|\psi\rangle = \mathbf{1}\,|\psi\rangle = \Big(\sum_i |f_i\rangle\langle f_i|\Big)\,|\psi\rangle = \sum_i |f_i\rangle\langle f_i|\,\psi\rangle, \tag{4.7}$$

so that if $\psi = \sum c_i f_i$, then $c_i = \langle f_i \mid \psi\rangle$. (4.8)

An operator such as $\hat{H}$ has an eigenvalue equation $\hat{H}\,|\psi\rangle = E\,|\psi\rangle$, where $\hat{H}$ is represented by a matrix in terms of the basis set $\{f_i\}$. Although it is not necessary, the $\{f_i\}$ are assumed to be orthogonal functions ($\langle f_i \mid f_j\rangle = \delta_{ij}$) and the matrix elements of $\hat{H}$ are given as

$$H_{ij} = \langle f_i|\,\hat{H}\,|f_j\rangle = \int f_i^* \hat{H} f_j\, d\tau. \tag{4.9}$$

In quantum mechanics the operators associated with observables are Hermitian ($\hat{A}^\dagger = \hat{A}$) so that the corresponding matrices, including $\hat{H}$, are also Hermitian ($\mathbf{H}^\dagger = \mathbf{H}$). The solution of Schrödinger's equation thus requires that the orthogonal eigenvectors and the real eigenvalues of the Hermitian Hamiltonian matrix $\mathbf{H}$ be determined.

The solution of the secular equation

$$|\mathbf{H} - E\mathbf{1}| = 0 \tag{4.10}$$

provides a set of n energies, while the associated n eigenvectors can be determined from the corresponding set of homogeneous equations. These eigenvectors can be used as the columns of a unitary matrix $\mathbf{X}$ and the eigenvalue equation written as

$$\mathbf{HX} = \mathbf{XE}, \tag{4.11}$$

or

$$\mathbf{E} = \mathbf{X}^{-1}\mathbf{HX}. \tag{4.12}$$

The matrix $\mathbf{X}$ corresponds to a coordinate transformation of the original basis functions $\{f_i\}$,

$$\mathbf{f} = \mathbf{Xf}', \tag{4.13}$$

or

$$\mathbf{f}' = \mathbf{X}^{-1}\mathbf{f}. \tag{4.14}$$

The matrix **E** has the energy eigenvalues along the diagonal and zeros elsewhere. In the representation provided by the new set of basis functions the Hamiltonian is diagonal, that is,

$$\langle f_i' | \hat{H} | f_j' \rangle = E_i \, \delta_{ij}. \tag{4.15}$$

The exact and the approximate solutions for the case of a 2×2 Hamiltonian matrix are very useful as an example and for simple applications. Let **H** be a 2-dimensional matrix represented by

$$\mathbf{H} = \begin{pmatrix} E_1^0 & V \\ V & E_2^0 \end{pmatrix} \tag{4.16}$$

in terms of the basis set $\{|f_1\rangle, |f_2\rangle\}$. Then the solution of the secular equation,

$$(E_1^0 - E)(E_2^0 - E) - V^2 = 0, \tag{4.17}$$

leads to the eigenvalues

$$E = \frac{E_1^0 + E_2^0}{2} \pm \frac{((E_1^0 - E_2^0)^2 + 4V^2)^{1/2}}{2}. \tag{4.18}$$

The transformation matrix **X** that diagonalizes **H** can be represented as a rotation of the basis functions, namely

$$\mathbf{X} = \begin{pmatrix} \cos\theta & \sin\theta \\ -\sin\theta & \cos\theta \end{pmatrix}. \tag{4.19}$$

The angle θ of this orthogonal matrix is chosen in order to satisfy the condition that $H_{12}' = H_{21}' = 0$ for the transformed Hamiltonian,

$$\mathbf{H}' = \mathbf{X}^{-1}\mathbf{H}\mathbf{X}. \tag{4.20}$$

This condition requires that

$$\tan 2\theta = \frac{-2V}{E_1^0 - E_2^0}. \tag{4.21}$$

The new basis functions are hence given by

$$\mathbf{f} = \mathbf{X}\mathbf{f}' \qquad \text{or} \qquad \mathbf{f}' = \mathbf{X}^{-1}\mathbf{f} \qquad \text{or} \qquad (\mathbf{f}')^t = (\mathbf{f})^t\mathbf{X}, \tag{4.22}$$

or

$$|f_1'\rangle = \cos\theta \, |f_1\rangle - \sin\theta \, |f_2\rangle \tag{4.23a}$$

$$|f_2'\rangle = \sin\theta \, |f_1\rangle + \cos\theta \, |f_2\rangle. \tag{4.23b}$$

In terms of this new basis set the transformed Hamiltonian $\mathbf{H}'$ is diagonal:

$$\mathbf{H}' = \begin{pmatrix} E_1 & 0 \\ 0 & E_2 \end{pmatrix}. \tag{4.24}$$

The two energies E_1 and E_2 are the two solutions of the secular equation (4.17).

Perturbation theory is also often used to solve a spectroscopic problem approximately. In the 2×2 example just given, the Hamiltonian is written as a sum of a zeroth-order term plus an interaction term, so that

$$\mathbf{H} = \mathbf{H}^{(0)} + \mathbf{H}^{(1)}, \tag{4.25}$$

with

$$\mathbf{H}^{(0)} = \begin{pmatrix} E_1^0 & 0 \\ 0 & E_2^0 \end{pmatrix} \tag{4.26}$$

and

$$\mathbf{H}^{(1)} = \begin{pmatrix} 0 & V \\ V & 0 \end{pmatrix} \tag{4.27}$$

with

$$\hat{H}^{(0)}\psi_n^{(0)} = E_n^{(0)}\psi_n^{(0)}, \tag{4.28}$$

and

$$V = \langle f_1 | \hat{H} | f_2 \rangle. \tag{4.29}$$

According to perturbation theory the energy for the nth eigenvalue is given by the zeroth-order energy plus an infinite sum of successive corrections, that is,

$$E_n = E_n^{(0)} + E_n^{(1)} + E_n^{(2)} + \cdots, \tag{4.30}$$

in which

$$E_n^{(1)} = \langle \psi_n^{(0)} | \hat{H}^{(1)} | \psi_n^{(0)} \rangle, \tag{4.31}$$

and

$$E_n^{(2)} = \sum_{m \neq n} \frac{\langle \psi_n^{(0)} | \hat{H}^{(1)} | \psi_m^{(0)} \rangle \langle \psi_m^{(0)} | \hat{H}^{(1)} | \psi_n^{(0)} \rangle}{E_n^{(0)} - E_m^{(0)}} = \sum_{m \neq n} \frac{|\langle \psi_n^{(0)} | \hat{H}^{(1)} | \psi_m^{(0)} \rangle|^2}{E_n^{(0)} - E_m^{(0)}} \tag{4.32}$$

and higher order contributions are similarly defined.

A similar expansion for the wavefunctions is used, namely

$$\psi_n = \psi_n^{(0)} + \psi_n^{(1)} + \cdots, \tag{4.33}$$

in which the first-order correction is given by

$$\psi_n^{(1)} = \sum_{m \neq n} \frac{\langle \psi_m^{(0)} | \hat{H}^{(1)} | \psi_n^{(0)} \rangle}{E_n^{(0)} - E_m^{(0)}} | \psi_m^{(0)} \rangle, \tag{4.34}$$

while higher order corrections are defined analogously. Since the diagonal elements of $\mathbf{H}^{(1)}$ are zero in this example, the first-order correction to the energy $E^{(1)} = 0$, while the second-order correction to the energy is given by

$$E_1^{(2)} = \frac{\langle \psi_1^{(0)} | \hat{H}^{(1)} | \psi_2^{(0)} \rangle \langle \psi_2^{(0)} | \hat{H}^{(1)} | \psi_1^{(0)} \rangle}{E_1^{(0)} - E_2^{(0)}} = \frac{V^2}{E_1^{(0)} - E_2^{(0)}} = -\frac{V^2}{\Delta E} \tag{4.35}$$

and

$$E_2^{(2)} = \frac{V^2}{E_2^{(0)} - E_1^{(0)}} = \frac{V^2}{\Delta E}. \tag{4.36}$$

The corresponding wavefunctions are

$$\psi_1 = \psi_1^{(0)} + \frac{V}{E_1^{(0)} - E_2^{(0)}} \psi_2^{(0)} = \psi_1^{(0)} - \frac{V}{\Delta E} \psi_2^{(0)} \tag{4.37}$$

and

$$\psi_2 = \psi_2^{(0)} + \frac{V}{E_2^{(0)} - E_1^{(0)}} \psi_1^{(0)} = \psi_2^{(0)} + \frac{V}{\Delta E} \psi_1^{(0)}. \tag{4.38}$$

The effect of the interaction

$$V = \langle \psi_2^{(0)} | \hat{H} | \psi_1^{(0)} \rangle \tag{4.39}$$

mixes the wavefunctions and shifts the energy levels E_1^0 and E_2^0 in opposite directions by the amount $+V^2/\Delta E$ for E_2^0 and $-V^2/\Delta E$ for E_1^0, as shown in Figure 4.1. The

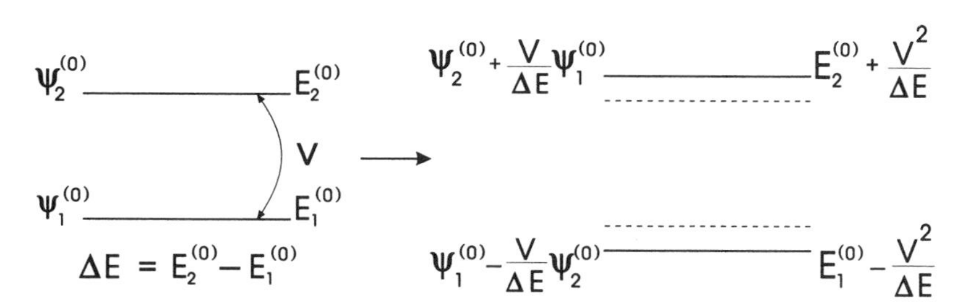

Figure 4.1. The interaction of two states using second-order perturbation theory.

degree to which the two wavefunctions can mix depends on both the magnitude of V and the initial energy difference $\Delta E = E_2^0 - E_1^0$ as indicated by equations (4.37) and (4.38).

The perturbation theory result can be compared to the exact result for the 2-level system by expanding the square root in the expression for the exact solution (4.18), first by rewriting it in the form

$$E_{\pm} = (E_1^{(0)} + E_2^{(0)})/2 \pm \Delta E\left(1 + \frac{4V^2}{\Delta E^2}\right)^{1/2} \Big/ 2 \tag{4.40}$$

and then expanding the square root to obtain

$$E_{\pm} = (E_1^{(0)} + E_2^{(0)})/2 \pm \Delta E\left(1 + 4\frac{V^2}{2(\Delta E)^2} + \frac{(1/2)(-1/2)}{2!}\left(\frac{4V^2}{\Delta E^2}\right)^2 + \cdots\right) \Big/ 2. \tag{4.41}$$

The two energies are therefore

$$E_2 = E_+ = E_2^{(0)} + \frac{V^2}{\Delta E} - \frac{V^4}{\Delta E^3} + \cdots \tag{4.42}$$

and

$$E_1 = E_- = E_1^{(0)} - \frac{V^2}{\Delta E} + \frac{V^4}{\Delta E^3} + \cdots. \tag{4.43}$$

Comparing equation (4.35) with equations (4.42) and (4.43) indicates that second-order perturbation theory gives accurate results if the terms $\pm V^4/\Delta E^3$ are negligible.

Born–Oppenheimer Approximation

The central approximation in molecular spectroscopy is the separation of electronic and nuclear motion. The nonrelativistic molecular Hamiltonian is given by

$$\hat{H} = \frac{-\hbar^2}{2}\sum_\alpha \frac{\nabla_\alpha^2}{M_\alpha} - \frac{\hbar^2}{2m_e}\sum_i \nabla_i^2 + \sum_\alpha \sum_{\beta>\alpha} \frac{Z_\alpha Z_\beta e^2}{4\pi\varepsilon_0 r_{\alpha\beta}} - \sum_\alpha \sum_i \frac{Z_\alpha e^2}{4\pi\varepsilon_0 r_{i\alpha}} + \sum_{i>j} \frac{e^2}{4\pi\varepsilon_0 r_{ij}} \tag{4.44}$$

$$\hat{H} = \hat{T}_N + \hat{T}_e + \hat{V}_{NN} + \hat{V}_{eN} + \hat{V}_{ee} \tag{4.45}$$

in which Greek subscripts in equation (4.44) refer to the nuclei in a molecule, while Roman subscripts refer to the electrons. The various terms in this expression are: the nuclear kinetic energy

$$\hat{T}_N = \frac{-\hbar^2}{2}\sum_\alpha \frac{\nabla_\alpha^2}{M_\alpha}, \tag{4.46}$$

the electronic kinetic energy

$$\hat{T}_e = \frac{-\hbar^2}{2m_e}\sum_i \nabla_i^2, \tag{4.47}$$

the nuclear–nuclear repulsion energy

$$\hat{V}_{NN} = \sum_\alpha \sum_{\beta>\alpha} \frac{Z_\alpha Z_\beta e^2}{4\pi\varepsilon_0 r_{\alpha\beta}}, \tag{4.48}$$

the electron–nuclear attraction energy

$$V_{eN} = -\sum_\alpha \sum_i \frac{Z_\alpha e^2}{4\pi\varepsilon_0 r_{i\alpha}}, \tag{4.49}$$

and the electron–electron repulsion energy

$$V_{ee} = \sum_i \sum_{j>i} \frac{e^2}{4\pi\varepsilon_0 r_{ij}}. \tag{4.50}$$

The Schrödinger equation $\hat{H}\psi = E\psi$ with this Hamiltonian is much easier to write than it is to solve. The solution of the appropriate Schrödinger equation can in principle explain all of chemistry and spectroscopy. The first step in solving the Schrödinger equation is to invoke the Born–Oppenheimer approximation.[1]

The Coulombic forces acting on the nuclei and on the electrons are similar in magnitude, but the electrons are much lighter. The electrons therefore move much faster than the nuclei and, as a consequence, the electronic motion can be separated from the nuclear motion. The electronic structure is solved by "clamping" the nuclei at fixed positions and solving the purely electronic equation

$$\hat{H}_{el}\psi_{el} = E_{el}\psi_{el}$$

in which (4.51)

$$\hat{H}_{el} = \hat{T}_e + \hat{V}_{eN} + \hat{V}_{ee}.$$

If the nuclei are fixed in space then

$$\hat{V}_{NN} = \sum_\alpha \sum_{\beta>\alpha} \frac{Z_\alpha Z_\beta e^2}{4\pi\varepsilon_0 r_{\alpha\beta}} \tag{4.52}$$

is just a number that can be added in at the end to form the total electronic energy

$$U = E_{el} + V_{NN}. \tag{4.53}$$

The separation of the Schrödinger equation into electronic and nuclear motion (vibration–rotation) parts means that ψ can be approximated as the product function

$$\psi \approx \psi_{el}\chi_N, \tag{4.54}$$

and that two equations

$$(\hat{H}_{el} + \hat{V}_{NN})\psi_{el}(\mathbf{r}_i; \mathbf{r}_\alpha) = U(\mathbf{r}_\alpha)\psi_{el}(\mathbf{r}_i; \mathbf{r}_\alpha) \tag{4.55}$$

$$[\hat{T}_N + U(\mathbf{r}_\alpha)]\chi_N(\mathbf{r}_\alpha) = E_N\chi_N(\mathbf{r}_\alpha) \tag{4.56}$$

now need to be solved. In the first equation (4.55) the value of the total electronic energy depends in a parametric way on the particular nuclear positions, $\mathbf{r}_\alpha$. Clamping the nuclei at different positions will result in a different numerical value for U and a different function for ψ_{el}. As the nuclei move, the electrons move so quickly that the $U(\mathbf{r}_\alpha)$ derived from equation (4.55) serves as the potential energy for the nuclear motion (4.56). The vibrational and rotational motions can also be approximately separated in equation (4.56). Although the vibration–rotation separation is similar to the Born–Oppenheimer approximation, it is a separate approximation that is not part of the Born–Oppenheimer separation of nuclear and electronic motion.

The terms neglected in the Born–Oppenheimer approximation can be examined by substituting the equation

$$\psi = \psi_{el}(\mathbf{r}_i; \mathbf{r}_\alpha)\chi_N(\mathbf{r}_\alpha) \tag{4.57}$$

into the full Schrödinger equation and remembering that χ_N depends only upon the nuclear coordinates ($\mathbf{r}_\alpha$), while ψ_{el} depends upon the electronic coordinates ($\mathbf{r}_i$) and, parametrically, also upon the nuclear coordinates so that

$$-\frac{\hbar^2}{2}\sum_\alpha \frac{\nabla_\alpha^2}{M_\alpha}\psi_{el}\chi_N - \frac{\hbar^2}{2m_e}\sum_i \nabla_i^2\psi_{el}\chi_N + V\psi_{el}\chi_N = E\psi_{el}\chi_N. \tag{4.58}$$

By employing the identity

$$\nabla^2 fg = g\nabla^2 f + 2\nabla f \cdot \nabla g + \nabla^2 g, \tag{4.59}$$

we see that equation (4.58) becomes

$$-\frac{\hbar^2}{2}\sum_\alpha \chi_N \frac{\nabla_\alpha^2}{M_\alpha}\psi_{el} - \hbar^2 \sum_\alpha \frac{(\nabla_\alpha \psi_{el})}{M_\alpha} \cdot (\nabla_\alpha \chi_N)$$

$$-\frac{\hbar^2}{2}\sum_\alpha \psi_{el}\frac{\nabla_\alpha^2\chi_N}{M_\alpha} - \frac{\hbar^2}{2m_e}\sum_i \chi_N \nabla_i^2\psi_{el} + V\psi_{el}\chi_N = E\psi_{el}\chi_N. \tag{4.60}$$

By neglecting the first and second terms in equation (4.60), the remaining terms can be separated to yield equations (4.55) and (4.56). The neglect of these two terms is equivalent to neglecting the first and second derivatives of the electronic wavefunction with respect to the nuclear coordinates $\nabla_\alpha \psi_{el}$ and $\nabla_\alpha^2 \psi_{el}$. Indeed, the first-order correction (diagonal correction) for the effects of the breakdown of the Born–Oppenheimer approximation requires that these derivatives be evaluated and used to deduce the energy correction from $E^{(1)} = \langle \psi^{(0)} | \hat{H}^{(1)} | \psi^{(0)} \rangle$, in which $\psi^{(0)}$ is the Born–Oppenheimer wavefunction and $\hat{H}^{(1)}$ represents the two neglected terms.

Symmetry of the Hamiltonian

The application of symmetry in quantum mechanics makes use of a key theorem: if the operators for two observables $\hat{A}$ and $\hat{B}$ commute, then it is possible to find a common set of orthogonal eigenfunctions. In mathematical terms if $[\hat{A}, \hat{B}] = 0$, then $\hat{A}\psi = a\psi$ and $\hat{B}\psi = b\psi$, where $[\hat{A}, \hat{B}] = \hat{A}\hat{B} - \hat{B}\hat{A}$, and the ψ's in the two eigenvalue equations are the same. This abstract theorem has far-reaching consequences. For example, if the two observables are the total energy and the square of the total angular momentum, then it can be proved that $[\hat{H}, \hat{J}^2] = 0$. Therefore a set of common wavefunctions can be found for the two equations $\hat{H}\psi_{nJ} = E\psi_{nJ}$ and $\hat{J}^2\psi_{nJ} = E_J\psi_{nJ}$ and, most importantly, the energies and wavefunctions of the system can be labeled with J: $\{E_{nJ}\}$, $\{\psi_{nJ}\}$. The "good quantum number" J is very useful in characterizing the eigenstates of molecular or atomic systems.

An "almost good quantum number" is associated with an observable of the system that "almost" commutes with the Hamiltonian: $\hat{H}\psi_n = E_n\psi_n$ and $\hat{A}\psi_n \approx a\psi_n$ (i.e., the wavefunction of the system is an approximate eigenfunction of some other observable). For example, spin-orbit coupling can couple the spin $\mathbf{S}$ and orbital angular momentum $\mathbf{L}$ in an atom. Although J ($\mathbf{J} = \mathbf{L} + \mathbf{S}$) is still a good quantum number, L and S are only approximate quantum numbers so that, for example, $\mathbf{S}^2\psi \approx S(S+1)\hbar^2\psi$. Although spin is no longer a good quantum number, it is almost good for light atoms and it is still useful to speak of, for example, triplet states.

The Hamiltonian of a system $\hat{H} = \hat{T} + \hat{V}$ has certain symmetry properties. For example, the kinetic energy part of the Hamiltonian always has the symmetry of a sphere K_h because the Laplacian operator ∇^2 is invariant under all reflections and rotations that contain the origin. This can be verified by applying the transformation operators, such as $\hat{O}_{\hat{C}_\theta}$, to the Laplacian. The potential energy part of the Hamiltonian therefore displays the particular point group symmetry of the molecule.

Since the operators $\hat{O}_R$ leave the Hamiltonian unchanged, they must commute with the Hamiltonian, that is, $\hat{O}_R\hat{H}f = \hat{H}\hat{O}_R f$ or $[\hat{H}, \hat{O}_R] = 0$. In fact it is possible to show that $\hat{O}_R$ commutes with $\hat{T}_e$, $\hat{T}_N$, and $\hat{V}$ individually. This means that a common set of eigenfunctions for $\hat{H}\psi = E\psi$ and $\hat{O}_R\psi = a\psi$ can be found. The wavefunctions can therefore be classified by their behavior with respect to the set of symmetry operators $\{\hat{O}_R\}$. The wavefunctions have the same symmetry properties as the irreducible representations, which can thus be used to label the wavefunctions. For example, if the molecule has a center of symmetry, then $\hat{H}_{el}\psi = E_{el}\psi$

C_i	$\hat{E}$	$\hat{i}$
A_g	1	1
A_u	1	−1

and $\hat{i}\psi_{el} = \pm\psi_{el}$ have a common set of eigenfunctions. Thus ψ_{el} behaves either like the A_g row or the A_u row of the character table. The electron wavefunction is either even or odd so that g and u can be used to classify the wavefunctions as $\psi_{el,g}$ or $\psi_{el,u}$.

For degenerate wavefunctions the symmetry operator $\hat{O}_R$ changes one wavefunction into a linear combination of the members of the degenerate set. If there is an n-fold degeneracy, that is, if

$$\hat{H}\psi_1 = E\psi_1$$

$$\hat{H}\psi_2 = E\psi_2$$

$$\vdots$$

$$\hat{H}\psi_n = E\psi_n,$$

then $\{\psi_1, \psi_2, \ldots, \psi_n\}$ will form an n-dimensional function space spanned by the n orthogonal wavefunctions. Thus the action of $\hat{O}_R$ on a single member ψ_i of the set of n degenerate wavefunctions can be represented by

$$\hat{O}_R\psi_i = \sum_j \psi_j D_{ji}(\hat{R}), \tag{4.61}$$

in which $\mathbf{D}(\hat{R})$ is the matrix representation of $\hat{R}$ in the n-dimensional wavefunction space. The new wavefunction

$$\psi' = \sum_i c_i\psi_i \tag{4.62}$$

produced by the action of a symmetry operator is also a solution of the Schrödinger equation having the same energy eigenvalue. The n degenerate wavefunctions form a basis for the matrix representation of the point group operations of the molecule. For example, the electronic wavefunctions of the NH_3 molecule might be totally symmetric with respect to the six symmetry operators (and have A_1 symmetry), might behave like the A_2 line of the character table, or might be doubly degenerate (with E symmetry). The electronic states of ammonia are thus said to belong to the A_1, A_2, or E representations. Basis functions that possess these symmetries are said to be symmetry-adapted basis functions.

Projection Operators

Projection operators are useful in generating functions of the proper symmetry for the solution of a molecular problem. Since a wavefunction must belong to a particular irreducible representation, it is very helpful to construct solutions of the correct symmetry for the problem. Consider a set of n_ν orthogonal functions $\{f_1^\nu, f_2^\nu, \ldots, f_{n_\nu}^\nu\}$ that belongs to a function space that forms the νth irreducible representation in a point group. The result of operating with the operator $\hat{O}_R$ on an arbitrary member of this set of functions can be written as

$$\hat{O}_R f_j^\nu = \sum_i f_i^\nu D_{ij}^\nu(\hat{R}), \tag{4.63}$$

in which $\Gamma^\nu = \{\mathbf{D}^\nu(\hat{R})\}$ is the unitary matrix representation. Let us define the projection operator

$$\hat{P}_{ij}^\mu = \sum_R D_{ij}^\mu(\hat{R})^* \hat{O}_R \tag{4.64}$$

and apply $\hat{P}_{ij}^\mu$ to a member of the function space $\{f^\nu\}$. We obtain the result

$$\begin{aligned}\hat{P}_{ij}^\mu f_k^\nu &= \sum_R D_{ij}^\mu(\hat{R})^* \hat{O}_R f_k^\nu = \sum_R D_{ij}^\mu(\hat{R})^* \sum_m D_{mk}^\nu(\hat{R}) f_m^\nu \\ &= \sum_m \left[\sum_R D_{ij}^\mu(\hat{R})^* D_{mk}^\nu(\hat{R})\right] f_m^\nu \\ &= \sum_m \left[\frac{g}{n_\nu}\delta_{\mu\nu}\delta_{im}\delta_{jk}\right] f_m^\nu = \frac{g}{n_\nu}\delta_{\nu\mu}\delta_{jk} f_i^\nu\end{aligned} \tag{4.65}$$

by employing the Great Orthogonality Theorem. Notice that if $\mu \neq \nu$, then $\hat{P}_{ij}^\mu f_k^\nu = 0$; if $j \neq k$, then $\hat{P}_{ij}^\mu f_k^\nu = 0$; but $\hat{P}_{ij}^\mu f_j^\mu = (g/n_\mu) f_i^\mu$. This means that if one member of a set of basis functions belonging to an irreducible representation is known, then it is possible by using projection operators to generate all other members of that representation. The only catch with this type of a projection operator is that the representation matrices $\{\mathbf{D}^\mu(\hat{R})\}$ are needed, and not just their traces or characters.

A simpler, but still useful, projection operator can be defined using only the characters of the matrix representation. Let the projection operator $\hat{P}^\mu$ be defined as

$$\hat{P}^\mu = \sum_{i=1}^{n_\mu} \hat{P}_{ii}^\mu = \sum_{i=1}^{n_\mu} \sum_R D_{ii}^\mu(\hat{R})^* \hat{O}_R = \sum_R \sum_{i=1}^{n_\mu} D_{ii}^\mu(\hat{R})^* \hat{O}_R \tag{4.66}$$

or

$$\hat{P}^\mu = \sum_R \chi^\mu(\hat{R})^* \hat{O}_R \tag{4.67}$$

since the summation over the diagonal elements of $\mathbf{D}^\mu(\hat{R})$ simply represents the

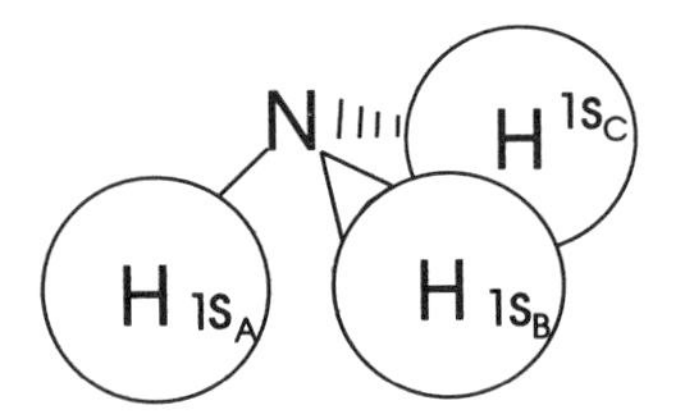

Figure 4.2. The three 1s hydrogen functions in the NH_3 molecule.

character of $\mathbf{D}^{\mu}(\hat{R})$. Thus we see that when $\hat{P}^{\mu}$ acts on a member of the νth function space, we obtain

$$\begin{aligned}\hat{P}^{\mu} f_j^{\nu} &= \sum_i \hat{P}_{ii}^{\mu} f_j^{\nu} = \sum_{i=1}^{n_\mu} \sum_R D_{ii}^{\mu}(\hat{R}) \hat{O}_R f_j^{\nu} \\ &= \sum_{i=1}^{n_\mu} \sum_R D_{ii}^{\mu}(\hat{R}) \sum_m D_{mj}^{\nu}(\hat{R}) f_m^{\nu} = \sum_{i=1}^{n_\mu} \sum_{m=1}^{n_\nu} \frac{g}{n_\mu} \delta_{\mu\nu}\, \delta_{im}\, \delta_{ij} f_m^{\nu} \\ &= \frac{g}{n_\mu} \delta_{\mu\nu} f_j^{\nu}. \end{aligned} \tag{4.68}$$

If $\mu \neq \nu$, then $\hat{P}^{\mu} f_j^{\nu} = 0$, and if $\mu = \nu$ and $i = j$, then $\hat{P}^{\mu} f_j^{\nu} = (g/n_\mu) f_j^{\nu}$.

At first sight the $\hat{P}^{\mu}$ operators do not seem to be very useful. Notice, however, that $\hat{P}^{\mu}$ annihilates all functions, or parts of functions, that do not belong to the μth irreducible representation and leaves behind a function that possesses the correct symmetry.

For example, consider a set of 1s functions on the hydrogen atoms of NH_3 (Figure 4.2) and suppose that a linear combination of the three atomic hydrogen orbitals is needed to make approximate molecular orbitals to bond with the nitrogen atomic orbitals. Individually the three hydrogen orbitals do not have the correct symmetry since, for example, $\hat{C}_3(1s_A) = 1s_C$. Projection operators can, however, be used to construct a set of symmetry-adapted linear combinations of the hydrogen 1s orbitals. As we have seen earlier the appropriate symmetry point group is C_{3v}, so that the projection operators

$$\hat{P}^{A_1} = 1 \cdot \hat{O}_E + 1 \cdot \hat{O}_{C_3} + 1 \cdot \hat{O}_{C_3^{-1}} + 1 \cdot \hat{O}_{\sigma'} + 1 \cdot \hat{O}_{\sigma''} + 1 \cdot \hat{O}_{\sigma'''}, \tag{4.69a}$$

$$\hat{P}^{A_2} = 1 \cdot \hat{O}_E + 1 \cdot \hat{O}_{C_3} + 1 \cdot \hat{O}_{C_3^{-1}} - 1 \cdot \hat{O}_{\sigma'} - 1 \cdot \hat{O}_{\sigma''} - 1 \cdot \hat{O}_{\sigma'''}, \tag{4.69b}$$

$$\hat{P}^{E} = 2 \cdot \hat{O}_E - 1 \cdot \hat{O}_{C_3} - 1 \cdot \hat{O}_{C_3^{-1}} + 0 \cdot \hat{O}_{\sigma'} + 0 \cdot \hat{O}_{\sigma''} + 0 \cdot O_{\sigma'''}, \tag{4.69c}$$

can be applied to a typical 1s function, say $1s_A$, to give

$$\hat{P}^{A_1} 1s_A = 1s_A + 1s_C + 1s_B + 1s_A + 1s_C + 1s_B, \tag{4.70a}$$

$$\hat{P}^{A_2} 1s_A = 1s_A + 1s_C + 1s_B - 1s_A - 1s_C - 1s_B = 0, \tag{4.70b}$$

$$\hat{P}^{E} 1s_A = 2(1s_A) - 1s_C - 1s_B. \tag{4.70c}$$

If the hydrogenic basis functions are orthogonal to one another, then we obtain the orthonormalized symmetry-adapted basis functions

$$\phi_1 = \frac{1}{\sqrt{3}}(1s_A + 1s_B + 1s_C), \tag{4.71a}$$

with A_1 symmetry and

$$\phi_2 = \frac{1}{\sqrt{6}}[(2(1s_A) - 1s_B - 1s_C)] \tag{4.71b}$$

with E symmetry. Notice that there is a missing linear combination since the application of a single $\hat{P}^E$ operator generates only a single function of E symmetry and to complete the E representation a second basis function is necessary. This second function can be generated by applying $\hat{P}^E$ to another atomic orbital, say the $1s_B$ orbital,

$$\hat{P}^E 1s_B = 2(1s_B) - 1s_C - 1s_A. \tag{4.72}$$

This second function can then be made orthogonal to the original E function by subtracting out the overlapping part using the Gram-Schmidt procedure (see the fifth section of Chapter 3). The new function has the form

$$\begin{aligned}\phi_3' &= \frac{1}{\sqrt{6}}[2(1s_B) - 1s_C - 1s_A] - \frac{1}{\sqrt{6}}\langle\phi_2|\, 2(1s_B) - 1s_C - 1s_A\rangle\phi_2 \\ &= \frac{1}{\sqrt{6}}[2(1s_B) - 1s_C - 1s_A] + \phi_2/2 \\ &= \frac{1}{\sqrt{6}}\left[\frac{3}{2}(1s_B) - \frac{3}{2}(1s_C)\right]\end{aligned} \tag{4.73}$$

which becomes upon normalizing,

$$\phi_2 = \frac{1}{\sqrt{2}}(1s_B - 1s_C). \tag{4.74}$$

The functions ϕ_1, ϕ_2, and ϕ_3 have the appropriate symmetry for the NH_3 molecule. Although these simple functions might not be the best functions for a realistic calculation of the electronic energy of NH_3, they are the approximate functions to use within the simple linear combination of atomic orbitals model. Since the final electronic wavefunction must belong to the A_1, A_2, or E irreducible representations, the calculation is simplified if the initial trial wavefunctions also belong to one of these representations.

Direct Product Representations

The total wavefunction is often written as a product $\psi = \psi_{el}\chi_N$, where ψ_{el} and χ_N each belong to particular irreducible representations of the molecular point group. To which representation does the product $\psi = \psi_e\chi_N$ belong? In order to determine this let $\{f_1^\nu f_2^\nu \cdots f_{n_\nu}^\nu\}$ be a set of n_ν functions belonging to the νth representation Γ^ν, while $\{f_1^\mu \cdots f_{n_\mu}^\mu\}$ belong to Γ^μ. A new function space with $n_\nu \times n_\mu$ members can be formed by taking all possible products

$$\begin{matrix} f_1^\nu f_1^\mu & f_1^\nu f_2^\mu & \cdots & f_1^\nu f_{n_\mu}^\mu \\ \vdots & & & \vdots \\ f_{n_\nu}^\nu f_1^\mu & \cdots & \cdots & f_{n_\nu}^\nu f_{n_\mu}^\mu. \end{matrix}$$

These product functions form a new representation $\Gamma^{\nu\otimes\mu} = \Gamma^\nu \otimes \Gamma^\mu$, where the symbol $\otimes$ is used to represent the direct product in order to distinguish the operation from ordinary multiplication. A new set of matrices, $n_\nu n_\mu \times n_\nu n_\mu$ in dimension is formed by taking direct products of the matrix representatives in Γ^ν and Γ^μ, that is, $\mathbf{D}^{\nu\otimes\mu} = \mathbf{D}^\nu \otimes \mathbf{D}^\mu$. For example, the direct product of two 2×2 matrices $\mathbf{A}$ and $\mathbf{B}$ is represented by

$$\mathbf{A}\otimes\mathbf{B} = \begin{pmatrix} A_{11}\mathbf{B} & A_{12}\mathbf{B} \\ A_{21}\mathbf{B} & A_{22}\mathbf{B} \end{pmatrix}$$

$$= \begin{pmatrix} A_{11}B_{11} & A_{11}B_{12} & A_{12}B_{11} & A_{12}B_{12} \\ A_{11}B_{21} & A_{11}B_{22} & A_{12}B_{21} & A_{12}B_{22} \\ A_{21}B_{11} & A_{21}B_{12} & A_{22}B_{11} & A_{22}B_{12} \\ A_{21}B_{21} & A_{21}B_{22} & A_{22}B_{21} & A_{22}B_{22} \end{pmatrix}. \tag{4.75}$$

The characters of the direct product matrices

$$\sum_i D_{ii}^{\nu\otimes\mu}(\hat{R}) = \sum_i^{n_\nu}\sum_j^{n_\mu} D_{ii}^\nu(\hat{R})D_{jj}^\mu(\hat{R}) = \Big(\sum_i D_{ii}^\nu(\hat{R})\Big)\Big(\sum_j D_{jj}^\mu(\hat{R})\Big) = \chi^\nu(\hat{R})\chi^\mu(\hat{R}) \tag{4.76}$$

are just the product of the characters of the individual matrix representations in Γ^μ and Γ^ν. Of course, these direct product representations are reducible in terms of the irreducible representations of the point group, that is,

$$\Gamma^{\nu\otimes\mu} = \Gamma^\nu \otimes \Gamma^\mu = a_1\Gamma^1 \oplus a_2\Gamma^2 \oplus \cdots = \oplus \sum a_i\Gamma^i, \tag{4.77}$$

in which the a_i are determined from

$$a_i = \frac{1}{g} \sum_R \chi^{\nu \otimes \mu}(\hat{R})[\chi^i(\hat{R})]^*. \tag{4.78}$$

For example, consider the product $E \otimes E$ obtained for the product wavefunction $\psi = \psi_{el,E}\chi_{N,E}$. The appropriate characters are given by

C_{3v}	$\hat{E}$	$2\hat{C}_3$	$3\hat{\sigma}_v$
A_1	1	1	1
A_2	1	1	−1
E	2	−1	0
$\chi^{E\otimes E}$	4	1	0

and the use of equation (4.78) leads to the results

$$a_{A_1} = \frac{1}{6}(4(1) + 2(1)(1)) = 1,$$

$$a_{A_2} = \frac{1}{6}(4(1) + 2(1)(1)) = 1,$$

$$a_E = \frac{1}{6}(4(2) + 2(-1)(1)) = 1,$$

so that the direct product representation is decomposed as

$$\Gamma^{E\otimes E} = \Gamma^{A_1} \oplus \Gamma^{A_2} \oplus \Gamma^{E}. \tag{4.79}$$

Integrals and Selection Rules

The intensity of an electronic transition is proportional to the square of a transition dipole moment, that is,

$$I \propto \left| \int \psi_f^* \boldsymbol{\mu} \psi_i \, d\tau \right|^2, \tag{4.80}$$

so that integrals of the type

$$\int \psi_i^* \mathrm{x} \psi_j \, d\tau, \tag{4.81}$$

or, more generally,

$$\int \psi_i^* \, \mathrm{f} \, \psi_j \, d\tau, \tag{4.82}$$

with f as an arbitrary function, are of interest. The integrand is a product of three functions, each of which belongs to a particular irreducible representation. What is the overall symmetry of the integrand? If ψ_i belongs to the Γ^μ irreducible representation, ψ_j to Γ^ν, and f to the Γ^λ irreducible representation, then the triple product $\psi_i^* \mathrm{f} \psi_j$ belongs to the direct product representation

$$\Gamma^\mu \otimes \Gamma^\lambda \otimes \Gamma^\nu = \oplus \sum a_i \Gamma^i \tag{4.83}$$

that can be reduced to the direct sum of the irreducible representations of the point group. If this reduction does not contain the A_1 irreducible (totally symmetric) representation, then the integral over all space is exactly zero. This is just a generalization of the fact that the integral over all space of an odd function, $f(-x) = -f(x)$, is zero when integrated over all space, or

$$\int_{-\infty}^{\infty} f_{\text{odd}}(x)\,dx = 0.$$

The proof of the assertion that a nonzero integral

$$\int \psi_i^* \mathrm{f} \psi_j \, d\tau$$

must have an integrand that belongs to a direct product representation that contains A_1 requires the use of the projection operator

$$\hat{P}^{A_1} = \sum_R \chi^{A_1}(\hat{R})\hat{O}_R = \sum_R \hat{O}_R. \tag{4.84}$$

If $\psi_i^* \mathrm{f} \psi_j$ does not contain a function of A_1 symmetry, then

$$\hat{P}^{A_1}[\psi_i^* \mathrm{f} \psi_j] = 0. \tag{4.85}$$

The $\hat{O}_R$ symmetry operators have no effect on the integral

$$\int \hat{O}_R[\psi_i^* \mathrm{f} \psi_j]\,d\tau = \int \psi_i^* \mathrm{f} \psi_j \, d\tau, \tag{4.86}$$

since the integration is over all space. Thus, summing over all R, we find that

$$\begin{aligned} g\int \psi_i^* \mathrm{f} \psi_j \, d\tau &= \sum_R \int \hat{O}_R[\psi_i^* \mathrm{f} \psi_j]\,d\tau \\ &= \int \hat{P}^{A_1}[\psi_i^* \mathrm{f} \psi_j]\,d\tau, \end{aligned} \tag{4.87}$$

and hence

$$\int \psi_i^* \mathrm{f} \psi_j \, d\tau = \frac{1}{g}\int \hat{P}^{A_1}[\psi_i^* \mathrm{f} \psi_j]\,d\tau. \tag{4.88}$$

Thus, if $\hat{P}^{A_1}[\psi_i^* \mathrm{f} \psi_j] = 0$, then the integral must vanish, and it is a necessary (but not sufficient) condition that the integrand contain a function of A_1 symmetry in order that the integral not vanish.

An important application of this rule (in addition to deriving selection rules) is in the construction of Hamiltonian matrices. It is possible to let f be an operator such as

$\hat{H}$, which belongs to the A_1 irreducible representation since the Hamiltonian is unchanged under all symmetry operations. Therefore

$$H_{ij} = \int \psi_i^* \hat{H} \psi_j \, d\tau \tag{4.89}$$

will be nonzero depending on the symmetry properties of ψ_i and ψ_j. The number of times that the direct product representation contains an irreducible representation of A_1 symmetry is given by

$$\begin{aligned} a_{A_1} &= \frac{1}{g} \sum_R \chi^{\mu \otimes \nu}(\hat{R}) \chi^{A_1}(\hat{R})^* \\ &= \frac{1}{g} \sum_R \chi^{\mu \otimes \nu}(\hat{R}) \cdot 1 = \frac{1}{g} \sum_R \chi^{\mu}(\hat{R}) \chi^{\nu}(\hat{R}) \\ &= \frac{1}{g} (g \delta_{\mu\nu}) = \delta_{\mu\nu}, \end{aligned} \tag{4.90}$$

since Γ^μ and Γ^ν are irreducible representations. This means that matrix elements between functions belonging to different irreducible representations ($\mu \neq \nu$) will be identically zero. The Hamiltonian becomes block diagonal, with each block corresponding to a different irreducible representation. Each block can now be diagonalized separately since there are no matrix elements connecting blocks of different symmetry

$$\mathbf{H} = \left(\begin{array}{c|c|c} A_1 \text{ block} & 0 & 0 \\ \hline 0 & \mu\text{th block} & 0 \\ \hline 0 & 0 & \nu\text{th block} \end{array} \right). \tag{4.91}$$

Problems

1. Given the matrices **A** and **B** as

$$\mathbf{A} = \begin{pmatrix} -\frac{1}{3} & \sqrt{\frac{2}{3}} & \frac{\sqrt{2}}{3} \\ \sqrt{\frac{2}{3}} & 0 & \frac{1}{\sqrt{3}} \\ \frac{\sqrt{2}}{3} & \frac{1}{\sqrt{3}} & -\frac{2}{3} \end{pmatrix} \qquad \mathbf{B} = \begin{pmatrix} \frac{5}{3} & \frac{1}{\sqrt{6}} & -\frac{1}{3\sqrt{2}} \\ \frac{1}{\sqrt{6}} & \frac{3}{2} & \frac{1}{2\sqrt{3}} \\ -\frac{1}{3\sqrt{2}} & \frac{1}{2\sqrt{3}} & \frac{11}{6} \end{pmatrix}.$$

Show that **A** and **B** commute. Find their eigenvalues and eigenvectors, and obtain a unitary transformation matrix **U** that diagonalizes both **A** and **B**.

2. Obtain eigenvalues to second order and eigenvectors to first order of the matrix

$$\mathbf{H} = \begin{pmatrix} 1 & 2\alpha & 0 \\ 2\alpha & 2+\alpha & 3\alpha \\ 0 & 3\alpha & 3+2\alpha \end{pmatrix}$$

using the small parameter α.

3. A particle of mass m is confined to an infinite potential box with potential

$$V(x) = \begin{cases} \infty, & x<0, \quad x>L, \\ k\left(1-\dfrac{x}{L}\right), & 0 \le x \le L. \end{cases}$$

Calculate the ground and fourth excited-state energies of the particle in this box using first-order perturbation theory. Obtain the ground and fourth excited-state wavefunctions to first order, and sketch their appearance. How do they differ from the corresponding unperturbed wavefunctions?

4. A matrix representation of the Hamiltonian for a 2-dimensional system is given by $\mathbf{H} = \mathbf{H}^{(0)} + \mathbf{H}^{(1)}$, with

$$\mathbf{H}^{(0)} = \begin{pmatrix} 2 & 1 \\ 1 & 2 \end{pmatrix} \qquad \mathbf{H}^{(1)} = \alpha\begin{pmatrix} 1 & -1 \\ -1 & 2 \end{pmatrix}, \qquad \alpha \ll 1.$$

(a) Obtain eigenvectors to first order and eigenvalues to second order for the problem

$$\mathbf{H}\mathbf{x}_k = \lambda_k \mathbf{x}_k, \qquad k = 1, 2,$$

using perturbation theory.

(b) For comparison solve the problem exactly, first for $\mathbf{H}^{(0)}$ and then for $\mathbf{H}$.

5. Consider the Hamiltonian matrix constructed in the $\{\phi_i\}$ basis (α, β are real numbers):

$$\mathbf{H} = \begin{pmatrix} -\alpha & \beta & 0 \\ \beta & -\alpha & \beta \\ 0 & \beta & 2\alpha \end{pmatrix}.$$

Obtain the eigenvalues, their corresponding eigenvectors, and the unitary transformation that brings $\mathbf{H}$ to diagonal form. (*Hint*: There is a trigonometric solution to certain cubic equations.)

6. Write out the characters for the following direct products and then determine which irreducible representations they decompose into:

(a) $A_1 \otimes A_1$, $A_2 \otimes A_2$, $A_2 \otimes E$, $E \otimes E$, $E \otimes E \otimes E$ for C_{3v}.

(b) $A_{2g} \otimes A_{2g}$, $E_g \otimes E_g$, $T_{2g} \otimes T_{2g}$, $T_{1u} \otimes T_{2g}$, $E_u \otimes T_{1u}$ for O_h.

(c) $E'' \otimes A_2''$, $A_2' \otimes A_2''$, $E' \otimes E''$, $E'' \otimes E''$ for D_{3h}.

7. Consider the transition dipole moment integral

$$\int \psi_1^* \boldsymbol{\mu} \psi_0 \, d\tau.$$

(a) For the C_{2v} point group, if ψ_0 belongs to the A_1 irreducible representation and μ_z, μ_x, and μ_y have A_1, B_1, and B_2 symmetry, what are the possible symmetries of ψ_1 in order to make the integral nonzero?

(b) Repeat (a) for D_{6h} where μ_z and (μ_x, μ_y) have A_{2u} and E_{1u} symmetry, and ψ_0 has A_{1g} symmetry.

(c) Repeat (a) for T_d where (μ_x, μ_y, μ_z) have T_2 symmetry and ψ_0 has E symmetry.

8. Show that the eigenvalues and eigenvectors of a symmetric 2×2 matrix (4.16) are given by equations (4.18), (4.21), and (4.23).

Reference

1. Born, M. and Oppenheimer, R. *Ann. Phys.* **84,** 457 (1927).

General References

Bishop, D. M. *Group Theory and Chemistry,* Dover, New York, 1993.

Cohen-Tannoudji, C., Diu, B., and Laloë, F. *Quantum Mechanics,* Vols 1 and 2, Wiley, New York, 1977.

Fischer, G. *Vibronic Coupling,* Academic Press, London, 1984.

Hamermesh, M. *Group Theory and Its Application to Physical Problems,* Dover, New York, 1989.

Tinkham, M. *Group Theory and Quantum Mechanics,* McGraw-Hill, New York, 1964.

Chapter 5
Atomic Spectroscopy

Introduction

Historically, atomic spectroscopy was developed before molecular spectroscopy. The discovery of the Fraunhofer absorption lines in the spectrum of the sun and the observation by Herschel of the colors emitted when metal salts are introduced into flames occurred in the early 1800's. It was not, however, until the 1850's that Kirchoff and Bunsen clearly established that each atom had a characteristic spectral signature. These ideas led to the identification of the elements rubidium and cesium by emission spectroscopy and the discovery of helium in the sun in advance of its isolation on earth.

Atomic spectroscopy was used simply as a diagnostic tool in these early measurements, although Balmer noted mathematical regularities in the spectrum of the hydrogen atom in 1885. It was not until the work of Bohr in 1913 that the spectrum of the hydrogen atom was explained. The Bohr model was unable to account for the spectra of atoms with more than one electron and was soon superseded by the development of quantum mechanics in the 1920's. In fact the desire to explain atomic spectra was one of the primary motivations for the development of quantum mechanics.

When the hydrogen molecule is excited in an electrical discharge a regular series of atomic hydrogen emission lines is observed (Figure 5.1). The line positions seem to converge to a limit for the Balmer series. This pattern repeats itself in other regions of the spectrum, for example, in the near-infrared (Paschen) series and the vacuum ultraviolet (Lyman) series. These lines are customarily labeled with the series name and with a Greek letter to indicate the member of the series (Figure 5.2). For example Balmer α (H_α) and Balmer β (H_β) denote the first and second members of the Balmer series at 15,233 cm^{-1} (6562.7 Å) and 20,565 cm^{-1} (4861.3 Å), respectively.

Balmer discovered that the wavelengths of the series that now bears his name could be represented by the empirical formula

$$\lambda = \frac{3645.6\, n^2}{n^2 - n_0^2}\ \text{Å}, \qquad n = 3, 4, 5, \ldots;\ n_0 = 2 \tag{5.1}$$

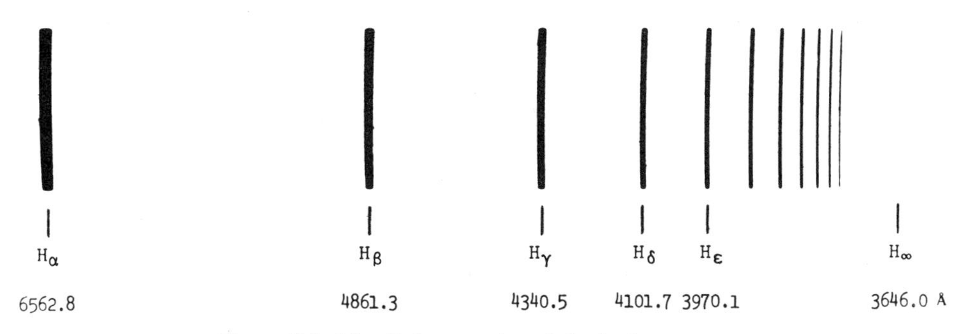

Figure 5.1. The Balmer series of the hydrogen atom.

where 3645.6 Å is the series limit. In terms of cm^{-1} the formula becomes

$$\tilde{\nu} = 109{,}678\left(\frac{1}{2^2} - \frac{1}{n^2}\right) \text{cm}^{-1}, \qquad n = 3, 4, \ldots$$

$$= R_H\left(\frac{1}{2^2} - \frac{1}{n^2}\right), \tag{5.2}$$

with R_H called the Rydberg constant. The other series were found to obey formulas similar to the Balmer formula (5.1) with each transition given as the difference between two terms, but with $n_0 = 1, 3, 4, 5, \ldots$.

Remarkably, the spectra of the alkali atoms provided similar patterns in both emission and absorption. Although the emission and absorption spectra had some lines in common, the emission spectra were more complex. The emission spectra could also be organized into series that were given the names sharp (S), principal (P), diffuse (D), and fundamental (F). The names sharp and diffuse were based on the appearance of the lines, while the principal series appeared in both absorption and emission. The fundamental series was thought to be more fundamental because it occurred to the red (longer wavelength) of the others and was most like the hydrogen series. Moreover, simple formulas similar to equation (5.2) were found to represent the various series of lines, namely

$$\tilde{\nu} = T - \frac{R}{(n - \delta)^2}, \tag{5.3}$$

in which T is the series limit, R is the Rydberg constant, and n is an integer. Unfortunately, a small non-integer quantum defect δ had to be introduced, since the use of integer quantum numbers could not reproduce the series of alkali line positions. Elements other than hydrogen and the alkalis had even more complex spectra.

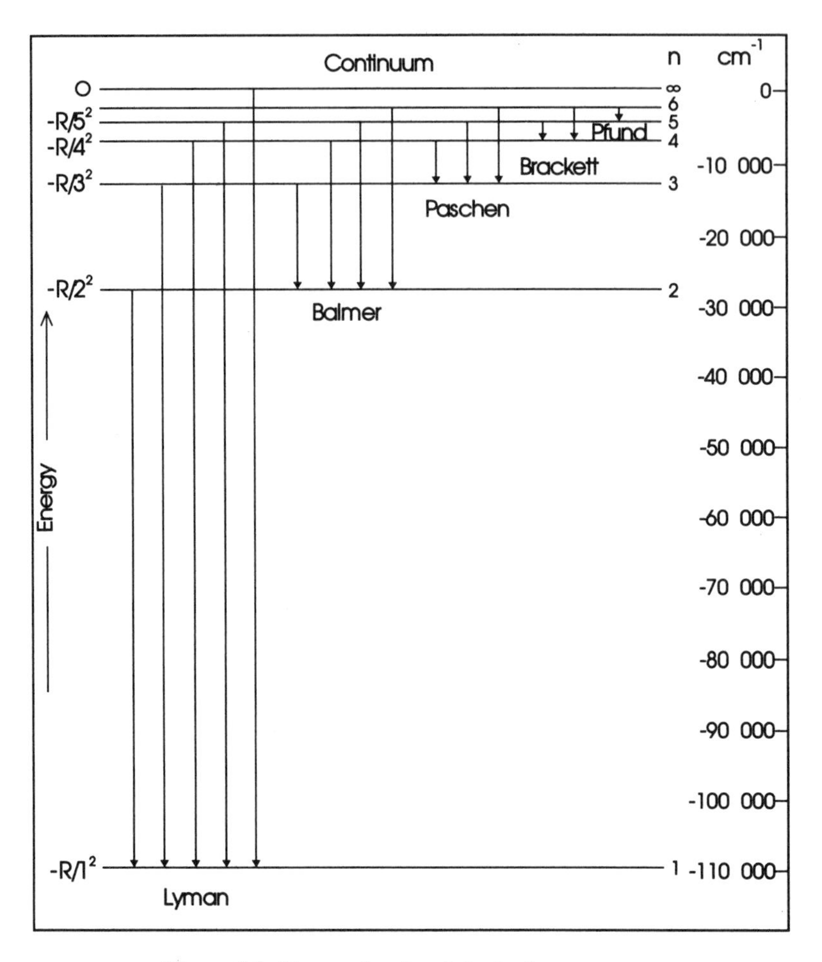

Figure 5.2. Energy levels of the hydrogen atom.

Angular Momentum

The interpretation of atomic spectra is closely related to the concept of angular momentum. The classical angular momentum $\mathbf{L} = \mathbf{r} \times \mathbf{p}$ can be transformed into a quantum mechanical operator by the usual substitution

$$\hat{p}_x = -i\hbar \frac{\partial}{\partial x}, \qquad \hat{p}_y = -i\hbar \frac{\partial}{\partial y}, \qquad \hat{p}_z = -i\hbar \frac{\partial}{\partial z}.$$

This leads to the following expressions for the components and square of $\hat{L}$ in

Cartesian coordinates, as well as in terms of the polar angle θ and the azimuthal angle ϕ of the electron:

$$\hat{L}_x = -i\hbar\left(y\frac{\partial}{\partial z} - z\frac{\partial}{\partial y}\right) = i\hbar\left(\sin\phi\frac{\partial}{\partial\theta} + \cot\theta\cos\phi\frac{\partial}{\partial\phi}\right) \tag{5.4}$$

$$\hat{L}_y = -i\hbar\left(z\frac{\partial}{\partial x} - x\frac{\partial}{\partial z}\right) = i\hbar\left(-\cos\phi\frac{\partial}{\partial\theta} + \cot\theta\sin\phi\frac{\partial}{\partial\phi}\right) \tag{5.5}$$

$$\hat{L}_z = -i\hbar\left(x\frac{\partial}{\partial y} - y\frac{\partial}{\partial x}\right) = -i\hbar\frac{\partial}{\partial\phi} \tag{5.6}$$

$$\hat{L}^2 = \hat{L}_x^2 + \hat{L}_y^2 + \hat{L}_z^2 = -\hbar^2\left[\frac{1}{\sin^2\theta}\frac{\partial^2}{\partial\phi^2} + \frac{1}{\sin\theta}\left(\frac{\partial}{\partial\theta}\sin\theta\frac{\partial}{\partial\theta}\right)\right]. \tag{5.7}$$

With the use of the definition for the commutator, $[\hat{A}, \hat{B}] = \hat{A}\hat{B} - \hat{B}\hat{A}$, one obtains

$$[\hat{L}_x, \hat{L}_y] = i\hbar\hat{L}_z, \qquad [\hat{L}_y, \hat{L}_z] = i\hbar\hat{L}_x, \qquad [\hat{L}_z, \hat{L}_x] = i\hbar\hat{L}_y, \tag{5.8}$$

$$[\hat{L}^2, \hat{L}_x] = [\hat{L}^2, \hat{L}_y] = [\hat{L}^2, \hat{L}_z] = 0, \tag{5.9}$$

so that a simultaneous set of eigenfunctions for $\hat{L}^2$ and $\hat{L}_z$ can be found. These eigenfunctions are the spherical harmonics $Y_{LM}(\theta, \phi)$ where

$$\hat{L}^2 Y_{LM}(\theta, \phi) = \hbar^2 L(L+1) Y_{LM}(\theta, \phi), \tag{5.10}$$

and

$$\hat{L}_z Y_{LM}(\theta, \phi) = M\hbar Y_{LM}(\theta, \phi). \tag{5.11}$$

The Y_{LM} can be further separated into a product of two functions

$$Y_{LM} = \Theta(\theta)\Phi(\phi), \tag{5.12}$$

where $\Theta(\theta)$ is an associated Legendre function and

$$\Phi(\phi) = \frac{e^{iM\phi}}{(2\pi)^{1/2}}. \tag{5.13}$$

Table 5.1. Spherical Harmonics

$$Y_{0,0}=\sqrt{\frac{1}{4\pi}}$$

$$Y_{1,0}=\sqrt{\frac{3}{4\pi}}\cos\theta$$

$$Y_{1,\pm1}=\mp\sqrt{\frac{3}{8\pi}}\sin\theta e^{\pm i\phi}$$

$$Y_{2,0}=\sqrt{\frac{5}{16\pi}}(3\cos^2\theta-1)$$

$$Y_{2,\pm1}=\mp\sqrt{\frac{15}{8\pi}}\sin\theta\cos\theta e^{\pm i\phi}$$

$$Y_{2,\pm2}=\sqrt{\frac{15}{32\pi}}\sin^2\theta e^{\pm 2i\phi}$$

$$Y_{LM}(\theta,\phi)=(-1)^M\left[\frac{(2L+1)(L-M)!}{4\pi(L+M)!}\right]^{1/2}P_L^M(\cos\theta)e^{iM\phi},\qquad M\geq 0;$$

$$Y_{L,-M}=(-1)^M Y_{LM}^*$$

$$P_L^M(x)=(1-x^2)^{|M/2|}\frac{d^{|M|}}{dx^{|M|}}P_L(x);$$

$$P_L(x)=\frac{1}{2^L L!}\frac{d^L}{dx^L}(x^2-1)^L.$$

$$\text{Normalization: }\int_0^{2\pi}\int_0^{\pi}Y_{LM}^*Y_{LM}\sin\theta\,d\theta\,d\phi=1.$$

The first few spherical harmonics for $L\leq 2$ as well as a formula for generating Y_{LM} for $l>2$ are provided in Table 5.1.

A simple geometric interpretation of the $\hat{L}^2$ and $\hat{L}_z$ operators is of a quantized vector of length $[L(L+1)]^{1/2}\hbar$ units precessing about the z-axis so that L_x and L_y have undefined values, but a definite projection of $M\hbar$ units along the z-axis (see below). The z-axis is arbitrarily chosen as the reference axis appropriate to the experimental situation in which observations are made.

The raising and lowering operators are defined as

$$\hat{L}_+=\hat{L}_x+i\hat{L}_y \tag{5.14}$$

and

$$\hat{L}_-=\hat{L}_x-i\hat{L}_y \tag{5.15}$$

because they raise and lower the M value for a given value of L. Specifically, their actions on Y_{LM} are represented by

$$\hat{L}_{\pm} Y_{LM} = \hbar[L(L+1) - M(M \pm 1)]^{1/2} Y_{LM\pm 1}. \tag{5.16}$$

The spherical harmonics $Y_{LM} = |LM\rangle$ are a convenient set of basis functions to construct a matrix representation of the angular momentum operators. In this basis $\hat{L}^2$ and $\hat{L}_z$ are represented by diagonal matrices with matrix elements

$$\langle LM| \hat{L}^2 |L'M'\rangle = L(L+1)\hbar^2 \delta_{LL'} \delta_{MM'} \tag{5.17}$$

$$\langle LM| \hat{L}_z |L'M'\rangle = M\hbar \delta_{LL'} \delta_{MM'}. \tag{5.18}$$

The $\hat{L}_+$ and $\hat{L}_-$ operators are represented by non-Hermitian matrices with matrix elements

$$\langle LM| \hat{L}_{\pm} |L'M'\rangle = [L(L+1) - M'(M' \pm 1)]^{1/2} \hbar \delta_{LL'} \delta_{M,M'\pm 1}.$$

Since $\hat{L}_+$ and $\hat{L}_-$ do not correspond to any observables, they do not require Hermitian representations. For example, $\mathbf{L}_+$ and $\mathbf{L}_-$ matrices can be used to construct $\mathbf{L}_x$ and $\mathbf{L}_y$ matrices from the relationships

$$\hat{L}_x = \frac{\hat{L}_+ + \hat{L}_-}{2} \tag{5.19}$$

and

$$\hat{L}_y = \frac{-i(\hat{L}_+ - \hat{L}_-)}{2}. \tag{5.20}$$

For the specific case when $L = 1$, $M = 1, 0, -1$, let the three basis functions be

$$\begin{pmatrix} 1 \\ 0 \\ 0 \end{pmatrix} = |1, 1\rangle, \quad \begin{pmatrix} 0 \\ 1 \\ 0 \end{pmatrix} = |1, 0\rangle, \quad \begin{pmatrix} 0 \\ 0 \\ 1 \end{pmatrix} = |1, -1\rangle.$$

In terms of these basis vectors the matrix representations of $\hat{L}^2$, $\hat{L}_z$, $\hat{L}_+$, $\hat{L}_-$, $\hat{L}_x$, and $\hat{L}_y$ are

$$\hat{\mathbf{L}}^2 = 2\hbar^2 \begin{pmatrix} 1 & 0 & 0 \\ 0 & 1 & 0 \\ 0 & 0 & 1 \end{pmatrix}, \tag{5.21}$$

$$\hat{\mathbf{L}}_z = \hbar \begin{pmatrix} 1 & 0 & 0 \\ 0 & 0 & 0 \\ 0 & 0 & -1 \end{pmatrix}, \tag{5.22}$$

$$\hat{\mathbf{L}}_+ = \sqrt{2}\,\hbar \begin{pmatrix} 0 & 1 & 0 \\ 0 & 0 & 1 \\ 0 & 0 & 0 \end{pmatrix}, \tag{5.23}$$

$$\hat{\mathbf{L}}_- = \sqrt{2}\,\hbar \begin{pmatrix} 0 & 0 & 0 \\ 1 & 0 & 0 \\ 0 & 1 & 0 \end{pmatrix}, \tag{5.24}$$

$$\hat{\mathbf{L}}_x = \frac{\sqrt{2}}{2}\hbar \begin{pmatrix} 0 & 1 & 0 \\ 1 & 0 & 1 \\ 0 & 1 & 0 \end{pmatrix}, \tag{5.25}$$

and

$$\hat{\mathbf{L}}_y = \frac{\sqrt{2}}{2} i\hbar \begin{pmatrix} 0 & -1 & 0 \\ 1 & 0 & -1 \\ 0 & 1 & 0 \end{pmatrix}. \tag{5.26}$$

Electron orbital angular momentum in the hydrogen atom depends on the θ and ϕ polar coordinates of the electron. In this case, L is restricted to integral values but, in general, angular momentum can assume half-integral values as well. For example, the spin angular momentum of an electron is found to be $\frac{1}{2}\hbar$. In this case, the angular momentum of the electron is *defined* in terms of the commutation relations (5.8) and the associated matrices. The letter S is used to designate spin angular momentum. More generally, the preceding equations must be transformed by $L \to J$ and $M \to M_J$, where J and M_J always symbolize the total angular momentum (spin and/or orbital) in a system (exclusive of nuclear spin) and its projection on the laboratory z-axis.

For the simple spin $\frac{1}{2}$ case let

$$|\alpha\rangle = \begin{pmatrix} 1 \\ 0 \end{pmatrix} = \left| S = \frac{1}{2}, M_s = \frac{1}{2} \right\rangle, \qquad |\beta\rangle = \begin{pmatrix} 0 \\ 1 \end{pmatrix} = \left| S = \frac{1}{2}, M_s = -\frac{1}{2} \right\rangle.$$

The corresponding matrix representations are

$$\hat{\mathbf{J}}^2 = \frac{3}{4}\hbar^2\begin{pmatrix}1 & 0\\0 & 1\end{pmatrix}, \tag{5.27}$$

$$\hat{\mathbf{J}}_z = \frac{\hbar}{2}\begin{pmatrix}1 & 0\\0 & -1\end{pmatrix} = \frac{\hbar}{2}\boldsymbol{\sigma}_z, \tag{5.28}$$

$$\hat{\mathbf{J}}_+ = \hbar\begin{pmatrix}0 & 1\\0 & 0\end{pmatrix}, \tag{5.29}$$

$$\hat{\mathbf{J}}_- = \hbar\begin{pmatrix}0 & 0\\1 & 0\end{pmatrix}, \tag{5.30}$$

$$\hat{\mathbf{J}}_x = \frac{\hbar}{2}\begin{pmatrix}0 & 1\\1 & 0\end{pmatrix} = \frac{\hbar}{2}\boldsymbol{\sigma}_x, \tag{5.31}$$

and

$$\hat{\mathbf{J}}_y = \frac{\hbar}{2}\begin{pmatrix}0 & -i\\i & 0\end{pmatrix} = \frac{\hbar}{2}\boldsymbol{\sigma}_y, \tag{5.32}$$

where $\boldsymbol{\sigma}_x$, $\boldsymbol{\sigma}_y$, and $\boldsymbol{\sigma}_z$ are the Pauli spin matrices.

Matrix representations of operators are vital in spectroscopy because they provide a quantitative description of the system. The Hamiltonian operator $\hat{H}$ for the system is expressed in terms of various operators such as the spin and orbital angular momentum operators. To transform the Schrödinger equation into a matrix equation, we choose a basis set and evaluate the matrix elements of $\hat{H}$. Finally, the matrix form of the Schrödinger equation in this basis, $\hat{\mathbf{H}}\psi = E\psi$, is solved by transforming $\mathbf{H}$ to diagonal form in order to find the eigenvalues $\{E_n\}$ and the associated eigenvectors $\{\psi_n\}$.

The Hydrogen Atom and One-Electron Spectra

The energy level structure of the hydrogen atom and hydrogenlike ions can be explained by solving the Schrödinger equation

$$\frac{-\hbar^2}{2\mu}\nabla^2\psi - \frac{Ze^2\psi}{4\pi\varepsilon_0 r} = E\psi. \tag{5.33}$$

This differential equation is most easily solved in terms of the spherical polar coordinates. Thus, the wavefunction is written as $\psi = \psi(r, \theta, \phi)$ and the Schrödinger equation becomes

$$\frac{-\hbar^2}{2\mu}\left(\frac{\partial^2\psi}{\partial r^2} + \frac{2}{r}\frac{\partial\psi}{\partial r} + \frac{1}{r^2}\frac{\partial^2\psi}{\partial\theta^2} + \frac{1}{r^2}\cot\theta\frac{\partial\psi}{\partial\theta} + \frac{1}{r^2\sin^2\theta}\frac{\partial^2\psi}{\partial\phi^2}\right) - \frac{Ze^2\psi}{4\pi\varepsilon_0 r} = E\psi, \tag{5.34}$$

or

$$\frac{-\hbar^2}{2mr^2}\frac{\partial}{\partial r}r^2\frac{\partial\psi}{\partial r}+\frac{1}{2mr^2}\hat{L}^2\psi-\frac{Ze^2\psi}{4\pi\varepsilon_0 r}=E\psi. \tag{5.35}$$

This form of the Schrödinger equation is used because the partial differential equation can then be separated into three ordinary differential equations involving r, θ, and ϕ alone. Boundary conditions force quantization, and the energy eigenvalues are

$$E_n=\frac{-\mu Z^2(e^2/4\pi\varepsilon_0)^2}{2n^2\hbar^2}=\frac{-R}{n^2}, \qquad n=1, 2, 3, \ldots, \tag{5.36}$$

in which $R = R_H = 109{,}677.4212\ \text{cm}^{-1}$ for the hydrogen atom. Note that most tables (Appendix A) report R_∞, appropriate for a stationary, infinitely heavy nucleus, rather than R_H. In fact R_∞ and R_H are related by the relationship

$$R_H=\frac{R_\infty}{1+m_e/m_p}, \tag{5.37}$$

which is obtained from the definition of the reduced mass of two particles, namely $\mu = m_1 m_2/(m_1+m_2)$. The solution of the Schrödinger equation yields three quantum numbers: the principal quantum number n, the azimuthal quantum number l, and the magnetic quantum number m, that can only assume the values

$$n=1, 2, 3, \ldots, \infty,$$

$$l=0, 1, \ldots, n-1,$$

and

$$m=0, \pm 1, \ldots, \pm l.$$

The l values of $0, 1, 2, 3, \ldots$ are usually labeled *s, p, d, f, g, h,* etc., for the historic reasons that were touched upon in the introduction to this chapter. The wavefunction is the product of a radial part and an angular part

$$\psi = R_{nl}(r)Y_{lm}(\theta, \phi),$$

in which the radial part $R_{nl}(r)$ is an associated Laguerre function and $Y_{lm}(\theta, \phi)$ is a spherical harmonic. A few of the Y_{lm} and R_{nl} functions are listed in Tables 5.1 and 5.2. The constant a_0 is the Bohr radius, which can be expressed in terms of fundamental constants (Appendix A) as

$$a_0=\frac{4\pi\varepsilon_0\hbar^2}{m_e e^2}=0.529177249\ \text{Å}.$$

Note that if μ replaces m_e in the definition of the Bohr radius, then one obtains a_H, the actual Bohr radius of the hydrogen atom.

Table 5.2. Radial Functions of the Hydrogen Atom

$$R_{10} = \left(\frac{Z}{a}\right)^{3/2} 2e^{-Zr/a}$$

$$R_{20} = \left(\frac{Z}{2a}\right)^{3/2} 2\left(1 - \frac{Zr}{2a}\right)e^{-Zr/2a}$$

$$R_{21} = \left(\frac{Z}{2a}\right)^{3/2} \frac{2}{\sqrt{3}}\left(\frac{Zr}{2a}\right)e^{-Zr/2a}$$

$$R_{30} = \left(\frac{Z}{3a}\right)^{3/2} 2\left[1 - 2\frac{Zr}{3a} + \frac{2}{3}\left(\frac{Zr}{3a}\right)^2\right]e^{-Zr/3a}$$

$$R_{31} = \left(\frac{Z}{3a}\right)^{3/2} \frac{4\sqrt{2}}{3}\left(\frac{Zr}{3a}\right)\left(1 - \frac{1}{2}\frac{Zr}{3a}\right)e^{-Zr/3a}$$

$$R_{32} = \left(\frac{Z}{3a}\right)^{3/2} \frac{2\sqrt{2}}{3\sqrt{5}}\left(\frac{Zr}{3a}\right)^2 e^{-Zr/3a}$$

Normalization: $\int_0^\infty R_{nl}^* R_{nl} r^2 \, dr = 1$

The angular parts of the hydrogen eigenfunctions Y_{lm} are complex when $|m| > 0$. These complex functions are not very useful when one tries to visualize the shape of the orbitals in real space. Since the energy does not depend on the magnetic quantum number m, the wavefunctions are degenerate and any linear combination of them is also a solution to the Schrödinger equation. Instead, the linear combinations

$$\frac{Y_{l|m|} + Y_{l-|m|}}{\sqrt{2}} \quad \text{and} \quad \frac{Y_{l|m|} - Y_{l-|m|}}{i\sqrt{2}} \tag{5.38}$$

are used when plotting the orbitals in real space. The real forms of some of the hydrogen orbitals are listed in Table 5.3 and their plots are illustrated in Figure 5.3.

The orbital angular momentum operators $\hat{l}^2$ and $\hat{l}_z$ commute with the hydrogenic Hamiltonian

$$[\hat{H}, \hat{l}_z] = [\hat{H}, \hat{l}^2] = 0, \tag{5.39}$$

where, from now on, the customary notation of lowercase letters for one-electron properties and uppercase letters for many-electron properties is adopted. Simultaneous eigenfunctions can be constructed for the three equations

$$\hat{H}\psi_{nlm} = E_n\psi_{nlm}, \tag{5.40}$$

$$\hat{l}^2\psi_{nlm} = l(l+1)\hbar^2\psi_{nlm}, \tag{5.41}$$

Table 5.3. Some of the Real Hydrogen Wavefunctions

$$\psi_{1s} = \frac{1}{\pi^{1/2}}\left(\frac{Z}{a}\right)^{3/2} e^{-Zr/a}$$

$$\psi_{2s} = \frac{1}{4(2\pi)^{1/2}}\left(\frac{Z}{a}\right)^{3/2}\left(2-\frac{Zr}{a}\right)e^{-Zr/2a}$$

$$\psi_{2p_z} = \frac{1}{4(2\pi)^{1/2}}\left(\frac{Z}{a}\right)^{5/2} re^{-Zr/2a}\cos\theta$$

$$\psi_{2p_x} = \frac{1}{4(2\pi)^{1/2}}\left(\frac{Z}{a}\right)^{5/2} re^{-Zr/2a}\sin\theta\cos\phi$$

$$\psi_{2p_y} = \frac{1}{4(2\pi)^{1/2}}\left(\frac{Z}{a}\right)^{5/2} re^{-Zr/2a}\sin\theta\sin\phi$$

$$\psi_{3s} = \frac{1}{81(3\pi)^{1/2}}\left(\frac{Z}{a}\right)^{3/2}\left(27-18\frac{Zr}{a}+2\frac{Z^2r^2}{a^2}\right)e^{-Zr/3a}$$

$$\psi_{3p_z} = \frac{2^{1/2}}{81\pi^{1/2}}\left(\frac{Z}{a}\right)^{5/2}\left(6-\frac{Zr}{a}\right)re^{-Zr/3a}\cos\theta$$

$$\psi_{3p_x} = \frac{2^{1/2}}{81\pi^{1/2}}\left(\frac{Z}{a}\right)^{5/2}\left(6-\frac{Zr}{a}\right)re^{-Zr/3a}\sin\theta\cos\phi$$

$$\psi_{3p_y} = \frac{2^{1/2}}{81\pi^{1/2}}\left(\frac{Z}{a}\right)^{5/2}\left(6-\frac{Zr}{a}\right)re^{-Zr/3a}\sin\theta\sin\phi$$

$$\psi_{3d_{z^2}} = \frac{1}{81(6\pi)^{1/2}}\left(\frac{Z}{a}\right)^{7/2} r^2e^{-Zr/3a}(3\cos^2\theta-1)$$

$$\psi_{3d_{xz}} = \frac{2^{1/2}}{81\pi^{1/2}}\left(\frac{Z}{a}\right)^{7/2} r^2e^{-Zr/3a}\sin\theta\cos\theta\cos\phi$$

$$\psi_{3d_{yz}} = \frac{2^{1/2}}{81\pi^{1/2}}\left(\frac{Z}{a}\right)^{7/2} r^2e^{-Zr/3a}\sin\theta\cos\theta\sin\phi$$

$$\psi_{3d_{x^2-y^2}} = \frac{1}{81(2\pi)^{1/2}}\left(\frac{Z}{a}\right)^{7/2} r^2e^{-Zr/3a}\sin^2\theta\cos 2\phi$$

$$\psi_{3d_{xy}} = \frac{1}{81(2\pi)^{1/2}}\left(\frac{Z}{a}\right)^{7/2} r^2e^{-Zr/3a}\sin^2\theta\sin 2\phi$$

and

$$\hat{l}_z\psi_{nlm} = m\hbar\psi_{nlm}. \tag{5.42}$$

The complex form of the wavefunctions listed in Tables 5.1 and 5.2 satisfy these equations and the corresponding quantum numbers n, l, and m are used to label the wavefunctions ψ_{nlm}.

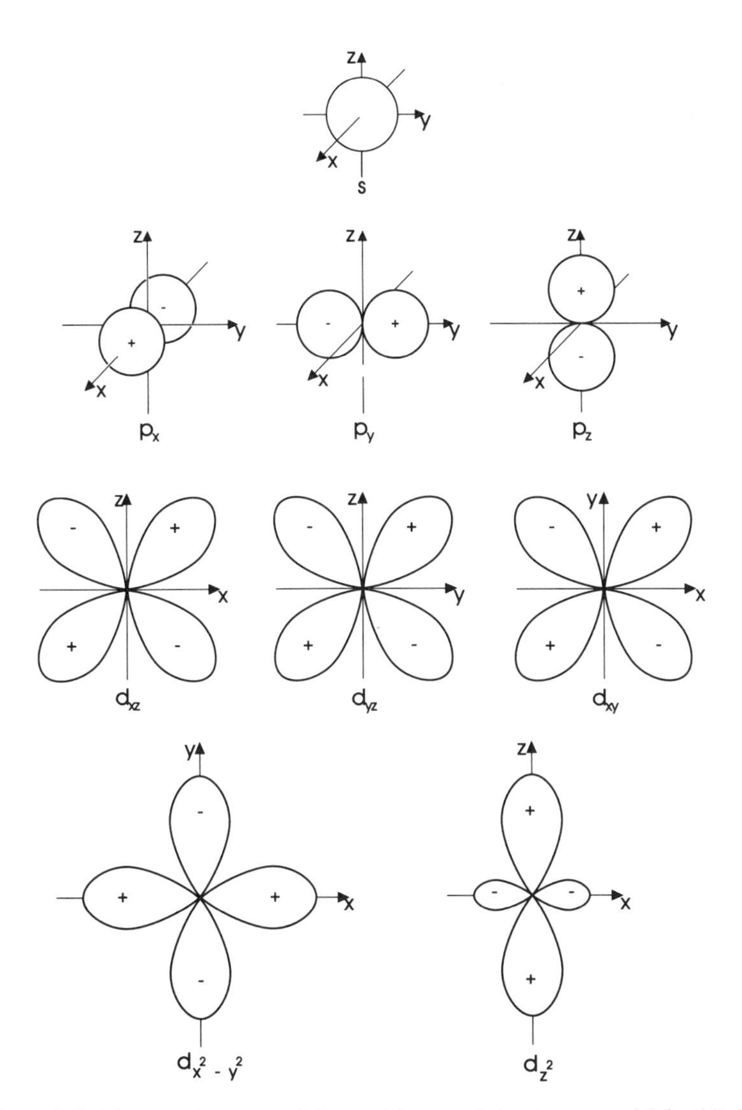

Figure 5.3. The angular part of the real forms of the 1*s*, 2*p*, and 3*d* orbitals.

Vector Model

Since angular momenta are so widely used in spectroscopy, it is useful to have a simple pictorial model (Figure 5.4). This model summarizes the mathematical results of quantum mechanics. An angular momentum **J** is represented in this picture by a vector of length $[J(J+1)]^{1/2}\hbar$ units. While **J** has a definite projection $M_J\hbar$ along the laboratory z-axis, the components along the x- and y-axes do not have definite values. This means that the vector **J** is inclined at an angle

$$\theta = \cos^{-1}\frac{M_J}{\sqrt{J(J+1)}} \tag{5.43}$$

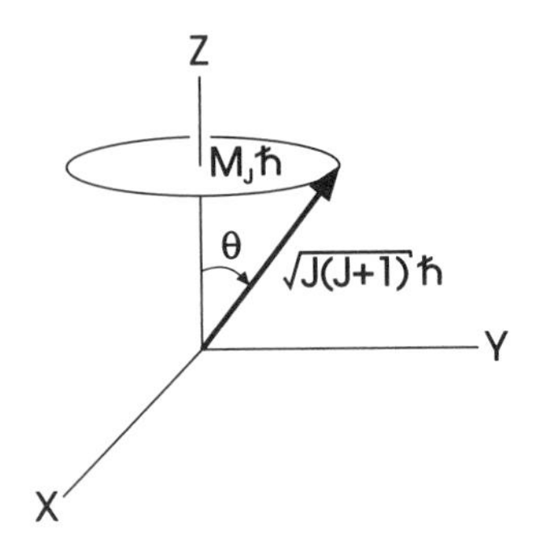

Figure 5.4. The vector model for angular momentum.

with respect to the z-axis and precesses at a constant angular velocity. This precessional motion ensures that the x- and y-components of **J** have undetermined values (until an external measurement forces one of these components to have a definite value). The different values of M_J thus correspond to different spatial orientations of **J**. If space is isotropic (i.e., there are no electric or magnetic fields present), then the energy cannot depend on the orientation of the total angular momentum in space so that there must be a $(2J+1)$-fold degeneracy corresponding to all allowed M_j values.

Spin-Orbit Coupling

Because the electron is a charged particle, the orbital motion of an electron produces a current. Associated with this current is a magnetic field that affects the orientation of the magnetic moment of the electron due to the presence of electron spin. This phenomenon is known as spin-orbit coupling and is responsible for the so-called "fine structure" in the spectrum of the hydrogen atom.

The strength of the magnetic field at the electron is proportional to **l**, while the magnitude of the electronic magnetic moment is proportional to **s**, given by

$$\boldsymbol{\mu}_s = -g_e \mu_B \mathbf{s}$$

where g_e is a numerical constant and μ_B is the Bohr magneton (see the section on Zeeman effects later in this chapter). Since the energy of interaction of a magnetic moment with a magnetic flux density **B** is

$$E = -\boldsymbol{\mu} \cdot \mathbf{B} \tag{5.44}$$

the spin-orbit coupling is written as

$$\hat{H}_{\text{so}} = \xi(r)\hat{\mathbf{l}} \cdot \hat{\mathbf{s}}. \tag{5.45}$$

Here $\xi(r)$ is defined by

$$\xi(r) = \frac{1}{2\mu^2c^2}\frac{1}{r}\frac{\partial V}{\partial r} = \frac{1}{2\mu^2c^2}\left(\frac{Ze^2}{4\pi\varepsilon_0 r^3}\right) \tag{5.46}$$

in which

$$V = -\frac{Ze^2}{4\pi\varepsilon_0 r} \tag{5.47}$$

is the potential energy due to the Coulombic attraction between the electron and the nucleus. A detailed derivation of equations (5.45) and (5.46) requires the use of relativistic quantum electrodynamics,[1] which is beyond the level of this book.

The inclusion of spin-orbit coupling transforms the eigenvalue equation to

$$[\hat{H}^{(0)} + \xi(r)\hat{\mathbf{l}}\cdot\hat{\mathbf{s}}]\psi = E\psi, \tag{5.48}$$

where $\hat{H}^{(0)}$ is the simple hydrogenic Hamiltonian. Provided that the spin-orbit coupling term is small, first-order perturbation theory can be applied to solve equation (5.48) approximately. Letting

$$\hat{H}' = \xi(r)\hat{\mathbf{l}}\cdot\hat{\mathbf{s}} \tag{5.49}$$

gives

$$E = E^{(0)} + \int \psi^{(0)*}\hat{H}'\psi^{(0)}\,d\tau = -\frac{R_H}{n^2} + \int \psi_n^{(0)*}\xi(r)\hat{\mathbf{l}}\cdot\hat{\mathbf{s}}\psi_n^{(0)}\,d\tau. \tag{5.50}$$

Degenerate perturbation theory must be used, however, to find the correct $\psi_n^{(0)}$ since spin-orbit coupling removes some of the $\mathbf{l}$ and $\mathbf{s}$ degeneracy in the hydrogen atom.

The quantum numbers l and s are no longer good when spin-orbit coupling is taken into account since the operators $\hat{l}_z$ and $\hat{s}_z$ do not commute with the Hamiltonian because of the $\xi(r)\hat{\mathbf{l}}\cdot\hat{\mathbf{s}}$ term. However, the total angular momentum

$$\mathbf{j} = \mathbf{l} + \mathbf{s} \tag{5.51}$$

is still a constant of motion so that $\hat{j}^2$ and $\hat{j}_z$ must commute with the Hamiltonian. The matrix elements of the $\xi(r)\hat{\mathbf{l}}\cdot\hat{\mathbf{s}}$ term can be evaluated by using the simple direct product basis set

$$\psi = |l, m_l\rangle\,|s, m_s\rangle. \tag{5.52}$$

This basis set is referred to as the uncoupled representation because $\mathbf{l}$ and $\mathbf{s}$ are not coupled to give $\mathbf{j}$. For example, consider the set of $2p$ functions for the hydrogen atom: $l = 1$, $s = \frac{1}{2}$, which yield a total of six basis functions, namely

$$|l = 1, m_l = 1\rangle\,|s = \tfrac{1}{2}, m_s = \pm\tfrac{1}{2}\rangle,$$

$$|l = 1, m_l = 0\rangle\,|s = \tfrac{1}{2}, m_s = \pm\tfrac{1}{2}\rangle,$$

and

$$|l=1, m_l=-1\rangle\,|s=\tfrac{1}{2}, m_s=\pm\tfrac{1}{2}\rangle.$$

The diagonal matrix elements of

$$\hat{H} = \hat{H}^{(0)} + \xi(r)\hat{\mathbf{l}}\cdot\hat{\mathbf{s}} = (\hat{H}^{(0)} + \xi(r)\hat{l}_z\hat{s}_z) + \frac{\xi(r)(\hat{l}_+\hat{s}_- + \hat{l}_-\hat{s}_+)}{2} \tag{5.53}$$

are

$$\begin{aligned}&\langle l=1, m_l|\,\langle s=\tfrac{1}{2}, m_s|\,\hat{H}\,|s=\tfrac{1}{2}, m_s\rangle\,|l=1, m_l\rangle\\ &= E_{2p}^{(0)} + \zeta_{2p}m_l m_s\end{aligned} \tag{5.54}$$

where $E_{2p}^{(0)} = -R_H/4$ and

$$\zeta_{2p} = \hbar^2\int R_{2p}^*(r)\xi(r)R_{2p}(r)r^2\,dr. \tag{5.55}$$

The off-diagonal matrix elements are due to the $\hat{l}_+\hat{s}_-$ term which connects basis functions with the same $m_j = m_l + m_s = \frac{3}{2}, \frac{1}{2}, -\frac{1}{2}, -\frac{3}{2}$, resulting in a matrix with mostly zeros. Diagonalizing the 2×2 blocks yields the energy levels $E_{2p}^{(0)} - \zeta_{2p}$, $E_{2p}^{(0)} + \zeta_{2p}/2$ from the Hamiltonian matrix

$$\mathbf{H} = \begin{pmatrix} E^{(0)}+\zeta/2 & 0 & 0 & 0 & 0 & 0 \\ 0 & E^{(0)}-\zeta/2 & \sqrt{2}\,\zeta & 0 & 0 & 0 \\ 0 & \sqrt{2}\,\zeta & E^{(0)} & 0 & 0 & 0 \\ 0 & 0 & 0 & E^{(0)} & \sqrt{2}\,\zeta & 0 \\ 0 & 0 & 0 & \sqrt{2}\,\zeta & E^{(0)}-\zeta/2 & 0 \\ 0 & 0 & 0 & 0 & 0 & E^{(0)}+\zeta/2 \end{pmatrix} \begin{matrix} |m_l=1, m_s=\frac{1}{2}\rangle \\ |1, -\frac{1}{2}\rangle \\ |0, \frac{1}{2}\rangle \\ |0, -\frac{1}{2}\rangle \\ |-1, \frac{1}{2}\rangle \\ |-1, -\frac{1}{2}\rangle. \end{matrix}$$

An alternate and easier way to solve the problem involves the use of the coupled basis functions $|lsjm_j\rangle$ where j and m_j are good quantum numbers. For the $2p$ functions $\mathbf{j} = \mathbf{l} + \mathbf{s}$, corresponding to $j = \frac{3}{2}$ and $j = \frac{1}{2}$. Notice that the two values of j correspond to a vector addition of $\mathbf{l}$ and $\mathbf{s}$ (see Figure 5.5). Note that the usual (confusing)

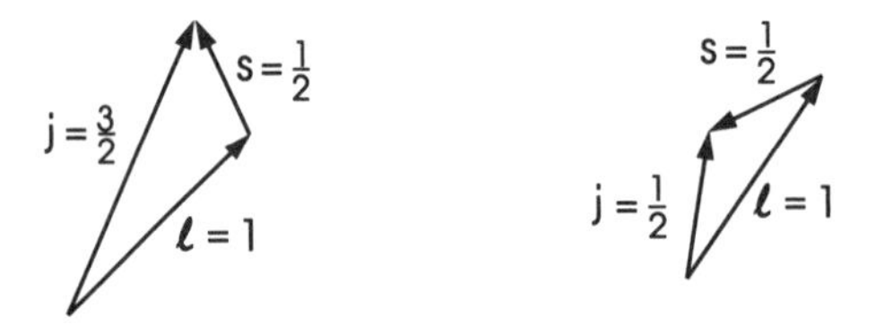

Figure 5.5. Vector addition of **l** and **s**.

shorthand notation of using $l = 1$, $s = \frac{1}{2}$, and $j = \frac{3}{2}$ to represent the lengths of vectors is used. In the coupled basis there are again six $2p$ functions:

$$|2p, j = \tfrac{3}{2}, m_j = \tfrac{3}{2}\rangle,$$

$$|2p, j = \tfrac{3}{2}, m_j = \tfrac{1}{2}\rangle,$$

$$|2p, j = \tfrac{3}{2}, m_j = -\tfrac{1}{2}\rangle,$$

$$|2p, j = \tfrac{3}{2}, m_j = -\tfrac{3}{2}\rangle,$$

$$|2p, j = \tfrac{1}{2}, m_j = \tfrac{1}{2}\rangle,$$

and

$$|2p, j = \tfrac{1}{2}, m_j = -\tfrac{1}{2}\rangle.$$

In this coupled, basis set the Hamiltonian matrix is already diagonal since

$$\hat{j}^2 = (\hat{\mathbf{l}} + \hat{\mathbf{s}}) \cdot (\hat{\mathbf{l}} + \hat{\mathbf{s}}) = \hat{\mathbf{l}}^2 + \hat{\mathbf{s}}^2 + 2\hat{\mathbf{l}} \cdot \hat{\mathbf{s}} \tag{5.56}$$

commutes with the Hamiltonian. The spin-orbit operator can be expressed in terms of $\hat{j}^2$, $\hat{l}^2$, and $\hat{s}^2$ as

$$\xi \hat{\mathbf{l}} \cdot \hat{\mathbf{s}} = \frac{\xi(\hat{j}^2 - \hat{l}^2 - \hat{s}^2)}{2}. \tag{5.57}$$

First-order perturbation theory gives

$$\begin{aligned} E &= E^{(0)} + \int (\psi_{nlm}^{(0)})^* \xi(r) \hat{\mathbf{l}} \cdot \hat{\mathbf{s}} \psi_{nlm}^{(0)} \, d\tau \\ &= E^{(0)} + \frac{1}{2} \int (\psi_{nlm}^{(0)})^* \xi(r) (\hat{j}^2 - \hat{l}^2 - \hat{s}^2) \psi_{nlm}^{(0)} \, d\tau \\ &= E^{(0)} + \frac{\zeta_{nl}}{2} [j(j+1) - l(l+1) - s(s+1)], \end{aligned} \tag{5.58}$$

using the definition of ζ, equation (5.55). When $s = \frac{1}{2}$, then $j = l \pm \frac{1}{2}$ and the energy levels are

$$E_{nl} = E^{(0)} + \frac{(\zeta_{nl})l}{2} \quad \text{for} \quad j = l + \frac{1}{2} \tag{5.59a}$$

or

$$E_{nl} = E^{(0)} - \frac{\zeta_{nl}(l+1)}{2} \quad \text{for} \quad j = l - \frac{1}{2}. \tag{5.59b}$$

For the case of the hydrogen $2p$ states ($l = 1$) the energy levels are

$$E = E^{(0)} + \frac{\zeta_{2p}}{2} \quad \text{for} \quad 2p_{3/2} \tag{5.60a}$$

and

$$E = E^{(0)} - \zeta_{2p} \quad \text{for} \quad 2p_{1/2}, \tag{5.60b}$$

where the good quantum number j is not used to label the wavefunctions. Notice that the $2p$ energy levels do not depend on whether the coupled or the uncoupled basis set is chosen to construct the Hamiltonian matrix. This is due to the fact that the coupled $|j_1 j_2 J M_J\rangle$ and uncoupled $|j_1 m_1\rangle |j_2 m_2\rangle$ basis sets are related by a linear transformation

$$|j_1, j_2 J, M_J\rangle = \sum_{m_1=-j_1}^{j_1} \sum_{m_2=-j_2}^{j_2} \langle j_1, j_2; m_1, m_2 \,|\, J M_J\rangle \, |j_1 m_1\rangle \, |j_2 m_2\rangle, \tag{5.61}$$

in which the coupling coefficients $\langle j_1, j_2; m_1 m_2 \,|\, JM\rangle$ are known as the Clebsch–Gordan coefficients. For the $2p$ hydrogen atom orbitals the transformation is

$$|2p_{3/2} M_J = \tfrac{3}{2}\rangle = |m_l = 1\rangle \, |m_s = \tfrac{1}{2}\rangle, \tag{5.62a}$$

$$|2p_{3/2} M_J = \tfrac{1}{2}\rangle = (\tfrac{2}{3})^{1/2} \, |m_l = 0\rangle \, |m_s = \tfrac{1}{2}\rangle + (\tfrac{1}{3})^{1/2} \, |m_l = 1\rangle \, |m_s = -\tfrac{1}{2}\rangle, \tag{5.62b}$$

$$|2p_{3/2} M_J = -\tfrac{1}{2}\rangle = (\tfrac{1}{2})^{1/3} \, |m_l = -1\rangle \, |m_s = \tfrac{1}{2}\rangle + (\tfrac{2}{3})^{1/2} \, |m_l = 0\rangle \, |m_s = -\tfrac{1}{2}\rangle, \tag{5.62c}$$

$$|2p_{3/2} M_J = -\tfrac{3}{2}\rangle = |m_l = -1\rangle \, |m_s = -\tfrac{1}{2}\rangle, \tag{5.62d}$$

$$|2p_{1/2} M_J = \tfrac{1}{2}\rangle = (\tfrac{1}{3})^{1/2} \, |m_l = 0\rangle \, |m_s = \tfrac{1}{2}\rangle - (\tfrac{2}{3})^{1/2} \, |m_l = 1\rangle \, |m_s = -\tfrac{1}{2}\rangle, \tag{5.62e}$$

and

$$|2p_{1/2} M_J = -\tfrac{1}{2}\rangle = (\tfrac{2}{3})^{1/2} \, |m_l = -1\rangle \, |m_s = \tfrac{1}{2}\rangle - (\tfrac{1}{3})^{1/2} \, |m_l = 0\rangle \, |m_s = -\tfrac{1}{2}\rangle. \tag{5.62f}$$

The transformation can be derived by using the operators

$$\hat{j}_{\pm} = \hat{l}_{\pm} + \hat{s}_{\pm} \tag{5.63}$$

and the orthogonality of the wavefunctions.

Many-Electron Atoms

The nonrelativistic Schrödinger equation for an N-electron atom with a nucleus of charge Z at the origin is

$$\left(\frac{-\hbar^2}{2m_e} \sum_{i=1}^{N} \nabla_i^2 - \sum_{i=1}^{N} \frac{Ze^2}{4\pi\varepsilon_0 r_i} + \sum_{i,j>i}^{N} \frac{e^2}{4\pi\varepsilon_0 r_{ij}}\right)\psi = E\psi. \tag{5.64}$$

By invoking the orbital approximation, the wavefunction is represented by a Slater determinant, namely

$$\psi = (N!)^{-1/2} \begin{vmatrix} \phi_1(1)\alpha(1) & \phi_1(2)\alpha(2) & \cdots & \phi_1(N)\alpha(N) \\ \phi_1(1)\beta(1) & \phi_1(2)\beta(2) & \cdots & \phi_1(N)\beta(N) \\ \phi_2(1)\alpha(1) & \phi_2(2)\alpha(2) & \cdots & \phi_2(N)\alpha(N) \\ \vdots & \vdots & & \vdots \\ \phi_{N/2}(1)\beta(1) & \phi_{N/2}(2)\beta(2) & \cdots & \phi_{N/2}(N)\beta(N) \end{vmatrix}$$

$$\equiv |\phi_1 \bar{\phi}_1 \phi_2 \cdots \bar{\phi}_{N/2}|. \tag{5.65}$$

In shorthand notation a bar represents a β or spin-down ($m_s = -\frac{1}{2}$) electron, while the absence of a bar represents an α or spin-up ($m_s = +\frac{1}{2}$) electron. The Slater determinant automatically satisfies the Pauli exclusion principle since the interchange of any two columns, which corresponds to the exchange of two electrons, changes the sign of the determinant. Since electrons are fermions the Pauli exclusion principle requires that

$$\hat{P}_{12}\psi = -\psi \tag{5.66}$$

hold for the exchange of two electrons, where $\hat{P}_{12}$ is the permutation operator that exchanges the coordinates of two electrons.

The Pauli exclusion principle requires that each orbital ϕ_i can contain, at most, two electrons with opposite spin. The orbitals are approximated by the product function

$$\phi_i = R(r_i)Y_{lm}(\theta_i, \phi_i).$$

By choosing this form we are assuming that the orbitals possess a hydrogenlike angular shape, but the radial functions need not be the associated Laguerre polynomials of the hydrogen atom. Instead, the radial functions associated with each ϕ_i are usually determined by minimizing the total energy of the atom using the variational principle.

The configuration of a multi-electron atom is constructed by placing electrons in the lowest energy orbitals in accordance with the Aufbau principle. For example, the lowest energy configuration of the Li atom is $(1s)^2 2s$, which corresponds to the Slater determinant

$$\psi = 6^{-1/2} \begin{vmatrix} 1s(1)\alpha(1) & 1s(2)\alpha(2) & 1s(3)\alpha(3) \\ 1s(1)\beta(1) & 1s(2)\beta(2) & 1s(3)\beta(3) \\ 2s(1)\alpha(1) & 2s(2)\alpha(2) & 2s(3)\alpha(3) \end{vmatrix} = |1s\overline{1s}2s|. \tag{5.67}$$

The tasks of either calculating atomic energy levels and wavefunctions or experimentally measuring energy-level differences by atomic spectroscopy are active areas of research. Calculation or measurement of atomic energy levels can be a complex task, but the labeling of atomic energy levels using the theory of angular momentum coupling is relatively straightforward.

All the various orbital and spin angular momenta of an atom must add vectorially to make **J** the total angular momentum, which must remain a constant of the motion. For light atoms where spin-orbit coupling is small, it is convenient to use the Russell–Saunders coupling scheme. A coupling scheme is no more than a prescription that describes the order in which angular momenta are coupled. In the Russell–Saunders scheme the orbital angular momenta of all electrons are coupled to give a total orbital angular momentum of the atom, that is,

$$\hat{\mathbf{L}} = \hat{\mathbf{l}}_1 + \hat{\mathbf{l}}_2 + \cdots + \hat{\mathbf{l}}_N = \sum_{i=1}^{N} \hat{\mathbf{l}}_i. \tag{5.68}$$

Similarly, for electron spin,

$$\hat{\mathbf{S}} = \sum_{i=1}^{N} \hat{\mathbf{s}}_i \tag{5.69}$$

and the total angular momentum is given by their vector sum, namely

$$\hat{\mathbf{J}} = \hat{\mathbf{L}} + \hat{\mathbf{S}}. \tag{5.70}$$

The convention of capital letters for a "total," many-electron angular momentum and lowercase letters for an individual or one-electron angular momentum is used.

As required by equations (5.68), (5.69), and (5.70), the individual operator components add to give the total component, for example

$$\hat{L}_z = \hat{l}_{z1} + \hat{l}_{z2} + \cdots + \hat{l}_{zN}, \tag{5.71}$$

$$\hat{S}_z = \hat{s}_{z1} + \hat{s}_{z2} + \cdots + \hat{s}_{zN}, \tag{5.72}$$

and

$$\hat{L}^2 = \hat{L}_x^2 + \hat{L}_y^2 + \hat{L}_z^2, \tag{5.73}$$

$$\hat{S}^2 = \hat{S}_x^2 + \hat{S}_y^2 + \hat{S}_z^2. \tag{5.74}$$

The components of $\hat{L}$ and $\hat{S}$ all commute with the nonrelativistic Hamiltonian, equation (5.64). The commutation of the spin operators with the atomic Schrödinger equation is obvious since there are no spin variables present in equation (5.64). For the multi-electron atom $\hat{L}_z$ and $\hat{H}$ are

$$\hat{L}_z = -i\hbar\left(\frac{\partial}{\partial\phi_1} + \frac{\partial}{\partial\phi_2} + \cdots + \frac{\partial}{\partial\phi_N}\right) \tag{5.75}$$

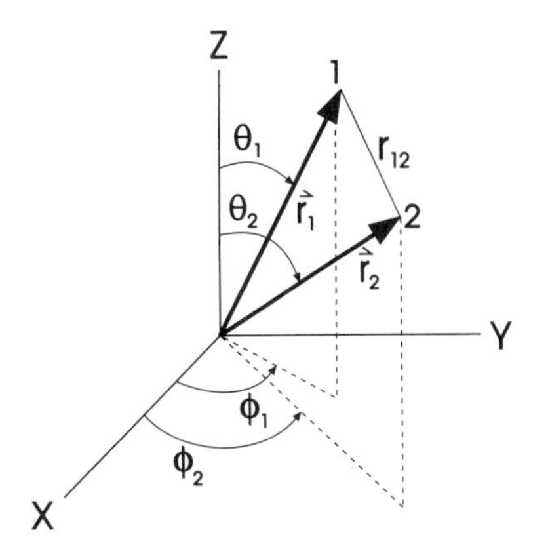

Figure 5.6. Polar coordinates of two electrons in an atom.

and

$$\hat{H} = \frac{-\hbar^2}{2m_e} \sum_i \nabla_i^2 - \sum_i \frac{Ze^2}{4\pi e_0 r_i} + \sum_{i,j>i} \frac{e^2}{4\pi \varepsilon_0 r_{ij}} \tag{5.76}$$

with

$$\nabla^2 = \frac{1}{r^2} \frac{\partial}{\partial r} r^2 \frac{\partial}{\partial r} + \frac{1}{r^2 \sin\theta} \frac{\partial}{\partial \theta}\left(\sin\theta \frac{\partial}{\partial \theta}\right) + \frac{1}{r^2 \sin^2\theta} \frac{\partial^2}{\partial \phi^2}. \tag{5.77}$$

$\hat{L}_z$ commutes with the kinetic energy operator because the ϕ_i variables appear as simple second derivatives in the Laplacian and as simple first derivatives in $\hat{L}_z$. The Coulomb attraction term is only a function of r_i, so it also commutes with $\hat{L}_z$. Finally, the electron–electron repulsion term is an implicit function of ϕ_i because of the presence of r_{ij}. Consideration of Figure 5.6, however, leads one to the conclusion that r_{12} depends only on $\phi_1 - \phi_2$ so

$$\begin{aligned} \hat{L}_z\left(\frac{1}{r_{12}}\right) &= -i\hbar\left(\frac{\partial}{\partial \phi_1} + \frac{\partial}{\partial \phi_2} + \cdots\right)\frac{1}{r_{12}} \\ &= -i\hbar\left(\frac{\partial(1/r_{12})}{\partial(\phi_1 - \phi_2)}\right)\left[\frac{\partial(\phi_1 - \phi_2)}{\partial \phi_1} + \frac{\partial(\phi_1 - \phi_2)}{\partial \phi_2}\right] + \cdots \\ &= 0 \end{aligned} \tag{5.78}$$

using the chain rule. Thus $\hat{L}_z$ commutes with $\sum e^2/r_{ij}$ since all of the azimuthal angles occur as differences, $\phi_i - \phi_j$.

Notice that individual electron quantum numbers m_l are not defined since

$$[\hat{l}_{zi}, \hat{H}] \neq 0, \tag{5.79}$$

and hence

$$\hat{l}_{zi}\psi \neq m_{li}\hbar\psi, \tag{5.80}$$

but the total M_L is a good quantum number in the absence of spin-orbit coupling since

$$\hat{L}_z\psi = M_L\hbar\psi, \tag{5.81}$$

with

$$M_L = m_{l1} + m_{l2} + \cdots + m_{lN}. \tag{5.82}$$

Analogous equations hold true for electron spin so that

$$\hat{S}_z\psi = M_S\hbar\psi \tag{5.83}$$

and

$$M_S = m_{s1} + m_{s2} + \cdots + m_{sN}. \tag{5.84}$$

Because an atom is spherically symmetric, the orientation of the z-axis is arbitrary so that if

$$[\hat{L}_z, \hat{H}] = 0,$$

then

$$[\hat{L}_x, \hat{H}] = 0 \quad \text{and} \quad [\hat{L}_y, \hat{H}] = 0.$$

Further, if

$$[\hat{L}_z^2, \hat{H}] = 0, \quad [\hat{L}_x^2, \hat{H}] = 0, \quad \text{and} \quad [\hat{L}_y^2, \hat{H}] = 0,$$

then

$$[\hat{L}^2, \hat{H}] = 0$$

by the properties of commutators. A comparison between the properties of multi-electron atoms and one-electron atoms are summarized in Table 5.4.

The operators $\hat{H}, \hat{L}^2, \hat{S}^2, \hat{L}_z, \hat{S}_z$ all commute with each other so that the wavefunction ψ is a simultaneous eigenfunction of all five operators. The corresponding quantum numbers E_n, L, S, M_L, and M_S can be used to label the wavefunctions $\psi = |nLM_LSM_s\rangle$. In the absence of external electric fields, magnetic fields and spin-orbit coupling the energy levels associated with ψ possess a $(2S+1)$-fold degeneracy due to the different M_S states and a $(2L+1)$-fold degeneracy due to the different M_L states. It

Table 5.4. One-electron and Multi-electron Atoms

Multi-electron Atoms	One-electron Atoms
$\hat{L}^2\psi = L(L+1)\hbar^2\psi$	$\hat{l}^2\psi = l(l+1)\hbar^2\psi$
$\hat{L}_z\psi = M_L\hbar\psi$	$\hat{l}_z\psi = m_l\hbar\psi$
$\hat{S}^2\psi = S(S+1)\hbar^2\psi$	$\hat{s}^2\psi = s(s+1)\hbar^2\psi$
$\hat{S}_z\psi = M_S\hbar\psi$	$\hat{s}_z\psi = m_s\hbar\psi$
$L = 0,\ 1,\ 2,\ 3,\ 4,\ 5$	$l = 0,\ 1,\ 2,\ 3,\ 4,\ 5$
$S\ \ P\ \ D\ \ F\ \ G\ \ H$	$s\ \ p\ \ d\ \ f\ \ g\ \ h$

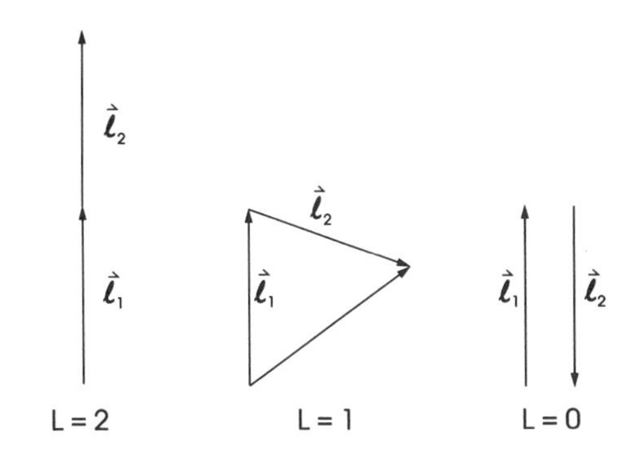

Figure 5.7. Vector addition of two $l = 1$ angular momenta.

is therefore convenient to label these energy levels by the term symbol ${}^{2S+1}L$, which gives rise to a total degeneracy

$$g = (2L + 1)(2S + 1). \tag{5.85}$$

The angular momenta $\mathbf{L}$ and $\mathbf{S}$ are vector quantities, made up of individual $\mathbf{l}_i$ and $\mathbf{s}_i$ vectors. When any two angular momenta, for example $\mathbf{l}_1$ and $\mathbf{l}_2$, are coupled together, they must be added vectorially. The possible values of the quantum number L are then given by $l_1 + l_2, l_1 + l_2 - 1, \ldots, |l_1 - l_2|$ from the vector coupling rules. The vector coupling $\mathbf{l}_1$ and $\mathbf{l}_2$ can be visualized with the aid of vector coupling diagrams, as shown in Figure 5.7 for the case of $l_1 = 1$ and $l_2 = 1$.

Consider the carbon atom with configuration $1s^2 2s^2 2p^2$. What terms arise from this configuration? The filled orbitals such as $1s^2$ or $2s^2$ have no net spin or orbital angular momentum, hence they can be ignored. The possible terms can be derived by considering all possible distributions of the two p electrons among the six spin-orbitals $p_1\alpha, p_0\alpha, p_{-1}\alpha, p_1\beta, p_0\beta, p_{-1}\beta$, in a manner consistent with the Pauli exclusion principle (Table 5.5). These possible states are often referred to as microstates and, in fact, correspond to individual Slater determinants.

A microstate designated as $|1, \bar{0}|$ means $m_{l1} = 1$, $m_{s1} = \frac{1}{2}$, $m_{l2} = 0$, $m_{s2} = -\frac{1}{2}$ with the orbitals arranged in some standard (arbitrary) order. Starting with the maximum values of M_L and M_S one deduces that there are nine microstates corresponding to ${}^3P$, five

Table 5.5. Slater Determinants for the Configuration p^2

	$M_S = 1$	0	-1
$M_L = 2$	—	$\lvert 1, \bar{1}\rvert$	—
$M_L = 1$	$\lvert 1, 0\rvert$	$\lvert 1, \bar{0}\rvert$ $\lvert \bar{1}, 0\rvert$	$\lvert \bar{1}, \bar{0}\rvert$
$M_L = 0$	$\lvert 1, -1\rvert$	$\lvert 1, -\bar{1}\rvert$ $\lvert \bar{1}, -1\rvert$ $\lvert 0, \bar{0}\rvert$	$\lvert \bar{1}, -\bar{1}\rvert$
$M_L = -1$	$\lvert 0, -1\rvert$	$\lvert -1, \bar{0}\rvert$ $\lvert -\bar{1}, 0\rvert$	$\lvert \bar{0}, -\bar{1}\rvert$
$M_L = -2$	—	$\lvert -1, -\bar{1}\rvert$	—

microstates corresponding to ^1D and one microstate corresponding to ^1S, consistent with the $(2S+1)(2L+1)$-fold total degeneracy of each Russell–Saunders term.

The microstate $|1, \bar{1}|$ clearly belongs to ^1D, while $|1, 0|$ belongs to ^3P, but to which terms do $|1, \bar{0}|$ and $|\bar{1}, 0|$ belong? Neither one is a proper eigenfunction of $\hat{L}^2$ and $\hat{L}_z$. The correct linear combination of determinants can be deduced, however, by the application of the lowering operator $\hat{L}_-$ to $|^1D, M_L = 2\rangle = |1, \bar{1}|$,

$$\hat{L}_- |1, \bar{1}| = (\hat{l}_{1-} + \hat{l}_{2-}) |1, \bar{1}| \tag{5.86}$$

$$[L(L+1) - M_L(M_L - 1)]^{1/2} |^1D, M_L = 1\rangle$$

$$= [l_1(l_1+1) - m_{l1}(m_{l1} - 1)]^{1/2} |0, \bar{1}| + [l_2(l_2+1) - m_{l2}(m_{l2} - 1)]^{1/2} |1, \bar{0}| \tag{5.87}$$

$$2 |^1D, M_L = 1\rangle = \sqrt{2} |0, \bar{1}| + \sqrt{2} |1, \bar{0}| \tag{5.88}$$

$$|^1D, M_L = 1\rangle = \frac{1}{\sqrt{2}} [|0, \bar{1}| + |1, \bar{0}|]$$

$$= \frac{1}{\sqrt{2}} [|1, \bar{0}| - |\bar{1}, 0|], \tag{5.89}$$

where the final determinant has been put in standard order. Similarly, the state

$$|^3P, M_L = 1, M_S = 0\rangle = \frac{1}{\sqrt{2}} [|1, \bar{0}| + |\bar{1}, 0|] \tag{5.90}$$

is orthogonal to $|^1D, M_L = 1, M_S = 0\rangle$.

The different terms arising from a configuration have different energies because of the electron–electron repulsion terms in the Hamiltonian. If

$$\hat{H}_0 = \frac{-\hbar^2}{2m_e} \sum \nabla_i^2 - \sum_i \frac{Ze^2}{4\pi\varepsilon_0 r_i} \tag{5.91}$$

were the only terms present in the Hamiltonian, then the orbital approximation would be exact because the electronic coordinates would be separable. Since $\hat{H}_0$ is comprised of one-electron, hydrogenlike operators, the wavefunction associated with $\hat{H}_0$ is a product of one-electron orbitals. The electron–electron repulsion term

$$\hat{H}_{ee} = \sum_{i, j>i} \frac{e^2}{4\pi\varepsilon_0 r_{ij}} \tag{5.92}$$

in the complete Hamiltonian prevents the separation of the total wavefunction into a

Table 5.6. Atomic Terms Arising from the p^n and d^n Configurations

p^1: ^2P	d^1, d^9: ^2D
p^2: ^1S, ^1D, ^3P	d^2, d^8: ^1S, ^1D, ^1G, ^3P, ^3F
p^3: ^2P, ^2D, ^4S	d^3, d^7: ^2P, ^2D (twice), ^2F, ^2G, ^2H, ^4P, ^4F
p^4: ^1S, ^1D, ^3P	d^4, d^6: ^1S (twice), ^1D (twice), ^1F, ^1G (twice), ^1I, ^3P (twice) ^3D, ^3F (twice), ^3G, ^3H, ^5D
p^5: ^2P	d^5: ^2S, ^2P, ^2D (three times), ^2F (twice), ^2G (twice), ^2H, ^2I, ^4P, ^4D, ^4F, ^4G, ^6S

product of one-electron orbitals. Nevertheless, it is still conceptually useful to retain the orbital approximation.

A set of empirical rules first proposed by Hund in 1927 is useful in predicting the lowest energy term arising from a configuration. Hund's first rule states that the term with the highest multiplicity $2S+1$ lies lowest in energy. If this rule does not select a unique term, then Hund's second rule comes into play: of terms of the same (maximum) multiplicity, the term with the highest L value lies lowest in energy. For example, Hund's rules predict that from a p^2 configuration the ^3P term lies lower in energy than the ^1D and ^1S terms. Experimentally the ground state of the carbon atom is indeed found to be ^3P. Hund's third rule is given below in the discussion of spin-orbit coupling.

The terms for many common configurations with equivalent electrons are given in Table 5.6. Notice that for the p^n configurations the terms are the same for the pairs, p^1, p^5 and p^2, p^4. This is because an electron (e.g., p^1) in a subshell (charge $-e$) has the same term as a hole (charge $+e$) in a full subshell (e.g., p^5). Similarly the terms arising from d^n and d^{10-n} are the same.

The enumeration of all the microstates arising from a d^5 configuration is quite a task (f^7 is worse!), but the direct application of Hund's rules will give the lowest energy term without having to determine all remaining terms. As a pictorial representation of the method, $2l+1$ boxes are drawn to represent the different orbitals. Each box is labeled with an m_l value and the electrons are placed into the boxes to maximize $M_L = \sum m_l$ and $M_S = \sum m_s$. The term that has these maximum M_L and M_S values can then be written down by inspection. For example d^4 has a diagram (Figure 5.8) corresponding to $M_L(\max) = 2$ and $M_S(\max) = 2$ arising from a ^5D term.

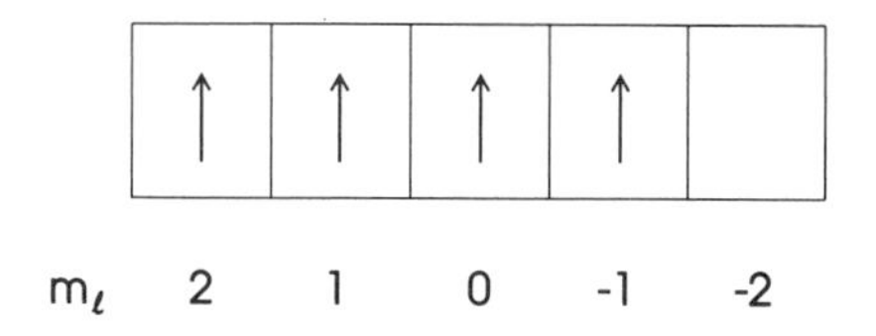

Figure 5.8. The microstate of d^4 with maximum M_S and M_L values.

Table 5.7. Names for the First Ten Multiplicities

$2S+1$	Name	$2S+1$	Name
1	Singlet	6	Sextet
2	Doublet	7	Septet
3	Triplet	8	Octet
4	Quartet	9	Nonet
5	Quintet	10	Decet

The $2S+1$ multiplicity of a term such as ^5D is read as "quintet" (Table 5.7) rather than as "five."

A term symbol also indicates the parity of an atomic state by the presence of a small following superscript "o" for an odd-parity term. The behavior of an atomic wavefunction under the operation of inversion can be determined from the parity of the orbitals since

$$\hat{E}^*\psi = \hat{E}^* |1s\overline{1s}2s \cdots| = (-1)^{\Sigma l_i}\psi \tag{5.93}$$

where the inversion operator $\hat{E}^*$ inverts the coordinates of all of the electrons through the origin. The parity of a state is determined simply by adding all the l values of the configuration from which the state arises. Thus, for example, the lowest energy term of nitrogen with a configuration $1s^22s^22p^3$ is written as $^4S^{\circ}$.

The effects of spin-orbit coupling must be incorporated into the Hamiltonian of the multi-electron atom. The spin-orbit term in the Hamiltonian is just the sum of the one-electron terms,

$$\hat{H}_{\text{so}} = \sum \xi(r_i)\hat{\mathbf{l}}_i \cdot \hat{\mathbf{s}}_i. \tag{5.94}$$

This form is not very convenient because of the presence of the individual $\mathbf{l}_i$ and $\mathbf{s}_i$ angular momenta. For a given term, however, an equivalent $\hat{H}_{\text{so}}$ can be derived using the properties of vector operators[2] (Wigner–Eckart theorem):

$$\hat{H}_{\text{so}} = \zeta \hat{\mathbf{L}} \cdot \hat{\mathbf{S}}. \tag{5.95}$$

The numerical factor $\zeta = \zeta(S, L)$ (typically with units of cm^{-1}) is referred to as the spin-orbit coupling constant of the L, S term. This form of the spin-orbit interaction is only applicable within a single, isolated ^{2S+1}L term and assumes that there are no interactions with other terms.

When $\hat{H}_{so} = \zeta \hat{\mathbf{L}} \cdot \hat{\mathbf{S}}$ is added to the atomic Hamiltonian (5.64), the components of **L** and **S** no longer commute with $\hat{H}$. The components of the total angular momentum $\mathbf{J} = \mathbf{L} + \mathbf{S}$ commute with $\hat{H}$, however, as do $\hat{L}^2$ and $\hat{S}^2$. The set of commuting observables is now $\{\hat{H}, \hat{L}^2, \hat{S}^2, \hat{J}^2, \hat{J}_z\}$ rather than the set $\{\hat{H}, \hat{L}^2, \hat{L}_z, \hat{S}^2, \hat{S}_z\}$ used in the absence of the spin-orbit coupling term, $\zeta \mathbf{L} \cdot \mathbf{S}$ (Table 5.8). In fact, as ζ increases the various terms begin to interact with each other since the full spin-orbit Hamiltonian $\sum \xi(r) \hat{\mathbf{l}}_i \cdot \hat{\mathbf{s}}_i$ has additional matrix elements with $\Delta L = 0, \pm 1$ and $\Delta S = 0, \pm 1$. The individual terms can no longer be considered isolated when spin-orbit coupling becomes large. This means that the true wavefunctions are no longer eigenfunctions of $\hat{L}^2$ and $\hat{S}^2$. Nevertheless

$$\hat{L}^2 \psi \approx L(L+1)\hbar^2 \psi \qquad \text{and} \qquad \hat{S}^2 \psi \approx S(S+1)\hbar^2 \psi,$$

so it is useful to retain the approximate quantum numbers L and S. The term symbol ^{2S+1}L is often used for heavy atoms with large spin-orbit coupling, but then J is always added as a subscript, $^{2S+1}L_J$.

The permissible values of J determined by vector coupling of L and S are $L+S$, $L+S-1, \ldots, |L-S|$. For example, the states of the p^2 configuration are $^3P_2$, $^3P_1$, $^3P_0$, $^1D_2$, and $^1S_0$ from the ^3P, ^1D, and ^1S terms. The energy separation of the $J=2$, $J=1$, and $J=0$ levels of the ^3P term (referred to as the multiplet splitting) is easily determined from the relationship

$$\mathbf{L} \cdot \mathbf{S} = \frac{\hat{J}^2 - \hat{L}^2 - \hat{S}^2}{2} \tag{5.96}$$

derived from

$$\hat{J}^2 = (\mathbf{L} + \mathbf{S}) \cdot (\mathbf{L} + \mathbf{S}) = \hat{L}^2 + \hat{S}^2 + 2\mathbf{L} \cdot \mathbf{S}. \tag{5.97}$$

If L and S are nearly good quantum numbers (in an isolated term), then perturbation

Table 5.8. Eigenvalue Equations for the Multi-electron Atom when Spin-Orbit Coupling is Included

$\hat{H} = \hat{H}_0 + \hat{H}_{ee}$	$\hat{H} = \hat{H}_0 + \hat{H}_{ee} + \zeta \mathbf{L} \cdot \mathbf{S}$	$\hat{H} = \hat{H}_0 + \hat{H}_{ee} + \sum \xi(r_i) \mathbf{l}_i \cdot \mathbf{s}_i$
$\hat{H}\psi = E\psi$	$\hat{H}\psi = E\psi$	$\hat{H}\psi = E\psi$
$\hat{L}^2\psi = L(L+1)\hbar^2\psi$	$\hat{L}^2\psi = L(L+1)\hbar^2\psi$	$\hat{L}^2\psi \approx L(L+1)\hbar^2\psi$
$\hat{L}_z\psi = M_L\hbar\psi$	$\hat{S}^2\psi = S(S+1)\hbar^2\psi$	$\hat{S}^2\psi \approx S(S+1)\hbar^2\psi$
$\hat{S}^2\psi = S(S+1)\hbar^2\psi$	$\hat{J}^2\psi = J(J+1)\hbar^2\psi$	$\hat{J}^2\psi = J(J+1)\hbar^2\psi$
$\hat{S}_z\psi = M_S\hbar\psi$	$\hat{J}_z\psi = M_J\hbar\psi$	$\hat{J}_z\psi = M_J\hbar\psi$

theory gives

$$\langle H_{so}\rangle = \zeta\langle \mathbf{L}\cdot\mathbf{S}\rangle = \zeta\langle nJM_JLS|\,\mathbf{L}\cdot\mathbf{S}\,|nJM_JLS\rangle$$

$$= \frac{\zeta[J(J+1) - L(L+1) - S(S+1)]}{2}. \tag{5.98}$$

The intervals are given by

$$E_{J+1} - E_J = \frac{\zeta[(J+1)(J+2) - J(J+1)]}{2} = \zeta(J+1). \tag{5.99}$$

This is the Landé interval rule: the spin-orbit splitting between sequential J levels in a term is proportional to the larger of the J values. Whether the level with the largest value of J lies highest in energy or lowest in energy is determined by the sign of ζ (Figure 5.9). If $\zeta > 0$, then the term is said to be regular (as for C), and if $\zeta < 0$, then the term is inverted (as for O). Hund's third rule predicts whether the lowest energy term will be regular or inverted. If the ground term arises from an electron configuration for which the valence electrons make up a less than half-filled subshell (e.g., C), then the lowest energy term will be regular, while if the configuration is more than half-filled (e.g., O), then the lowest energy term will be inverted. Both C and O have ground ^3P terms arising from $1s^22s^22p^2$ and $1s^22s^22p^4$ configurations, respectively; therefore the ^3P term of C is regular, while the ^3P term of O is inverted (Figure 5.9). If

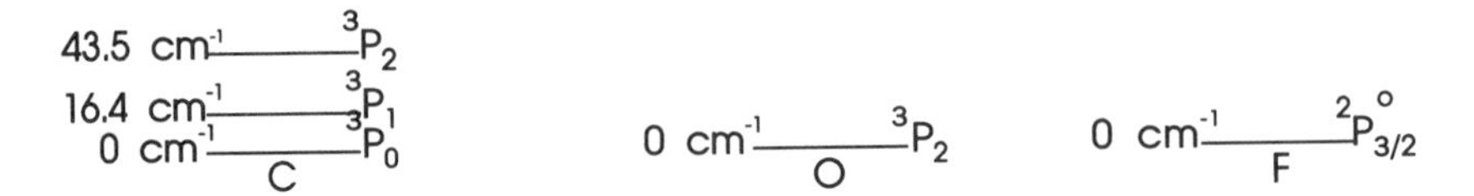

Figure 5.9. The multiplet splittings for the lowest energy term of the C, O, and F atoms.

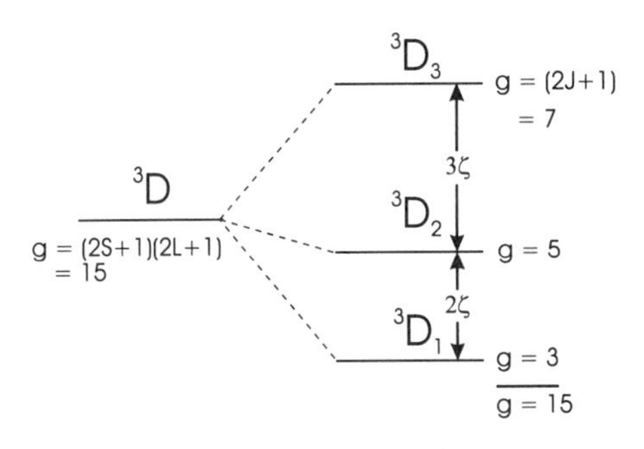

Figure 5.10. Degeneracy in the ^3D term.

an incomplete subshell is exactly half full (e.g., p^3), then the lowest energy term is always S and no spin-orbit splittings are possible.

The total degeneracy $2J + 1$ of a level J arises from the M_J degeneracy. For a term such as ^3D the total degeneracy $g = (2L + 1)(2S + 1) = 5(3) = 15$. The presence of spin-orbit coupling partially lifts this degeneracy and gives rise to the levels $^3D_3$, $^3D_2$, and $^3D_1$; the total degeneracy remains 15 (Figure 5.10).

In summary, if only the hydrogenlike terms $\hat{H}_0$ (equation (5.91)) are retained in the atomic Hamiltonian, then the terms in a given configuration are degenerate (Figure 5.11). If the electron–electron repulsion term $\hat{H}_{ee}$ (equation (5.92)) is added, then the orbital approximation begins to break down and the different terms in the configuration separate (Figure 5.11). Finally, when the spin-orbit term $\hat{H}_{so}$ (equation (5.95)) is added, the degeneracy of the levels in a term is lifted.

Selection Rules

Explaining the appearance of a spectrum requires detailed knowledge of the energy-level structure and selection rules that govern transitions between levels. To

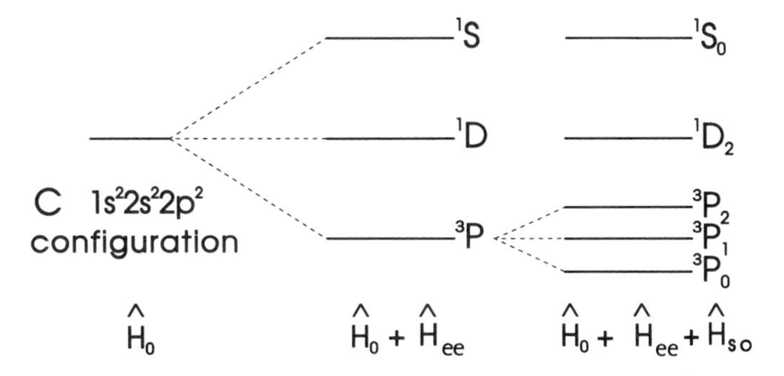

Figure 5.11. The qualitative effect of the $\hat{H}_0$, $\hat{H}_{ee}$, and $\hat{H}_{so}$ terms on the energy level pattern of the $1s^22s^22p^2$ configuration of the C atom.

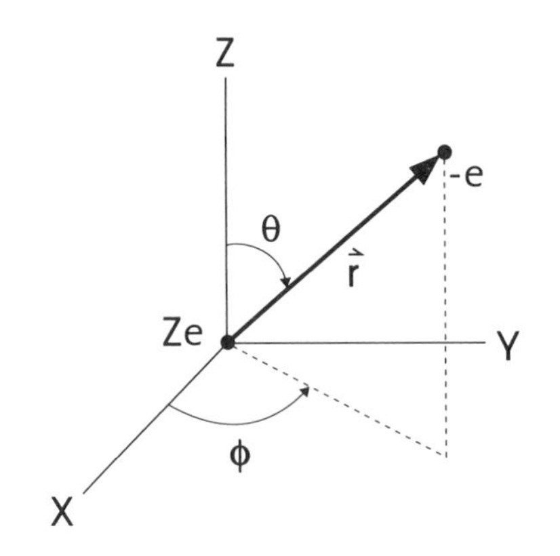

Figure 5.12. Coordinate system for the hydrogen atom.

begin, the one-electron selection rules of the hydrogen atom[3] are determined from the transition moment integral (Chapter 1)

$$\mathbf{M} = \int \psi_f^* \boldsymbol{\mu} \psi_i \, d\tau. \tag{5.100}$$

Because the parity of an atomic wavefunction is determined by l and the dipole moment operator is of odd parity, the selection rule for one-electron atomic transitions is that Δl must be odd. More restrictive selection rules can be derived by considering the atomic wavefunctions. With the nucleus at the origin of the coordinate system (Figure 5.12), $\boldsymbol{\mu} = -e\mathbf{r}$ and

$$\mathbf{M} = -e \iiint R_{n'l'}^*(r) Y_{l'm'}^*(\theta, \phi) \mathbf{r} R_{nl}(r) Y_{lm}(\theta, \phi) r^2 \sin\theta \, dr \, d\theta \, d\phi. \tag{5.101}$$

Each coordinate can be considered separately with

$$\mathbf{r} = x\hat{\mathbf{i}} + y\hat{\mathbf{j}} + z\hat{\mathbf{k}} = r(\sin\theta\cos\phi\hat{\mathbf{i}} + \sin\theta\sin\phi\hat{\mathbf{j}} + \cos\theta\hat{\mathbf{k}}). \tag{5.102}$$

For the z component we have

$$M_z = -eN \int_0^\infty r^3 R_{n'l'} R_{nl} \, dr \int_0^\pi P_{l'}^{m'}(\theta) \cos\theta \, P_l^m(\theta) \sin\theta \, d\theta \int_0^{2\pi} e^{-im'\phi} e^{im\phi} \, d\phi, \tag{5.103}$$

so that for $M_z \neq 0$, $m' = m$ or $\Delta m = 0$. Also from the properties of associated Legendre polynomials, we can show that

$$\cos\theta \, P_l^m(\theta) = \frac{(l - m + 1)P_{l+1}^m + (l + m)P_{l-1}^m}{2l + 1} \tag{5.104}$$

and from the orthogonality of these polynomials, as expressed by

$$\int_0^{\pi} P_{l'}^{m'} P_l^m \sin\theta \, d\theta = \delta_{l'l}\delta_{m'm}, \tag{5.105}$$

it is required that $\Delta l = \pm 1$. For the x component of the transition moment, we find

$$M_x = -eN \int_0^{\infty} r^3 R_{n'l'} R_{nl} \, dr \int_0^{\pi} P_{l'}^{m'} \sin\theta \, P_l^m \sin\theta \, d\theta \int_0^{2\pi} e^{-im'\phi} \cos\phi \, e^{im\phi} \, d\phi \tag{5.106}$$

but since

$$\cos\phi = \frac{e^{i\phi} + e^{-i\phi}}{2}, \tag{5.107}$$

we see that $\Delta m = \pm 1$ for nonzero M_x. In addition we can show that

$$\sin\theta \, P_l^{m-1} = \frac{P_{l+1}^m - P_{l-1}^m}{2l+1}, \tag{5.108}$$

so that once again $\Delta l = \pm 1$. The y component of $\mathbf{M}$ gives the same selection rules as does M_x. Further, no restriction exists for n' and n'' so that $\Delta n = 0, \pm 1, \pm 2, \ldots$ transitions are possible.

Thus single-photon, electric-dipole-allowed selection rules in hydrogenic atoms are $\Delta l = \pm 1$, $\Delta m = 0, \pm 1$, and $\Delta n =$ any integer. In reality, restrictions on Δn do exist for $l' \leftarrow l$ transitions, due to the decreasing overlap between the $R_{n'l'}$ and R_{nl} radial wavefunctions with increasing Δn. Values of the radial part of equation (5.103) and of the transition probabilities are tabulated, for example, in Condon and Shortley.[4]

Selection rules for multi-electron atoms are much more difficult to derive than are the one-electron selection rules, and consequently, only the results will be quoted here. Within the orbital approximation only a single electron can make a jump from one orbital to another, with $\Delta l = \pm 1$ during an electronic transition. All electrons other than the one making the transition remain in their original orbitals.

The parity selection rule of even $\leftrightarrow$ odd applies to multi-electron atoms. The parity of a multi-electron atomic state can easily be determined by evaluating $(-1)^{\sum l_i}$. This parity selection rule, often referred to as the *Laporte rule*, remains valid in all cases for electric-dipole transitions, even when the l_i are no longer good quantum numbers.

The selection rule for J is $\Delta J = 0, \pm 1$, but $J' = 0 \not\leftrightarrow J'' = 0$ also always remains valid for one-photon, electric-dipole-allowed transitions. The selection rules $\Delta L = 0, \pm 1$ and $\Delta S = 0$ for L and S only remain valid for small spin-orbit coupling. For the heavier elements the Russell–Saunders coupling scheme is no longer useful because the large spin-orbit coupling allows mixing between terms with different L and S values so that these selection rules break down. For example, the Hg $^3P_1 - {}^1S_0$ transition at 253.7 nm becomes quite strong, while the analogous transition for He is very weak.

It would seem that the multi-electron selection rule $\Delta L = 0$ conflicts with the one-electron selection rule $\Delta l = \pm 1$. The Ti atom transition $^3F^{\circ}(3d^24s4p) - {}^3F(3d^24s^2)$ is an example that illustrates that a $\Delta L = 0$, $\Delta l = 1$ transition is possible.

Atomic Spectra

The alkali atoms Li, Na, K, Rb, and Cs, as well as hydrogen, are the prototypes for one-electron atom transitions. In Figure 5.13 the energy levels and transitions of K are displayed. In atomic spectroscopy, energy-level diagrams are called *Grotrian diagrams.*

The $^2P_{3/2}$–$^2S_{1/2}$ (5890 Å) and $^2P_{1/2}$–$^2S_{1/2}$ (5896 Å) transitions of Na are known as the Na *D* lines. The letter designation was made by Fraunhofer when he first observed

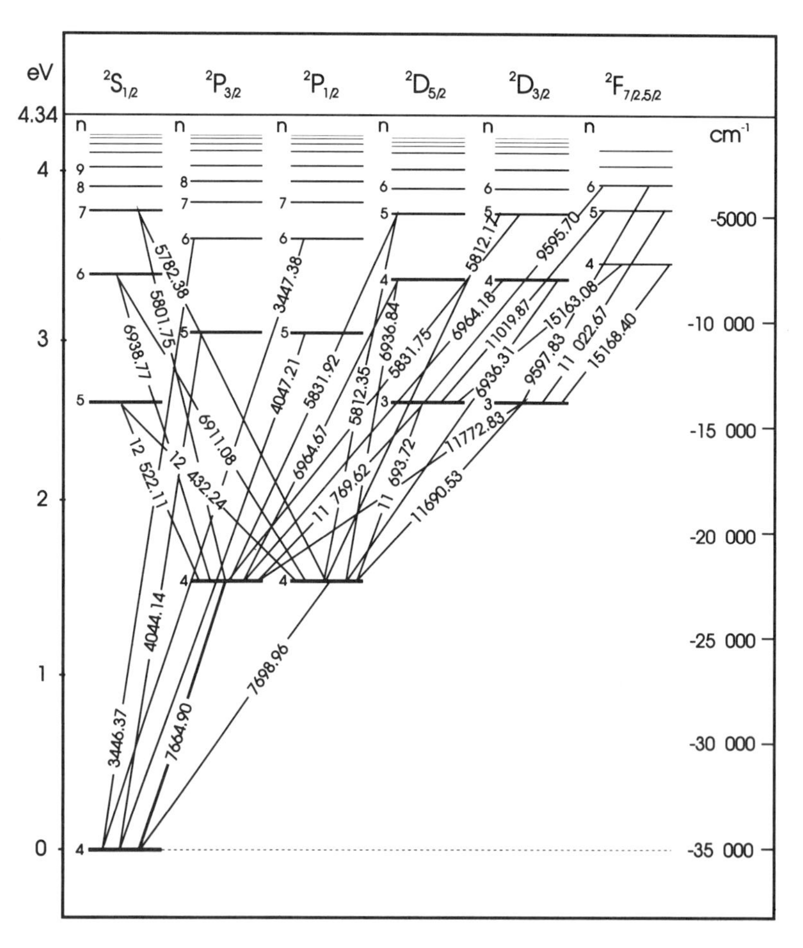

Figure 5.13. Energy-level diagram for the K atom.

them in the sun. The corresponding transitions of potassium are at 7644.90 Å and 7698.96 Å (Figure 5.13). They are examples of "resonance lines" because they originate from the ground state and are strongly allowed transitions. They are also the so-called "persistent lines" used for analytical purposes because they are readily observed in emission even when the K atom concentration is very low. There are several series of potassium transitions in Figure 5.13 that converge to the ionization limit of 4.34 eV for K.

He (Figure 5.14) and Ca (Figure 5.15) are examples of atoms with two valence electrons. The spectra are organized into separate singlet and triplet manifolds of states with only weak intercombination transitions connecting them.

Spin-orbit interaction causes many atomic lines to split into multiplets such as the Na *D* lines or the six-line pattern of the Ca ^3D–^3P transition at 442.5–445.6 nm (Figure 5.16). The splitting into multiplets is called *fine structure.*

The transition elements, lanthanides, and actinides have very complex energy-level patterns because of the many terms and levels arising from open *d* and *f* subshells.

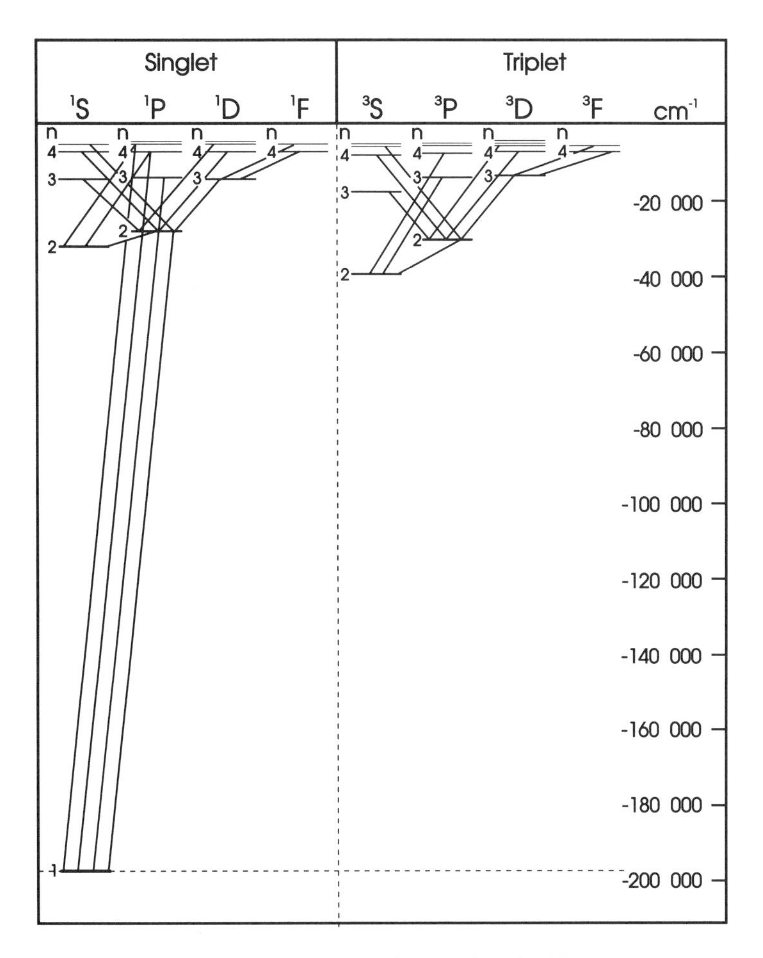

Figure 5.14. Energy-level diagram for the He atom.

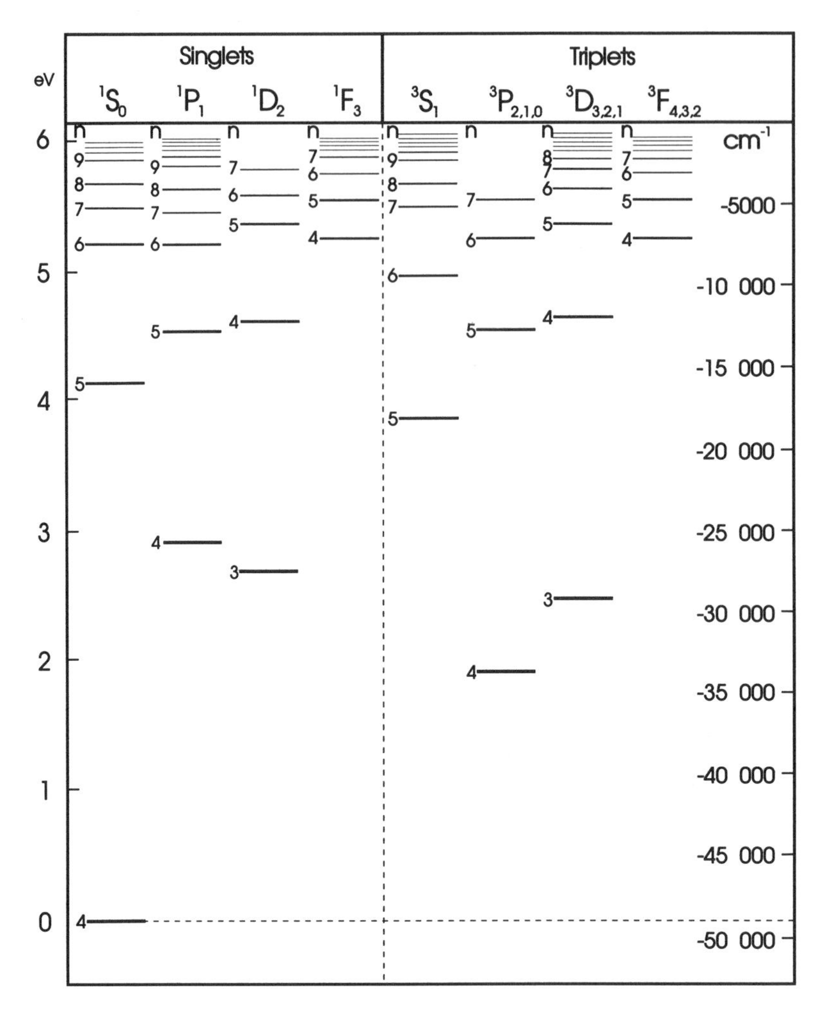

Figure 5.15. Energy-level diagram for the Ca atom.

Hyperfine Structure

The presence of nuclear spin produces further splittings in the lines of many elements. Since the splittings caused by electron spin are called fine structure, the much smaller splittings due to the nuclear spin are called *hyperfine structure.*

The nuclear spin **I** couples with **J** to form the total angular momentum **F** via vector coupling, namely

$$\mathbf{F} = \mathbf{J} + \mathbf{I}. \tag{5.109}$$

When nuclear spin is present, the only strictly good quantum number is F. Splittings due to hyperfine structure are relatively small, typically less than 1 cm^{-1}, so J remains a nearly good quantum number. The selection rules for F are the same as for J, $\Delta F = 0, \pm 1$, but $F' = 0 \not\leftrightarrow F'' = 0$, and $\Delta M_F = 0, \pm 1$.

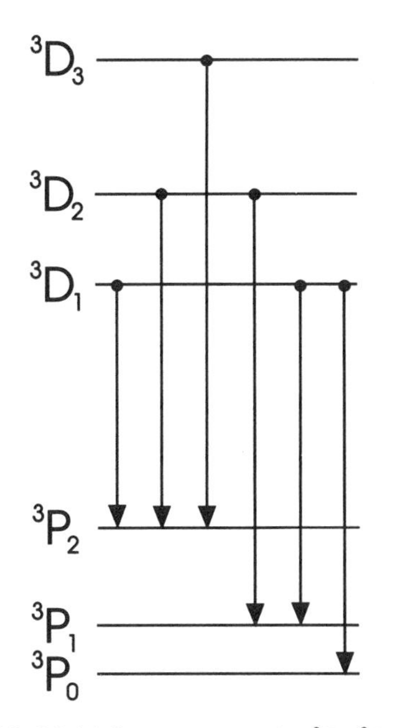

Figure 5.16. Multiplet structure of a ^3D–^3P transition.

For example, consider the energy-level pattern[5] for the ^{87}Rb $5^2P_{3/2}$–$5^2S_{1/2}$ transition (one of the resonance lines) of ^{87}Rb at 7800 Å[5] displayed in Figure 5.17. ^{87}Rb has a nuclear spin of 3/2. The separation between the two lowest hyperfine levels in the ground electronic state of Rb is used for a frequency standard. The similar ground-state hyperfine transition at 9192.631770 GHz[6] in Cs ($I = 7/2$) is used in an atomic clock to provide the national time standard in many countries.

Hydrogen Atom

The hydrogen atom continues to fascinate scientists. For example, precise frequency measurements in the hydrogen atom spectrum have led to a refinement in the value of the Rydberg constant as well as practical applications such as a hydrogen maser (microwave laser) that can also serve as an atomic clock. The Rydberg constant is the most accurately known fundamental physical constant,[7] with a value $R_\infty =$ 109737.3156841 cm^{-1}.

When split-orbit coupling is included,[8] the Lyman α line is comprised of two components, $2^2P_{3/2}$–$1^2S_{1/2}$ and $2^2P_{1/2}$–$1^2S_{1/2}$ (Figure 5.18). Dirac's relativistic model of the hydrogen atom, however, predicts that the $2^2P_{1/2}$ and $2^2S_{1/2}$ levels have the same energy. In fact Lamb and Retherford experimentally determined that these two levels are split by about 1058 MHz. The theory of quantum electrodynamics, developed by Feynman and others, is able to account for this Lamb shift.

In addition to spin-orbit coupling, hyperfine structure[8] is also present in the

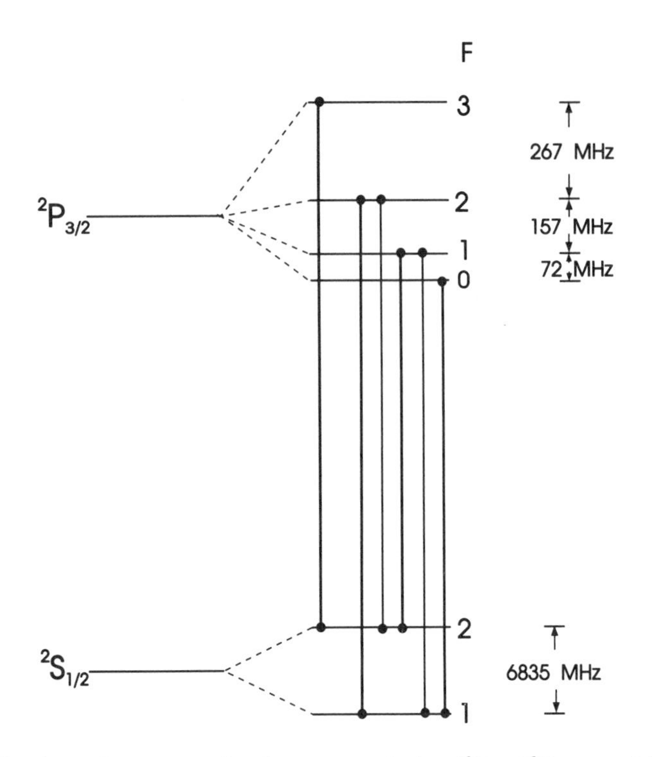

Figure 5.17. The hyperfine energy-level pattern of the $5^2P_{3/2}$–$5^2S_{1/2}$ transition of ^{87}Rb near 7800 Å.

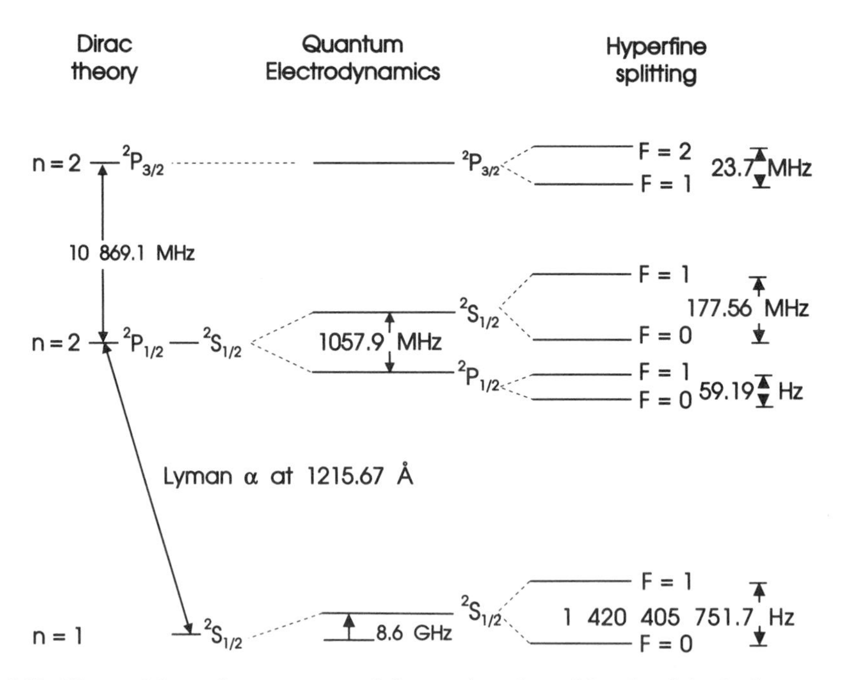

Figure 5.18. Fine and hyperfine structure of the $n = 1$ and $n = 2$ levels of the hydrogen atom.

spectrum of the hydrogen atom because the proton has a nuclear spin of 1/2. Hyperfine structure doubles all the energy levels. In the ground state the $F=1$–$F=0$ splitting is 1420 MHz, which corresponds to a wavelength of 21 cm. This 21-cm radiation was first detected in interstellar space, leading to the development of radio astronomy. The hydrogen maser also oscillates on this 1420-MHz hyperfine transition, which is electric-dipole forbidden but magnetic-dipole allowed.

Zeeman Effect

Associated with the various angular momenta of atoms are magnetic moments. The theory begins with orbital angular momentum since there is a classical analogy. The orbit of a negatively charged electron around the nucleus is equivalent to a small current loop that creates a magnetic moment. The constant of proportionality between the orbital angular momentum $\mathbf{L}$ and the magnetic moment $\boldsymbol{\mu}$ is γ, the magnetogyric ratio, and we have

$$\boldsymbol{\mu}_L = \gamma \mathbf{L}. \tag{5.110}$$

The value of γ is $-e/(2m_e)$ in SI units or $-e/(2m_e c)$ in cgs units. The magnitude of $\mathbf{L}$ is

$$|\mathbf{L}| = [L(L+1)]^{1/2}\hbar \tag{5.111}$$

so that

$$|\boldsymbol{\mu}_L| = -\left(\frac{e\hbar}{2m_e}\right)\sqrt{L(L+1)} = -\mu_B\sqrt{L(L+1)}, \tag{5.112}$$

where it is convenient to define the Bohr magneton, μ_B as

$$\mu_B = \frac{e\hbar}{2m_e} \tag{5.113}$$

in SI units. The Bohr magneton is the size of the magnetic moment that one unit of orbital angular momentum produces in an atom.

Other angular momenta such as spin also have associated magnetic moments, but there are no classical analogies to draw upon. In the case of the electron it is possible to write, by analogy with the orbital angular momentum, the expression

$$\boldsymbol{\mu}_S = g_e \gamma \mathbf{S} = g_e\left(\frac{-\mu_B}{\hbar}\right)\mathbf{S}, \tag{5.114}$$

in which the numerical factor g_e is defined by this equation. The g_e value for a single free electron is found to be 2.0023. The negative sign of the magnetogyric ratio indicates that $\boldsymbol{\mu}_S$ and **S** point in opposite directions (notice that $\boldsymbol{\mu}_L$ and **L** also point in opposite directions).

Similarly, for nuclear spin a nuclear moment $\boldsymbol{\mu}_I$ can be described by the equation

$$\boldsymbol{\mu}_I = \gamma_I \mathbf{I} = g_I \frac{\mu_N}{\hbar} \mathbf{I} \tag{5.115}$$

in which γ_I is the nuclear magnetogyric ratio and μ_N is the basic unit for nuclear magnetic moments, called the *nuclear magneton.* For nuclei the g_I values can be either positive or negative. The nuclear magneton

$$\mu_N = \frac{e\hbar}{2m_p} \tag{5.116}$$

is defined by analogy with the Bohr magneton, but the mass of the proton (m_p) is used rather than the mass of an electron. This means that (Appendix A)

$$\frac{\mu_N}{\mu_B} = \frac{m_e}{m_p} = \frac{1}{1837}, \tag{5.117}$$

so that nuclear magnetic moments are typically two to three orders of magnitude smaller than are electric moments, since $g_I \sim 1$ ($g_I = 5.585$ for ^1H).

When an atom is placed in a magnetic field, the magnetic moments interact with the field and, as a consequence, an interaction energy term

$$\hat{H}' = -\boldsymbol{\mu} \cdot \mathbf{B} \tag{5.118}$$

must be included in the Hamiltonian. This is called the *Zeeman interaction Hamiltonian* and leads to the Zeeman effect. When $S = 0$, the Zeeman effect is called "normal," and when $S \neq 0$ it is called "anomalous" since, historically, the Zeeman effect was discovered before electron spin was known.

The normal Zeeman effect applies to singlet states, for which only $\boldsymbol{\mu}_L$ is present. If **B** is aligned along the laboratory z-axis, then the Zeeman Hamiltonian takes the form

$$\hat{H}' = -\boldsymbol{\mu}_L \cdot \mathbf{B} = -\gamma \hat{L}_z B_z. \tag{5.119}$$

Simple perturbation theory gives the Zeeman energy as

$$\begin{aligned} E &= \langle LM_L| -\gamma \hat{L}_z B_z |LM_L\rangle \\ &= -\gamma \hbar M_L B = \mu_B B M_L. \end{aligned} \tag{5.120}$$

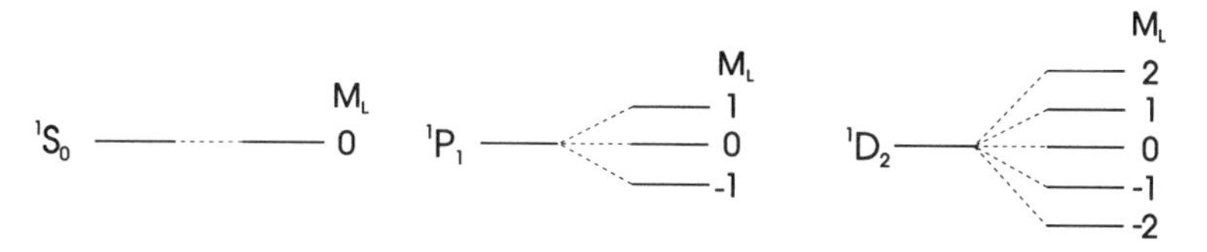

Figure 5.19. The Zeeman pattern for ^1S, ^1P, and ^1D states.

The energy levels are thus split into the $2L+1$ states (Figure 5.19) characterized by individual M_L values. The energy of each state varies linearly with the strength of the magnetic field. The interval between adjacent M_L values is $\mu_L B$, which is independent of L. The selection rules for electronic transitions are $\Delta M_L = 0$ for light polarized along z and parallel to $\mathbf{B}$, $\Delta M_L = \pm 1$ for light polarized perpendicular to $\mathbf{B}$.

When a magnetic field is applied to atomic transitions between singlet ($S=0$) states, the atomic line is split into three components (Figure 5.20). The line coinciding with the zero field position is a $\Delta M = 0$ transition, and the line shifted to higher frequency is a $\Delta M = +1$ transition, while the line shifted to lower frequency is a $\Delta M = -1$ transition. Classically, one can view the orbital magnetic moment as precessing (Figure 5.21) at the Larmor frequency around the direction of the applied field. The Larmor frequency is given by $h\nu_L = \mu_B B$ or

$$\nu_L = \frac{\mu_B B}{h}. \tag{5.121}$$

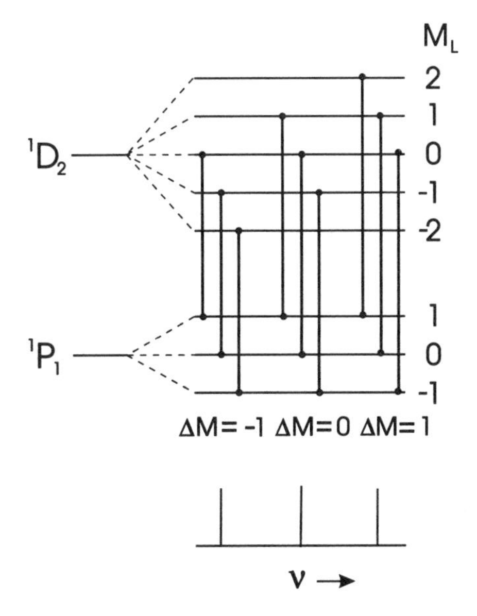

Figure 5.20. The three-line pattern of the "normal" Zeeman effect.

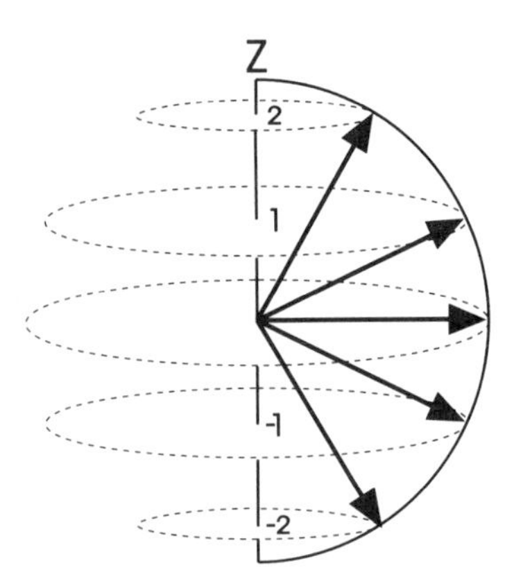

Figure 5.21. The precessional motion of $L = 2$ with the possible $M_L = 2, 1, 0, -1,$ and -2 values.

The case with $S \neq 0$ is more complex, since **L** and **S** couple first to give **J**, while $\boldsymbol{\mu}_S$ and $\boldsymbol{\mu}_L$ interact to give the corresponding $\boldsymbol{\mu}_J$. By analogy with $\boldsymbol{\mu}_S$ and $\boldsymbol{\mu}_L$ one writes

$$\boldsymbol{\mu}_J = g_J \gamma \mathbf{J} = -g_J \frac{\mu_B}{\hbar} \mathbf{J} \tag{5.122a}$$

so that the problem reduces to finding an expression for g_J. According to the classical vector coupling picture **L** and **S** precess rapidly about **J**, so that only those components of **L** and **S** parallel to **J** contribute to the total magnetic moment (Figure 5.22). The total magnetic moment $\boldsymbol{\mu}_J$ can be written, according to what has been previously said, as

$$\begin{aligned} \boldsymbol{\mu}_J &= \left[\boldsymbol{\mu}_S \cdot \frac{\mathbf{J}}{|\mathbf{J}|^2} + \frac{\boldsymbol{\mu}_L \cdot \mathbf{J}}{|\mathbf{J}|^2} \right] \mathbf{J} \\ &= \left[\gamma(g_s \mathbf{S} + \mathbf{L}) \cdot \frac{\mathbf{J}}{|\mathbf{J}|^2} \right] \mathbf{J}, \end{aligned} \tag{5.122b}$$

from which we see that the appropriate g_J-factor is

$$\begin{aligned} g_J &= (g_s \mathbf{S} + \mathbf{L}) \cdot \frac{\mathbf{J}}{|\mathbf{J}|^2} = \frac{(g_s \mathbf{S} + \mathbf{L}) \cdot (\mathbf{L} + \mathbf{S})}{|\mathbf{J}|^2} \\ &= \frac{g_s(\mathbf{L} \cdot \mathbf{S} + |\mathbf{S}|^2) + \mathbf{L} \cdot \mathbf{S} + |\mathbf{L}|^2}{|\mathbf{J}|^2}. \end{aligned} \tag{5.123}$$

Notice that g_J is not a simple number but is technically a scalar operator that will give rise to different numerical values for different levels. Now since

$$J^2 = (\mathbf{L} + \mathbf{S}) \cdot (\mathbf{L} + \mathbf{S}) = L^2 + S^2 + 2\mathbf{L} \cdot \mathbf{S} \tag{5.124}$$

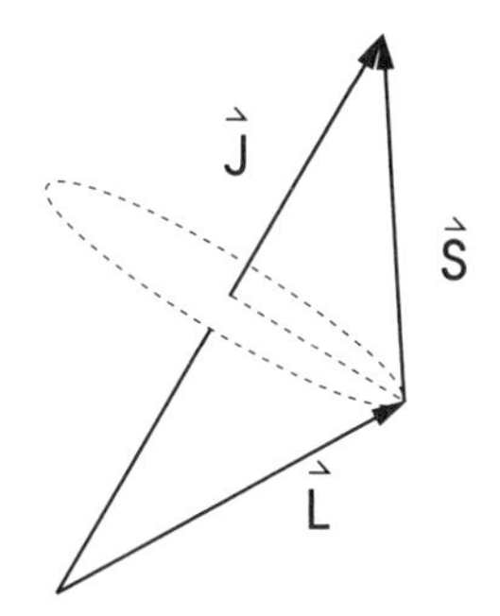

Figure 5.22. The rapid precessional motion of **L** and **S** about **J**.

we can solve for $\mathbf{L}\cdot\mathbf{S}$ in terms of J^2, L^2, and S^2

$$\mathbf{L}\cdot\mathbf{S}=\frac{J^2-L^2-S^2}{2}$$

to obtain for g_J the expression

$$g_J=\frac{g_s(J^2-L^2+S^2)+(J^2-S^2+L^2)}{2J^2}. \tag{5.125}$$

Because $g_S\approx 2$, the expression for g_J can also be written in the form

$$g_J\approx\frac{(3J^2+S^2-L^2)}{2J^2}. \tag{5.126}$$

In the corresponding energy eigenvalue expression (analogous to equation (5.120)), we would replace J^2, L^2, and S^2 by their magnitudes $J(J+1)$, $L(L+1)$, and $S(S+1)$ to obtain

$$g_J=\frac{3J(J+1)+S(S+1)-L(L+1)}{2J(J+1)}=1+\frac{J(J+1)+S(S+1)-L(L+1)}{2J(J+1)}. \tag{5.127}$$

By applying a magnetic field **B** along the z-axis of an atom, one obtains an energy splitting of

$$E_{M_J}=\langle\hat{H}'\rangle=-\langle\boldsymbol{\mu}_J\cdot\boldsymbol{B}\rangle=g_J\frac{\mu_B}{\hbar}\langle\mathbf{J}\cdot\boldsymbol{B}\rangle=g_J\frac{\mu_B B}{\hbar}\langle J_z B_z\rangle=g_J\mu_B M_J B, \tag{5.128}$$

with g_J given by equation (5.127). This "anomalous" Zeeman effect has proved to be

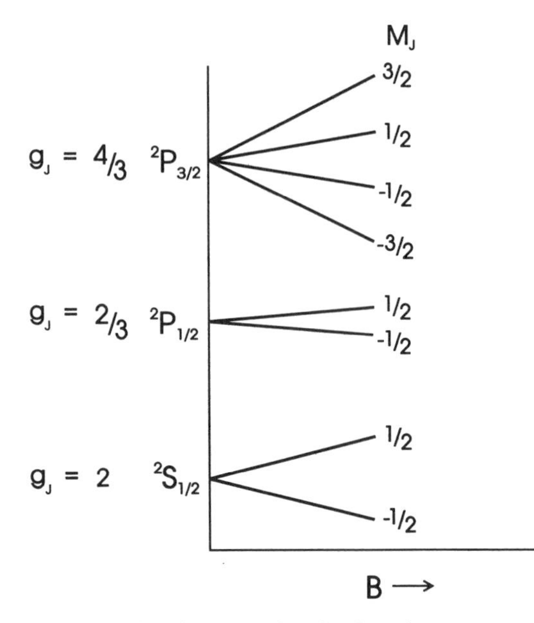

Figure 5.23. Zeeman energy-level pattern for the first three states of an alkali atom.

of great value in atomic spectroscopy, since the number of components is related to the J value, and the measured g_J value allows L, S, and J to be determined (Figure 5.23).

Paschen–Back Effect

As the applied magnetic field becomes very large, the splitting between the M_J components becomes larger than the splitting between the spin-orbit components. The coupling of **L** and **S** with the magnetic field is then stronger than the spin-orbit coupling. Spin-orbit coupling breakdown then occurs because of the decoupling of **L** and **S** from **J** by the magnetic field. As the magnetic field increases, the complex pattern of the anomalous effect is replaced by the simple three-line pattern of the normal Zeeman effect. This is called the *Paschen–Back effect.*

Stark Effect

The application of an electric field to an atom is called the *Stark effect.* In this case the atomic levels split into their $L + 1$ $|M_J|$ values and the $(2J + 1)$-degeneracy is only partly removed. While the Stark effect is widely used in molecular spectroscopy to obtain dipole moments, it is rarely used in atomic spectroscopy. The application of an electric field to the $n = 2$ states of the hydrogen atom results in a space-fixed dipole moment in the laboratory coordinate system (see Problem 17).

Problems

1. Positronium is an atom-like system formed from an electron and a positron. Predict the energy-level pattern and the wavelengths of some of the electronic transitions of positronium.
2. Show that the spherical harmonics obey the parity equation

$$E^*Y_{L,M_L} = Y_{L,M_L}(\pi - \theta, \phi + \pi) = (-1)^L Y_{L,M_L},$$

where E^* inverts all of the coordinates through the origin.
3. (a) Plot the R_{10}, R_{20}, and R_{21} radial functions for the hydrogen atom.
   (b) Plot the angular part of the real form of the hydrogen function for the $2s$, $2p_x$, $2p_y$, and $2p_z$ orbitals. What are the shapes of the orbitals?
4. Derive the spin matrices for $\hat{J}^2$, $\hat{J}_x$, $\hat{J}_y$, and $\hat{J}_z$ for the case $S = 3/2$.
5. Verify the relationship

$$(\boldsymbol{\sigma} \cdot \mathbf{a})(\boldsymbol{\sigma} \cdot \mathbf{b}) = \mathbf{a} \cdot \mathbf{b} + i\boldsymbol{\sigma} \cdot \mathbf{a} \times \mathbf{b}$$

where the components of $\boldsymbol{\sigma}$ are the Pauli matrices, and **a** and **b** are any two vector operators that commute with $\boldsymbol{\sigma}$.
6. (i) Construct the table of microstates and derive the terms for the atomic configurations
   (a) p^3
   (b) d^2
   (ii) For (a) and (b), what are the lowest energy terms and energy levels?
7. Without using microstates, derive the ground-state terms and energy levels for the transition elements of the third row (Sc through Zn) of the periodic table. (Remember Cr and Cu are exceptions to the regular Aufbau filling of electrons into orbitals.)
8. For the configurations
   (i) $ns\, n's$
   (ii) $ns\, n'd$
   (iii) $np\, n'p\, n''p$
   (a) What are the possible terms?
   (b) For each term, what are the possible energy levels?
9. For the He^+ atom calculate the vacuum wavelength and wavenumber of the transition corresponding to the red Balmer H_α transition in the H atom.
10. The spectrum of He^+ contains transitions at 303.780 Å, 256.317 Å, 243.027 Å, and 237.331 Å. Assign principal quantum numbers to these transitions.
11. (a) What are $\langle r \rangle$ and $\langle 1/r \rangle$ for the $1s$ orbital of hydrogen?
    (b) What is the transition dipole moment in debye for the $2p_z \leftarrow 1s$ transition of hydrogen?

12. The air wavelengths for the Balmer series are 6562.72 Å and 6562.852 Å (H_α), 4861.33 Å (H_β), 4340.47 Å (H_γ), and 4101.74 Å (H_δ). Derive a value for the Rydberg constant R_∞. Why are two wavelengths listed for H_α?
13. Plot the angular dependence of the square magnitude of the spherical harmonics $|Y_{l,m_l}|^2$ for $l = 0$, 1, and 2.
14. (a) For the hydrogen atom in $n = 2$ evaluate the ζ_{2p} integral.
    (b) Calculate the splitting in cm^{-1} for the $^2P_{3/2}$–$^2P_{1/2}$ interval for $n = 2$ of H.
15. In the atomic spectrum of neutral Ca there is a normal multiplet of six lines at 0, 14, 36, 106, 120, and 158 cm^{-1} above the lowest frequency line of the multiplet. What are the quantum numbers of the states involved in the transition?
16. For the two Na D lines calculate the spectral patterns for emission lines at a magnetic field strength of 0.25 tesla (T). What are the Zeeman splittings of the lines in cm^{-1}?
17. An electric field along the laboratory z-axis is applied to the hydrogen atom.
    (a) If the interaction energy is represented by $\hat{H}' = eEz = eEr \cos\theta$, evaluate the Hamiltonian matrix for $n = 2$ using the complex form of the hydrogen wavefunctions. (Ignore the effects of the Lamb shift, fine, and hyperfine structure.) Parity considerations simplify the problem.
    (b) What is the energy-level pattern?
    (c) For an applied electric field of 1000 V, what are the energy splittings in cm^{-1}?
18. The following wavenumbers are listed in Moore's tables for the n^2P^o–3^2S transitions of Na:

n	J	Wavenumber (cm^{-1})
5	0.5	35,040.27
5	1.5	35,042.79
6	0.5	37,296.51
6	1.5	37,297.76
7	0.5	38,540.40
7	1.5	38,541.14
8	0.5	39,298.54
8	1.5	39,299.01
9	0.5	39,794.53
9	1.5	39,795.00
10	0.5 and 1.5	40,137.23

    (a) Correct the line positions for the effect of spin-orbit coupling and determine ζ for the excited n^2P terms of Na.
    (b) Devise an extrapolation procedure to determine the ionization potential and the quantum defect for this Rydberg series.
19. On the basis of first-order perturbation theory the hyperfine structure of the ground electronic state of the H atom involves the interaction of the spins of the electron and proton with one another, and with any applied magnetic fields. It is

possible to integrate out the spatial coordinates and to consider the system as two spins $S = I = 1/2$ governed by the spin Hamiltonian

$$\hat{H}_{\text{spin}} = \frac{b_F}{\hbar^2}\hat{I}\cdot\hat{S} + \frac{k_S}{\hbar}\hat{S}_z + \frac{k_I}{\hbar}\hat{I}_z \equiv \hat{H}_{\text{hfs}} + \hat{H}_{\text{Zeeman}},$$

in which b_F, k_S, and k_I are given by

$$b_F = \frac{2\mu_0}{3} g_e\mu_B g_I\mu_I\,|\psi_{1s}(0)|^2$$

$$k_S = g_e\mu_B B_0$$

$$k_I = -g_I\mu_N B_0$$

and g_e, g_I, μ_B, μ_N are the g-factors and magnetons for the electron and the proton. The spin Hamiltonian can be split into two parts, $b_F\hat{I}\cdot\hat{S}/\hbar^2$ (referred to as the hyperfine structure (hfs) Hamiltonian), and $(k_S\hat{S}_z + k_I\hat{I}_z)/\hbar$ (referred to as the Zeeman Hamiltonian). SI units are used and $\mu_0 = 4\pi\times 10^{-7}\,\text{N A}^{-2}$ is the permeability of vacuum.

(a) Calculate the values of b_F, k_S, and k_I (the latter two as multiples of the field strength B_0) for the hydrogen 1s state.

(b) Now consider an isolated H atom (with no applied magnetic field). Show that the matrix of $\hat{H}_{\text{hfs}}$ with respect to the $|m_S m_I\rangle$ basis is

$$\mathbf{H}_{\text{hfs}} = \frac{b_F}{4}\begin{pmatrix} 1 & 0 & 0 & 0 \\ 0 & -1 & 2 & 0 \\ 0 & 2 & -1 & 0 \\ 0 & 0 & 0 & 1 \end{pmatrix}.$$

Find the energies and eigenstates in this basis and construct the matrix $\mathbf{X}$ that diagonalizes $\mathbf{H}_{\text{hfs}}$. What will be the eigenstates $|FM_F\rangle$ of $\hat{H}_{\text{hfs}}$ in terms of the $|m_S m_I\rangle$ states? Give a discussion of this in terms of vector coupling.

(c) Determine (in terms of b_F, k_S, k_I) the matrices with elements $\langle m'_S m'_I|\,\hat{H}_{\text{spin}}\,|m''_S m''_I\rangle$ and $\langle F'M'_F|\,\hat{H}_{\text{spin}}\,|F''M''_F\rangle$ in the general case when an applied field B_0 is present.

(d) From the results of part (c) show how the zero-field $|FM_F\rangle$ levels split in a weak magnetic field. In this case it is necessary to treat the magnetic field as a perturbation, namely

$$\hat{H}^{(0)} = \frac{b_F}{\hbar^2}\hat{I}\cdot\hat{S}, \qquad \hat{H}^{(1)} = \frac{k_S}{\hbar}\hat{S}_z + \frac{k_I}{\hbar}\hat{I}_z.$$

Give a plot of the splitting of these levels as calculated earlier for fields B_0 from 0 to 0.2 T (put your energy scale in MHz).

(e) Determine the energy levels in a strong magnetic field of 1 T, regarding the hyperfine interaction as a small perturbation, that is,

$$\hat{H}^{(0)} = \frac{k_S}{\hbar}\hat{S}_z + \frac{k_I}{\hbar}\hat{I}_z, \qquad \hat{H}^{(1)} = \frac{b_F}{\hbar^2}\hat{I}\cdot\hat{S}.$$

In this case show explicitly that the first-order perturbation spin functions are

$$\psi_1^{(1)} = \phi_1^{(0)}, \qquad \psi_4^{(1)} = \phi_4^{(0)},$$

$$\psi_2^{(1)} = \phi_2^{(0)} + \frac{b_F}{2(g_e\mu_B B_0 + g_I\mu_N B_0)}\phi_3^{(0)},$$

$$\psi_3^{(1)} = \phi_3^{(0)} - \frac{b_F}{2(g_e\mu_B B_0 + g_I\mu_N B_0)}\phi_2^{(0)},$$

while the second-order energies corresponding to these four functions are

$$E_1 = \frac{1}{2}g_e\mu_B B_0 - \frac{1}{2}g_I\mu_N B_0 + \frac{1}{4}b_F,$$

$$E_2 = \frac{1}{2}g_e\mu_B B_0 + \frac{1}{2}g_I\mu_N B_0 - \frac{1}{4}b_F + \frac{b_F^2}{4(g_e\mu_B B_0 + g_I\mu_N B_0)},$$

$$E_3 = -\frac{1}{2}g_e\mu_B B_0 - \frac{1}{2}g_I\mu_N B_0 - \frac{1}{4}b_F - \frac{b_F^2}{4(g_e\mu_B B_0 + g_I\mu_N B_0)},$$

$$E_4 = -\frac{1}{2}g_e\mu_B B_0 + \frac{1}{2}g_I\mu_N B_0 + \frac{1}{4}b_F.$$

The electron spin resonance (ESR) spectrum for the hydrogen atoms has only *two* equally intense lines, because the magnetic moment of the proton is too small to contribute to the intensity, and because the mixing of the $|m_S m_I\rangle$ states in the strong field is small. Show explicitly with numerical results that this is indeed the case for the problem that you are considering. Calculate the splitting of the two ESR lines in MHz, and compare your result with the experimentally observed value of 1420.4 MHz. What is the corresponding wavelength? How could you use this calculation to substantiate the existence of interstellar clouds of atomic hydrogen?

20. Consider the $2p^1 3d^1$ electron configuration of an atom.
    (a) What terms are possible?
    (b) Construct wavefunctions for one M_L component of the total $L = 2$ states for all possible total S values.

References

1. Mizushima, M. *Quantum Mechanics of Atomic Spectra and Atomic Structure,* Benjamin, New York, 1970, p. 207.
2. Tinkham, M. *Group Theory and Quantum Mechanics,* McGraw-Hill, New York, 1964, pp. 132 and 183.
3. Woodgate, G. K. *Elementary Atomic Structure,* Oxford University Press, Oxford, 1980, p. 46.
4. Condon, E. U. and Shortley, G. H. *The Theory of Atomic Spectra,* Cambridge University Press, Cambridge, England, 1970, pp. 133 and 136.
5. Belin, G. and Svanberg, S. Phys. Scr. **4,** 269 (1971).
6. Corney, A. *Atomic and Laser Spectroscopy,* Oxford University Press, Oxford, 1977, p. 706.
7. Andreae, T., et al. Phys. Rev. Lett. **69,** 1923 (1992) and references therein.
8. Cohen-Tannoudji, C., Diu, B. and Laloë, F. *Quantum Mechanics,* Vol. 2, Chap. XII, Wiley, New York, 1977.

General References

Bashkin, S. and Stoner, J. O. *Atomic Energy Levels and Grotrian Diagrams,* Vols. I, II, III, North-Holland, Amsterdam, 1975, 1978, 1981.

Bethe, H. A. and Salpeter, E. E. *Quantum Mechanics of One- and Two-Electron Atoms,* Springer-Verlag, Berlin, 1957.

Brink, D. M. and Satchler, G. R. *Angular Momentum,* Oxford University Press, Oxford, 1979.

Condon, E. U. and Odabasi, H. *Atomic Structure,* Cambridge University Press, Cambridge, England, 1980.

Condon, E. U. and Shortley, G. H. *The Theory of Atomic Spectra,* Cambridge University Press, Cambridge, England, 1970.

Corney, A. *Atomic and Laser Spectroscopy,* Oxford University Press, Oxford, 1979.

Edmonds, A. R. *Angular Momentum in Quantum Mechanics,* Princeton University Press, Princeton, N.J., 1974.

Fuhr, J. R., Martin, G. A., and Wiese, W. L. Atomic Transition Probabilities, Iron Through Nickel, J. Phys. Chem., Ref. Data **17,** Suppl. 4 (1988).

Gerloch, M. *Orbitals, Terms and States,* Wiley, Chichester, England, 1986.

Heckmann, P. H. and Träbert, E. *Introduction to the Spectroscopy of Atoms,* North-Holland, Amsterdam, 1989.

Harrison, G. R. *MIT Wavelength Tables,* M.I.T. Press, Cambridge, Mass. 1969.

Herzberg, G. *Atomic Spectra and Atomic Structure,* Dover, New York, 1945.

Kelley, R. L. *Atomic and Ionic Spectrum Lines Below 2000 Å,* Parts I, II, and III, J. Phys. Chem., Ref. Data. **16,** Suppl. 1 (1987).

King, G. W. *Spectroscopy and Molecular Structure,* Holt, Rinehart & Winston, New York, 1964.

Levine, I. N. *Quantum Chemistry,* 4th ed., Prentice-Hall, Englewood Cliffs, N.J., 1991.
Mitchell, A. C. G. and Zemansky, M. W. *Resonance Radiation and Excited Atoms,* Cambridge University Press, Cambridge, England, 1971.
Mizushima, M. *Quantum Mechanics of Atomic Spectra and Atomic Structure,* Benjamin, New York, 1970.
Moore, C. E. *Atomic Energy Levels,* Vols. I, II, and III (NSRDS-NBS35), U.S. Government Printing Office, Washington, D.C., 1971.
Pilar, F. L. *Elementary Quantum Chemistry,* 2nd ed., McGraw-Hill, New York, 1990.
Reader, J. and Corliss, C. H. Line Spectra of the Elements, in *CRC Handbook of Chemistry and Physics,* 73rd ed., D. R. Lide, Ed., CRC Press, Boca Raton, Fla., 1992.
Rose, M. E. *Elementary Theory of Angular Momentum,* Wiley, New York, 1957.
Sobelman, I. I. *Atomic Spectra and Radiative Transitions,* 2nd ed., Springer-Verlag, Berlin, 1992.
Sugar, J. and Corliss, C. Atomic Energy Levels of the Iron-Period Elements, J. Phys. Chem., Ref. Data. **14,** Suppl. 2 (1985).
Tinkham, M. *Group Theory and Quantum Mechanics,* McGraw-Hill, New York, 1964.
Woodgate, G. K. *Elementary Atomic Structure,* 2nd ed., Oxford University Press, Oxford, 1983.
Zare, R. N. *Angular Momentum,* Wiley, New York, 1988.

Chapter 6
Rotational Spectroscopy

Rotation of Rigid Bodies

The classical mechanics of rotational motion of a rigid body remains a relatively mysterious subject compared to that for linear motion. In order to dispel some of the mystery it is useful to note the extensive correspondence between linear motion of a point particle of mass m and rotational motion of the same particle (Figure 6.1 and Table 6.1). For simplicity the vector natures of most of the quantities are suppressed. The correspondences between the analogous linear and angular quantities in Table 6.1 are quite striking. The linear and angular variables are related by various equations,

$$\theta = \frac{x}{r}$$

$$\omega = \frac{v}{r} \qquad (\boldsymbol{\omega} \times \mathbf{r} = \mathbf{v})$$

$$a = \frac{v^2}{r} \qquad (\text{constant } \omega)$$

$$L = rp \qquad (\mathbf{L} = \mathbf{r} \times \mathbf{p})$$

$$T = rF \qquad (\mathbf{T} = \mathbf{r} \times \mathbf{F})$$

where the full vector forms are listed in parentheses. For a single particle, $\boldsymbol{\omega}$ and $\mathbf{L}$ are vectors that point out of the plane of the rotation. In this case, the $\boldsymbol{\omega}$ and $\mathbf{L}$ vectors point in the same direction (Figure 6.2). If an extended object is rotating, then $\mathbf{L}$ and $\boldsymbol{\omega}$ need not point in the same direction (Figure 6.3) and $\mathbf{I}$ is represented by a symmetric 3×3 matrix. This behavior is represented mathematically by the matrix product

$$\mathbf{L} = \mathbf{I}\boldsymbol{\omega} \tag{6.1}$$

written explicitly as

$$\begin{pmatrix} L_x \\ L_y \\ L_z \end{pmatrix} = \begin{pmatrix} I_{xx} & I_{xy} & I_{xz} \\ I_{xy} & I_{yy} & I_{yz} \\ I_{xz} & I_{yz} & I_{zz} \end{pmatrix} \begin{pmatrix} \omega_x \\ \omega_y \\ \omega_z \end{pmatrix}. \tag{6.2}$$

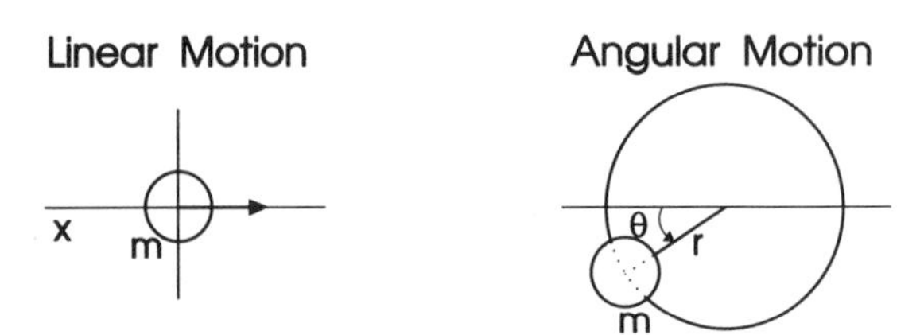

Figure 6.1. Linear and circular motion of a particle of mass m.

Table 6.1. The Correspondence between Linear and Angular Motion

Linear Motion		Angular Motion
	Position	
Distance, x		Angle, θ
	Velocity	
Velocity, $v = \dot{x} = dx/dt$		Angular velocity, $\omega = \dot{\theta} = d\theta/dt$
	Acceleration	
Acceleration, $a = \ddot{x} = d^2x/dt^2$		Angular acceleration, $\alpha = \ddot{\theta} = d^2\theta/dt^2$
	Mass	
Mass, m		Moment of inertia, $I = mr^2$
	Momentum	
Linear momentum, $p = mv$		Angular momentum, $L = I\omega$
	Kinetic energy	
$E_k = \frac{1}{2}mv^2 = p^2/2m$		$E_k = \frac{1}{2}I\omega^2 = L^2/2I$
	Force	
Force, F		Torque, T
	Newton's second law	
$F = ma = dp/dt$		$T = I\alpha = dL/dt$

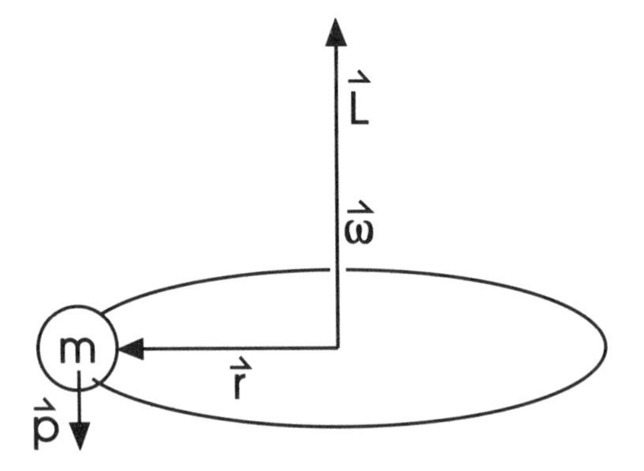

Figure 6.2. The circular motion of a particle of mass m.

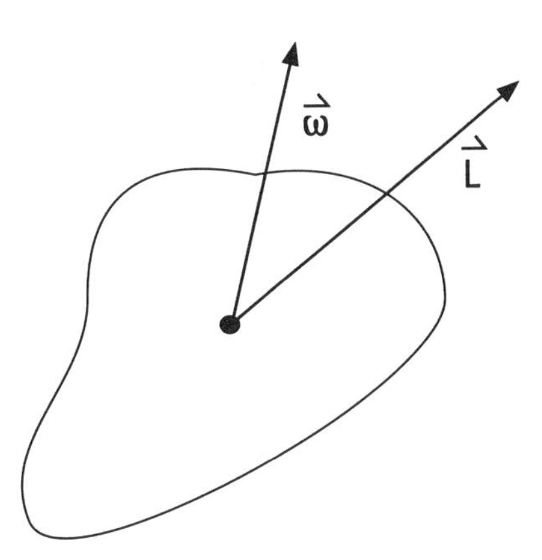

Figure 6.3. For an extended object $\boldsymbol{\omega}$ and $\mathbf{L}$ can point in different directions.

The matrix $\mathbf{I}$ is called the *moment of inertia tensor* in classical mechanics.

The derivation of the form of the moment of inertia tensor for a collection of nuclei rotating together requires the use of some vector identities and the definition of angular momentum. Consider a collection of nuclei of mass m_α located at positions $\mathbf{r}_\alpha$ relative to the origin in a Cartesian coordinate system (Figure 6.4) and all rotating with angular velocity $\boldsymbol{\omega}$, so that angular momentum is given by

$$\mathbf{L} = \sum_\alpha \mathbf{r}_\alpha \times \mathbf{p}_\alpha = \sum_\alpha m_\alpha \mathbf{r}_\alpha \times (\boldsymbol{\omega} \times \mathbf{r}_\alpha) \tag{6.3}$$

in which

$$\boldsymbol{\omega}_\alpha = \boldsymbol{\omega} \qquad \text{and} \qquad \mathbf{p}_\alpha = m_\alpha \mathbf{v} = m_\alpha \boldsymbol{\omega} \times \mathbf{r}.$$

The cross product identity

$$\mathbf{P} \times (\mathbf{Q} \times \mathbf{R}) = \mathbf{Q}(\mathbf{P} \cdot \mathbf{R}) - \mathbf{R}(\mathbf{P} \cdot \mathbf{Q}) \tag{6.4}$$

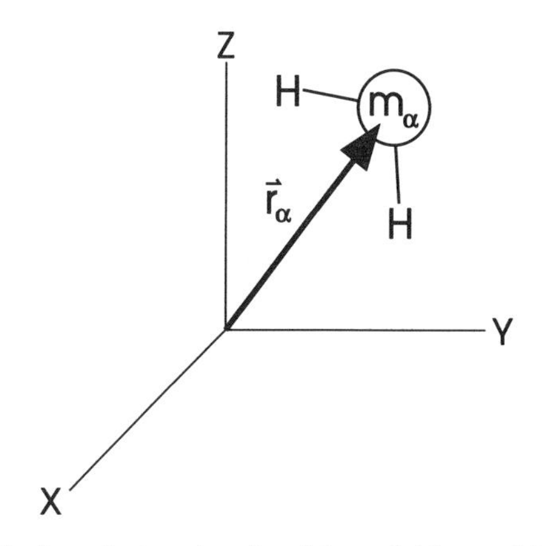

Figure 6.4. A typical molecule with nuclei located by $\mathbf{r}_\alpha$ vectors.

gives

$$\mathbf{L} = \sum_\alpha m_\alpha(\boldsymbol{\omega}(\mathbf{r}_\alpha \cdot \mathbf{r}_\alpha) - \mathbf{r}_\alpha(\mathbf{r}_\alpha \cdot \boldsymbol{\omega}))$$

$$= \sum_\alpha m_\alpha[\boldsymbol{\omega}(x_\alpha^2 + y_\alpha^2 + z_\alpha^2) - \mathbf{r}_\alpha(x_\alpha \omega_x + y_\alpha \omega_y + z_\alpha \omega_z)]. \tag{6.5}$$

Writing out the vector components gives

$$\begin{aligned}\mathbf{L} = \sum_\alpha m_\alpha[&\omega_x(x_\alpha^2 + y_\alpha^2 + z_\alpha^2)\hat{\mathbf{e}}_1 + \omega_y(x_\alpha^2 + y_\alpha^2 + z_\alpha^2)\hat{\mathbf{e}}_2 + \omega_z(x_\alpha^2 + y_\alpha^2 + z_\alpha^2)\hat{\mathbf{e}}_3 \\ &- x_\alpha^2\omega_x\hat{\mathbf{e}}_1 - x_\alpha y_\alpha \omega_y \hat{\mathbf{e}}_1 - x_\alpha z_\alpha \omega_z \hat{\mathbf{e}}_1 \\ &- y_\alpha x_\alpha \omega_x \hat{\mathbf{e}}_2 - y_\alpha^2 \omega_y \hat{\mathbf{e}}_2 - y_\alpha z_\alpha \omega_z \hat{\mathbf{e}}_2 \\ &- z_\alpha x_\alpha \omega_x \hat{\mathbf{e}}_3 - z_\alpha y_\alpha \omega_y \hat{\mathbf{e}}_3 - z_\alpha^2 \omega_z \hat{\mathbf{e}}_3],\end{aligned} \tag{6.6}$$

which can also be expressed in matrix form as

$$\begin{pmatrix} L_x \\ L_y \\ L_z \end{pmatrix} = \begin{pmatrix} \sum m_\alpha(y_\alpha^2 + z_\alpha^2) & -\sum m_\alpha x_\alpha y_\alpha & -\sum m_\alpha x_\alpha z_\alpha \\ -\sum m_\alpha y_\alpha x_\alpha & \sum m_\alpha(x_\alpha^2 + z_\alpha^2) & -\sum m_\alpha y_\alpha z_\alpha \\ -\sum m_\alpha z_\alpha x_\alpha & -\sum m_\alpha z_\alpha y_\alpha & \sum m_\alpha(x_\alpha^2 + y_\alpha^2) \end{pmatrix} \begin{pmatrix} \omega_x \\ \omega_y \\ \omega_z \end{pmatrix}. \tag{6.7}$$

Let us now identify the diagonal matrix elements of the matrix **I** as

$$I_{xx} = \sum_\alpha m_\alpha(y_\alpha^2 + z_\alpha^2) = \sum_\alpha m_\alpha r_{x,\perp}^2, \tag{6.8a}$$

$$I_{yy} = \sum_\alpha m_\alpha(x_\alpha^2 + z_\alpha^2) = \sum_\alpha m_\alpha r_{y,\perp}^2, \tag{6.8b}$$

$$I_{zz} = \sum_\alpha m_\alpha(x_\alpha^2 + y_\alpha^2) = \sum_\alpha m_\alpha r_{z,\perp}^2. \tag{6.8c}$$

These elements are referred to as the *moments of inertia.* Similarly, let us identify the nondiagonal matrix elements as

$$I_{xy} = -\sum_\alpha m_\alpha x_\alpha y_\alpha, \tag{6.9a}$$

$$I_{xz} = -\sum_\alpha m_\alpha x_\alpha z_\alpha, \tag{6.9b}$$

$$I_{yz} = -\sum_\alpha m_\alpha y_\alpha z_\alpha. \tag{6.9c}$$

These elements are referred to as *products of inertia.* Notice that the moment of inertia with respect to an axis involves the squares of the perpendicular distances of the masses from that axis, for example, $r_{x,\perp}^2$ from the x-axis.

In classical mechanics the motion of a collection of objects can be broken into the center of mass translational motion (see below) and the rotational motion about the center of mass. If a rigid rotor is assumed, the 3N-6 internal vibrations are ignored. The natural origin for the molecular coordinate system is the center of mass of the molecule.

The location of the center of mass (given by a vector $\mathbf{R}$) for a system of total mass

$$M = \sum_{\alpha} m_{\alpha} \tag{6.10}$$

made up of a collection of particles is given by

$$M\mathbf{R} = \sum_{\alpha} m_{\alpha}\mathbf{r}_{\alpha}. \tag{6.11}$$

If the origin of the coordinate system is at the center of mass, then $\mathbf{R} = 0$ and

$$\sum_{\alpha} m_{\alpha}\mathbf{r}_{\alpha} = 0. \tag{6.12}$$

The moment of inertia tensor is a real symmetric matrix, so it is always possible to find an orthogonal transformation matrix $\mathbf{X}$ that transforms the moment of inertia tensor $\mathbf{I}$ (equation (6.7)) into diagonal form. The matrix $\mathbf{X}$ represents a rotation of the coordinate system, which can be written as

$$\mathbf{r}' = \mathbf{X}^{-1}\mathbf{r} \qquad \text{or} \qquad \mathbf{r} = \mathbf{X}\mathbf{r}'. \tag{6.13}$$

The columns of the matrix $\mathbf{X}$ are made up of the normalized eigenvectors of $\mathbf{I}$.

As discussed in Chapter 3, the diagonalized matrix $\mathbf{I}'$ is related to the original matrix $\mathbf{I}$ by a similarity transformation, that is,

$$\mathbf{I}\mathbf{X} = \mathbf{X}\mathbf{I}' \tag{6.14}$$

or

$$\mathbf{X}^{-1}\mathbf{I}\mathbf{X} = \mathbf{I}'. \tag{6.15}$$

The $\mathbf{I}'$ matrix is a diagonal matrix whose elements are the eigenvalues of $\mathbf{I}$. This new coordinate system is called the *principal axis system* and $\mathbf{I}'$ has the form

$$\mathbf{I}' = \begin{pmatrix} I_{x'x'} & 0 & 0 \\ 0 & I_{y'y'} & 0 \\ 0 & 0 & I_{z'z'} \end{pmatrix}. \tag{6.16}$$

In most work the use of the principal axis system is assumed so that the primes will be dropped and $I_x = I_{x'x'}$, $I_y = I_{y'y'}$, and $I_z = I_{z'z'}$.

In the principal axis system we write

$$\begin{pmatrix} L_x \\ L_y \\ L_z \end{pmatrix} = \begin{pmatrix} I_x & 0 & 0 \\ 0 & I_y & 0 \\ 0 & 0 & I_z \end{pmatrix} \begin{pmatrix} \omega_x \\ \omega_y \\ \omega_z \end{pmatrix} \tag{6.17}$$

or $L_x = I_x\omega_x$, $L_y = I_y\omega_y$, and $L_z = I_z\omega_z$.

The kinetic energy expression also has the very simple form

$$\begin{aligned} E_k = T &= \tfrac{1}{2}\boldsymbol{\omega}^t \mathbf{I} \boldsymbol{\omega} \\ &= \tfrac{1}{2}(\omega_x \omega_y \omega_z) \begin{pmatrix} I_x\omega_x \\ I_y\omega_y \\ I_z\omega_z \end{pmatrix} \\ &= \frac{1}{2} I_x\omega_x^2 + \frac{1}{2} I_y\omega_y^2 + \frac{1}{2} I_z\omega_z^2 \\ &= \frac{L_x^2}{2I_x} + \frac{L_y^2}{2I_y} + \frac{L_z^2}{2I_z}. \end{aligned} \tag{6.18}$$

The x-, y-, and z-axes are chosen by some set of geometrical conventions. For example, the z-axis of a molecule is always chosen to be the highest order axis of rotational symmetry, and the x-axis is out of the plane for a planar molecule. For example, the moments of inertia for the H_2O molecule (Figure 6.5) are

$$I_z = 2m_H f^2, \tag{6.19}$$

$$I_y = m_O h^2 + 2m_H g^2, \tag{6.20}$$

and

$$I_x = I_z + I_y = m_O h^2 + 2m_H(g^2 + f^2). \tag{6.21}$$

For any planar molecule the out-of-plane moment of inertia is equal to the sum of the two in-plane moments of inertia. There is another labeling scheme for the axes in a molecule based upon the magnitude of the moments of inertia. In this case, the axes are labeled a, b, and c with

$$I_A \leq I_B \leq I_C \tag{6.22}$$

so that I_C is always the largest moment of inertia and I_A is the smallest. The a-, b-, and c-axes are chosen in order to ensure that this inequality holds.

For example, using $r = 0.958$ Å, $\theta = 104.5°$, $m_H = 1.00$ atomic mass unit (amu) and $m_O = 16.00$ amu for H_2O, results in $f = 0.7575$ Å, $g = 0.5213$ Å, and $h = 0.0652$ Å using

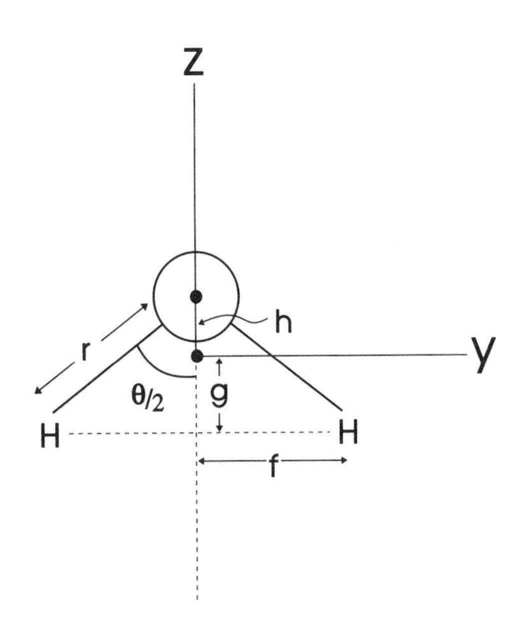

Figure 6.5. The H_2O molecule.

the center of mass definition (6.12). Thus from equations (6.19), (6.20), and (6.21) we obtain

$$I_z = 1.148 \text{ amu Å}^2 (= I_B)$$

$$I_y = 0.6115 \text{ amu Å}^2 (= I_A)$$

$$I_x = 1.760 \text{ amu Å}^2 (= I_C)$$

and $z = b$, $y = a$, and $x = c$. There are six possible ways that (x, y, z) can be mapped into (a, b, c) depending on the particular values of the moments of inertia. The x-, y-, z-axes are picked by a customary set of rules, such as z is along the highest axis of rotational symmetry, but a, b, and c are chosen to make equation (6.22) true.

Molecules can be classified on the basis of the values of the three moments of inertia. The five cases are as follows:

1. Linear molecules, $I_B = I_C$, $I_A = 0$; for example, HCN (Figure 6.6).
2. Spherical tops, $I_A = I_B = I_C$; for example, SF_6 and CH_4 (Figure 6.7)
3. Prolate symmetric tops, $I_A < I_B = I_C$; for example, CH_3Cl (Figure 6.8).
4. Oblate symmetric tops, $I_A = I_B < I_C$; for example, BF_3 (Figure 6.9).
5. Asymmetric tops, $I_A < I_B < I_C$; for example, H_2O (Figure 6.10).

Group theory can be used to classify the rotational properties of molecules. The

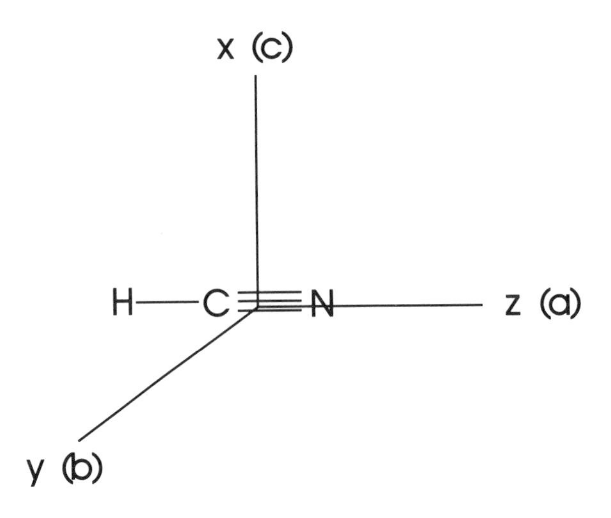

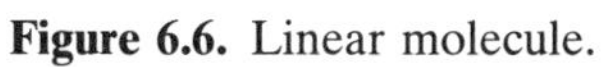

Figure 6.6. Linear molecule.

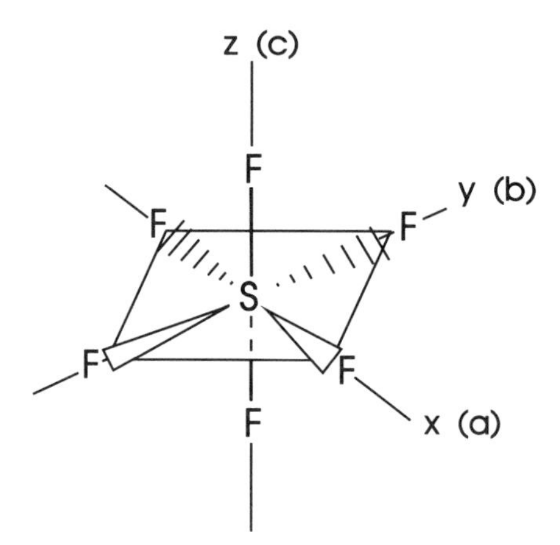

Figure 6.7. Spherical top.

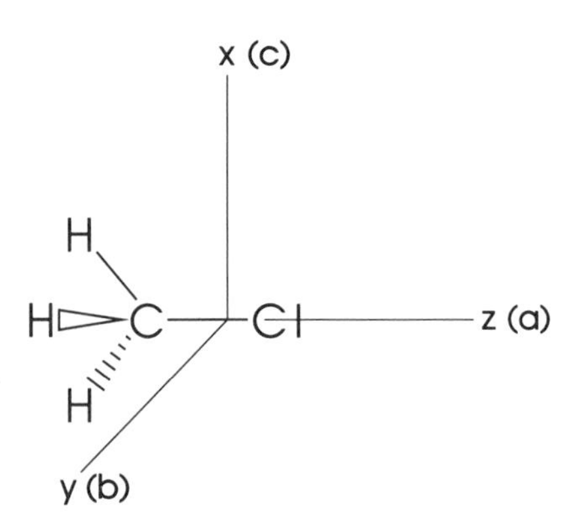

Figure 6.8. Prolate symmetric top.

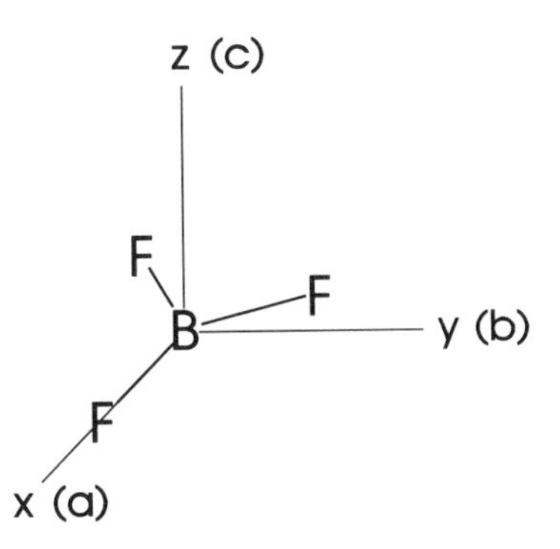

Figure 6.9. Oblate symmetric top.

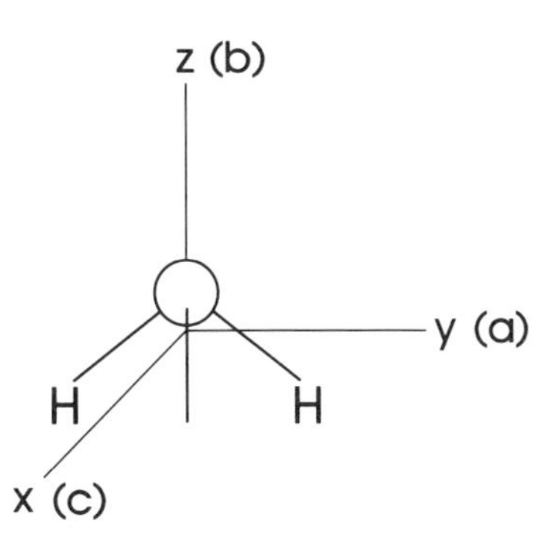

Figure 6.10. Asymmetric top.

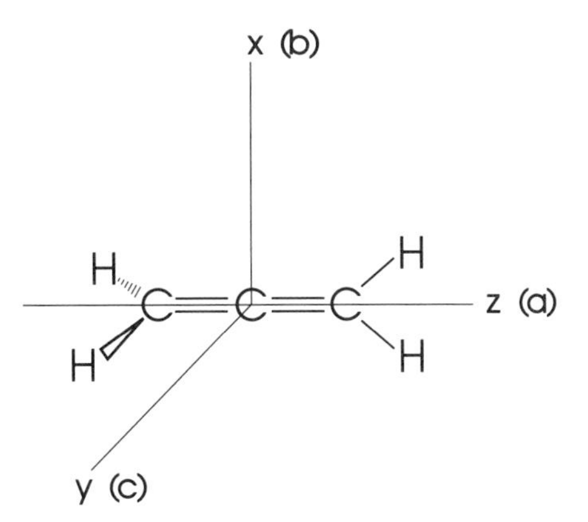

Figure 6.11. Allene, a symmetric top.

spherical tops (O_h, T_d, and I_h point groups) and linear molecules ($C_{\infty v}$ and $D_{\infty h}$) are readily recognized. All symmetric tops have a C_n-axis, with n greater than 2. For example, the symmetric tops CH_3Cl and benzene have C_3- and C_6-axes, while the asymmetric top H_2O has only a C_2-axis. But what about allene (Figure 6.11)? By symmetry allene has $I_B = I_C$, and hence it must be a prolate symmetric top. Allene has only a C_2-axis, but it does have an S_4-axis. The complete rule is, therefore, all molecules with a C_n- ($n > 2$) or an S_4-axis are symmetric tops. Note that the presence of an S_n-axis with $n > 4$ implies the presence of a C_n-axis, $n > 2$, so this case need not be explicitly stated.

The symmetry properties of a molecule are also helpful in locating the principal axes. For example, if there is a C_n-axis with $n > 1$, then one of the principal axes lies along it (e.g., H_2O). Any molecule with a plane of symmetry has one of the principal inertial axes perpendicular to the plane (e.g., H_2O).

Pure Rotational Spectroscopy of Diatomic and Linear Molecules

For a rigid linear molecule with no net orbital and spin angular momentum the classical expression for the rotational kinetic energy is

$$\begin{aligned} E_k = T &= \frac{1}{2}I_x\omega_x^2 + \frac{1}{2}I_y\omega_y^2 + \frac{1}{2}I_z\omega_z^2 \\ &= \frac{1}{2}I_x\omega_x^2 + \frac{1}{2}I_y\omega_y^2 \\ &= \frac{J_x^2}{2I} + \frac{J_y^2}{2I} = \frac{J^2}{2I} \end{aligned} \tag{6.23}$$

since $I_z = 0$, $I_x = I_y = I$ for a linear molecule and the customary symbol J is used to represent the total angular momentum (exclusive of nuclear spin) (Figure 6.12). For a rigid rotor in isotropic (field free) space the rotational Hamiltonian for a linear molecule is

$$\hat{H} = \frac{\hat{J}^2}{2I}. \tag{6.24}$$

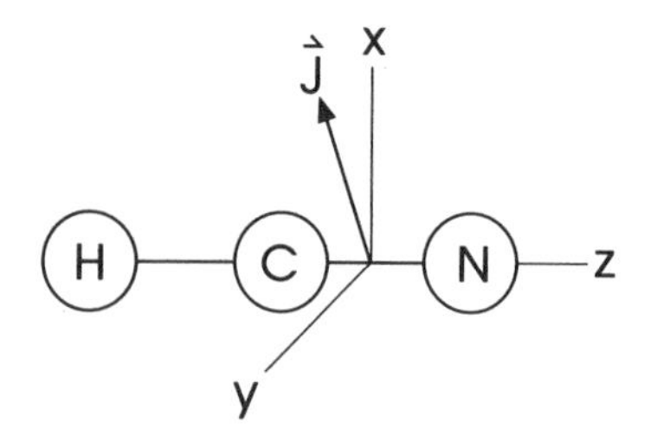

Figure 6.12. Rotational angular momentum in a linear molecule.

The Schrödinger equation can be solved immediately, since ψ must be one of the spherical harmonics, $\psi_{LM} = Y_{JM}$. The specific Schrödinger equation for this case is

$$\frac{\hat{J}^2\psi}{2I} = E\psi, \tag{6.25}$$

so that

$$\frac{\hat{J}^2\psi}{2I} = \frac{J(J+1)\hbar^2\psi}{2I} = BJ(J+1)\psi. \tag{6.26}$$

Thus we see that $F(J)$ is

$$F(J) = BJ(J+1) \tag{6.27}$$

in which

$$B = \frac{\hbar^2}{2I} = \frac{h^2}{8\pi^2 I}(\text{joules}). \tag{6.28}$$

In spectroscopy it is customary to use $F(J)$ to represent the rotational energy expression and the value of B is usually given in MHz or cm^{-1} rather than in joules. Since $E = h\nu = hc/\lambda = hc\tilde{\nu}$, the value of B is

$$B = \frac{h}{8\pi^2 I}(\text{Hz}) = \frac{h}{8\pi^2 I} \times 10^{-6}\,(\text{MHz}), \tag{6.29}$$

or

$$B = \frac{h}{8\pi^2 cI}(\text{m}^{-1}) = \frac{h}{8\pi^2 cI} \times 10^{-2}\,(\text{cm}^{-1}). \tag{6.30}$$

Convenient explicit expressions for B are

$$B = \frac{16.8576314}{I(\text{amu Å}^2)}\,\text{cm}^{-1}; \tag{6.31}$$

and

$$B = \frac{505{,}379.07}{I(\text{amu Å}^2)}\,\text{MHz}. \tag{6.32}$$

For a diatomic molecule we have

$$I = \mu r^2, \tag{6.33}$$

where μ is the reduced mass

$$\mu = \frac{m_A m_B}{m_A + m_B}. \tag{6.34}$$

The use of a single symbol B to represent a number that may be in units of joules, MHz, or cm^{-1} is an unfortunate but common practice. This convention will nonetheless be followed in this book.

Selection Rules

The intensity of a pure rotational transition is determined by the transition dipole moment

$$\mathbf{M} = \int \psi_{J'M'} \boldsymbol{\mu} \psi_{JM} \, d\tau. \tag{6.35}$$

For a linear molecule the wavefunction ψ_{JM} can be written explicitly as

$$\psi_{JM} = Y_{JM}(\theta, \phi) = \Theta_{JM}(\theta) e^{iM\phi}/\sqrt{2\pi} \tag{6.36}$$

and the dipole moment is oriented along the internuclear axis of the molecule, so that its components in the laboratory axis system can be expressed in the form

$$\begin{aligned} \boldsymbol{\mu} &= \mu_x \hat{\mathbf{e}}_1 + \mu_y \hat{\mathbf{e}}_2 + \mu_z \hat{\mathbf{e}}_3 \\ &= \mu_0 (\sin\theta \cos\phi \hat{\mathbf{e}}_1 + \sin\theta \sin\phi \hat{\mathbf{e}}_2 + \cos\theta \hat{\mathbf{e}}_3). \end{aligned} \tag{6.37}$$

so that equation (6.35) becomes

$$\begin{aligned} \mathbf{M} = \frac{\mu_0}{2\pi} \Bigg[&\hat{\mathbf{e}}_1 \int_0^{2\pi} \int_0^{\pi} \Theta_{J'M'} e^{-iM'\phi} \sin\theta \cos\phi \Theta_{JM} e^{iM\phi} \sin\theta \, d\theta \, d\phi \\ &+ \hat{\mathbf{e}}_2 \int_0^{2\pi} \int_0^{\pi} \Theta_{J'M'} e^{-iM'\phi} \sin\theta \sin\phi \Theta_{JM} e^{iM\phi} \sin\theta \, d\theta \, d\phi \\ &+ \hat{\mathbf{e}}_3 \int_0^{2\pi} \int_0^{\pi} \Theta_{J'M'} e^{-iM'\phi} \cos\theta \Theta_{JM} e^{iM\phi} \sin\theta \, d\theta \, d\phi \Bigg]. \end{aligned} \tag{6.38}$$

If we now employ the relationship $\cos\phi = (e^{i\phi} + e^{-i\phi})/2$ and a recursion relationship for the associated Legendre polynomials, namely

$$(2l+1) z P_l^m(z) = (l + |m|) P_{l-1}^m(z) + (l - |m| + 1) P_{l+1}^m(z), \tag{6.39}$$

where $z = \cos\theta$ and $\Theta_{JM}(\theta) = N P_l^m(\cos\theta)$, the selection rules $\Delta M = 0, \pm 1$ and $\Delta J = \pm 1$ are obtained.

In addition, if the molecule has no permanent dipole moment ($\mu_0 = 0$), then there are no allowed transitions. Thus, symmetric molecules O=C=O, Cl—Cl, H—C≡C—H have no pure rotational transitions, if only one-photon electric-dipole selection rules are considered. Molecules such as O_2 ($X\,^3\Sigma_g^-$) undergo weakly allowed magnetic-dipole pure rotational transitions. Molecules such as H—C≡C—D or H—D, where the center of mass is displaced from the center of charge when the molecule is vibrating, possess a small dipole moment (8×10^{-4} D for HD[1]) and also undergo weak rotational transitions.

The selection rule $\Delta J = \pm 1$ for a linear molecule results in transitions with frequencies

$$\begin{aligned} \nu_{J+1 \leftarrow J} &= F(J') - F(J'') \\ &= B(J+1)(J+2) - BJ(J+1) \\ &= 2B(J+1). \end{aligned} \tag{6.40}$$

Customarily, transitions are written with the upper state, indicated by primes (J'), first and the lower state, indicated by double primes (J''), second with an arrow to indicate absorption $J' \leftarrow J''$ or emission $J' \rightarrow J''$. The first transition $J = 1 \leftarrow 0$ occurs at $2B$, and the other transitions are spaced by multiples of $2B$ from one another (Figure 6.13). This is illustrated by the pure rotational transitions of hot HF (Figure 6.14) and the far-infrared absorption spectrum of CO (Figure 6.15).

The intensity of a rotational transition is determined both by the dipole moment and the population difference between the two levels (Chapter 1). The rotational populations can be calculated from statistical thermodynamics. If the total concentration of molecules is N, then the concentration of molecules N_J with the rotational quantum number J is

$$N_J = N(2J+1)\frac{e^{-BJ(J+1)/kT}}{q_r} = NP_J, \tag{6.41}$$

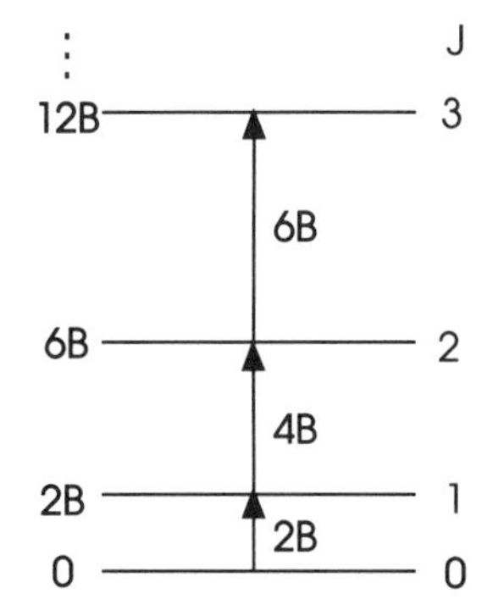

Figure 6.13. Transitions of a linear molecule.

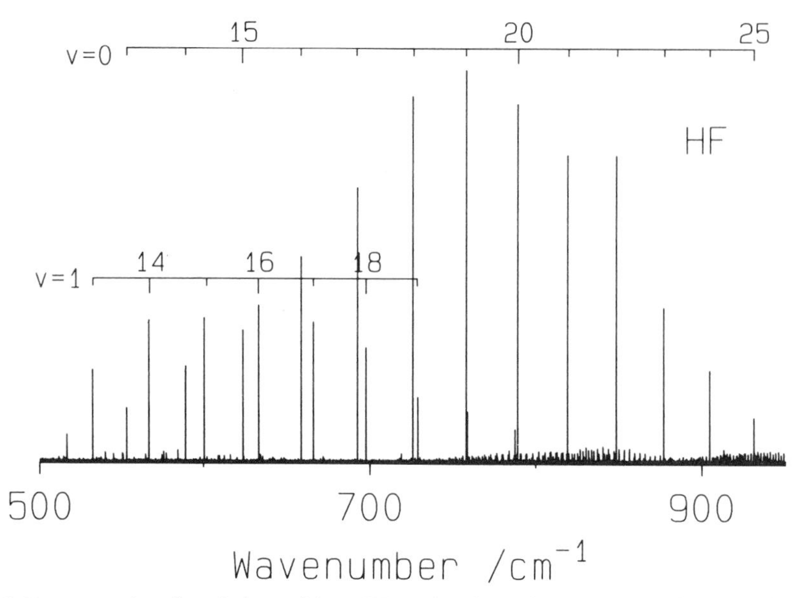

Figure 6.14. Pure rotational emission of hot HF molecules. The spectrum also contains weaker lines due to H_2O and LiF molecules.

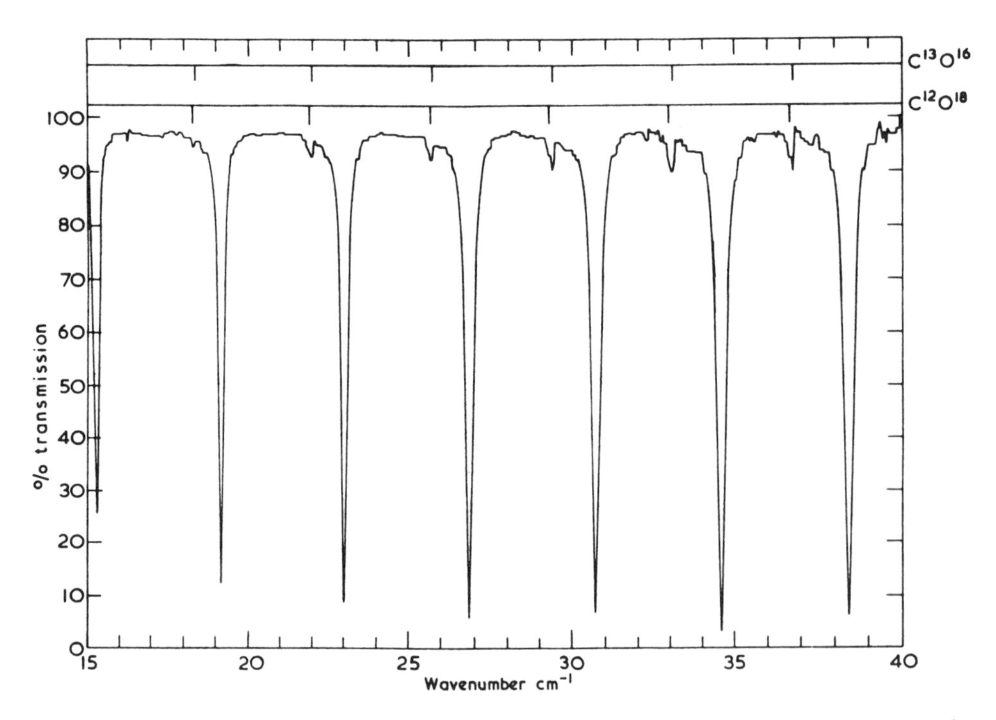

Figure 6.15. Far-infrared absorption spectrum of CO showing $J = 4 \leftarrow 3$ at 15.38 cm^{-1} to $J = 10 \leftarrow 9$ at 38.41 cm^{-1}.

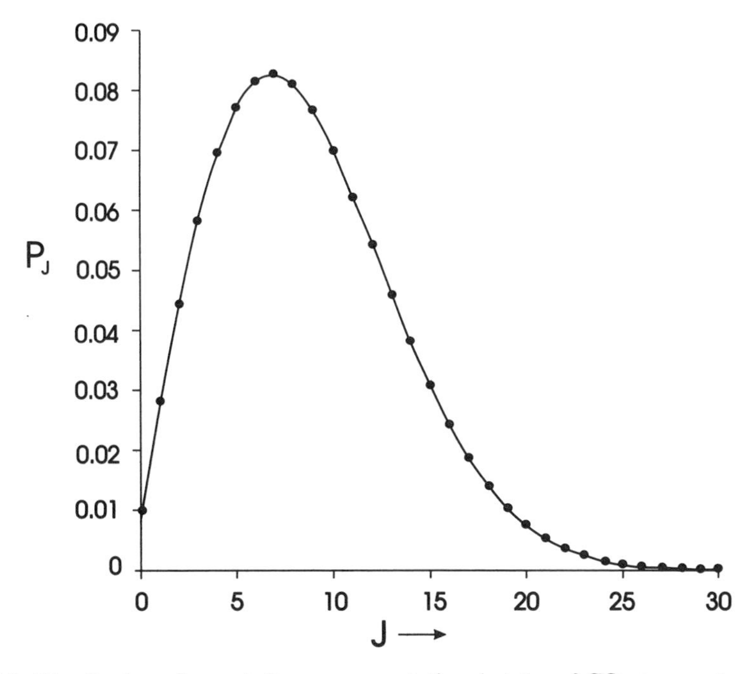

Figure 6.16. Distribution of population among rotational states of CO at room temperature.

where q_r is the rotational partition function

$$q_r = \sum_J (2J+1)e^{-BJ(J+1)/kT} \approx \frac{kT}{\sigma B}, \tag{6.42}$$

where σ, the symmetry number, is equal to 2 or 1 for a symmetric or nonsymmetric molecule, respectively. The expression (6.41) assumes that only the ground vibrational and electronic states are populated at temperature T. This distribution is plotted in Figure 6.16 for CO ($B = 1.9225\,\text{cm}^{-1}$) at room temperature (298 K). The rotational state with maximum population $J_{\max}$ is determined by setting $dN_J/dJ = 0$ and solving for J. This gives

$$J_{\max} = \left(\frac{kT}{2B}\right)^{1/2} - \frac{1}{2}. \tag{6.43}$$

For CO at room temperature the state with maximum population has a J value of 7.

Centrifugal Distortion

A molecule is not strictly a rigid rotor. As a molecule rotates, the atoms experience a centrifugal force in the rotating molecular reference frame that distorts the

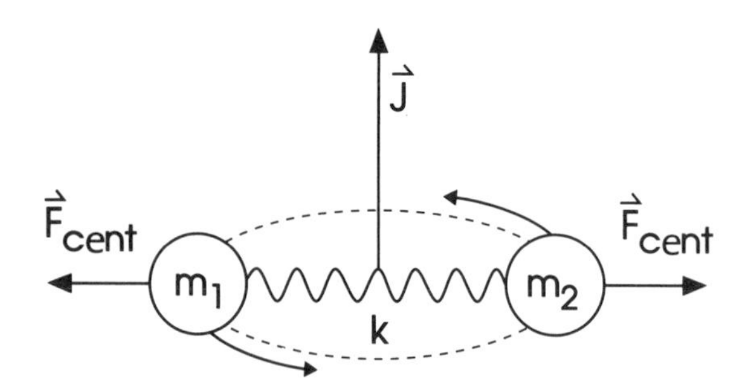

Figure 6.17. Nonrigid diatomic rotor with m_1 and m_2 connected by a spring.

internuclear positions (Figure 6.17). For a diatomic molecule one can obtain an expression for the stretching of the internuclear separation r by allowing the bond to stretch from r_e to r_c under the action of the centrifugal force

$$F_c = \frac{\mu v^2}{r} = \mu\omega^2 r = \frac{J^2}{\mu r^3}. \tag{6.44}$$

The centrifugal force is balanced by the Hooke's law restoring force

$$F_r = k(r_e - r_c) \tag{6.45}$$

in the bond, and after some algebra (Problem 14) one finds that

$$F(J) = BJ(J+1) - D[J(J+1)]^2 = [B - DJ(J+1)]J(J+1). \tag{6.46}$$

The constant D is called the *centrifugal distortion constant* and, in fact, there are additional higher order distortion corrections that lead to the rotational energy expression

$$F(J) = BJ(J+1) - D[J(J+1)]^2 + H[J(J+1)]^3 + L[J(J+1)]^4 + M[J(J+1)]^5 + \cdots. \tag{6.47}$$

A useful expression for D is given by the Kratzer relationship (Problem 14)

$$D = \frac{4B_e^3}{\omega_e^2}, \tag{6.48}$$

in which ω_e is the equilibrium vibration frequency. The negative sign in front of D in equations (6.46) and (6.47) has been introduced in order to make D a positive number. Equation (6.47) applies to both diatomic and linear polyatomic molecules.

Centrifugal distortion increases the internuclear separation r, which decreases the

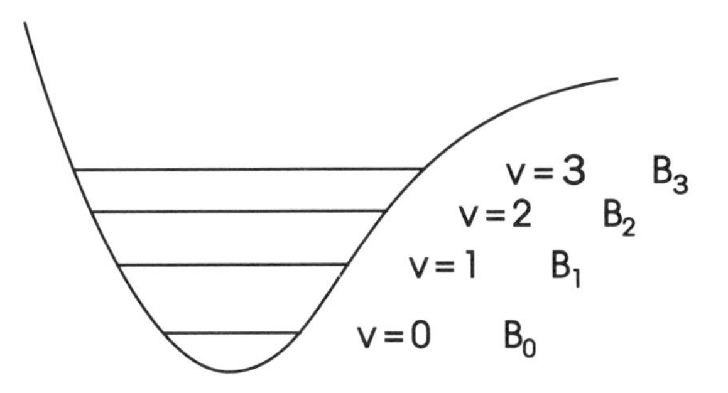

Figure 6.18. Each vibrational level of a diatomic molecule has its own rotational constant B_v.

effective $B_{\text{eff}} = B - DJ(J+1)$ of a pure rotational transition, so that the transition frequency can be written as

$$\nu_{J+1 \leftarrow J} = F(J+1) - F(J) = 2B(J+1) - 4D(J+1)^3 = 2[B - 2D(J+1)^2](J+1). \quad (6.49)$$

For example, the values of B and D for CO in the vibrational ground state are[2] $B(v=0) = 57.6359683$ GHz, $D(v=0) = 0.1835055$ MHz, $H(v=0) = 1.725 \times 10^{-7}$ MHz, and $L(v=0) = 3.1 \times 10^{-13}$ MHz.

The rotational constant also depends on the vibrational and electronic state (Figure 6.18). For a diatomic molecule, as v increases the molecule spends more of its time at large r where the potential energy curve is flatter (Figure 7.5). Thus, the average internuclear separation $\langle r \rangle$ increases with v while

$$B_v = \frac{h^2}{8\pi^2\mu}\left\langle \frac{1}{r^2} \right\rangle \quad (6.50)$$

decreases. This vibrational dependence is customarily parameterized[3] by the equations

$$B_v = B_e - \alpha_e(v + \tfrac{1}{2}) + \gamma_e(v + \tfrac{1}{2})^2 + \cdots \quad (6.51)$$

and

$$D_v = D_e + \beta_e(v + \tfrac{1}{2}) + \cdots. \quad (6.52)$$

The rotational energy level expression also becomes dependent on v, namely

$$F_v(J) = B_v J(J+1) - D_v[J(J+1)]^2 + \cdots. \quad (6.53)$$

At room temperature the pure rotational spectrum of a small molecule will not

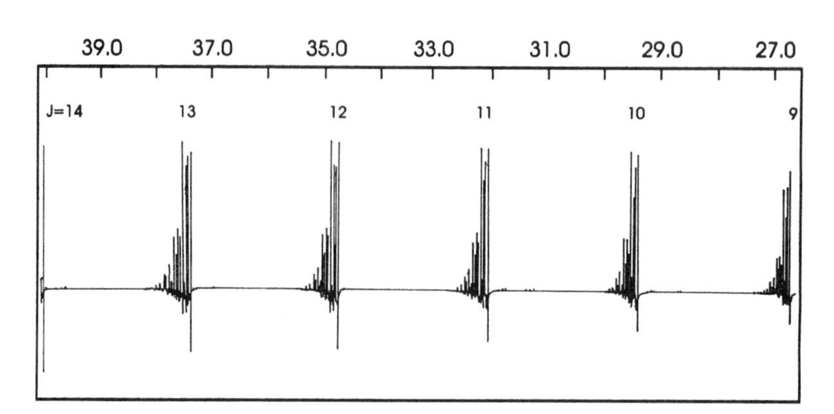

Figure 6.19. The microwave spectrum of H—C≡C—C≡C—C≡N showing vibrational satellites.

usually display the effects of vibration because the excited vibrational energy levels have little population. For a more floppy molecule with low-frequency vibrations, "vibrational satellites" appear in the pure rotational spectrum (Figure 6.19) since each vibrational level has its own set of rotational constants. Including the effects of centrifugal distortion and the vibrational dependence of the rotational constants results in transition frequencies given by

$$\begin{aligned}\nu_{J+1\leftarrow J} &= B_v(J+1)(J+2) - D_v[(J+1)(J+2)]^2 \\ &\quad - B_vJ(J+1) + D_v[J(J+1)]^2 \\ &= 2B_v(J+1) - 4D_v(J+1)^3.\end{aligned} \tag{6.54}$$

Vibrational Angular Momentum

The total angular momentum **J** in a linear molecule is given by

$$\mathbf{J} = \mathbf{R} + \mathbf{L} + \mathbf{S} + \mathbf{l} \tag{6.55}$$

where **R**, **L**, **S**, and **l** are the rotational, electronic orbital, spin, and vibrational angular momenta, respectively. In spectroscopy it is customary to associate different standard symbols[4] with different types of angular momenta. Most common molecules ($O_2(X\,^3\Sigma_g^-)$ and $NO(X\,^2\Pi)$ are exceptions) have no unpaired spins or electronic orbital angular momenta ($\mathbf{L} = \mathbf{S} = 0$) and only **l** needs to be considered in addition to **R**. In recent years the sensitivity of pure rotational spectroscopy has improved, particularly with the development of submillimeter wave technology, so that

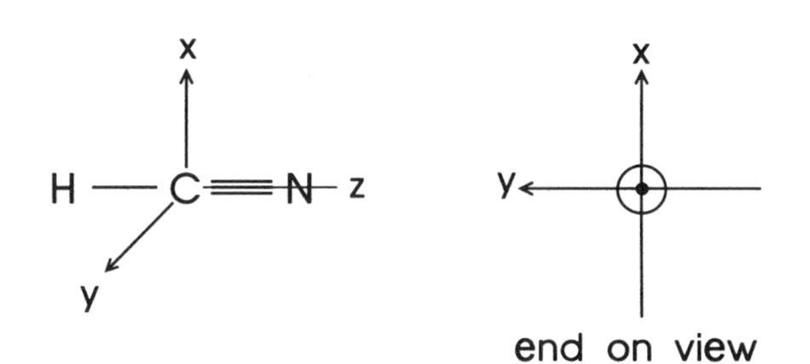

Figure 6.20. The doubly degenerate bending mode of a linear molecule.

microwave spectroscopy of free radicals and ions,[5] often with **L** and **S** not equal to zero, is now an important area of research. However, it is beyond the scope of this book.

Vibrationally excited linear polyatomic molecules can display the effects of vibrational angular momentum. A molecule like H—C≡N or H—C≡C—Cl has doubly degenerate bending modes, since the molecule could bend in plane or out of plane (Figure 6.20). For example, HCN has $3N - 5 = 4$ vibrational modes with

$\nu_1(\sigma^+)$	$3311\ \text{cm}^{-1}$	for the H—C stretching mode
$\nu_2(\pi)$	$713\ \text{cm}^{-1}$	for the H—C≡N bending mode
$\nu_3(\sigma^+)$	$2097\ \text{cm}^{-1}$	for the C≡N stretching mode

with ν_2 being a doubly degenerate bending mode.

The degenerate bending mode ν_2 is modeled by a two-dimensional harmonic oscillator[6] with a Hamiltonian given by

$$\hat{H} = \frac{-\hbar^2}{2\mu}\left(\frac{\partial^2}{\partial x^2} + \frac{\partial^2}{\partial y^2}\right) + \frac{1}{2}k(x^2 + y^2), \tag{6.56}$$

in which μ and k are the effective mass and force constant, respectively. The x and y parts are separable, so the Schrödinger equation is solved by writing the wavefunctions as

$$\psi(x, y) = \psi_{\text{HO}}(x)\psi_{\text{HO}}(y) \tag{6.57}$$

and splitting the total energy into two parts as

$$\begin{aligned} E &= h\nu(v_x + \tfrac{1}{2}) + h\nu(v_y + \tfrac{1}{2}) \\ &= h\nu(v + 1), \qquad v = 0, 1, 2, \ldots, \end{aligned} \tag{6.58}$$

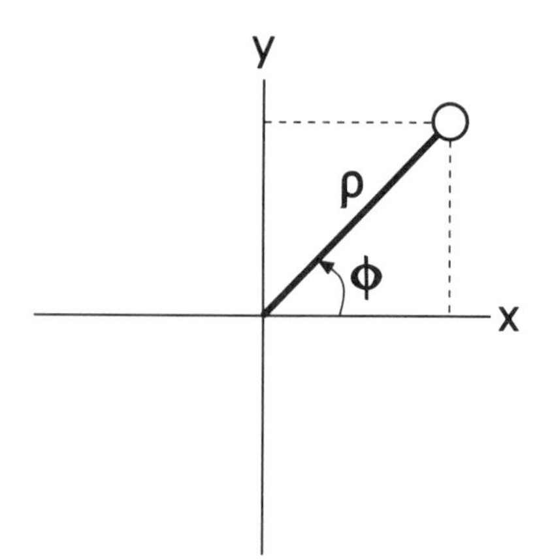

Figure 6.21. Plane polar coordinates.

where the level v has a degeneracy of $v+1$. In general for the d-dimensional harmonic oscillator

$$E = h\nu\left(v + \frac{d}{2}\right), \qquad v = 0, 1, 2, \ldots, \tag{6.59}$$

with $d = 1, 2, 3, \ldots$, depending on the number of degenerate oscillators, each contributing $h\nu/2$ of zero-point energy.

The 2-dimensional harmonic oscillator Hamiltonian[6] can be converted to plane polar coordinates in which $\rho = (x^2 + y^2)^{1/2}$ and $\phi = \tan^{-1}(y/x)$, see Figure 6.21. The Hamiltonian becomes

$$\hat{H} = \frac{-\hbar^2}{2\mu}\left[\frac{1}{\rho}\frac{\partial}{\partial\rho}\left(\rho\frac{\partial}{\partial\rho}\right) + \frac{1}{\rho^2}\frac{\partial^2}{\partial\phi^2}\right] + \frac{1}{2}k\rho^2. \tag{6.60}$$

In this coordinate system the problem is also separable and results in a wavefunction

$$\psi_{vl} = R_{vl}(\rho)e^{il\phi} \tag{6.61}$$

where l is a new quantum number associated with vibrational angular momentum of $\pm\,|l|\,\hbar$. The operator for vibrational angular momentum about z is

$$\hat{p}_z = -i\hbar\frac{\partial}{\partial\phi} \tag{6.62}$$

since

$$\hat{p}_z\psi_{vl} = l\hbar\psi_{vl}. \tag{6.63}$$

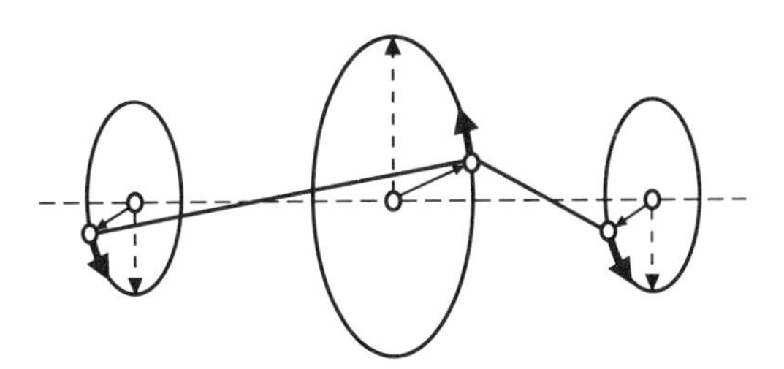

Figure 6.22. Classical picture of vibrational angular momentum.

The possible values of $|l|$ are $v, v-2, \ldots, 0$ or 1.

Following the usual custom in spectroscopy, a single positive value of $|l|$ is used although $\pm |l|$ are possible. The double degeneracy for each value of l is associated with clockwise or counterclockwise motion of the nuclei in a linear molecule (Figure 6.22). As before, the total degeneracy for the level v is $v + 1$. Classically, the two oscillators in the x and y directions can be phased such that the nuclei execute circular motion of small amplitude about the z-axis. In quantum mechanics this motion is quantized and only $\pm l\hbar$ units of angular momentum are possible. Sometimes Greek letters are used to designate vibrational angular momentum (in analogy to the use of Σ, Π, Δ, and so forth, to represent $\Lambda = 0, 1, 2, \ldots$ for the component of the orbital angular momentum about the internuclear axis of a diatomic molecule; see Chapter 9) and l is often written as a superscript, v_2^l (Figure 6.23).

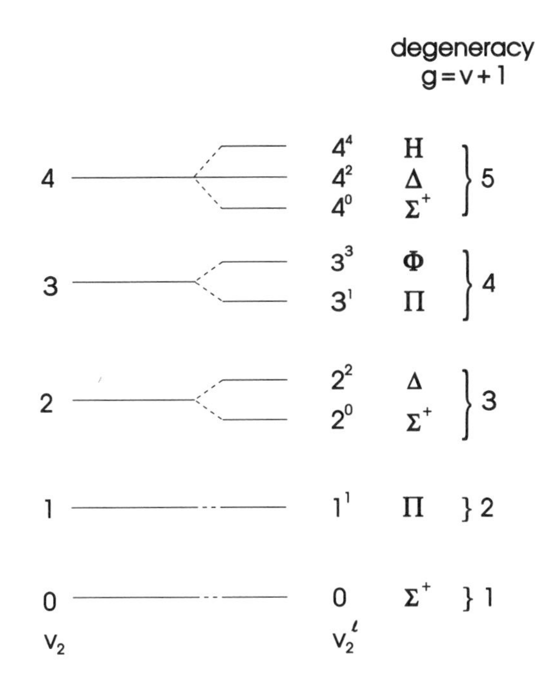

Figure 6.23. Vibrational energy-level pattern for the bending mode of a linear molecule.

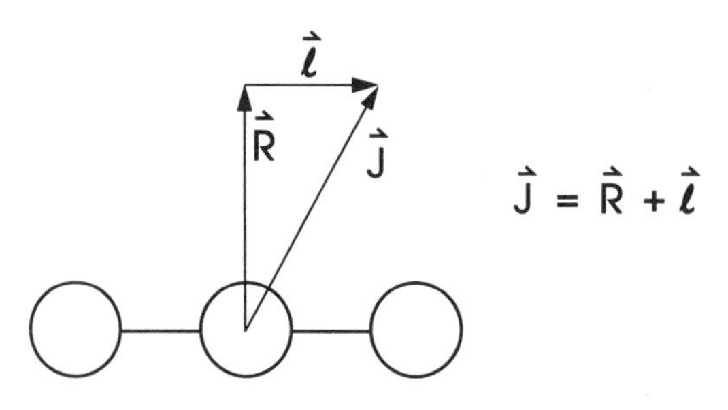

Figure 6.24. The rotational **R** and vibrational **l** angular momenta couple to give **J**.

Although the different l values for a given v are degenerate for the 2-dimensional harmonic oscillator, they become split if the oscillator is anharmonic. Since real molecules are always anharmonic oscillators, the different $|l|$ values are split by typically a few cm^{-1}. The twofold degeneracy for each l value $\pm |l|$ remains in the nonrotating molecule (Figure 6.23).

When only vibrational and rotational angular momenta (Figure 6.24) are present, we have

$$\mathbf{J} = \mathbf{R} + \mathbf{l}. \tag{6.64}$$

The possible values of the quantum number J are $|l|$, $|l| + 1 \ldots$, since a vector cannot be shorter than one of its components (Figure 6.25).

There is a different rotational constant for each vibrational level, which is customarily expressed as

$$B_v = B_e - \sum \alpha_i \left(v_i + \frac{d_i}{2} \right). \tag{6.65}$$

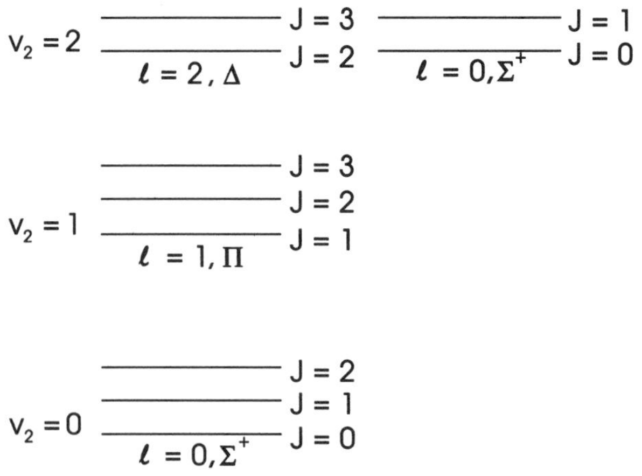

$v_2 = 2$: $l = 2, \Delta$ — J = 3, J = 2; $l = 0, \Sigma^+$ — J = 1, J = 0

$v_2 = 1$: $l = 1, \Pi$ — J = 3, J = 2, J = 1

$v_2 = 0$: $l = 0, \Sigma^+$ — J = 2, J = 1, J = 0

Figure 6.25. Rotational structure (not to scale) of the first few bending vibrational and rotational energy levels of a linear triatomic molecule.

For example, the vibrational dependence of the rotational constant for BeF_2 is[7]

$$B_{v_1v_2v_3} = 0.235356 - 0.000794(v_1 + \tfrac{1}{2}) + 0.001254(v_2 + 1) - 0.002446(v_3 + \tfrac{1}{2})\ \text{cm}^{-1}.$$

This B_e value gives an $r_e = 1.372973$ Å for the Be—F bond length while the B_{000} value of 0.234990 cm^{-1} gives an r_0 value of 1.374042 Å.

Symmetric Tops

The classical energy-level expression for a rigid symmetric top is

$$E = \frac{J_a^2}{2I_A} + \frac{J_b^2}{2I_B} + \frac{J_c^2}{2I_C} \tag{6.66}$$

$$= \frac{J_a^2}{2I_A} + \frac{1}{2I_B}(J_b^2 + J_c^2) \qquad \text{(prolate top)}$$

or

$$= \frac{1}{2I_B}(J_a^2 + J_b^2) + \frac{J_c^2}{2I_C} \qquad \text{(oblate top)}. \tag{6.67}$$

For simplicity the treatment will be limited to a prolate top, but the results also apply to an oblate top by interchanging the labels a and c. Since

$$J_a^2 + J_b^2 + J_c^2 = J^2 \tag{6.68}$$

or

$$J_b^2 + J_c^2 = J^2 - J_a^2 \tag{6.69}$$

then

$$E = \frac{1}{2I_b} J^2 + \left(\frac{1}{2I_a} - \frac{1}{2I_b}\right) J_a^2. \tag{6.70}$$

The corresponding quantum mechanical Hamiltonian to equation (6.70) is

$$\hat{H} = \frac{1}{2I_B} \hat{J}^2 + \left(\frac{1}{2I_A} - \frac{1}{2I_B}\right) \hat{J}_a^2. \tag{6.71}$$

The solution of the symmetric top Schrödinger equation requires a small digression into quantum mechanics.

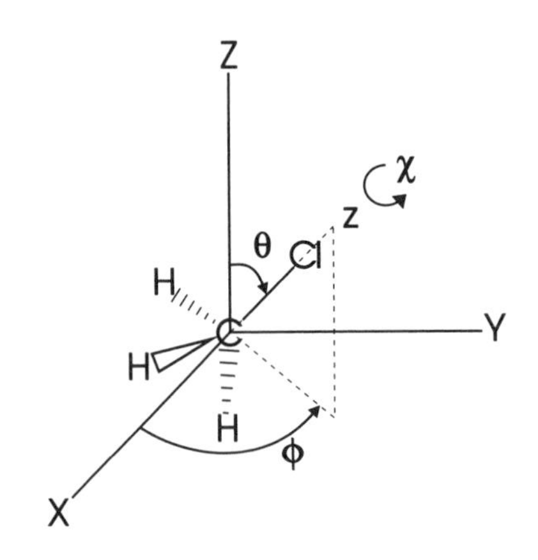

Figure 6.26. The laboratory and molecular coordinate systems for CH_3Cl.

Molecule and Space-Fixed Angular Momenta

The symmetric top molecule is described in two coordinate systems, the space-fixed laboratory system, X, Y, Z, and the molecular coordinate system, x, y, z (or a, b, c), both with origins at the center of mass (Figure 6.26). The orientation of the molecular system relative to the laboratory system is described by three Euler angles, θ, ϕ, and χ, defined in different ways by various authors. Our convention[8] is illustrated in Figure 6.27. The angles θ and ϕ correspond to the usual polar angles of the molecular z-axis in the X, Y, Z frame, while χ describes the internal orientation of the molecule along the molecular z-axis. In the example of CH_3Cl, χ is the angle that describes the rotation of the CH_3 group around the molecular z-axis (Figure 6.26).

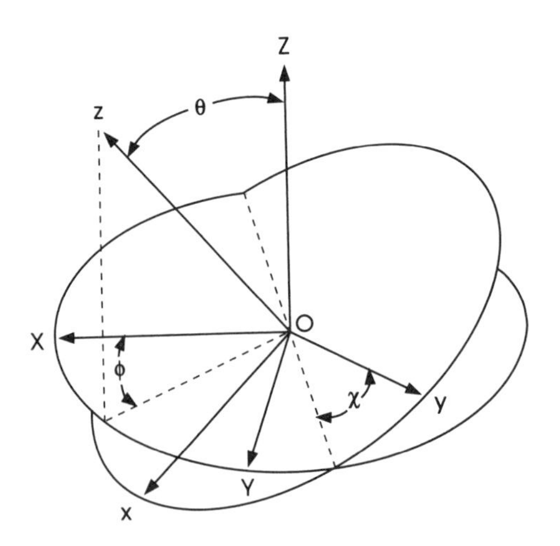

Figure 6.27. The Euler angles θ, ϕ, and χ that relate the space fixed coordinate system (X, Y, Z) to the molecular coordinate system (x, y, z).

The laboratory and molecular coordinate systems are related by the transformation matrix $\mathbf{S}$, that is,

$$\begin{pmatrix} x \\ y \\ z \end{pmatrix} = \mathbf{S}\begin{pmatrix} X \\ Y \\ Z \end{pmatrix} = \begin{pmatrix} \Phi_{xX} & \Phi_{xY} & \Phi_{xZ} \\ \Phi_{yX} & \Phi_{yY} & \Phi_{yZ} \\ \Phi_{zX} & \Phi_{zY} & \Phi_{zZ} \end{pmatrix}\begin{pmatrix} X \\ Y \\ Z \end{pmatrix}. \tag{6.72}$$

The elements of $\mathbf{S}$ are just the direction cosines of vector algebra (Figure 6.28) with

$$\begin{aligned} x &= \Phi_{xX}X + \Phi_{xY}Y + \Phi_{xZ}Z \\ &= \cos\alpha_1 X + \cos\beta_1 Y + \cos\gamma_1 Z \\ &= \hat{\mathbf{x}}\cdot\hat{\mathbf{X}}X + \hat{\mathbf{x}}\cdot\hat{\mathbf{Y}}Y + \hat{\mathbf{x}}\cdot\hat{\mathbf{Z}}Z \end{aligned} \tag{6.73}$$

and so forth, where $\hat{\mathbf{x}}, \hat{\mathbf{y}}, \hat{\mathbf{z}}$, and $\hat{\mathbf{X}}, \hat{\mathbf{Y}}, \hat{\mathbf{Z}}$ are sets of unit vectors for the molecular and laboratory coordinate systems, respectively.

The $\mathbf{S}$ matrix can be derived using the description of the Euler angles as rotations about axes (Figure 6.27):

1. Rotate X and Y by an angle ϕ about Z into X' and Y',
2. Rotate X' and Z by an angle θ about Y' into X'' and z,
3. Rotate X'' and Y' by an angle χ about z into x and y.

Thus,

$$\mathbf{S} = \begin{pmatrix} \cos\chi & \sin\chi & 0 \\ -\sin\chi & \cos\chi & 0 \\ 0 & 0 & 1 \end{pmatrix}\begin{pmatrix} \cos\theta & 0 & -\sin\theta \\ 0 & 1 & 0 \\ \sin\theta & 0 & \cos\theta \end{pmatrix}\begin{pmatrix} \cos\phi & \sin\phi & 0 \\ -\sin\phi & \cos\phi & 0 \\ 0 & 0 & 1 \end{pmatrix}$$

$$= \begin{pmatrix} \cos\theta\cos\phi\cos\chi - \sin\phi\sin\chi & \cos\theta\sin\phi\cos\chi + \cos\phi\sin\chi & -\sin\theta\cos\chi \\ -\cos\theta\cos\phi\sin\chi - \sin\phi\cos\chi & -\cos\theta\sin\phi\sin\chi + \cos\phi\cos\chi & \sin\theta\sin\chi \\ \sin\theta\cos\phi & \sin\theta\sin\phi & \cos\theta \end{pmatrix}, \tag{6.74}$$

where $\mathbf{S}$ is an orthogonal matrix ($\mathbf{S}^{-1} = \mathbf{S}^t$), since $\mathbf{S}$ represents a rotational transformation of the coordinate system. The angular momenta can be measured in the laboratory frame ($\hat{J}_X, \hat{J}_Y, \hat{J}_Z$) or in the molecular frame ($\hat{J}_x, \hat{J}_y, \hat{J}_z$) with

$$\hat{J}_x = \Phi_{xX}\hat{J}_X + \Phi_{xY}\hat{J}_Y + \Phi_{xZ}\hat{J}_Z, \tag{6.75}$$

$$\hat{J}_y = \Phi_{yX}\hat{J}_X + \Phi_{yY}\hat{J}_Y + \Phi_{yZ}\hat{J}_Z, \tag{6.76}$$

$$\hat{J}_z = \Phi_{zX}\hat{J}_X + \Phi_{zY}\hat{J}_Y + \Phi_{zZ}\hat{J}_Z. \tag{6.77}$$

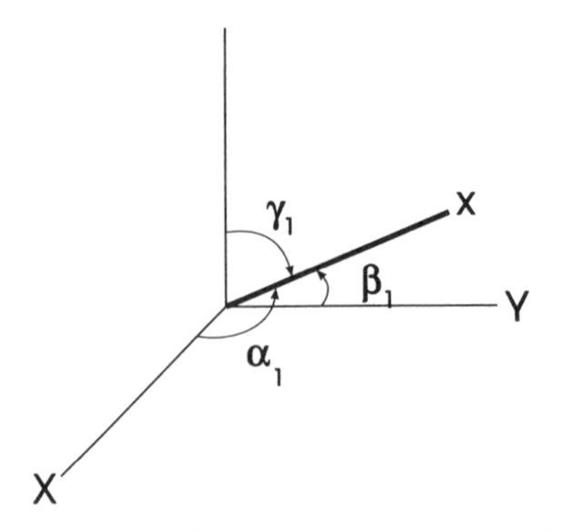

Figure 6.28. The direction cosine angles for the molecular x-axis in the laboratory frame.

Using the matrix elements of **S** and the operator expressions for $\hat{J}_X, \hat{J}_Y$, and $\hat{J}_Z$ one obtains[8] a set of expressions for $\hat{J}_x$, $\hat{J}_y$, and $\hat{J}_z$, namely

$$\hat{J}_x = -i\hbar\left[\frac{-\cos\chi}{\sin\theta}\frac{\partial}{\partial\phi} + \frac{\cos\chi\cos\theta}{\sin\theta}\frac{\partial}{\partial\chi} + \sin\chi\frac{\partial}{\partial\theta}\right] \tag{6.78}$$

$$\hat{J}_y = -i\hbar\left[\frac{\sin\chi}{\sin\theta}\frac{\partial}{\partial\phi} - \frac{\sin\chi\cos\theta}{\sin\theta}\frac{\partial}{\partial\chi} + \cos\chi\frac{\partial}{\partial\theta}\right] \tag{6.79}$$

$$\hat{J}_z = -i\hbar\frac{\partial}{\partial\chi}. \tag{6.80}$$

The corresponding expressions in the laboratory frame[8] are

$$J_X = -i\hbar\left[\frac{-\cos\phi\cos\theta}{\sin\theta}\frac{\partial}{\partial\phi} + \frac{\cos\phi}{\sin\theta}\frac{\partial}{\partial\chi} - \sin\phi\frac{\partial}{\partial\theta}\right] \tag{6.81}$$

$$J_Y = -i\hbar\left[\frac{-\sin\phi\cos\theta}{\sin\theta}\frac{\partial}{\partial\phi} + \frac{\sin\phi}{\sin\theta}\frac{\partial}{\partial\chi} + \cos\phi\frac{\partial}{\partial\theta}\right] \tag{6.82}$$

$$J_Z = -i\hbar\frac{\partial}{\partial\phi}. \tag{6.83}$$

Checking the commutation relationships gives the surprising result

$$[\hat{J}_x, \hat{J}_y] = -i\hbar\hat{J}_z, \tag{6.84}$$

in contrast to the usual commutation relation

$$[\hat{J}_X, \hat{J}_Y] = i\hbar J_Z. \tag{6.85}$$

All of the commutation relationships have a minus sign in the molecular frame when compared to the corresponding equation in the laboratory frame. The "anomalous" commutation relationships in the molecular frame, equation (6.84), are due entirely to the direction cosine terms. Note that the molecular $\hat{J}_z$ operator commutes with $\hat{J}^2$ and that the space-fixed and molecular frame operators commute with each other, that is,

$$[J^2, \hat{J}_z] = [J_x^2 + \hat{J}_y^2 + \hat{J}_z^2, \hat{J}_z] = 0 \tag{6.86}$$

and

$$[\hat{J}_\alpha, \hat{J}_i] = 0; \qquad \alpha = X, Y, Z \quad i = x, y, z. \tag{6.87}$$

Consider now the rigid rotor symmetric top Hamiltonian

$$\hat{H} = \left(\frac{1}{2I_B}\right)\hat{J}^2 + \left(\frac{1}{2I_A} - \frac{1}{2I_B}\right)\hat{J}_z^2 \tag{6.88}$$

in the molecular frame. Clearly $J^2, \hat{J}_z$, and $\hat{J}_Z$ all commute with $\hat{H}$ so that a set of simultaneous eigenfunctions can be found, namely

$$\hat{H}\,|JKM\rangle = E\,|JKM\rangle, \tag{6.89}$$

$$\hat{J}^2\,|JKM\rangle = J(J+1)\hbar^2\,|JKM\rangle, \tag{6.90}$$

$$\hat{J}_Z\,|JKM\rangle = M_J\hbar\,|JKM\rangle, \tag{6.91}$$

$$\hat{J}_z\,|JKM\rangle = K\hbar\,|JKM\rangle, \tag{6.92}$$

in which $K\hbar$ is defined as the projection of $\hat{J}$ along the molecular z-axis. Furthermore, since

$$\hat{J}_Z = -i\hbar\frac{\partial}{\partial\phi},$$

and

$$\hat{J}_z = -i\hbar\frac{\partial}{\partial\chi},$$

the symmetric top wavefunctions $|JKM\rangle$ must have the form[9]

$$|JKM\rangle = \left[\frac{2J+1}{8\pi^2}\right]^{1/2} e^{iM\phi}\, d_{MK}^{(J)}(\theta)e^{iK\chi}. \tag{6.93}$$

The $d_{MK}^{(J)}$ functions are hypergeometric functions of $\sin^2(\theta/2)$ and are also related to the rotation matrices of angular momentum theory. The symmetric top functions are rarely listed since the explicit functions are not needed for calculations.

The anomalous commutation relationships in the molecular frame mean that $\hat{J}^+ = \hat{J}_x + i\hat{J}_y$ is a lowering operator and $\hat{J}^- = \hat{J}_x - i\hat{J}_y$ is a raising operator. (Note that $\hat{J}^+$ and $\hat{J}^-$ are in the molecular frame, but $\hat{J}_-$ and $\hat{J}_+$ are in the laboratory frame.) The effects of the raising and lowering operators on the symmetric top eigenfunctions are given by the equations

$$\hat{J}_+ |JKM\rangle = \hbar[J(J+1) - M(M+1)]^{1/2} |JKM+1\rangle, \quad (6.94)$$

$$\hat{J}_- |JKM\rangle = \hbar[J(J+1) - M(M-1)]^{1/2} |JKM-1\rangle, \quad (6.95)$$

$$\hat{J}^+ |JKM\rangle = \hbar[J(J+1) - K(K-1)]^{1/2} |JK-1M\rangle, \quad (6.96)$$

$$\hat{J}^- |JKM\rangle = \hbar[J(J+1) - K(K+1)]^{1/2} |JK+1M\rangle. \quad (6.97)$$

Returning to the symmetric top energy Hamiltonian, equation (6.88), one can solve the Schrödinger equation using the symmetric top wavefunctions

$$\begin{aligned}\hat{H}\psi &= \left[\frac{\hat{J}^2}{2I_B} + \left(\frac{1}{2I_A} - \frac{1}{2I_B}\right)\hat{J}_a^2\right] |JKM\rangle \\ &= \left[\frac{\hbar^2}{2I_B} J(J+1) + \left(\frac{\hbar^2}{2I_A} - \frac{\hbar^2}{2I_B}\right)K^2\right] |JKM\rangle \qquad \text{(prolate top)}\end{aligned} \quad (6.98a)$$

$$\begin{aligned}\hat{H}\psi &= \left(\frac{\hat{J}^2}{2I_B} + \left(\frac{1}{2I_C} - \frac{1}{2I_B}\right)\hat{J}_c^2\right) |JKM\rangle \\ &= \left[\frac{\hbar^2}{2I_B} J(J+1) + \left(\frac{\hbar^2}{2I_C} - \frac{\hbar^2}{2I_B}\right)K^2\right] |JKM\rangle \qquad \text{(oblate top)}\end{aligned} \quad (6.98b)$$

so

$$E_{JK} = BJ(J+1) + (A-B)K^2 \qquad \text{(prolate top)} \quad (6.99)$$

or

$$E_{JK} = BJ(J+1) + (C-B)K^2 \qquad \text{(oblate top).} \quad (6.100)$$

The rotational constants in energy units (joules) are defined by

$$A = \frac{h^2}{8\pi^2 I_A} \quad (6.101a)$$

$$B = \frac{h^2}{8\pi^2 I_B} \quad (6.101b)$$

$$C = \frac{h^2}{8\pi^2 I_C} \quad (6.101c)$$

and equations analogous to equations (6.29) to (6.32) are valid for nonlinear molecules. As expected, the energy-level expression is not a function of M so that the $2J+1$ M_J degeneracy remains. The degeneracy of states with the $2J+1$ possible values of K is partially lifted; however, states with $\pm K$ still have the same energy. The energy of the symmetric top is the same for clockwise and counterclockwise rotation around the molecular z-axis so that a twofold K degeneracy remains.

It is convenient to classify the energy levels of symmetric tops by the K quantum number. For a given K, $J \geq K$ and the energies have a simple linear molecule structure apart from a $(A - B)K^2$ (>0) or $(C - B)K^2$ (<0) offset (Figure 6.29). Note that levels of a given J value increase in energy with increasing K for a prolate top while they decrease in energy for an oblate top. For example CH_3I has $A = 5.11\ \text{cm}^{-1}$ and $B = 0.250\ \text{cm}^{-1}$ so[10]

$$E_{JK} = 0.250J(J+1) + 4.86K^2(\text{cm}^{-1}).$$

Rotational Spectra

The pure rotational spectra of symmetric tops are determined by the application of selection rules to the energy-level pattern in Figure 6.29. The derivation of selection rules for the symmetric top is somewhat involved since the transformation from the laboratory frame to the molecular frame needs to be considered. The selection rules are $\Delta J = \pm 1$, $\Delta M = 0, \pm 1$, and $\Delta K = 0$, and they result in very simple pure rotational spectra (Figures 6.30 and 6.31). The transitions are confined to lie within a K-stack so the transition frequencies are given by the diatomic expression, that is,

$$\nu_{J+1,K \leftarrow J,K} = 2B(J+1) \tag{6.102}$$

Centrifugal Distortion

As a molecule rotates, it also distorts under the effects of centrifugal forces resulting in an energy-level expression

$$F(J, K) = BJ(J+1) - D_J[J(J+1)]^2 + (A - B)K^2 - D_K K^4 - D_{JK}J(J+1)K^2, \tag{6.103}$$

where there are now three centrifugal distortion constants D_J, D_K, and D_{JK}. The transition frequencies are then given by

$$\begin{aligned} \nu_{J+1,K \leftarrow J,K} &= F(J+1, K) - F(J, K) \\ &= 2B(J+1) - 4D_J(J+1)^3 - 2D_{JK}(J+1)K^2. \end{aligned} \tag{6.104}$$

The constant D_{JK} splits out the transitions with different K for a given $J+1 \leftarrow J$ transition, as shown in Figure 6.32.

Asymmetric Tops

For an asymmetric top, $I_A \neq I_B \neq I_C$, the classical energy for a rigid rotor is given by

$$E = \frac{J_a^2}{2I_A} + \frac{J_b^2}{2I_B} + \frac{J_c^2}{2I_C}. \tag{6.105}$$

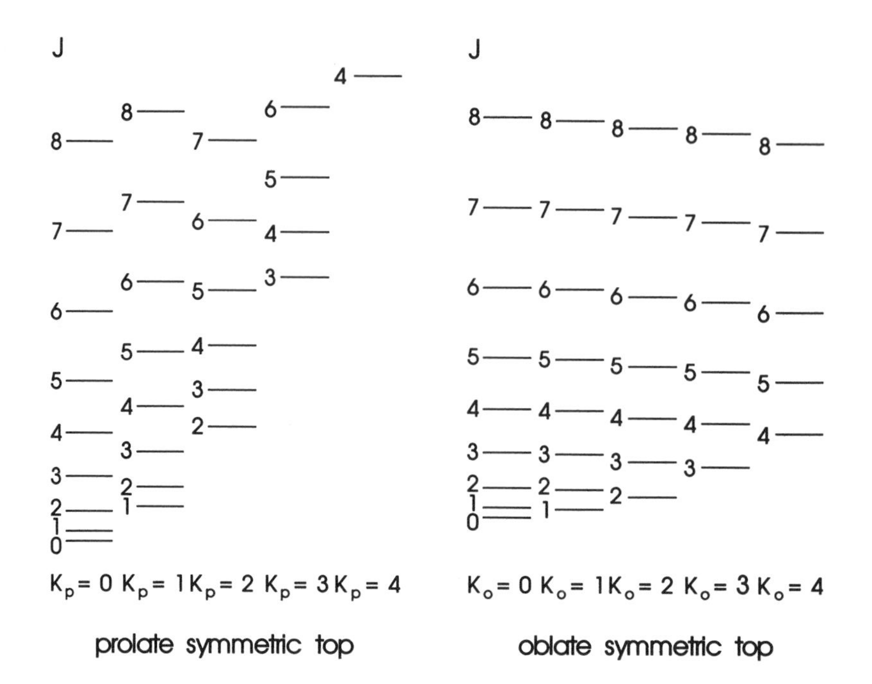

Figure 6.29. Energy levels of a prolate and an oblate symmetric top.

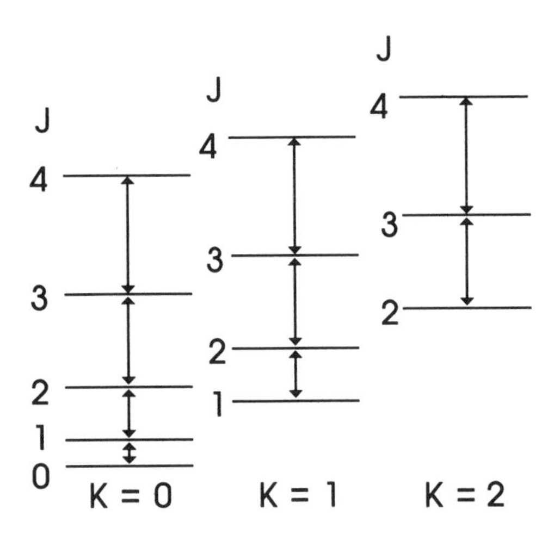

Figure 6.30. The allowed electric dipole transitions of a prolate symmetric top.

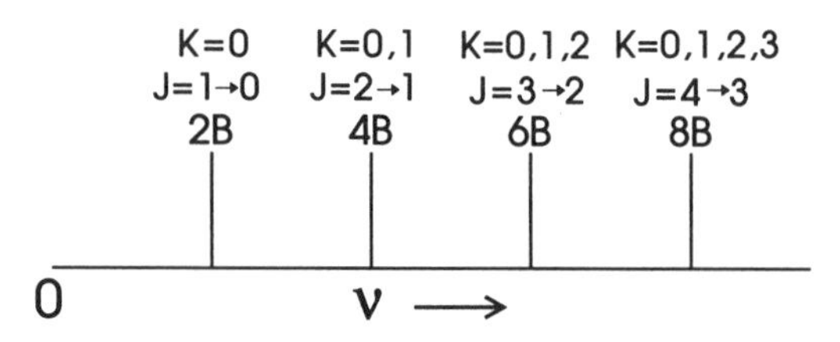

Figure 6.31. Pure rotational spectrum of a symmetric top.

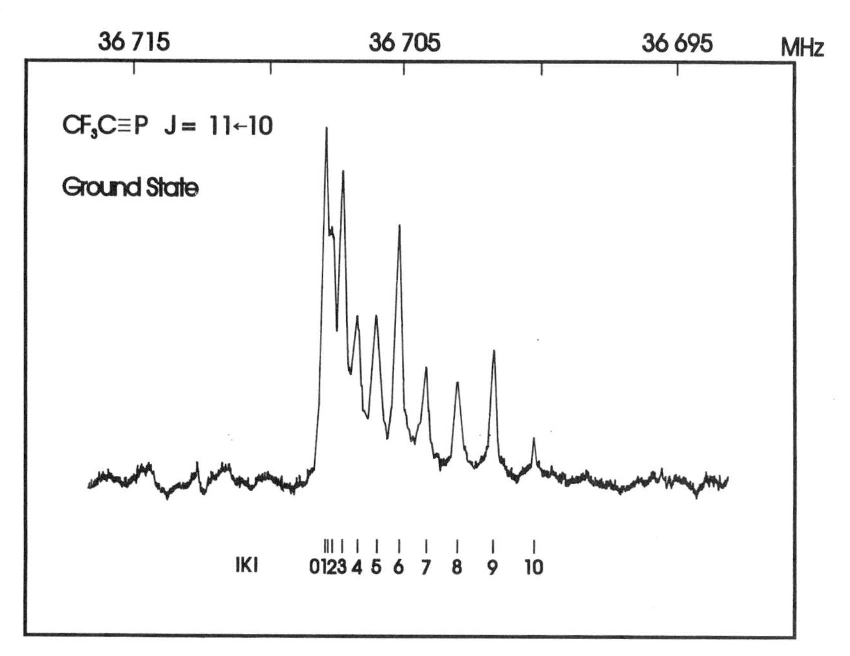

Figure 6.32. The 11 ← 10 transition of $CF_3C{\equiv}P$ showing K-structure. The intensity pattern is affected by nuclear spin statistics.

This results in the rigid asymmetric rotor Hamiltonian

$$\hat{H} = \frac{\hat{J}_a^2}{2I_A} + \frac{\hat{J}_b^2}{2I_B} + \frac{\hat{J}_c^2}{2I_C}. \tag{6.106}$$

The Schrödinger equation for the asymmetric top has no general analytical solutions and therefore must be solved numerically with the help of a computer. For certain special values of J, however, analytical solutions are available for the rigid rotor.

The asymmetric top Hamiltonian can be solved using a symmetric top basis set, changing the form of the terms in the Hamiltonian for convenience. Let

$$A = \frac{\hbar^2}{2I_A}, \tag{6.107a}$$

$$B = \frac{\hbar^2}{2I_B}, \tag{6.107b}$$

and

$$C = \frac{\hbar^2}{2I_C} \tag{6.107c}$$

so that

$$\hbar^2\hat{H} = A\hat{J}_a^2 + B\hat{J}_b^2 + C\hat{J}_c^2$$
$$= \left(\frac{A+B}{2}\right)(\hat{J}_a^2 + \hat{J}_b^2) + C\hat{J}_c^2 + \frac{(A-B)}{2}(\hat{J}_a^2 - \hat{J}_b^2)$$
$$= \left(\frac{A+B}{2}\right)\hat{J}^2 + \left(C - \frac{A+B}{2}\right)\hat{J}_c^2 + \left(\frac{A-B}{4}\right)[(\hat{J}^+)^2 + (\hat{J}^-)^2]. \quad (6.108)$$

The following symmetric top matrix elements[11] are useful

$$\langle JK|\, \hat{J}^2\, |JK\rangle = \hbar^2 J(J+1), \quad (6.109)$$
$$\langle JK|\, J_c^2\, |JK\rangle = \hbar^2 K^2, \quad (6.110)$$
$$\langle JK+2|\, (\hat{J}^-)^2\, |JK\rangle = \hbar^2[(J-K)(J+K+1)(J-K-1)(J+K+2)]^{1/2}, \quad (6.111)$$
$$\langle JK-2|\, (J^+)^2\, |JK\rangle = \hbar^2[(J+K)(J-K+1)(J+K-1)(J-K+2)]^{1/2}, \quad (6.112)$$

remembering that $\hat{J}^+$ is a lowering operator and $\hat{J}^-$ is a raising operator. With the symmetric top basis functions the asymmetric top Hamiltonian has matrix elements with $\Delta K = 0$ and $\Delta K = \pm 2$.

For example, the basis set for $J = 1$ has three members $|J = 1, K = 1\rangle$, $|1, 0\rangle$, and $|1, -1\rangle$, and the Hamiltonian is the 3×3 matrix

$$\mathbf{H} = \begin{array}{c} \\ \langle 1,1| \\ \langle 1,0| \\ \langle 1,-1| \end{array} \begin{array}{ccc} |1,1\rangle & |1,0\rangle & |1,-1\rangle \\ \end{array} \begin{pmatrix} C + \frac{A+B}{2} & 0 & \frac{A-B}{2} \\ 0 & A+B & 0 \\ \frac{A-B}{2} & 0 & C + \frac{A+B}{2} \end{pmatrix}. \quad (6.113)$$

The eigenvalues of this equation are easily determined by first exchanging the second and third rows and columns to give

$$\mathbf{H} = \begin{array}{c} \\ \langle 1,1| \\ \langle 1,-1| \\ \langle 1,0| \end{array} \begin{array}{ccc} |1,1\rangle & |1,-1\rangle & |1,0\rangle \\ \end{array} \begin{pmatrix} C + \frac{A+B}{2} & \frac{A-B}{2} & 0 \\ \frac{A-B}{2} & C + \frac{A+B}{2} & 0 \\ 0 & 0 & A+B \end{pmatrix} \quad (6.114)$$

and then solving the secular equation for the 2×2 block

$$\begin{vmatrix} C + \frac{A+B}{2} - \lambda & \frac{A-B}{2} \\ \frac{A-B}{2} & C + \frac{A+B}{2} - \lambda \end{vmatrix} = 0 \quad (6.115)$$

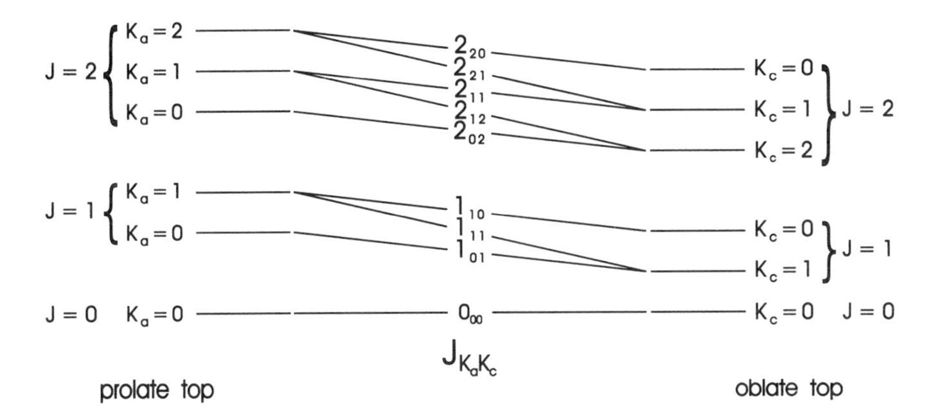

Figure 6.33. Prolate–oblate correlation diagram useful for labeling asymmetric top levels.

to give

$$\lambda = C + A$$

or

$$\lambda = C + B.$$

The three solutions for $J = 1$ are thus $A + B$, $A + C$, and $B + C$. Labeling of the energy levels is carried out by considering the correlation diagram that connects the energy levels of a prolate top with those of an oblate top in Figure 6.33 and the requirement that $I_A \leq I_B \leq I_C$.

The energy levels of prolate and oblate symmetric tops are

$$E_p = (A - B)K_a^2 + BJ(J + 1); \qquad E_o = (C - B)K_c^2 + BJ(J + 1).$$

Notice that by definition $(A - B) > 0$ and $(C - B) < 0$, so that for a given value of J the K_a levels increase in energy as K_a increases for a prolate top, while the K_c levels decrease in energy as K_c increases for an oblate top. As is typical in correlation diagrams, the energy levels are not to scale and the lines connecting the prolate levels with the oblate levels correspond to a hypothetical distortion of a molecule from a prolate to an oblate top. The levels are labeled by $J_{K_aK_c}$, where J is a good quantum number, but K_a and K_c are just labels for the asymmetric top. Clearly, K_a and K_c become good quantum numbers only in the prolate or oblate symmetric top limits. Note that the sum of K_a and K_c is J or $J + 1$.

It is sometimes convenient to define a label $\tau = K_a - K_c$, which runs from $\tau = +J$ to $\tau = -J$ in order of descending energy. The label τ emphasizes that for the asymmetric top there are $2J + 1$ distinct energy levels corresponding to the $2J + 1$ different possible values of τ or "K" for every J. The degree of asymmetry can be quantified by an

Table 6.2. Rigid Asymmetric Rotor Energy Levels for $J = 0, 1, 2, 3$

$J_{K_aK_c}$	J_τ	$F(J_\tau)$
0_{00}	0_0	0
1_{10}	1_1	$A + B$
1_{11}	1_0	$A + C$
1_{01}	1_{-1}	$B + C$
2_{20}	2_2	$2A + 2B + 2C + 2[(B - C)^2 + (A - C)(A - B)]^{1/2}$
2_{21}	2_1	$4A + B + C$
2_{11}	2_0	$A + 4B + C$
2_{12}	2_{-1}	$A + B + 4C$
2_{02}	2_{-2}	$2A + 2B + 2C - 2[(B - C)^2 + (A - C)(A - B)]^{1/2}$
3_{30}	3_3	$5A + 5B + 2C + 2[4(A - B)^2 + (A - C)(B - C)]^{1/2}$
3_{31}	3_2	$5A + 2B + 5C + 2[4(A - C)^2 - (A - B)(B - C)]^{1/2}$
3_{21}	3_1	$2A + 5B + 5C + 2[4(B - C)^2 + (A - B)(A - C)]^{1/2}$
3_{22}	3_0	$4A + 4B + 4C$
3_{12}	3_{-1}	$5A + 5B + 2C - 2[4(A - B)^2 + (A - C)(B - C)]^{1/2}$
3_{13}	3_{-2}	$5A + 2B + 5C - 2[4(A - C)^2 - (A - B)(B - C)]^{1/2}$
3_{03}	3_{-3}	$2A + 5B + 5C - 2[4(B - C)^2 + (A - B)(A - C)]^{1/2}$

asymmetry parameter ("Ray's asymmetry parameter") κ, which runs from -1 for a prolate top to $+1$ for an oblate top. The asymmetry parameter is defined as

$$\kappa = \frac{2B - A - C}{A - C}. \tag{6.116}$$

The asymmetric top labels $K_a = K_p$ and $K_c = K_o$ are sometimes called K_{-1} and K_{+1} because of the values of the asymmetry parameter for the prolate and oblate symmetric top limits. The notation $J_{K_pK_o}$ allows the three energy levels associated with $J = 1$ to be labeled $E(1_{10}) = A + B$, $E(1_{11}) = A + C$, and $E(1_{01}) = B + C$, since $I_A \leq I_B \leq I_C$ means $A \geq B \geq C$. The explicit energy level expressions for a rigid asymmetric rotor are provided in Table 6.2 for $J = 0, 1, 2$, and 3.

Selection Rules

The asymmetric top selection rules are more complicated than those of a linear molecule or a symmetric top. In general, an arbitrary molecule has three dipole moment components μ_a, μ_b, and μ_c along the principal axes (Figure 6.34). Each nonvanishing dipole moment component makes a certain set of transitions possible, and leads to a set of selection rules. The selection rules on J and M are $\Delta J = 0, \pm 1$ and $\Delta M = 0, \pm 1$.

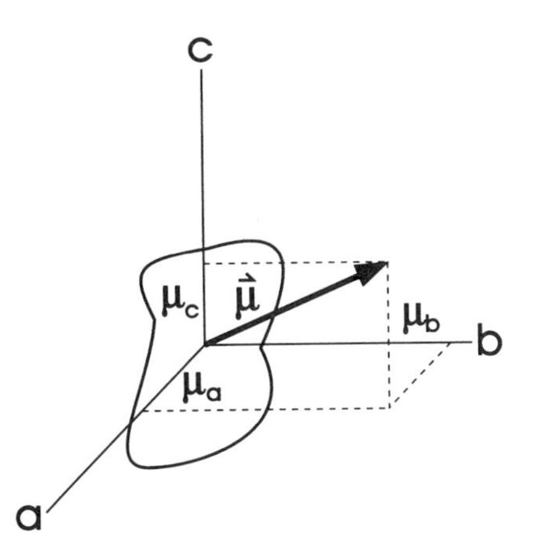

Figure 6.34. An arbitrary molecule has three components of the dipole moment in the principal axis system of the molecule.

a-Type Transitions

If $\mu_a \neq 0$ and $\mu_b = \mu_c = 0$, then a molecule such as H_2CO is said to obey a-type selection rules (Figure 6.35), $\Delta K_a = 0$ (± 2, ± 4) and $\Delta K_c = \pm 1$, (± 3, ± 5) in which the transitions in brackets are much weaker than the main ones. Thus, for example, the 1_{01}–0_{00} transition of formaldehyde is allowed, but the transitions 1_{10}–0_{00} and 1_{11}–0_{00} are forbidden because they require $\mu_c \neq 0$ and $\mu_b \neq 0$, respectively.

b-Type Transitions

If $\mu_b \neq 0$, then transitions with the selection rules

$$\Delta K_a = \pm 1(\pm 3, \ldots)$$

$$\Delta K_c = \pm 1(\pm 3, \ldots)$$

are allowed. The transitions with selection rules in brackets are much weaker.

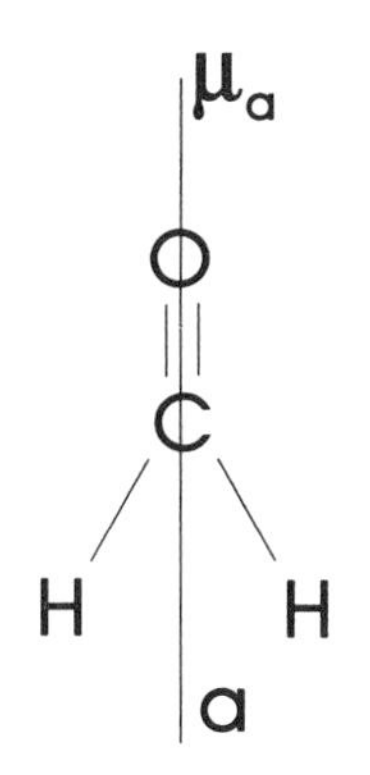

Figure 6.35. Formaldehyde has $\mu_a \neq 0$.

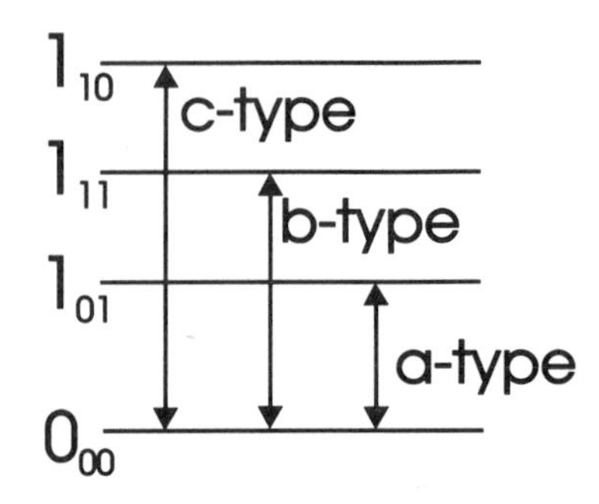

Figure 6.36. *a*-, *b*-, and *c*-type transitions.

c-Type Transitions

If $\mu_c \neq 0$, then transitions with the selection rules

$$\Delta K_a = \pm 1(\pm 3, \pm 5, \ldots)$$

$$\Delta K_c = 0(\pm 2, \pm 4, \ldots)$$

are allowed. The three possible types of transitions are illustrated in Figure 6.36.

Structure Determination

One of the main applications of molecular spectroscopy is the determination of molecular structures. The moments of inertia are related to bond lengths and bond angles. For a diatomic molecule the determination of r from B is simple, but each vibrational level has a different B_v value so that there are numerous corresponding r_v values (Figure 6.37). Each $B_v = (\hbar^2/2\mu)\langle v|\, 1/r^2\, |v\rangle$ corresponds to an average over a

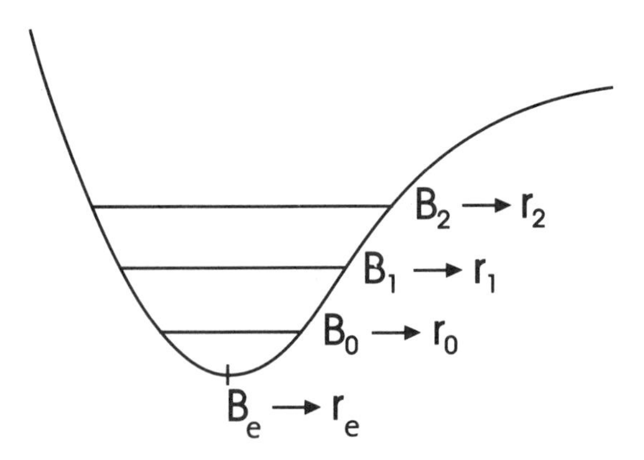

Figure 6.37. B_v and r_v values for a diatomic molecule.

different vibrational wavefunction. Perhaps the "best" value of r is r_e, which is computed by extrapolating B_v down to the bottom of the potential well, that is,

$$B_v = B_e - \alpha_e(v + \tfrac{1}{2}) + \gamma_e(v + \tfrac{1}{2})^2 + \cdots.$$

Given at least B_1 and B_0, B_e is easily computed as $B_0 - B_1 = \alpha_e$ and $B_e = B_0 + \alpha_e/2$, and hence r_e is determined. This is usually possible for diatomic molecules, so that the r_e values are customarily reported.

The determination of molecular structures in polyatomic molecules is much more difficult. The first problem is that there are now $3N - 6$ (or 5) α's for each A, B, and C:

$$A_v = A_e - \sum_{i=1}^{3N-6} \alpha_i^A\left(v_i + \frac{d_i}{2}\right)$$

$$B_v = B_e - \sum_{i=1}^{3N-6} \alpha_i^B\left(v_i + \frac{d_i}{2}\right)$$

$$C_v = C_e - \sum_{i=1}^{3N-6} \alpha_i^C\left(v_i + \frac{d_i}{2}\right).$$

The determination of so many α's is a tedious task even for a triatomic molecule, so that very few r_e structures are known. For most polyatomic molecules r_0 structures are computed from A_0, B_0, and C_0.

Another difficulty is that there are at most three moments of inertia (even for the largest molecules!), but usually more than three structural parameters need to be determined. Consider the case of formaldehyde (H_2CO) where the three structural parameters are r_{CO}, r_{CH}, and θ_{HCH}. At first sight it seems that there is no difficulty, since A, B, and C allow I_A, I_B, and I_C to be determined. Formaldehyde is a planar molecule, however, so that $I_C = I_A + I_B$ and three structural parameters need to be determined from only two independent moments of inertia.

In fact I_C is not exactly equal to $I_A + I_B$ because of several effects including vibrations. It is useful to define the inertial defect $\Delta = I_C - I_A - I_B$, which should be a small positive value (e.g., $\Delta = 0.05767$ amu Å^2 for formaldehyde[12]). Any deviation from the empirically expected value of Δ is taken as evidence of nonplanarity or fluxional behavior in the molecule.

The solution to the structural problem in formaldehyde lies in making use of data from isotopic molecules (Table 6.3). It is necessary to assume that all the isotopic

Table 6.3. Rotational Constants[12] for $H_2^{12}CO$ and $H_2^{13}CO$ (in MHz).

	$H_2^{12}CO$	$H_2^{13}CO$
A	281,970.572	281,993.257
B	38,836.0455	37,811.0886
C	34,002.2034	33,213.9789

variants of formaldehyde have the same r_{CO}, r_{CH}, and θ_{HCH} values. This is a good approximation because the potential surface is independent of nuclear mass. From the pure rotational spectrum of, for example, ^{13}C-substituted H_2CO a set of two additional independent moments of inertia can be derived. Now four independent moments of inertia (from the six total) are available to determine the three structural parameters by least squares fitting (Problem 8).

An r_0 structure is very useful, but some of the geometrical parameters, particularly C—H bond lengths, are not very reliable in that they lie far from equilibrium r_e values. A "better" structure can be derived by isotopic substitution using Kraitchman's equations[13] or other, more sophisticated, techniques. Kraitchman's equations allow the distance between the center of mass and the isotopically substituted atom to be calculated (see Gordy and Cook[13] or Domenicano and Hargittai). Structures determined by substituting a different isotope for every atom in a molecule (in turn, one at a time) are called r_s structures[13] (s stands for substitution). A complete r_s structure is very tedious to determine because a large number of isotopic forms of a molecule must be synthesized and rotationally analyzed. For example, for formaldehyde, HDCO, $H_2^{13}CO$, $H_2C^{18}O$, and H_2CO are needed. In the case of formaldehyde an equilibrium or r_e structure has been determined[12] with $r_e(CH) = 1.100$ Å, $r_e(CO) = 1.203$ Å, and $\theta_e(HCH) = 116°8'$.

Problems

1. Classify each of the following molecules as spherical, symmetric, or asymmetric top molecules:
   (a) CH_4
   (b) CH_3F
   (c) CH_3D
   (d) SF_6
   (e) SF_5Br
   (f) *trans*-SF_4Br_2
   (g) *cis*-SF_4Br_2
   (h) HCN
   (i) H_2S
   (j) C_5H_8, spiropentane.
2. (a) Show for a linear triatomic molecule made of atoms with masses m_1, m_2, and m_3 that

$$I = (1/M)(m_1 m_2 r_{12}^2 + m_1 m_3 r_{13}^2 + m_2 m_3 r_{23}^2)$$

   where M is the mass of the molecule.
   (b) The lowest frequency microwave transitions of $^1H^{12}C^{14}N$ and $^2H^{12}C^{14}N$ occur at 88,631 and 72,415 MHz, respectively. (These are for the ground vibrational state.) Calculate the bond distances in HCN.

3. A triatomic molecule has the formula A_2B. Its microwave spectrum shows strong lines at $15, 30, 45, \ldots$ MHz, and no other lines. Which of the following structures is (are) compatible with this spectrum?
   (a) linear AAB
   (b) linear ABA
   (c) bent AAB
   (d) bent ABA
4. For the $^{12}C^{32}S$ molecule the following millimeter wave pure rotational transitions have been observed (in MHz):

Transition	$v = 0$	$v = 1$	$v = 2$
$J = 1 - 0$	48,990.978	48,635.977	48,280.887
$J = 2 - 1$	97,980.950	97,270.980	96,560.800
$J = 3 - 2$	146,969.033	145,904.167	144,838.826
$J = 4 - 3$	195,954.226	194,534.321	193,113.957

   (a) For each vibrational level derive a set of rotational constants by fitting the data.
   (b) From the results of (a) derive an expression (by fitting) for the vibrational dependence of B.
   (c) From B_o calculate r_o; from B_e calculate r_e.
5. The F_2O molecule of C_{2v} symmetry has an O—F bond length of 1.405 Å and an FOF bond angle of 103.0°.
   (a) Calculate A, B, and C for F_2O.
   (b) Will the microwave spectrum of F_2O show a-, b-, or c-type transitions?
   (c) Predict the frequency of the $J = 1 - 0$ microwave transition.
6. For the BF_3 molecule of D_{3h} symmetry the B—F bond length is 1.310 Å. Calculate A, B, and C. What is the rotational energy-level expression?
7. The $J = 2 \leftarrow 1$ microwave absorption is observed near 42,723 MHz for $^{14}NF_3$ and 42,517 MHz for $^{15}NF_3$.
   (a) Derive the rotational constants for $^{14}NF_3$ and $^{15}NF_3$.
   (b) Determine the N—F bond length and the F—N—F bond angle.
8. The following is a complete list of observed transitions involving levels $J = 0, 1,$ and 2 for two isotopes of formaldehyde in their vibrational ground states:

$H_2^{12}C^{16}O$ (MHz)	$H_2^{13}C^{16}O$ (MHz)
71.14	—
4,829.66	4,593.09
14,488.65	13,778.86
72,837.97	71,024.80
140,839.54	137,449.97
145,602.98	141,983.75
150,498.36	146,635.69

(a) Assign these microwave transitions for both isotopomers. Assume that H_2CO belongs to the C_{2v} point group and estimate a molecular geometry using bond-length tables. Assign the spectrum by prediction of the expected rotational spectrum.

(b) What are A, B, and C for the two isotopic species? Since we have neglected centrifugal distortion, it will not be possible to fit all transitions exactly with only three rotational constants. Devise a procedure that gives a "best fit" to all lines.

(c) Explain why the inertial defect

$$\Delta = I_C - I_A - I_B$$

is a good test for planarity. Why does H_2CO appear not to be planar from the microwave spectrum?

(d) Obtain a best possible geometry for H_2CO using your A, B, C values for the two isotopes.

9. For CF_3I, $r_{CF} = 1.332$ Å, $r_{CI} = 2.134$ Å, $\theta_{FCF} = 108°$.
   (a) Calculate the rotational constants.
   (b) Predict the pattern (be quantitative) of the microwave spectrum.
10. The S_2O molecule is a bent triatomic molecule isovalent with ozone. The S—S bond is 1.884 Å long, the S—O bond is 1.465 Å long, and the SSO angle is 118.0°.
    (a) Locate the center of mass and set up the moment of inertia tensor. Pick the z-axis out of the plane and the x-axis parallel to the S—S bond.
    (b) Diagonalize the moment of inertia tensor to find I_A, I_B, and I_C.
    (c) Predict frequencies of the possible transitions from the 0_{00} rotational state.
11. The HCl molecule has a B_0 value of 10.4 cm^{-1}.
    (a) What are the J values of the levels with maximum population at 300 K and 2000 K?
    (b) Graph the populations of the J levels as a function of J for 300 K and 2000 K.
12. What is the moment of inertia of a cube of uniform density ρ and sides of length a?
13. For the HCl molecule with a B_0 value of 10.4 cm^{-1} and $J = 1$, treat the rotational motion classically.
    (a) What is the period of rotation?
    (b) What is the linear velocity of the H atom?
    (c) What is the angular momentum?
14. Derive equation (6.46) by showing that the total energy of the distorted molecule

$$E = \frac{J^2}{2\mu r_c^2} + \frac{1}{2}k(r_c - r_e)^2$$

gives rise to a centrifugal distortion term of approximately $E_{cd} = -DJ^4 = -\hbar^2 J^4/(2\mu^2 r_e^6 k)$ for a classical nonrigid rotor. Show that this term gives the Kratzer formula (6.48) for D.

15. Check the commutation relationships (6.84) and (6.85) using the differential form of the operators.
16. The rotational constants for the ground vibrational state of CH_3I are found to be $B = 0.25022\ \text{cm}^{-1}$, $A = 5.1739\ \text{cm}^{-1}$, $D_J = 2.09 \times 10^{-7}\ \text{cm}^{-1}$, $D_{JK} = 3.29 \times 10^{-6}\ \text{cm}^{-1}$, and $D_K = 87.6 \times 10^{-6}\ \text{cm}^{-1}$.
    (a) Predict the microwave spectrum of the $J = 1 \leftarrow 0$ and $J = 4 \leftarrow 3$ transitions. (Ignore the possibility of hyperfine structure.)
17. (a) Show that the rigid rotor Hamiltonian (6.108) is equivalent to equation (6.106).
    (b) Derive the matrix elements (6.111) and (6.112).
    (c) Construct the Hamiltonian matrix for $J = 2$ and derive the equations (for $J = 2$) in Table 6.2. It is helpful to transform the Hamiltonian to a new basis set defined by $|JK\pm\rangle = (|JK\rangle \pm |J - K\rangle)/\sqrt{2}$ via the Wang transformation matrix

$$\mathbf{U} = \frac{1}{\sqrt{2}} \begin{pmatrix} -1 & 0 & 0 & 0 & 1 \\ 0 & -1 & 0 & 1 & 0 \\ 0 & 0 & \sqrt{2} & 0 & 0 \\ 0 & 1 & 0 & 1 & 0 \\ 1 & 0 & 0 & 0 & 1 \end{pmatrix}.$$

18. Consider the HOD (partially deuterated water) molecule with bond length $r = 0.958$ Å and bond angle $\theta(\text{HOD}) = 104.5°$.
    (a) Find the moment of inertia tensor in amu Å^2 in any convenient coordinate system.
    (b) Find the principal axis moments of inertia, rotational constants (MHz), and transformation from the original axis system to the principal axis system.
19. For the symmetric top wavefunction $|JKM\rangle = Ce^{i\chi}e^{-2i\phi}\sin\theta(3\sin^2\theta + 2\cos\theta - 2)$, find J, K and M.
20. The application of an electric field to a molecular system partially lifts the M_J degeneracy. This Stark effect may be treated as a perturbation of the rotational energies. The perturbation Hamiltonian $H' = -\mu_z E_z$, where z is a lab frame coordinate and E_z is the electric field along the laboratory z-axis.
    (a) Show that there will be no first-order Stark effect for a linear molecule.
    (b) Develop a formula for the second-order Stark effect of a linear molecule.

References

1. McKellar, A. R. W., Johns, J. W. C., Majewski, W., and Rich, N. H. Can. J. Phys. **62,** 1673 (1984).
2. Varberg, T. D. and Evenson, K. M. Astrophys. J. **385,** 763 (1992).
3. Herzberg, G. *Spectra of Diatomic Molecules,* Van Nostrand Reinhold, New York, 1950.
4. Mulliken, R. S. J. Chem. Phys. **23,** 1997 (1955).

5. Hirota, E. *High Resolution Spectroscopy of Transient Molecules,* Springer-Verlag, Berlin, 1985.
6. Papousek, D. and Aliev, M.R. *Molecular Vibrational-Rotational Spectra,* Elsevier, Amsterdam, 1982, p. 57.
7. Frum, C. I., Engleman, R., and Bernath, P. F. J. Chem. Phys. **95,** 1435 (1991).
8. Zare, R. N. *Angular Momentum,* Wiley, New York, 1988, pp. 77–85.
9. ———. *Angular Momentum,* pp. 266–277.
10. Herzberg, G. *Electronic Spectra and Electronic Structure of Polyatomic Molecules,* Van Nostrand Reinhold, New York, 1966, p. 621.
11. Kroto, H. W. *Molecular Rotation Spectra,* Dover, New York, 1992, pp. 34–40.
12. Clouthier, D. J. and Ramsay, D. A. Annu. Rev. Phys. Chem. **34,** 31 (1983).
13. Gordy, W. and Cook, R. L. *Microwave Molecular Spectra,* 3rd ed., Wiley, New York, 1984, pp. 647–720.

General References

Domenicano, A. and Hargittai, I. *Accurate Molecular Structures,* Oxford University Press, Oxford, 1992.

Flygare, W. H. *Molecular Structure and Dynamics,* Prentice-Hall, Englewood Cliffs, N.J., 1978.

Goldstein, H. *Classical Mechanics,* 2nd ed., Addison-Wesley, Reading, Mass., 1980.

Gordy, W. and Cook, R. L. *Microwave Molecular Spectra,* 3rd ed., Wiley, N.Y., 1984.

Graybeal, J. D. *Molecular Spectroscopy,* McGraw-Hill, N.Y., 1988.

Herzberg, G. and Huber, K. P., *Constants of Diatomic Molecules,* Van Nostrand Reinhold, New York, 1979.

Hollas, J. M. *High Resolution Spectroscopy,* Butterworths, London, 1982.

Kroto, H. W. *Molecular Rotation Spectra,* Dover, N.Y., 1992.

Struve, W. S. *Fundamentals of Molecular Spectroscopy,* Wiley, N.Y., 1989.

Sugden, T. M. and Kenney, C. N. *Microwave Spectroscopy of Gases,* Van Nostrand, London, 1965.

Townes, C. H. and Schawlow, A. L. *Microwave Spectroscopy,* Dover, N.Y., 1975.

Wollrab, J. E. *Rotational Spectra and Rotational Structure,* Academic Press, N.Y., 1967.

Zare, R. N., *Angular Momentum,* Wiley, N.Y., 1988.

Chapter 7
Vibrational Spectroscopy

Diatomic Molecules

The solution of the Schrödinger equation for a diatomic molecule plays an important role in spectroscopy. The study of the vibration–rotation spectra of diatomic molecules is an area of spectroscopy with many practical applications. In addition the vibrational spectra of diatomics serve as models for polyatomic molecules.

Consider a diatomic molecule A–B rotating and vibrating in the laboratory coordinate system X, Y, Z (Figure 7.1). The motion of the two nuclei can always be exactly separated into a center-of-mass part and an internal part by using the internal coordinates

$$\mathbf{r} = \mathbf{r}_B - \mathbf{r}_A \tag{7.1a}$$

and the definition for the center-of-mass position,

$$\mathbf{R} = \frac{m_A\mathbf{r}_A + m_B\mathbf{r}_B}{m_A + m_B} = \frac{m_A\mathbf{r}_A + m_B\mathbf{r}_B}{M} \tag{7.1b}$$

with

$$M = m_A + m_B. \tag{7.2}$$

These two equations can be solved for $\mathbf{r}_A$ and $\mathbf{r}_B$ in terms of $\mathbf{R}$ and $\mathbf{r}$ to give

$$\mathbf{r}_A = \mathbf{R} - \frac{m_B}{M}\mathbf{r} \tag{7.3}$$

and

$$\mathbf{r}_B = \mathbf{R} + \frac{m_A}{M}\mathbf{r}. \tag{7.4}$$

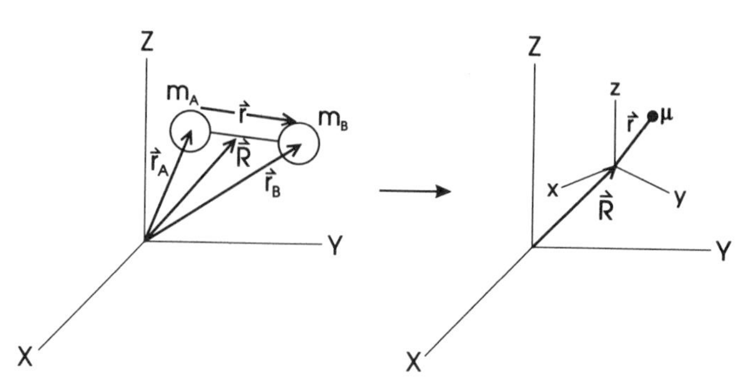

Figure 7.1. Center-of-mass transformation of a two-particle system.

The corresponding velocities are given by the time derivatives of $\mathbf{r}_A$ and $\mathbf{r}_B$, that is, $\dot{\mathbf{r}}_A$ and $\dot{\mathbf{r}}_B$. If the velocities obtained from equations (7.3) and (7.4) are substituted into the kinetic energy expression, it becomes

$$T = \frac{1}{2} m_A v_A^2 + \frac{1}{2} m_B v_B^2$$

$$T = \frac{1}{2} m_A \left(\dot{\mathbf{R}} - \frac{m_B \dot{\mathbf{r}}}{M} \right) \cdot \left(\dot{\mathbf{R}} - \frac{m_B \dot{\mathbf{r}}}{M} \right) + \frac{1}{2} m_B \left(\dot{\mathbf{R}} + \frac{m_A \dot{\mathbf{r}}}{M} \right) \cdot \left(\dot{\mathbf{R}} + \frac{m_A \dot{\mathbf{r}}}{M} \right), \tag{7.5}$$

which simplifies to

$$T = \frac{1}{2} M \, |\dot{\mathbf{R}}|^2 + \frac{1}{2} \mu \, |\dot{\mathbf{r}}|^2, \tag{7.6}$$

with μ the reduced mass

$$\mu = \frac{m_A m_B}{m_A + m_B}. \tag{7.7}$$

The kinetic energy has thus been split in equation (7.6) into an overall center-of-mass term and an equivalent one-particle (mass μ) term (Figure 7.1).

By expressing the kinetic energy in terms of the momentum rather than velocity, one obtains the classical Hamiltonian for the two-particle system,

$$H = \frac{p_A^2}{2m_A} + \frac{p_B^2}{2m_B} + V(r) = \frac{p_R^2}{2M} + \frac{p_r^2}{2\mu} + V(r), \tag{7.8}$$

in which the potential energy depends only upon the distance between the atoms. The center-of-mass contribution to the kinetic energy is ignored, since it only represents a shift in the total energy of the system. The quantum mechanical Hamiltonian for a vibrating rotor is therefore

$$\frac{-\hbar^2}{2\mu}\nabla^2\psi + V(r)\psi = E\psi. \tag{7.9}$$

Upon replacing the Cartesian coordinates x, y, and z for the location of the equivalent mass in equation (7.9) by the spherical polar coordinates r, θ, and ϕ, one obtains

$$\frac{-\hbar^2}{2\mu}\left[\frac{1}{r^2}\frac{\partial}{\partial r}r^2\frac{\partial\psi}{\partial r} + \frac{1}{r^2\sin\theta}\frac{\partial}{\partial\theta}\sin\theta\frac{\partial\psi}{\partial\theta} + \frac{1}{r^2\sin^2\theta}\frac{\partial^2\psi}{\partial\phi^2}\right] + V(r)\psi = E\psi, \tag{7.10}$$

or

$$\frac{-\hbar^2}{2\mu}\left[\frac{1}{r^2}\frac{\partial}{\partial r}r^2\frac{\partial\psi}{\partial r}\right] + \frac{1}{2\mu r^2}\hat{J}^2\psi + V(r)\psi = E\psi, \tag{7.11}$$

in which $\hat{J}^2$ is the operator representing the square of the angular momentum. Substitution of

$$\psi = R(r)Y_{JM}(\theta, \phi), \tag{7.12}$$

in which Y_{JM} is a spherical harmonic, into equation (7.11) yields the one-dimensional radial Schrödinger equation

$$\frac{-\hbar^2}{2\mu r^2}\frac{d}{dr}r^2\frac{dR}{dr} + \left(\frac{\hbar^2 J(J+1)}{2\mu r^2} + V(r)\right)R = ER. \tag{7.13}$$

Let us define

$$\frac{\hbar^2 J(J+1)}{2\mu r^2} = V_{\text{cent}} \tag{7.14}$$

as the centrifugal potential, and the sum

$$V(r) + V_{\text{cent}} = V_{\text{eff}} \tag{7.15}$$

is the effective potential. Only a specific functional form of $V(r)$ is needed in order to obtain the energy levels and wavefunctions of the vibrating rotor by solving equation (7.13). The substitution

$$S(r) = rR(r) \tag{7.16}$$

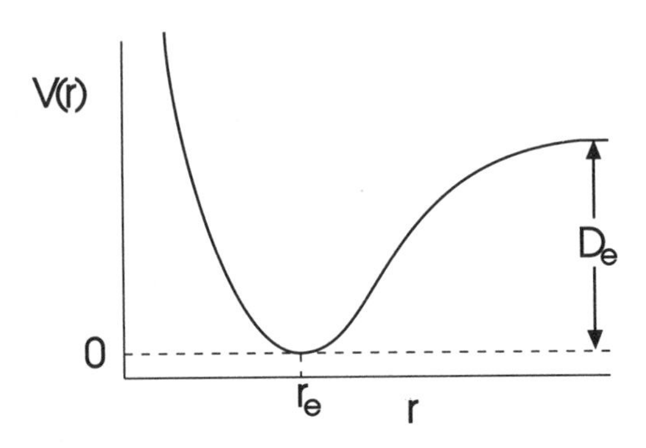

Figure 7.2. Potential energy curve for a diatomic molecule.

into equation (7.13) leads to the equation

$$\frac{-\hbar^2}{2\mu}\frac{d^2S}{dr^2} + \left[\frac{\hbar^2 J(J+1)}{2\mu r^2} + V(r)\right]S = ES. \tag{7.17}$$

In general $V(r) = E_{\text{el}}(r) + V_{NN}$ (Chapter 4) where E_{el} is obtained by solving the electronic Schrödinger equation

$$\hat{H}_{\text{el}}\psi = E_{\text{el}}\psi. \tag{7.18}$$

For the electronic Schrödinger equation (7.18) the energy depends on the particular value of r chosen for the calculation. As a result, E_{el} is a parametric function of r, so that no simple analytical form for $E_{\text{el}}(r)$ exists in general. Instead, considerable effort has been devoted to developing empirical forms for $V(r)$ (Figure 7.2). One of the most general forms, often denoted as the Dunham potential,[1] is a Taylor series expansion about r_e, namely

$$V(r) = V(r_e) + \left.\frac{dV}{dr}\right|_{r_e}(r - r_e) + \frac{1}{2}\left.\frac{d^2V}{dr^2}\right|_{r_e}(r - r_e)^2 + \cdots. \tag{7.19}$$

By setting $V(r_e) = 0$, the bottom of the well is arbitrarily chosen as the point of zero energy. In the expansion of $V(r)$ about its minimum at r_e, the first derivative is zero,

$$\left.\frac{dV}{dr}\right|_{r_e} = 0 \tag{7.20}$$

and therefore

$$V(r) = \frac{1}{2}k(r - r_e)^2 + \frac{1}{6}k_3(r - r_e)^3 + \frac{1}{24}k_4(r - r_e)^4 + \cdots \tag{7.21}$$

with

$$k = \left.\frac{d^2V}{dr^2}\right|_{r_e} \tag{7.22}$$

and

$$k_n = \left.\frac{d^nV}{dr^n}\right|_{r_e}. \tag{7.23}$$

By retaining only the leading term $\frac{1}{2}k(r - r_e)^2$ in this expansion for the nonrotating molecule ($J = 0$), one obtains the harmonic oscillator solutions

$$S = N_v H_v(\sqrt{\alpha}x)e^{(-\alpha x^2)/2}, \tag{7.24}$$

with

$$x = r - r_e, \qquad \alpha = \frac{\mu\omega}{\hbar}, \qquad N_v = \left[\frac{1}{2^v v!}\left(\frac{\alpha}{\pi}\right)^{1/2}\right]^{1/2}.$$

The functions $H_v(\alpha^{1/2}x)$ are the Hermite polynomials listed in Table 7.1. The corresponding eigenvalues for the nonrotating harmonic oscillator are

$$\begin{aligned} E_v &= h\nu(v + \tfrac{1}{2}), \qquad v = 0, 1, 2, \ldots \\ &= \hbar\omega(v + \tfrac{1}{2}) \end{aligned} \tag{7.25}$$

with

$$\omega = \left(\frac{k}{\mu}\right)^{1/2}, \qquad \nu = \frac{1}{2\pi}\left(\frac{k}{\mu}\right)^{1/2}. \tag{7.26}$$

Another popular choice for a simple form for the potential function is the Morse potential[2]

$$V(r) = D(1 - e^{-\beta(r-r_e)})^2. \tag{7.27}$$

The Morse potential, unlike the harmonic oscillator, asymptotically approaches a

Table 7.1. The Hermite Polynomials $H_v(x)$

$H_0 = 1$	$H_4 = 16x^4 - 48x^2 + 12$
$H_1 = 2x$	$H_5 = 32x^5 - 160x^3 + 120x$
$H_2 = 4x^2 - 2$	$H_6 = 64x^6 - 480x^4 + 720x^2 - 120$
$H_3 = 8x^3 - 12x$	$H_v(x) = (-1)^v e^{x^2} d^v/dx^v e^{-x^2}$

dissociation limit $V(r) = D$ as $r \to \infty$. Moreover, the Schrödinger equation can be solved analytically for the Morse potential. Specifically, one can show[2] that the eigenvalues for the Morse potential (plus the centrifugal term) can be written as (in cm^{-1})

$$\frac{E}{hc} = \omega_e(v + \tfrac{1}{2}) - \omega_e x_e(v + \tfrac{1}{2})^2 + B_e[J(J+1)] - D_e[J(J+1)]^2 - \alpha_e(v + \tfrac{1}{2})[J(J+1)] \quad (7.28)$$

with

$$\omega_e = \beta\left(\frac{Dh \times 10^2}{2\pi^2 c\mu}\right)^{1/2}, \quad (7.29)$$

$$\omega_e x_e = \frac{h\beta^2 \times 10^2}{8\pi^2 \mu c}, \quad (7.30)$$

$$B_e = \frac{h \times 10^{-2}}{8\pi^2 \mu r_e^2 c}, \quad (7.31)$$

$$D_e = \frac{4B_e^3}{\omega_e^2} \quad (7.32)$$

and

$$\alpha_e = \frac{6(\omega_e x_e B_e^3)^{1/2}}{\omega_e} - \frac{6B_e^2}{\omega_e}. \quad (7.33)$$

When using equations (7.28) to (7.33), some care with units is required since all spectroscopic constants and the Morse potential parameter β (equation (7.27)) are in cm^{-1}, while SI units are used for the physical constants. Note that in these equations D_e denotes the centrifugal distortion constant (equation (7.28)) as opposed to D, which denotes the dissociation energy, equation (7.27). The equation $D_e = 4B_e^3/\omega_e^2$, equation (7.32), applies to all realistic diatomic potentials and is known as the *Kratzer relationship.* The equation for α_e, equation (7.33), applies to the Morse potential and is often referred to as the *Pekeris relationship.* Notice that the vibrational energy expression

$$G(v) = \omega_e(v + \tfrac{1}{2}) - \omega_e x_e(v + \tfrac{1}{2})^2 \quad (7.34)$$

for the Morse oscillator has exactly two terms, and $G(v)$ is the customary symbol for the vibrational energy levels. In contrast, the rotational parts of the Morse oscillator energy-level equation (7.28) are only the leading terms of a series solution.

An even more general form than the Morse potential is the Dunham potential[1]

$$V(\xi) = a_0\xi^2(1 + a_1\xi + a_2\xi^2 + \cdots) \quad (7.35)$$

with

$$\xi = \frac{r - r_e}{r_e}. \tag{7.36}$$

The Dunham potential is just the Taylor series expansion (7.21) with some minor changes in notation such as

$$a_0 = \frac{kr_e^2}{2} = \frac{\omega_e^2}{4B_e}. \tag{7.37}$$

Although exact analytical forms for the wavefunctions and eigenvalues are impossible to derive for the Dunham potential, approximate analytical forms are relatively easy to obtain.

Dunham obtained an analytical expression for the energy levels of the vibrating rotor by using the first-order semiclassical quantization condition[3] from WKB (Wentzel–Kramers–Brillouin) theory, specifically

$$\left(\frac{2\mu}{\hbar^2}\right)^{1/2} \int_{r_-}^{r_+} \sqrt{E - V(r)}\, dr = \left(v + \frac{1}{2}\right)\pi, \tag{7.38}$$

in which r_- and r_+ are the classical inner and outer turning points for $V(r)$ at the energy E. The approximate wavefunctions are given by

$$\psi = A \exp\left(\pm i \left(\frac{2\mu}{\hbar^2}\right)^{1/2} \int [E - V(r)]^{1/2}\, dr\right), \tag{7.39}$$

and the energy levels are given by

$$E_{vJ} = \sum_{jk} Y_{jk}(v + \tfrac{1}{2})^j [J(J+1)]^k. \tag{7.40}$$

Dunham[1] was able to relate the coefficients Y_{jk} back to the potential energy parameters a_i by a series of equations listed, for example, in Townes and Schawlow.[4] The customary energy-level expressions[5]

$$F_v(J) = B_v J(J+1) - D_v[J(J+1)]^2 + H_v[J(J+1)]^3 + \cdots \tag{7.41}$$

$$G(v) = \omega_e(v + \tfrac{1}{2}) - \omega_e x_e(v + \tfrac{1}{2})^2 + \omega_e y_e(v + \tfrac{1}{2})^3 + \omega_e z_e(v + \tfrac{1}{2})^4 + \cdots \tag{7.42}$$

$$B_v = B_e - \alpha_e(v + \tfrac{1}{2}) + \gamma_e(v + \tfrac{1}{2})^2 + \cdots \tag{7.43}$$

$$D_v = D_e + \beta_e(v + \tfrac{1}{2}) + \cdots. \tag{7.44}$$

allow the relationships[4] between the Dunham Y_{jk} parameters and the conventional spectroscopic constants to be derived:

$$
\begin{array}{lll}
Y_{10} \approx \omega_e & Y_{20} \approx -\omega_e x_e & Y_{30} \approx \omega_e y_e \\
Y_{01} \approx B_e & Y_{11} \approx -\alpha_e & Y_{21} \approx \gamma_e \\
Y_{02} \approx -D_e & Y_{12} \approx -\beta_e & Y_{40} \approx \omega_e z_e. \\
Y_{03} \approx H_e & &
\end{array}
$$

The various isotopic forms of a molecule have different spectroscopic constants because the reduced mass is different. The pattern of isotopic mass dependence for a few of the spectroscopic constants can be seen in equations (7.29) to (7.33), that is, $\omega_e \propto \mu^{-1/2}$, $B_e \propto \mu^{-1}$, $\omega_e x_e \propto \mu^{-1}$, $D_e \propto \mu^{-2}$, and $\alpha_e \propto \mu^{-3/2}$. In general the isotopic dependence of the Dunham Y_{jk} constants is given by

$$Y_{jk} \propto (\mu^{-j/2})(\mu^{-k}) = \mu^{-(j+2k)/2}. \tag{7.45}$$

Defining a set of mass-independent constants as

$$U_{jk} = \mu^{(j+2k)/2} Y_{jk} \tag{7.46}$$

enables one to combine spectroscopic data from different isotopic forms of a molecule using the single equation

$$E_{vJ} = \sum_{jk} \mu^{-(j+2k)/2} U_{jk} (v + \tfrac{1}{2})^j [J(J+1)]^k. \tag{7.47}$$

When the Born–Oppenheimer approximation breaks down or the first-order WKB condition of equation (7.38) is inadequate, small correction terms[6] must be added to equation (7.47).

Wavefunctions for the Harmonic and Anharmonic Oscillators

The harmonic oscillator wavefunctions are given in Table 7.2 and are plotted in Figure 7.3. There are several notable features of these wavefunctions, including a finite probability density outside the walls of the confining potential. As the vibrational quantum number v increases, the probability for the oscillator being found near a classical turning point increases.

A diatomic molecule behaves like an anharmonic oscillator because the inner wall of a realistic potential is steeper than the harmonic oscillator potential, while the outer wall is much less steep than the harmonic oscillator (Figure 7.4). For small values of v,

Table 7.2. The Harmonic Oscillator Wavefunctions

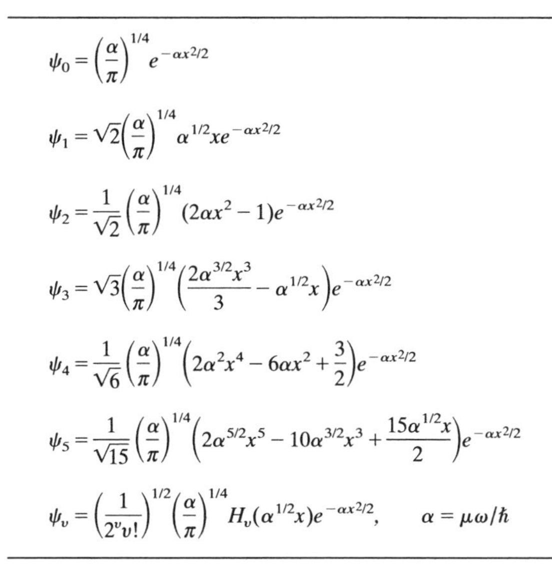

$$\psi_0 = \left(\frac{\alpha}{\pi}\right)^{1/4} e^{-\alpha x^2/2}$$

$$\psi_1 = \sqrt{2}\left(\frac{\alpha}{\pi}\right)^{1/4} \alpha^{1/2} x e^{-\alpha x^2/2}$$

$$\psi_2 = \frac{1}{\sqrt{2}}\left(\frac{\alpha}{\pi}\right)^{1/4} (2\alpha x^2 - 1) e^{-\alpha x^2/2}$$

$$\psi_3 = \sqrt{3}\left(\frac{\alpha}{\pi}\right)^{1/4} \left(\frac{2\alpha^{3/2}x^3}{3} - \alpha^{1/2}x\right) e^{-\alpha x^2/2}$$

$$\psi_4 = \frac{1}{\sqrt{6}}\left(\frac{\alpha}{\pi}\right)^{1/4} \left(2\alpha^2 x^4 - 6\alpha x^2 + \frac{3}{2}\right) e^{-\alpha x^2/2}$$

$$\psi_5 = \frac{1}{\sqrt{15}}\left(\frac{\alpha}{\pi}\right)^{1/4} \left(2\alpha^{5/2}x^5 - 10\alpha^{3/2}x^3 + \frac{15\alpha^{1/2}x}{2}\right) e^{-\alpha x^2/2}$$

$$\psi_v = \left(\frac{1}{2^v v!}\right)^{1/2} \left(\frac{\alpha}{\pi}\right)^{1/4} H_v(\alpha^{1/2}x) e^{-\alpha x^2/2}, \qquad \alpha = \mu\omega/\hbar$$

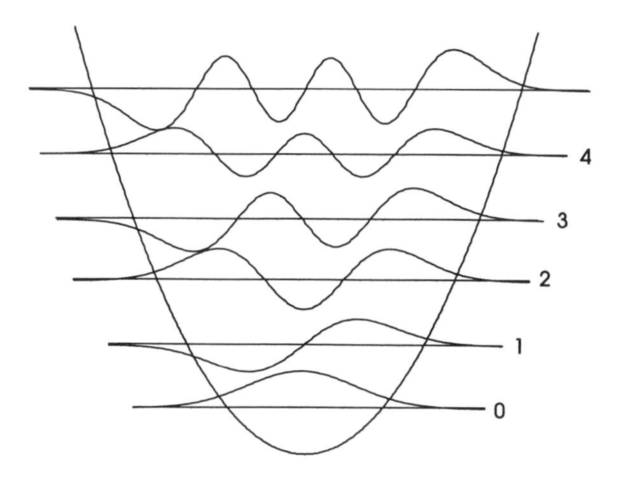

Figure 7.3. The harmonic oscillator wavefunctions.

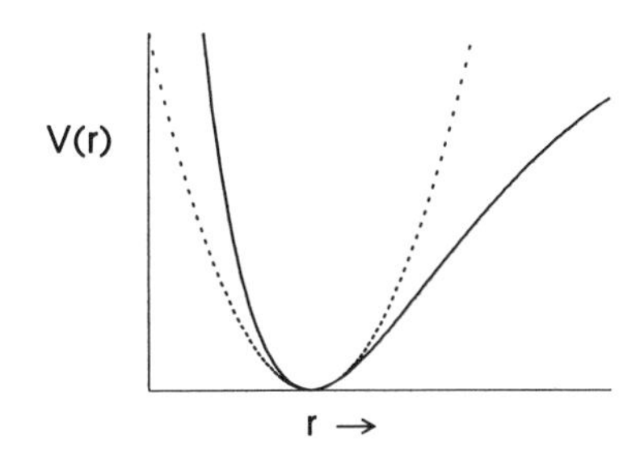

Figure 7.4. A harmonic oscillator potential (dots) as compared to a realistic diatomic potential (solid).

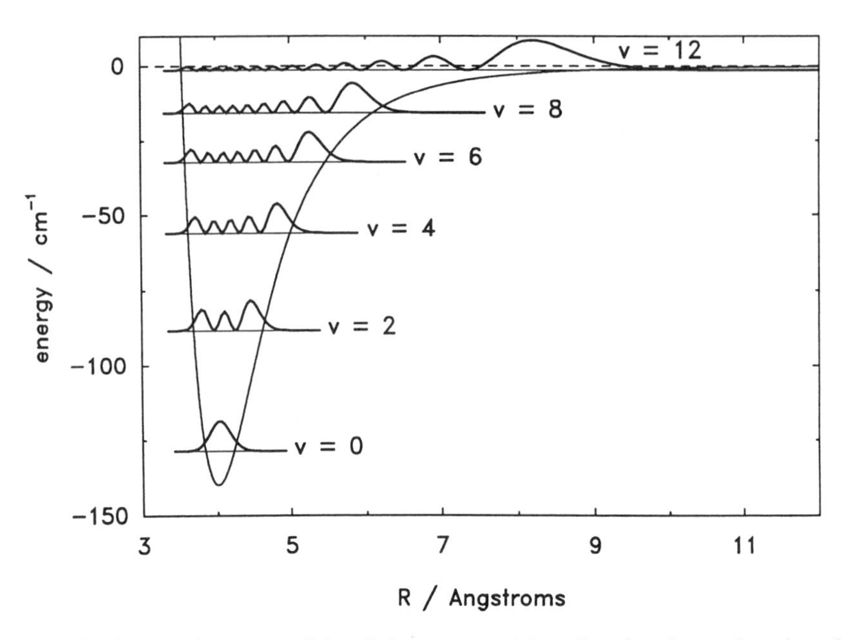

Figure 7.5. The internuclear potential and the square of the vibrational wavefunctions for Kr_2.

the harmonic oscillator model provides a good approximation and the differences between the harmonic and anharmonic oscillator wavefunctions are small. As v increases, however, the amplitude of the anharmonic potential's wavefunction increases at the outer turning point relative to its value at the inner turning point because the system spends most of its time at large r (Figure 7.5). The harmonic oscillator approximation is then no longer realistic.

Vibrational Selection Rules for Diatomic Molecules

In order to predict a spectrum from the energy levels, selection rules are required. The intensity of a vibrational transition is governed by the transition dipole moment integral

$$\mathbf{M}_{v'v''} = \int \psi'^{*}_{\text{vib}} \boldsymbol{\mu}(r) \psi''_{\text{vib}}\, dr, \tag{7.48}$$

in which single primes refer to the upper level of a transition and double primes to the lower. The dipole moment of a diatomic molecule is a function of r in which the functional dependence of $\boldsymbol{\mu}$ on r can be determined from Stark effect measurements, from the intensities of infrared bands or from *ab initio* calculations. As an example, the dipole moment function[7] calculated for the $X^2\Pi$ ground state of OH is illustrated in Figure 7.6.

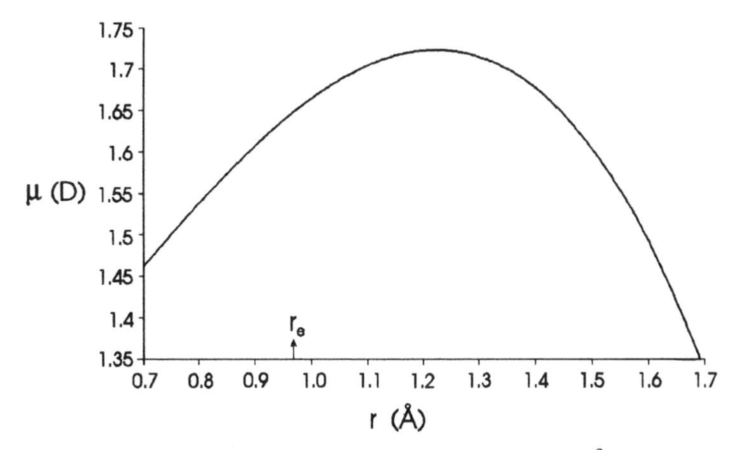

Figure 7.6. The *ab initio* dipole moment function of the $X^2\Pi$ state of OH.

Since any well-behaved function can be expanded as a Taylor series, let us expand $\boldsymbol{\mu}(r)$ about $r = r_e$ as

$$\boldsymbol{\mu} = \boldsymbol{\mu}_e + \left.\frac{d\boldsymbol{\mu}}{dr}\right|_{r_e}(r - r_e) + \frac{1}{2}\left.\frac{d^2\boldsymbol{\mu}}{dr^2}\right|_{r_e}(r - r_e)^2 + \cdots \tag{7.49}$$

so that

$$\mathbf{M}_{v'v''} = \boldsymbol{\mu}_e \int \psi'^{*}_{\text{vib}}\psi''_{\text{vib}}\,dr + \left.\frac{d\boldsymbol{\mu}}{dr}\right|_{r_e}\int \psi'^{*}_{\text{vib}}(r - r_e)\psi''_{\text{vib}}\,dr + \cdots. \tag{7.50}$$

The first term on the right-hand side of equation (7.50) is exactly zero because different vibrational wavefunctions of the potential curve are orthogonal to each other. The second term makes the dominant contribution to the intensity of most infrared fundamental transitions that thus depend on the value of the dipole moment derivative at the equilibrium distance, that is, upon $d\boldsymbol{\mu}/dr|_{r_e}$. More precisely, the intensity of a vibrational emission or absorption transition is given by

$$I \propto |\mathbf{M}_{v'v''}|^2 \propto \left|\frac{d\boldsymbol{\mu}}{dr}\right|^2_{r_e}. \tag{7.51}$$

This approximation neglects quadratic and higher power terms in equation (7.49) and assumes that the electrical dipole moment function is a linear function of r in the region close to $r = r_e$.

According to equation (6.38), the intensity of pure rotational transitions depend on $|\boldsymbol{\mu}|^2$, rather than on the square of the derivative (7.51), as in the case for vibrational transitions. Since homonuclear diatomic molecules such as Cl_2 have $\boldsymbol{\mu} = 0$ and $d\boldsymbol{\mu}/dr = 0$, they do not have electric dipole-allowed pure rotational or vibrational spectra. However, homonuclear diatomic molecules do have very weak electric quadrupole vibrational transitions which can be detected with very long path lengths.[8] These electric quadrupole transitions are about a factor of 10^{-6} weaker than typical electric dipole-allowed infrared transitions.

The intensity of an infrared vibrational transition also depends upon the value of the integral

$$\int \psi'^{*}_{\text{vib}}(r - r_e)\psi''_{\text{vib}}\, dr. \tag{7.52}$$

If harmonic oscillator wavefunctions (Table 7.2) are used to represent the wavefunctions in equation (7.52), and if the recursion relationship

$$2xH_n(x) = H_{n+1}(x) + 2nH_{n-1}(x), \tag{7.53}$$

between Hermite polynomials is employed, the result

$$\langle v'|\, x\, |v\rangle = \left(\frac{\hbar}{2m\omega}\right)^{1/2} [\sqrt{v+1}\, \delta_{v',v+1} + \sqrt{v}\, \delta_{v',v-1}] \tag{7.54}$$

is obtained for the integral (7.52) with $x = r - r_e$. The vibrational selection rule is thus $\Delta v = \pm 1$ for harmonic oscillator wavefunctions since $v' = v + 1$ or $v - 1$ in the Kronecker δ of equation (7.54).

If anharmonic wavefunctions are used, then transitions with $\Delta v = \pm 2, \pm 3, \ldots$ also become allowed because each anharmonic wavefunction can be represented by an expansion of harmonic oscillator wavefunctions

$$\psi_{\text{vib}} = \sum c_i \psi_{i,\text{HO}}. \tag{7.55}$$

Although this mechanical anharmonicity allows overtones to occur, their intensity drops off with increasing Δv. Typically an increase in Δv by one unit diminishes the intensity of the overtone band by a factor of 10 or 20 in infrared absorption spectroscopy. If the dipole moment function, equation (7.49), is not truncated after the linear term, then integrals of the type $\langle v'|\, (r - r_e)^2\, |v\rangle$ and $\langle v'|\, (r - r_e)^3\, |v\rangle$ are also present in the transition moment expression, equation (7.50). These terms give rise to matrix elements with $\Delta v = \pm 2, \pm 3, \ldots$ so that they also give overtones of appreciable intensity. The oscillator is said to be "electrically anharmonic" if terms higher than linear are used to represent $\boldsymbol{\mu}(r)$. Thus both electrical (equation (7.49)) and mechanical (equation (7.55)) anharmonicity terms contribute to the appearance of overtones in a spectrum.

The various types of infrared transitions have specific names associated with them

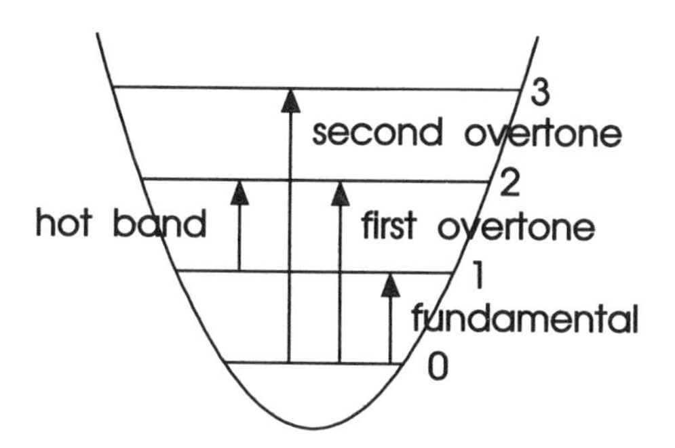

Figure 7.7. Names for infrared vibrational transitions.

(Figure 7.7). The $v = 1 \leftarrow 0$ transition is called the *fundamental,* while any transition with $v'' \neq 0$ is called a *hot band.* The name hot band originates from the experimental observation that the intensities of these bands increase upon heating the sample. The first overtone is the $v = 2 \leftarrow 0$ transition, the second overtone has $v = 3 \leftarrow 0$, and so on.

The mechanical anharmonicity of a diatomic oscillator results in an energy-level expression (7.42)

$$G(v) = \omega_e(v + \tfrac{1}{2}) - \omega_e x_e(v + \tfrac{1}{2})^2 + \omega_e y_e(v + \tfrac{1}{2})^3 + \omega_e z_e(v + \tfrac{1}{2})^4 + \cdots$$

so that a transition between vibrational levels characterized by $v + 1$ and v has an associated energy given by

$$\Delta G_{v+1/2} = G(v + 1) - G(v) = \omega_e - 2\omega_e x_e - 2\omega_e x_e v + \omega_e y_e(3v^2 + 6v + 13/4) + \cdots. \tag{7.56}$$

As an example, for $H^{35}Cl$ the vibrational energy expression is[9]

$$G(v) = 2990.946(v + \tfrac{1}{2}) - 52.8186(v + \tfrac{1}{2})^2 + 0.2244(v + \tfrac{1}{2})^3 - 0.0122(v + \tfrac{1}{2})^4 \text{ cm}^{-1}, \tag{7.57a}$$

while for $D^{35}Cl$ the expression is[9]

$$G(v) = 2145.163(v + \tfrac{1}{2}) - 27.1825(v + \tfrac{1}{2})^2 + 0.08649(v + \tfrac{1}{2})^3 - 0.00355(v + \tfrac{1}{2})^4 \text{ cm}^{-1}. \tag{7.57b}$$

Notice that the anharmonicity constants decrease rapidly in magnitude with $|\omega_e x_e| \gg |\omega_e y_e| \gg |\omega_e z_e|$.

Determination of Dissociation Energies from Spectroscopic Data

Under favorable circumstances it is possible to deduce the dissociation energy from spectroscopic data. Indeed, this is usually the most accurate of all methods for

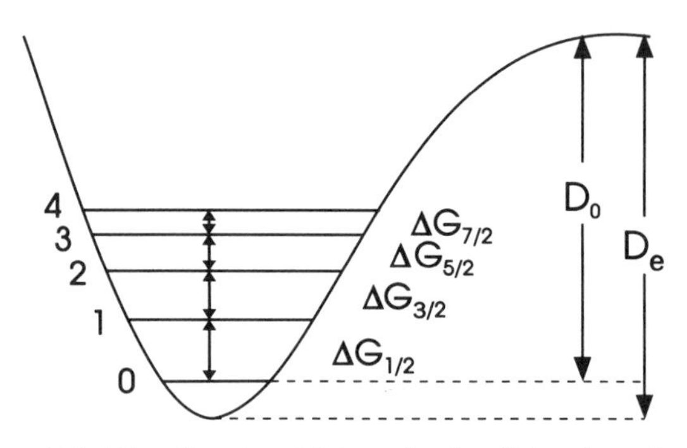

Figure 7.8. The vibrational intervals of a diatomic molecule.

determining dissociation energies. In principle, if all of the vibrational intervals $\Delta G_{v+1/2}$ are available, then the dissociation energy D_0 (measured from $v = 0$) is given by the sum of the intervals

$$D_0 = \sum_v \Delta G_{v+1/2} \tag{7.58}$$

as illustrated in Figure 7.8. Graphically this can be represented by a Birge–Sponer plot[10] of $\Delta G_{v+1/2}$ versus $v + \frac{1}{2}$ with the dissociation energy given by the area under the curve (see Figure 7.9). If the vibrational energy expression has only two terms

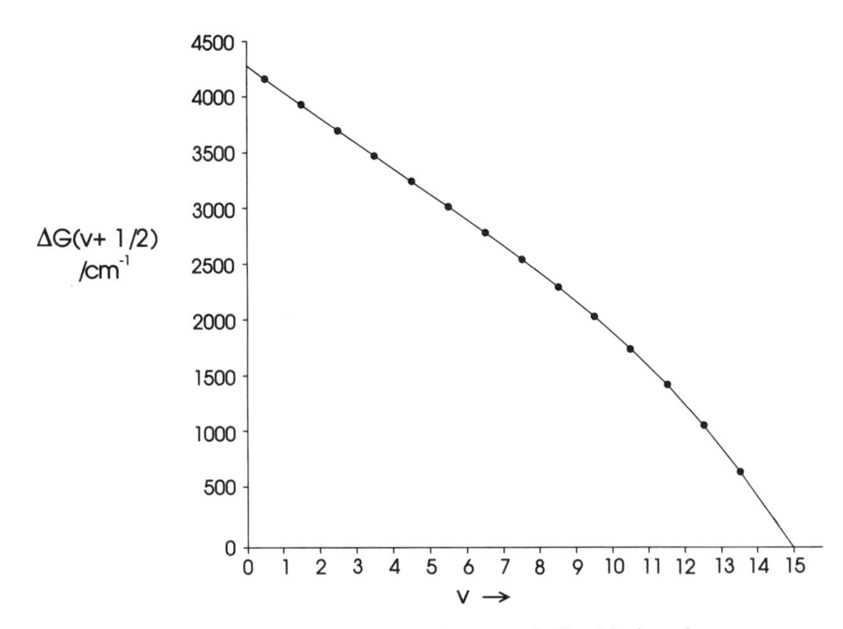

Figure 7.9. A Birge–Sponer plot for the ground state of H_2. Notice the curvature at high vibrational levels.

$G(v) = \omega_e(v + \frac{1}{2}) - \omega_e x_e(v + \frac{1}{2})^2$ (e.g., for the Morse oscillator), then $\Delta G_{v+1/2} = \omega_e - 2\omega_e x_e - 2\omega_e x_e v$. Thus the Birge–Sponer plot is linear over the entire range of v and the equilibrium dissociation energy D_e (Figure 7.8) is

$$D_e = \frac{\omega_e^2}{4\omega_e x_e}. \tag{7.59}$$

The dissociation energy of a Morse oscillator (equation (7.59)) is derived by substituting for β in equation (7.29) with the expression obtained from equation (7.30).

If all the vibrational levels of a molecule are known, then the simple application of equation (7.58) gives the dissociation energy. Only rarely, however, are all of the vibrational levels associated with a particular electronic state of a molecule known experimentally (e.g., H_2[11]). In practice an extrapolation from the last few observed levels to the unobserved dissociation limit (v_D) is necessary. A simple linear extrapolation has often been used, but this typically introduces considerable uncertainty into the exact location of the dissociation limit even when the extrapolation is short (Figure 7.10). The number v_D is the vibrational "quantum number" at dissociation and can be noninteger. It corresponds to the intercept of the Birge–Sponer curve (Figure 7.9) with the v-axis of the plot.

A more reliable procedure makes use of a Le Roy–Bernstein[12] plot in which the extrapolation is based on the theoretical long-range behavior of the potential. It has been shown that the vibrational spacings and other properties of levels lying near dissociation depend mainly on the long-range part of the potential, which is known to have the form

$$V(r) = D - \frac{C_n}{r^n} + \cdots, \tag{7.60}$$

in which D is the dissociation energy, n is an integer (typically 5 or 6 for a neutral molecule), and C_n is a constant. Substitution of equation (7.60) for $V(r)$ into the semiclassical quantization condition (equation (7.38)) followed by mathematical manipulation,[13] yields the approximate equation

$$(\Delta G_{v+1/2})^{(n-2)/(n+2)} = (v_D - v - \tfrac{1}{2})L(n, C_n), \tag{7.61}$$

in which $L(n, C_n)$ is a constant. A Le Roy–Bernstein plot of $(\Delta G_{v+1/2})^{(n-2)/(n+2)}$ versus v is a straight line at long range, so that linear extrapolation gives the dissociation limit marked by v_D in Figure 7.11. In essence the Le Roy–Bernstein procedure corrects for the curvature of the Birge–Sponer extrapolation (Figure 7.10) by making use of the known form (7.60) of the long-range interaction of two atoms.[13] For the case of the B state of I_2 this plot shows that the last bound vibrational level is $v' = 87$, which contrasts markedly with the uncertainty of the intercept on the conventional Birge–Sponer plot of Figure 7.10.

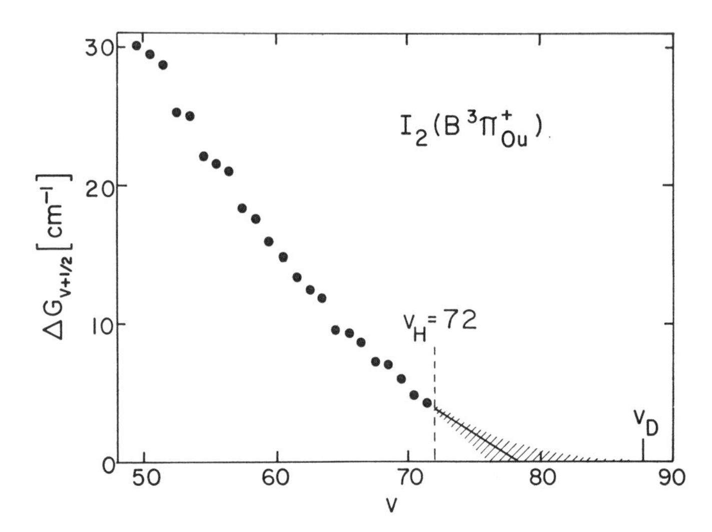

Figure 7.10. A Birge–Sponer plot for the B state of I_2. The highest observed vibrational level is $v_H = 72$.

The leading long-range interaction term C_n/r^n depends upon the nature of the two interacting atoms. All atom pairs have at least a C_6/r^6 term from the fluctuating induced dipole-induced dipole interaction. For I_2, however, the leading long-range term is C_5/r^5 (this is associated with the quadrupole–quadrupole interaction[13] between the two open-shell $^2P_{3/2}$ atoms). Indeed, the leading long-range interaction terms are known for all possible combinations of atoms.[13] For the B state of I_2 the Le Roy–Bernstein plot of $(\Delta G_{v+1/2})^{3/7}$ versus v predicts $v_D = 87.7$ from the old vibrational

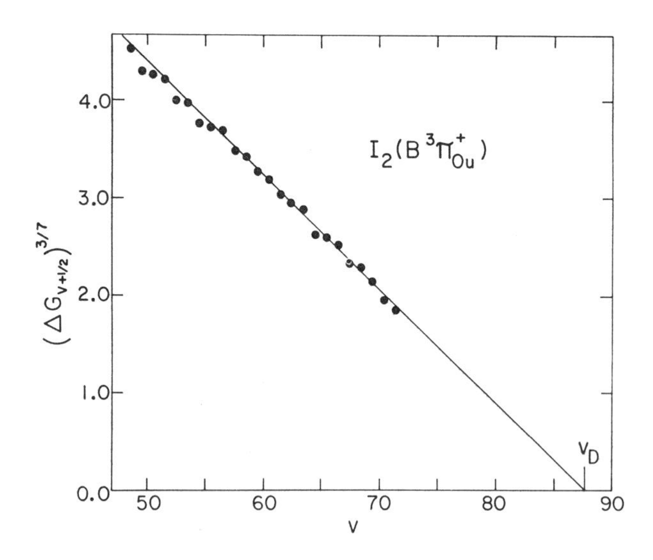

Figure 7.11. A Le Roy–Bernstein plot for the B state of I_2.

data of Brown[14] measured in 1931. More recent data,[15] including observations of levels up to $v = 82$, have confirmed this value of v_D.

Vibration–Rotation Transitions of Diatomic Molecules

In general, diatomic molecules vibrate and rotate at the same time, thus giving rise to vibration–rotation spectra. The selection rules are obtained simply by combining the pure rotational selection rules $\Delta J = \pm 1$ with the vibrational selection rules ($\Delta v = \pm 1$). The selection rules $\Delta J = \pm 1$ apply to molecules with no net spin or orbital angular momentum (i.e., $^1\Sigma^+$ states). For molecules such as O_2 ($X^3\Sigma_g^-$), NO ($X^2\Pi$), and free radicals in which **L** or **S** are nonzero, weak Q branches ($\Delta J = 0$) are also possible (Chapter 9).

Transitions are organized into branches on the basis of the change in J value. For one-photon, electric dipole-allowed transitions only $\Delta J = 0, \pm 1$ are possible, but for Raman transitions (Chapter 8), multiphoton transitions, magnetic dipole transitions, or electric quadrupole transitions, there are additional possibilities to be discussed later. Transitions with $\Delta J = -3, -2, -1, 0, 1, 2, 3$ are labeled N, O, P, Q, R, S, and T, respectively.

For a molecule such as HCl ($X^1\Sigma^+$) the spectrum contains only P and R branches. The energy of a given v, J level is

$$\begin{aligned} E_{vJ} &= G(v) + F(J) \\ &= \omega_e(v + \tfrac{1}{2}) - \omega_e x_e(v + \tfrac{1}{2})^2 + \omega_e y_e(v + \tfrac{1}{2})^3 + \cdots \\ &\quad + B_v J(J+1) - D_v[J(J+1)]^2 + \cdots \end{aligned} \tag{7.62}$$

so that (ignoring the effect of centrifugal distortion) the line positions for R and P transitions are given by

$$\nu_R(v', J+1 \leftarrow v'', J) = \nu_0 + 2B' + (3B' - B'')J + (B' - B'')J^2, \tag{7.63}$$

$$\nu_P(v', J-1 \leftarrow v'', J) = \nu_0 - (B' + B'')J + (B' - B'')J^2, \tag{7.64}$$

in which ν_0, the band origin, is $G(v') - G(v'')$. The P and R branch formulas can be combined into the single expression,

$$\nu = \nu_0 + (B' + B'')m + (B' - B'')m^2, \tag{7.65}$$

by defining $m = J + 1$ for the R branch and $m = -J$ for the P branch. By convention upper state quantum numbers and spectroscopic constants are labeled by single primes, while lower state quantum numbers and constants are labeled by double primes.

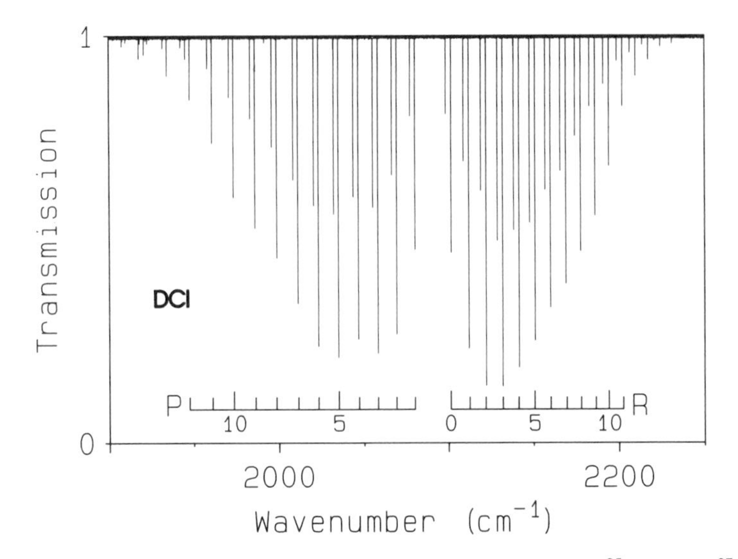

Figure 7.12. The fundamental vibration–rotation band of $D^{35}Cl$ and $D^{37}Cl$.

The fundamental band of HCl is the $v = 1 \leftarrow 0$ transition, and from equation (7.56) the band origin is given by

$$\nu_0 = \omega_e - 2\omega_e x_e + \tfrac{13}{4}\omega_e y_e + 5\omega_e z_e. \tag{7.66}$$

From expressions (7.57a) and (7.57b) we can obtain the band origin for $H^{35}Cl$ as 2885.977 cm^{-1}, while the band origin for $D^{35}Cl$ is 2091.061 cm^{-1}. The vibration–rotation transitions are illustrated in Figure 7.12 for the DCl infrared spectrum. The labeled peaks in Figure 7.12 are due to the more abundant $D^{35}Cl$ isotopomer, while the weaker satellite features are due to transitions of $D^{37}Cl$. For DCl the vibrational dependence of B is given by[9]

$$B_v = 5.448794 - 0.113291(v + \tfrac{1}{2}) + 0.0004589(v + \tfrac{1}{2})^2, \tag{7.67}$$

so that $B_0 = 5.392263$ cm^{-1} and $B_1 = 5.279890$ cm^{-1}. Thus, even for a light hydride, $B_0 \approx B_1$ and the consecutive lines in the P and R branches are spaced by about $2B$, equation (7.65). Notice that there is a gap at the band origin where a Q branch would be present if $\Delta J = 0$ were allowed. This "band gap" between the first lines $R(0)$ and $P(1)$ of the two branches is approximately $4B$.

Combination Differences

In general, a transition depends on both upper and lower state constants (equation (7.65)) so that each state cannot be treated independently. The differences between lines that share a common upper or lower level are known as *combination differences*

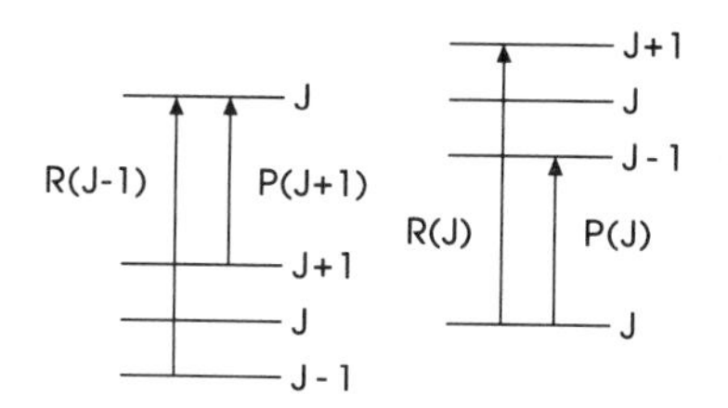

Figure 7.13. Ground-state and excited-state combination differences.

(Figure 7.13). These differences of line positions are very useful because they depend only on upper or lower state spectroscopic constants. The lower state combination differences (Figure 7.13) are

$$\begin{aligned}\Delta_2 F''(J) &= \nu[R(J-1)] - \nu[P(J+1)] \\ &= B''(J+1)(J+2) - B''(J-1)J \\ &= 4B''(J+\tfrac{1}{2}),\end{aligned} \tag{7.68}$$

while the upper state combination differences are

$$\Delta_2 F'(J) = 4B'(J+\tfrac{1}{2}). \tag{7.69}$$

In equations (7.68) and (7.69) the Δ indicates a difference between line positions represented by the standard $F(J)$ formulas, and the subscript 2 signifies that $\Delta J = 2$ for the differences. A plot of $\Delta_2 F(J)$ versus J yields approximately a straight line with a slope of $4B$ as shown in Figure 7.14 for HCl.[16] The slight curvature is due to the neglect of centrifugal distortion, which gives

$$\Delta_2 F(J) = (4B - 6D)(J+\tfrac{1}{2}) - 8D(J+\tfrac{1}{2})^3 \tag{7.70}$$

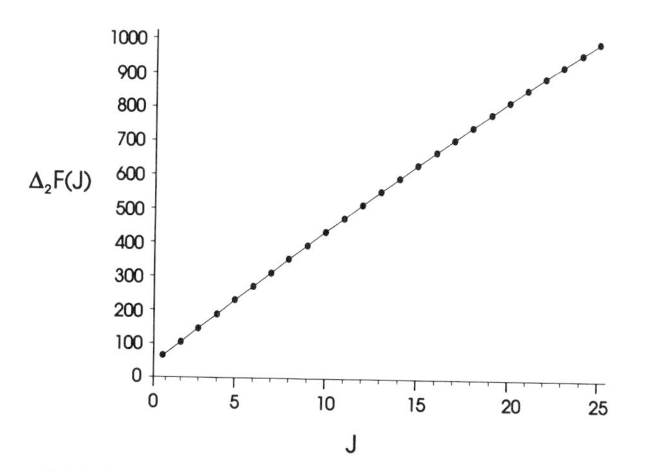

Figure 7.14. Ground-state combination differences for $v'' = 0$ of HCl.

when it is included. The combination differences thus allow the rotational constants of the upper and lower states to be determined independently.

Vibrational Motion of Polyatomic Molecules

Classical Mechanical Description

The classical Hamiltonian for the vibrational motion of a molecule (Figure 7.15) with N atoms is given by $H = T + V$ where the kinetic energy T is

$$\begin{aligned} T &= \frac{1}{2}\sum_i m_i\,|\mathbf{v}_i|^2 \\ &= \frac{1}{2}\sum_i m_i(\dot{x}_i^2 + \dot{y}_i^2 + \dot{z}_i^2), \end{aligned} \tag{7.71}$$

where the dot notation has been used for derivatives with respect to time, as for example, $\dot{x}_i \equiv dx_i/dt$.

This expression can be rewritten by introducing mass-weighted Cartesian displacement coordinates. Let

$$q_1 = (m_1)^{1/2}(x_1 - x_{1e}) \tag{7.72a}$$

$$q_2 = (m_1)^{1/2}(y_1 - y_{1e}) \tag{7.72b}$$

$$q_3 = (m_1)^{1/2}(z_1 - z_{1e}) \tag{7.72c}$$

$$q_4 = (m_2)^{1/2}(x_2 - x_{2e}) \tag{7.72d}$$

$$\vdots$$

$$q_{3N} = (m_N)^{1/2}(z_N - z_{Ne}), \tag{7.72e}$$

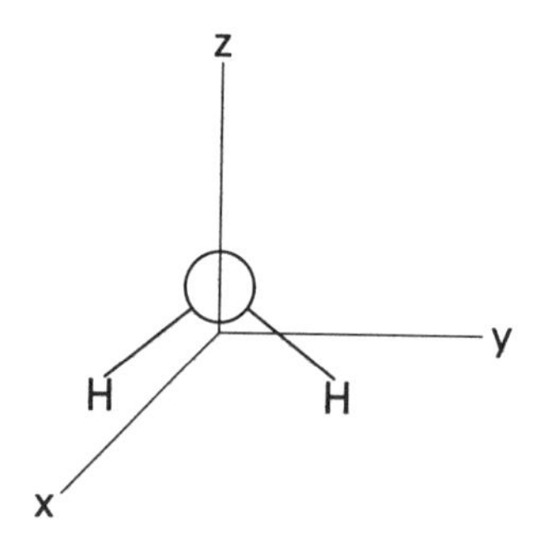

Figure 7.15. Coordinate system for a molecule with N (=3) atoms.

in which the q_i coordinate is proportional to the displacement from the equilibrium value. The set of equilibrium nuclear coordinates, $\{\mathbf{r}_{ie}\}$, describes the location of the nuclei at the absolute minimum in the potential energy. In terms of the q_i coordinates the kinetic energy of nuclear motion takes the especially simple form

$$T = \frac{1}{2} \sum_{i=1}^{3N} \dot{q}_i^2. \tag{7.73}$$

In general the potential energy $V(r_i)$ is a complicated function of the Cartesian coordinates of the atoms. Expanding the potential V in a Taylor series about the equilibrium nuclear positions using the mass-weighted Cartesian displacement coordinates gives

$$V = V(q_i = 0) + \sum_i \left.\frac{\partial V}{\partial q_i}\right|_{q_i=0} q_i + \frac{1}{2} \sum_i \sum_j \left.\frac{\partial^2 V}{\partial q_i\, \partial q_j}\right|_{q_i=0, q_j=0} q_i q_j + \cdots. \tag{7.74}$$

The potential energy is arbitrarily chosen to be zero at equilibrium, that is, $V(q_i = 0) = 0$. At the same time we also have by definition $(\partial V/\partial q_i)|_{q_i=0} = 0$ at equilibrium.

The present discussion is based on the harmonic approximation for the potential energy, according to which terms in the expansion (7.74) with powers greater than two are neglected. The second derivatives of the potential are force constants f_{ij} defined by

$$f_{ij} = \left.\frac{\partial^2 V}{\partial q_i\, \partial q_j}\right|_{q_i=0} \tag{7.75}$$

so that

$$2V = \sum_{i,j} f_{ij} q_i q_j. \tag{7.76}$$

The q_i are examples of generalized coordinates for which Newton's laws of motion are best formulated using Lagrange's equations. Lagrange's equations are equivalent to

$$\mathbf{F} = m\mathbf{a} \tag{7.77}$$

but are valid for any coordinates, not just the Cartesian coordinates implicit in equation (7.77). Lagrange's formulation of the classical equations of motion is based on the construction of the Lagrangian L,

$$L = T(\dot{q}_i) - V(q_i), \tag{7.78}$$

which is a function of the generalized coordinates and velocities q_i and $\dot{q}_i$. Newton's second law of motion, equation (7.77), is equivalent to Lagrange's equation

$$\frac{d}{dt}\left(\frac{\partial L}{\partial \dot{q}_i}\right)_{q_i} - \left(\frac{\partial L}{\partial q_i}\right)_{\dot{q}_i} = 0. \tag{7.79}$$

It is easy to verify this equivalence for a single particle moving in one dimension x, for a conservative system in which the potential is independent of time $[V \neq V(t)]$. In this case the Lagrangian is

$$L = \tfrac{1}{2}m\dot{x}^2 - V(x), \tag{7.80}$$

and Lagrange's equation (7.79) becomes

$$\frac{d}{dt}\frac{\partial}{\partial \dot{x}}\left(\frac{1}{2}m\dot{x}^2 - V\right) - \frac{\partial}{\partial x}\left(\frac{1}{2}m\dot{x}^2 - V\right) = 0. \tag{7.81}$$

Taking the derivatives with respect to $\dot{x}$ and x gives

$$\frac{d}{dt}m\dot{x} + \frac{\partial V}{\partial x} = 0, \tag{7.82}$$

or

$$-\frac{\partial V}{\partial x} = \frac{d(m\dot{x})}{dt} = \frac{dp}{dt} = m\ddot{x} = ma. \tag{7.83}$$

However, since force is also related to the potential function by

$$\mathbf{F} = -\nabla V \tag{7.84a}$$

we also have

$$F = -\frac{\partial V}{\partial x} = ma. \tag{7.84b}$$

As expected Lagrange's equation is equivalent to $F = ma$. In Lagrangian mechanics the generalized force is $(\partial L/\partial q_i)_{\dot{q}_i}$ and the generalized momentum is

$$\left(\frac{\partial L}{\partial \dot{q}_i}\right)_{q_i} = p_i. \tag{7.85}$$

Applying Lagrange's equation to the vibrations of polyatomic molecules gives

$$L = T - V = \frac{1}{2}\sum_{i=1}^{3N} \dot{q}_i^2 - \frac{1}{2}\sum_{i,j}^{3N} f_{ij}q_iq_j \tag{7.86}$$

but since

$$\frac{d}{dt}\left(\frac{\partial L}{\partial \dot{q}_i}\right)_{q_i} - \left(\frac{\partial L}{\partial q_i}\right)_{\dot{q}_i} = 0 \qquad i = 1, \ldots, 3N, \tag{7.87}$$

the equations of motion become

$$\ddot{q}_i + f_{ii}q_i + \sum_{j \neq i} f_{ij}q_j = 0, \tag{7.88}$$

or

$$\ddot{q}_i + \sum_j f_{ij}q_j = 0, \qquad i = 1, \ldots, 3N. \tag{7.89}$$

This is a set of $3N$ coupled second-order differential equations (7.89) with constant coefficients. Such a system of equations can be solved by assuming a solution of the form

$$q_i = A_i \cos(\lambda^{1/2}t + \phi). \tag{7.90}$$

Substitution of equation (7.90) into equation (7.89) converts the set of second-order differential equations into a set of $3N$ homogeneous linear equations

$$-\lambda A_i \cos(\lambda^{1/2}t + \phi) + \sum_j f_{ij}A_j \cos(\lambda^{1/2}t + \phi) = 0, \tag{7.91}$$

or

$$-\lambda A_i + \sum_j f_{ij}A_j = 0. \tag{7.92}$$

The set of $3N$ equations has only the trivial solution for the amplitudes $A_1 = A_2 = \cdots A_{3N} = 0$ unless the determinant of the coefficients is zero or

$$\begin{vmatrix} f_{11} - \lambda & f_{12} & \cdots & f_{1,3N} \\ f_{21} & f_{22} - \lambda & \cdots & \\ \vdots & & & \vdots \\ f_{3N,1} & \cdots & & f_{3N,3N} - \lambda \end{vmatrix} = 0. \tag{7.93}$$

The secular equation (7.93) is a polynomial of order $3N$ so that there exist $3N$ values of λ for which equation (7.93) is satisfied. The $3N \times 3N$ force constants can be arranged in a symmetric force constant matrix **f**, and the $3N$ values of λ are the eigenvalues of **f**. It turns out that six of the eigenvalues are zero for a nonlinear molecule, and five are zero for a linear molecule. Three degrees of freedom are associated with the translation (x, y, z) of the center of mass and three (or two for a linear molecule) with rotational motion of the molecule as a whole (θ, ϕ, χ). Since there is no restoring force acting on these degrees of freedom, their frequencies are zero.

Associated with each eigenvalue λ_i is a coordinate called a *normal mode coordinate*, Q_i. The normal modes represent a new set of coordinates related to the old q_j by a linear transformation,

$$Q_k = \sum_j l_{kj} q_j \tag{7.94}$$

or

$$\mathbf{Q} = \mathbf{l}\mathbf{q} \tag{7.95}$$

with $\mathbf{l}$ being a real orthogonal matrix ($\mathbf{l}^{-1} = \mathbf{l}'$) so that

$$\mathbf{q} = \mathbf{l}'\mathbf{Q}. \tag{7.96}$$

In matrix notation the original $3N$ differential equations (7.89) are written as

$$\ddot{\mathbf{q}} + \mathbf{f}\mathbf{q} = 0. \tag{7.97}$$

Substituting equation (7.96) gives

$$\mathbf{l}'\ddot{\mathbf{Q}} + \mathbf{f}\mathbf{l}'\mathbf{Q} = 0, \tag{7.98}$$

so that multiplication by $\mathbf{l}$ from the left gives

$$\ddot{\mathbf{Q}} + (\mathbf{l}\mathbf{f}\mathbf{l}')\mathbf{Q} = 0. \tag{7.99}$$

The transformation matrix $\mathbf{l}$ is chosen to diagonalize $\mathbf{f}$, that is,

$$\mathbf{l}\mathbf{f}\mathbf{l}' = \mathbf{\Lambda} \tag{7.100}$$

so that the eigenvalues of $\mathbf{f}$ are the diagonal elements of $\mathbf{\Lambda}$. Since $\mathbf{f}$ is a real symmetric matrix, there are $3N$ real eigenvalues. Furthermore, an orthogonal matrix can always be constructed from the orthonormal eigenvectors of $\mathbf{f}$. We see that the linear transformation has uncoupled the $3N$ equations so that now

$$\ddot{\mathbf{Q}} + \mathbf{\Lambda}\mathbf{Q} = 0 \tag{7.101}$$

or written out as a vector equation

$$\begin{pmatrix} \ddot{Q}_1 \\ \ddot{Q}_2 \\ \vdots \\ \ddot{Q}_{3N} \end{pmatrix} + \begin{pmatrix} \lambda_1 & & & 0 \\ & \lambda_2 & & \\ & & \ddots & \\ 0 & & & \lambda_{3N} \end{pmatrix} \begin{pmatrix} Q_1 \\ Q_2 \\ \vdots \\ Q_{3N} \end{pmatrix} = 0. \tag{7.102}$$

Expanding the matrix equation (7.102) gives

$$\ddot{Q}_1 + \lambda_1 Q_1 = 0 \tag{7.103a}$$

$$\ddot{Q}_2 + \lambda_2 Q_2 = 0 \tag{7.103b}$$

$$\vdots$$

$$\ddot{Q}_{3N} + \lambda_{3N} Q_{3N} = 0. \tag{7.103c}$$

Applying the same normal coordinate transformation to T and V gives

$$2T = \dot{\mathbf{q}}^t\dot{\mathbf{q}} = (\mathbf{l}^t\dot{\mathbf{Q}})^t(\mathbf{l}^t\dot{\mathbf{Q}}) = (\dot{\mathbf{Q}})^t\mathbf{l}\mathbf{l}^t(\dot{\mathbf{Q}}) = \dot{\mathbf{Q}}^t\dot{\mathbf{Q}} \tag{7.104}$$

or

$$T = \tfrac{1}{2}\sum \dot{Q}_i^2 \tag{7.105}$$

and

$$2V = \mathbf{q}^t\mathbf{f}\mathbf{q} = \mathbf{Q}^t\mathbf{l}\mathbf{f}\mathbf{l}^t\mathbf{Q} = \mathbf{Q}^t\boldsymbol{\Lambda}\mathbf{Q} \tag{7.106}$$

or

$$V = \tfrac{1}{2}\sum \lambda_i Q_i^2. \tag{7.107}$$

Thus, both the kinetic and potential energy terms of the Hamiltonian have no cross-terms that connect different coordinates. The system therefore behaves like a set of $3N-6$ (or $3N-5$) independent harmonic oscillators, each oscillating without interaction with the others. Naturally, a real system has cubic and quartic terms (and higher!) in the potential energy expansion. For a real molecule, the **l** matrix and the normal coordinates are still defined in the way outlined previously, using only the harmonic terms, but the **l** matrix transformation no longer completely uncouples the $3N$ differential equations. The anharmonic terms in the potential energy expansion are then said to couple the normal modes, so that the normal mode approximation is not completely valid.

Quantum Mechanical Description

With the classical Hamiltonian available, the transition to quantum mechanics is (deceptively) simple. First the classical Hamiltonian is written in terms of the generalized coordinates, the normal modes Q_i, and the associated generalized momenta P_i with

$$P_i = \left(\frac{\partial L}{\partial \dot{Q}_i}\right)_{Q_i} = \dot{Q}_i. \tag{7.108}$$

The resulting classical Hamiltonian is

$$H = \tfrac{1}{2}\sum P_i^2 + \tfrac{1}{2}\sum \lambda_i Q_i^2. \tag{7.109}$$

The classical Hamiltonian is converted to a quantum mechanical operator by making the usual substitutions $Q_i \to \hat{Q}_i$ and

$$P_i \to \hat{P}_i = -i\hbar \frac{\partial}{\partial Q_i}$$

which gives

$$\begin{aligned} \hat{H} &= -\frac{\hbar^2}{2}\sum_i \frac{\partial^2}{\partial Q_i^2} + \frac{1}{2}\sum_i \lambda_i \hat{Q}_i^2 \\ &= \sum \left[-\frac{\hbar^2}{2}\frac{\partial^2}{\partial Q_i^2} + \frac{1}{2}\lambda_i \hat{Q}_i^2 \right] \\ &= \sum \hat{H}_i. \end{aligned} \tag{7.110}$$

In terms of the normal coordinates the Hamiltonian operator equation (7.110) is just a sum of $3N-6$ independent harmonic oscillator Hamiltonians. Consequently, the total wavefunction ψ for the Schrödinger equation $\hat{H}\psi = E\psi$ is just a product of $3N-6$ (or $3N-5$) harmonic oscillator wavefunctions

$$\psi = \psi_1(Q_1)\psi_2(Q_2)\cdots\psi_{3N-6}(Q_{3N-6}) \tag{7.111}$$

with

$$\psi_i(Q_i) = N_{v_i} H_{v_i}(\xi_i) e^{-\xi_i^2/2} \quad \text{and} \quad \xi_i = Q_i\left(\frac{\lambda_i^{1/2}}{\hbar}\right)^{1/2}. \tag{7.112}$$

The six (or five) translations and rotations have already been removed since they are not genuine vibrational modes. The $3N-6$ (or $3N-5$) fictitious harmonic oscillators all have unit mass, since the actual atomic masses were already used to mass-weight the Cartesian displacement coordinates. The square of the angular frequency ($\omega = 2\pi\nu$) is $\omega^2 = k/m = \lambda$ so that $\omega = \lambda^{1/2}$. The total energy is thus the sum of the energies of $3N-6$ (or $3N-5$ for a linear molecule) harmonic oscillator energies, namely

$$\begin{aligned} E &= \hbar\omega_1(v_1 + \tfrac{1}{2}) + \cdots + \hbar\omega_{3N-6}(v_{3N-6} + \tfrac{1}{2}) \\ &= \sum \hbar\omega_i(v_i + \tfrac{1}{2}) \\ &= \sum \hbar\lambda_i^{1/2}(v_i + \tfrac{1}{2}). \end{aligned} \tag{7.113}$$

The preceding treatment is valid within the harmonic oscillator approximation and assumes that the molecule is not rotating. Real molecules, however, are rotating, anharmonic oscillators. The approximate vibrational wavefunction (7.111) is nevertheless a very good starting point for the description of true molecular vibrations. The effect of anharmonic terms in the potential can be thought of as mixing the various zeroth-order harmonic oscillator wavefunctions. The harmonic oscillator functions form a complete, orthogonal set of basis functions and are therefore a suitable expansion set.

The effect of rotation is much more difficult to handle because the vibrational and rotational motions are not separable. The use of a noninertial (accelerated) coordinate system (internal molecular coordinates) rotating with the molecule introduces Coriolis terms into the classical Hamiltonian. A discussion of Coriolis effects appears later in this chapter. Special techniques are then required to transform the classical Hamiltonian to an appropriate quantum mechanical form and, of course, there are additional terms in the Hamiltonian operator. The coupling of vibration with rotation introduces vibrational angular momentum and prevents the exact separation of vibrational and rotational motion. These additional terms can mix the normal-mode wavefunctions by Coriolis coupling and can mix the vibrational and rotational wavefunctions. Despite these problems the simple normal mode picture is a remarkably successful model. Only for highly excited modes, such as the fifth overtone of the OH stretching motion of H_2O, is a different, non-normal mode picture (the local mode approximation) commonly used.

Internal Coordinates

The use of force constants f_{ij} associated with mass-weighted Cartesian coordinates is very convenient mathematically, but they are difficult to associate with specific internal motions such as bond stretching. It would be preferable to describe the vibrational motion of a molecule in terms of the readily recognizable structural features, namely the bond lengths and angles. For example, in the water molecule three internal coordinates r_1, r_2, and θ are required to describe the relative positions of the atoms (i.e., Figure 7.16). There are also $3N - 6 = 3$ vibrational modes that must be related in some manner to changes in r_1, r_2, and θ. It is convenient to define the internal displacement coordinates Δr_1, Δr_2, and $\Delta\theta$ to correspond to bond-stretching and bond-bending motions. It is also convenient to use $(r_1 r_2)^{1/2}\Delta\theta = r\Delta\theta$ as the bending coordinate in order to make all internal coordinates have the same dimensions.

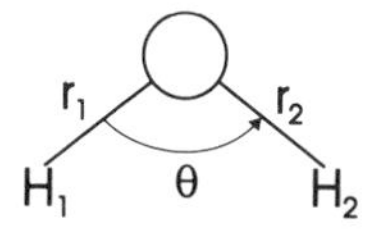

Figure 7.16. Internal coordinates of the H_2O molecule.

Symmetry Coordinates

Even more preferable than internal coordinates are symmetry coordinates, since the vibrational Hamiltonian is unchanged by the symmetry operations associated with the molecule. This also means that the vibrational wavefunctions can be labeled with the irreducible representations of the molecular point group. The best plan is to utilize as much symmetry information as possible.

Symmetry coordinates can be obtained by inspection or more systematically with the aid of projection operators (Chapter 4). From the character table all possible projection operators are constructed and applied to the internal coordinates until the required $3N - 6$ symmetry coordinates are generated. For example, using the H_2O molecule, the totally symmetric projection operator

$$\hat{P}_{A_1} = \sum_R (\chi_R^{A_1})^* \hat{O}_R = \hat{O}_E + \hat{O}_{C_2} + \hat{O}_{\sigma_v} + \hat{O}_{\sigma'_v} \tag{7.114}$$

when applied to Δr_1 gives

$$\begin{aligned}\hat{P}_{A_1}\Delta r_1 &= \Delta r_1 + \Delta r_2 + \Delta r_1 + \Delta r_2 \\ &= 2(\Delta r_1 + \Delta r_2).\end{aligned} \tag{7.115}$$

Similarly, the application of $\hat{P}_{B_2}$ to Δr_1 and $\hat{P}_{A_1}$ to $r\Delta\theta$ gives

$$\hat{P}_{B_2}\Delta r_1 = 2(\Delta r_1 - \Delta r_2) \tag{7.116}$$

$$\hat{P}_{A_1}(r\Delta\theta) = r\Delta\theta. \tag{7.117}$$

Let the three symmetry coordinates s_1, s_2, and s_3 be defined by

$$s_1 = \frac{\Delta r_1 + \Delta r_2}{\sqrt{2}} \qquad (A_1 \text{ symmetry}) \tag{7.118}$$

$$s_2 = r\Delta\theta \qquad (A_1 \text{ symmetry}) \tag{7.119}$$

$$s_3 = \frac{\Delta r_1 - \Delta r_2}{\sqrt{2}} \qquad (B_2 \text{ symmetry}). \tag{7.120}$$

The x-axis is defined to be out of plane (i.e., perpendicular to the plane of the water molecule). The three normalized symmetry coordinates s_1, s_2, and s_3 have symmetry appropriate to a molecule belonging to the C_{2v} point group.

Using internal coordinates the harmonic potential energy function of H_2O is

$$V = \tfrac{1}{2}f_{11}(\Delta r_1)^2 + \tfrac{1}{2}f_{22}(\Delta r_2)^2 + \tfrac{1}{2}f_{33}(r\Delta\theta)^2 + f_{12}(\Delta r_1)(\Delta r_2) + f_{13}(\Delta r_1)(r\Delta\theta) + f_{23}(\Delta r_2)(r\Delta\theta) \tag{7.121}$$

and $f_{11} = f_{22}$, $f_{13} = f_{23}$ by symmetry. This potential energy function (7.121) is derived by considering the three possible quadratic terms (f_{11}, f_{22}, and f_{33}) and the three possible cross-terms (f_{12}, f_{13}, f_{23}) obtained from the three internal coordinates. Converting to symmetry coordinates by using

$$\Delta r_1 = \frac{s_1 + s_3}{\sqrt{2}} \tag{7.122}$$

$$r\Delta\theta = s_2 \tag{7.123}$$

and

$$\Delta r_2 = \frac{s_1 - s_3}{\sqrt{2}} \tag{7.124}$$

gives

$$V(s_1, s_2, s_3) = \underbrace{\frac{(f_{11} + f_{12})s_1^2}{2}}_{\text{symmetric stretching}} + \underbrace{\frac{(f_{11} - f_{12})s_3^2}{2}}_{\text{antisymmetric stretching}} + \underbrace{\frac{f_{33}s_2^2}{2}}_{\text{bending}} + \underbrace{\sqrt{2}f_{13}s_1s_2}_{\text{stretch-bend interaction}} \tag{7.125}$$

Notice that there are no s_1s_3 terms present in equation (7.125) because s_1 and s_3 have different symmetry, so that no terms in the Hamiltonian can connect them (see Chapter 4). If a term such as s_1s_3 were present it would have $A_1 \otimes B_2 = B_2$ symmetry and thus would be changed by a symmetry operation such as $\hat{C}_2$ or $\hat{\sigma}_v$, contradicting the principle that the Hamiltonian is unchanged by the symmetry operations of the molecule.

The transformation to symmetry coordinates, equations (7.118) to (7.120), can be summarized by the matrix equation

$$\begin{pmatrix} s_1 \\ s_2 \\ s_3 \end{pmatrix} = \begin{pmatrix} \frac{1}{\sqrt{2}} & \frac{1}{\sqrt{2}} & 0 \\ 0 & 0 & 1 \\ \frac{1}{\sqrt{2}} & -\frac{1}{\sqrt{2}} & 0 \end{pmatrix} \begin{pmatrix} \Delta r_1 \\ \Delta r_2 \\ r\Delta\theta \end{pmatrix} \tag{7.126}$$

or

$$\mathbf{s} = \mathbf{Ur}. \tag{7.127}$$

Upon inversion equation (7.126) gives

$$\begin{pmatrix} \Delta r_1 \\ \Delta r_2 \\ r\Delta\theta \end{pmatrix} = \begin{pmatrix} \frac{1}{\sqrt{2}} & 0 & \frac{1}{\sqrt{2}} \\ \frac{1}{\sqrt{2}} & 0 & -\frac{1}{\sqrt{2}} \\ 0 & 1 & 0 \end{pmatrix} \begin{pmatrix} s_1 \\ s_2 \\ s_3 \end{pmatrix} \tag{7.128}$$

or

$$\mathbf{r} = \mathbf{U}^{-1}\mathbf{s} = \mathbf{U}^t\mathbf{s}, \tag{7.129}$$

in which **s** is the vector of symmetry coordinates, **r** is the vector of internal coordinates, and **U** is the orthogonal transformation matrix defined in equation (7.126).

In matrix notation the potential energy V is written as

$$\begin{aligned} V(\Delta r_1, \Delta r_2, r\Delta\theta) &= \tfrac{1}{2}(\Delta r_1, \Delta r_2, r\Delta\theta)\begin{pmatrix} f_{11} & f_{12} & f_{13} \\ f_{12} & f_{11} & f_{13} \\ f_{13} & f_{13} & f_{33} \end{pmatrix}\begin{pmatrix} \Delta r_1 \\ \Delta r_2 \\ r\Delta\theta \end{pmatrix} \\ &= \tfrac{1}{2}\mathbf{r}^t\mathbf{F}_r\mathbf{r}. \end{aligned} \tag{7.130}$$

For the H_2O molecule the potential energy is

$$\begin{aligned} V(s_1, s_2, s_3) &= \tfrac{1}{2}(s_1, s_2, s_3)\begin{pmatrix} f_{11}+f_{12} & \sqrt{2}f_{13} & 0 \\ \sqrt{2}f_{13} & f_{33} & 0 \\ 0 & 0 & f_{11}-f_{12} \end{pmatrix}\begin{pmatrix} s_1 \\ s_2 \\ s_3 \end{pmatrix} \\ &= \tfrac{1}{2}\mathbf{s}^t\mathbf{F}_s\mathbf{s} \end{aligned} \tag{7.131}$$

or

$$\begin{aligned} V &= \tfrac{1}{2}\mathbf{r}^t\mathbf{F}_r\mathbf{r} = \tfrac{1}{2}(\mathbf{U}^t\mathbf{s})^t\mathbf{F}_r\mathbf{U}^t\mathbf{s} \\ &= \tfrac{1}{2}\mathbf{s}^t(\mathbf{U}\mathbf{F}_r\mathbf{U}^t)\mathbf{s}. \end{aligned} \tag{7.132}$$

Thus, the internal coordinate force constant matrix is related to the symmetry coordinate force constant matrix by the equation

$$\mathbf{F}_s = \mathbf{U}\mathbf{F}_r\mathbf{U}^t. \tag{7.133}$$

The change of basis affects the force constant matrix via a similarity transformation. Let the symmetry-adapted force constants of H_2O be defined by the equation

$$V = \tfrac{1}{2}(s_1, s_2, s_3)\begin{pmatrix} F_{11} & F_{12} & 0 \\ F_{12} & F_{22} & 0 \\ 0 & 0 & F_{33} \end{pmatrix}\begin{pmatrix} s_1 \\ s_2 \\ s_3 \end{pmatrix} \tag{7.134}$$

with $F_{11} = f_{11} + f_{12}$, $F_{12} = \sqrt{2}f_{13}$ and $F_{33} = f_{11} - f_{12}$. The force constant matrix $\mathbf{F}_s$ is thus block factored into a 2×2 A_1 block and a 1×1 B_2 block. The capital F_{ij} are symmetry-adapted force constants, while lowercase f_{ij} are internal coordinate force constants.

The water molecule is described at the harmonic oscillator level by four force constants and three normal modes. There is not enough information in an infrared spectrum of H_2O to determine the force field since there are four unknowns to be derived from three vibrational frequencies. This is a general problem that gets worse as the molecule becomes larger and less symmetric. The only solution is to use vibrational frequencies from isotopic molecules such as D_2O and assume that the force constants and the geometry are the same, but that the nuclear masses are different.

The major problem with using internal coordinates or symmetry coordinates is that the kinetic energy operator becomes more complicated[17] than when Cartesian coordinates are used, that is,

$$\begin{aligned} T &= \tfrac{1}{2}\sum_{i=1}^{3N} \dot{q}_i^2 \\ &= \tfrac{1}{2}\sum_{i=1}^{3N-6}\sum_{j=1}^{3N-6} (G^{-1})_{ij}\dot{s}_i\dot{s}_j. \end{aligned} \tag{7.135}$$

When Cartesian coordinates are used, the classical kinetic energy operator contains no cross-terms between coordinates. When internal or symmetry coordinates are used, however, there are cross-terms $\dot{s}_i(G^{-1})_{ij}\dot{s}_j$ connecting the various coordinates. Of course, if symmetry coordinates are used, there can be no terms with $\dot{s}_i$ and $\dot{s}_j$ of different symmetry. This means that the T and V matrices have the same block structure and are factored into blocks belonging to the same irreducible representations. The matrix elements $(G^{-1})_{ij}$ or, more easily, the G_{ij} coefficients can be derived, given a molecular geometry.

The classical normal modes of vibration can be derived using symmetry coordinates in just the same way as for mass-weighted Cartesian displacement coordinates. The kinetic energy T is

$$T = \tfrac{1}{2}\dot{\mathbf{s}}^t\mathbf{G}^{-1}\dot{\mathbf{s}} \tag{7.136}$$

so that the classical Hamiltonian can be written as

$$H = \tfrac{1}{2}\dot{\mathbf{s}}^t\mathbf{G}^{-1}\dot{\mathbf{s}} + \tfrac{1}{2}\mathbf{s}^t\mathbf{F}\mathbf{s}. \tag{7.137}$$

A solution of the form

$$s_j = A_j \cos(\lambda^{1/2}t + \phi) \tag{7.138}$$

will be assumed. The Lagrangian is

$$L = T - V \tag{7.139}$$

and Lagrange's equations are

$$\frac{d}{dt}\left(\frac{\partial L}{\partial \dot{s}_i}\right)_{s_i} - \left(\frac{\partial L}{\partial s_i}\right)_{\dot{s}_i} = 0. \tag{7.140}$$

Using equations (7.136) and (7.137) in equation (7.140) results in

$$\frac{d}{dt}\sum_j (G^{-1})_{ij}\dot{s}_j - \left(-\sum_j F_{ij}s_j\right) = 0, \qquad i = 1, \ldots, 3N-6 \tag{7.141}$$

or

$$\sum_j (G^{-1})_{ij}\ddot{s}_j + \sum_j F_{ij}s_j = 0. \tag{7.142}$$

Substitution of the assumed solution (7.138) into equation (7.142) gives

$$\sum_j (G^{-1})_{ij}(-\lambda)A_j \cos(\lambda^{1/2}t + \phi) + \sum_j F_{ij}A_j \cos(\lambda^{1/2}t + \phi) = 0 \tag{7.143}$$

or

$$\sum_j (-\lambda(G^{-1})_{ij} + F_{ij})A_j = 0, \qquad i = 1, \ldots, 3N-6. \tag{7.144}$$

In matrix form equation (7.144) is written as

$$(-\lambda\mathbf{G}^{-1} + \mathbf{F})\mathbf{A} = 0. \tag{7.145}$$

Multiplying by **G** from the left gives

$$(-\lambda\mathbf{I} + \mathbf{GF})\mathbf{A} = 0. \tag{7.146}$$

Thus equation (7.146) is a set of $3N-6$ homogeneous linear equations that have a nontrivial solution if the determinant of the coefficients is zero, that is,

$$|\mathbf{GF} - \lambda\mathbf{I}| = 0. \tag{7.147}$$

This is the celebrated **GF** matrix solution for the vibrational modes of a polyatomic molecule.

The **G** matrix is easier to derive than $\mathbf{G}^{-1}$, but still requires some work. The technique of deriving **G** matrix elements is given in the classic book by Wilson, Decius, and Cross.[17] For example, the **G** matrix for H_2O is[18]

$$\mathbf{G} = \begin{pmatrix} \mu_H + \mu_O(1+\cos\theta) & -\sqrt{2}\mu_O \sin\theta & 0 \\ -\sqrt{2}\mu_O \sin\theta & 2(\mu_H + \mu_O - \mu_O\cos\theta) & 0 \\ 0 & 0 & \mu_H + \mu_O(1-\cos\theta) \end{pmatrix}$$

$$= \begin{pmatrix} 1.039 & -0.0858 & 0 \\ -0.0858 & 2.139 & 0 \\ 0 & 0 & 1.070 \end{pmatrix} \tag{7.148}$$

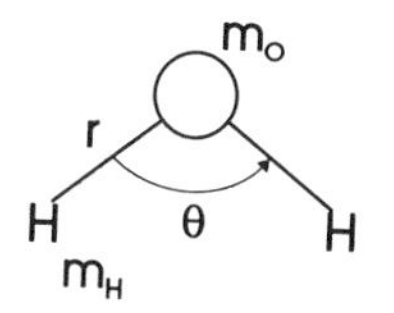

Figure 7.17. The H_2O molecule.

with $\mu_H = 1/m_H$, $\mu_O = 1/m_O$ for $m_H = 1.008$ amu, $m_O = 16$ amu, $\theta = 104°$, and $r =$ 0.958 Å (Figure 7.17).

A typical force constant analysis proceeds by selecting initial values for the force constants F_{11}, F_{12}, F_{22}, and F_{33}, and then calculating λ_1, λ_2, and λ_3, the three eigenvalues of the **GF** matrix for both H_2O and D_2O. The **G** matrix is different for H_2O and D_2O, but the **F** matrix is the same. The six calculated vibrational frequencies are then compared with the experimentally measured frequencies and the values of the four force constants are adjusted through iterative refinement to improve the agreement between observed and calculated frequencies. The iterative refinement involves a nonlinear least squares fitting procedure to minimize the sum of the squared deviations between the observed and calculated vibrational frequencies.

Force constants[18] for some bent XY_2 molecules are provided in Table 7.3 in units of millidynes/Å (1 millidyne/Å = 100 N/m).

The Symmetry of the Normal Modes of Vibration

As an example, consider the H_2O molecule. Three Cartesian coordinates are required to specify the position of each atom in space (Figure 7.18). Of the resulting nine degrees of freedom, however, three coordinates are required to locate the center of mass and three additional coordinates, for example, the Euler angles θ, ϕ, and χ, specify the orientation of the molecule in space. The spherical polar coordinates θ and ϕ specify how the molecular z-axis is oriented relative to the laboratory Z-axis. The Euler angle χ specifies the relative angular position of the plane containing the two hydrogen atoms and the O atom with respect to the molecular z-axis (Figure 7.19).

For a linear molecule the angle χ has no meaning and can be arbitrarily chosen. Therefore $3N - 5$ coordinates are necessary to describe the relative internal positions of the atoms in a linear molecule (Figure 7.20), but $3N - 6$ are needed for a nonlinear molecule.

Table 7.3. Force Constants in Millidyne/Å for Some XY_2 Molecules

V Term	Interaction	H_2O	F_2O	O_3	SO_2
$\frac{1}{2}f_{11}(\Delta r_1)^2$	f_{11} (X–Y stretch)	7.684	3.97	5.74	10.33
$f_{12}(\Delta r_1 \Delta r_2)$	f_{12} (stretch–stretch interaction)	−0.082	0.83	1.57	0.08
$\frac{1}{2}f_{33}(r\Delta\theta)^2$	f_{33} (bend)	0.707	0.70	1.26	0.82
$f_{13}\Delta r_1(r\Delta\theta)$	f_{13} (bend–stretch interaction)	0.169	0.15	0.39	0.23

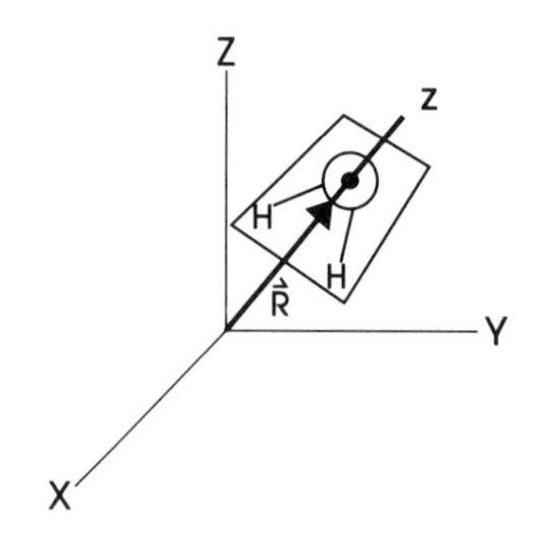

Figure 7.18. The H_2O molecule oriented in space relative to the center-of-mass vector **R**.

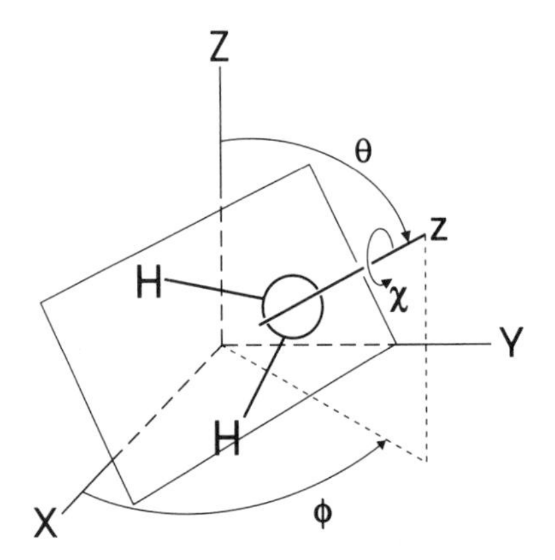

Figure 7.19. The orientation of the H_2O molecule in space, specified by the three Euler angles θ, ϕ, and χ.

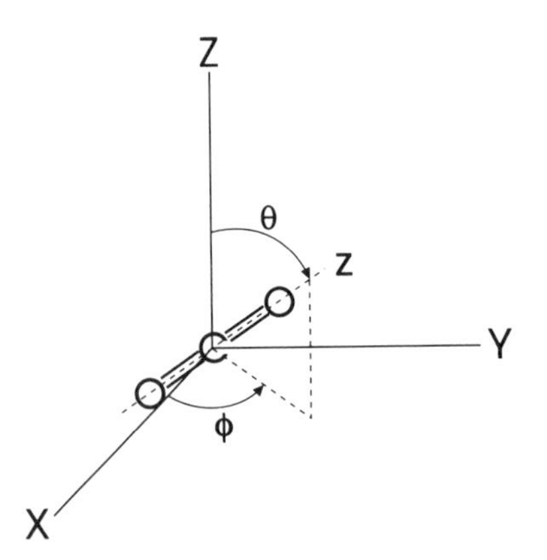

Figure 7.20. The orientation of a linear molecule in space, specified by the two angles θ and ϕ.

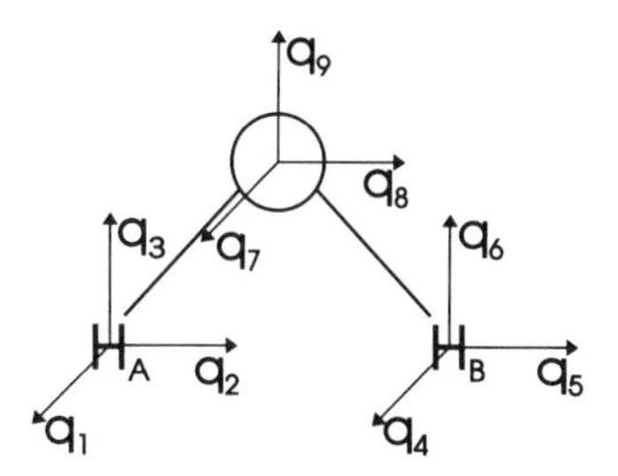

Figure 7.21. The nine mass-weighted Cartesian displacement coordinates for H_2O.

Following the methods outlined in Chapter 3, the set of $3N$ mass-weighted Cartesian displacement coordinates (Figure 7.21) can be used to construct a reducible representation for the group. For example, $\mathbf{D}(E)$ is given by

$$(q_1' \, q_2' \, q_3' \, q_4' \, q_5' \, q_6' \, q_7' \, q_8' \, q_9')$$

$$= (q_1 \, q_2 \, q_3 \, q_4 \, q_5 \, q_6 \, q_7 \, q_8 \, q_9) \begin{pmatrix} 1 & 0 & 0 & 0 & 0 & 0 & 0 & 0 & 0 \\ 0 & 1 & 0 & 0 & 0 & 0 & 0 & 0 & 0 \\ 0 & 0 & 1 & 0 & 0 & 0 & 0 & 0 & 0 \\ 0 & 0 & 0 & 1 & 0 & 0 & 0 & 0 & 0 \\ 0 & 0 & 0 & 0 & 1 & 0 & 0 & 0 & 0 \\ 0 & 0 & 0 & 0 & 0 & 1 & 0 & 0 & 0 \\ 0 & 0 & 0 & 0 & 0 & 0 & 1 & 0 & 0 \\ 0 & 0 & 0 & 0 & 0 & 0 & 0 & 1 & 0 \\ 0 & 0 & 0 & 0 & 0 & 0 & 0 & 0 & 1 \end{pmatrix}. \tag{7.149}$$

The effect of a $\hat{C}_2$ operation on the nine q's is shown in Figure 7.22. Notice that the $\hat{C}_2$ operation has left the atoms fixed, but has changed the coordinates, since we are working with the nine *displacement* coordinates. The matrix representation for $\hat{C}_2$, $\mathbf{D}(C_2)$ can be derived by inspection from Figure 7.22.

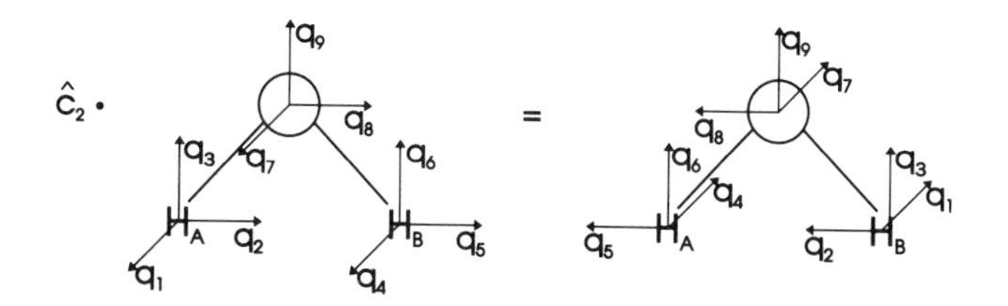

Figure 7.22. The effect of the $\hat{C}_2$ operation on H_2O.

$$(q_1'\, q_2'\, q_3'\, q_4'\, q_5'\, q_6'\, q_7'\, q_8'\, q_9') = (q_1\, q_2\, q_3\, q_4\, q_5\, q_6\, q_7\, q_8\, q_9) \begin{pmatrix} 0 & 0 & 0 & -1 & 0 & 0 & 0 & 0 & 0 \\ 0 & 0 & 0 & 0 & -1 & 0 & 0 & 0 & 0 \\ 0 & 0 & 0 & 0 & 0 & 1 & 0 & 0 & 0 \\ -1 & 0 & 0 & 0 & 0 & 0 & 0 & 0 & 0 \\ 0 & -1 & 0 & 0 & 0 & 0 & 0 & 0 & 0 \\ 0 & 0 & 1 & 0 & 0 & 0 & 0 & 0 & 0 \\ 0 & 0 & 0 & 0 & 0 & 0 & -1 & 0 & 0 \\ 0 & 0 & 0 & 0 & 0 & 0 & 0 & -1 & 0 \\ 0 & 0 & 0 & 0 & 0 & 0 & 0 & 0 & 1 \end{pmatrix}. \tag{6.150}$$

From Figure 7.23, the matrix representation of $\hat{\sigma}_v(xz)$ is

$$\mathbf{D}(\sigma_v(xz)) = \begin{pmatrix} 0 & 0 & 0 & 1 & 0 & 0 & 0 & 0 & 0 \\ 0 & 0 & 0 & 0 & -1 & 0 & 0 & 0 & 0 \\ 0 & 0 & 0 & 0 & 0 & 1 & 0 & 0 & 0 \\ 1 & 0 & 0 & 0 & 0 & 0 & 0 & 0 & 0 \\ 0 & -1 & 0 & 0 & 0 & 0 & 0 & 0 & 0 \\ 0 & 0 & 1 & 0 & 0 & 0 & 0 & 0 & 0 \\ 0 & 0 & 0 & 0 & 0 & 0 & 1 & 0 & 0 \\ 0 & 0 & 0 & 0 & 0 & 0 & 0 & -1 & 0 \\ 0 & 0 & 0 & 0 & 0 & 0 & 0 & 0 & 1 \end{pmatrix}. \tag{7.151}$$

From Figure 7.24, the matrix representation of $\hat{\sigma}(yz)$ is

$$\mathbf{D}(\sigma_v(yz)) = \begin{pmatrix} -1 & 0 & 0 & 0 & 0 & 0 & 0 & 0 & 0 \\ 0 & 1 & 0 & 0 & 0 & 0 & 0 & 0 & 0 \\ 0 & 0 & 1 & 0 & 0 & 0 & 0 & 0 & 0 \\ 0 & 0 & 0 & -1 & 0 & 0 & 0 & 0 & 0 \\ 0 & 0 & 0 & 0 & 1 & 0 & 0 & 0 & 0 \\ 0 & 0 & 0 & 0 & 0 & 1 & 0 & 0 & 0 \\ 0 & 0 & 0 & 0 & 0 & 0 & -1 & 0 & 0 \\ 0 & 0 & 0 & 0 & 0 & 0 & 0 & 1 & 0 \\ 0 & 0 & 0 & 0 & 0 & 0 & 0 & 0 & 1 \end{pmatrix} \tag{7.152}$$

The characters for the 9×9 representations are

$$\chi^{3N}(\hat{E}) = 9, \qquad \chi^{3N}(\hat{C}_2) = -1, \qquad \chi^{3N}(\hat{\sigma}_v(xz)) = 1, \qquad \chi^{3N}(\hat{\sigma}_v(yz)) = 3.$$

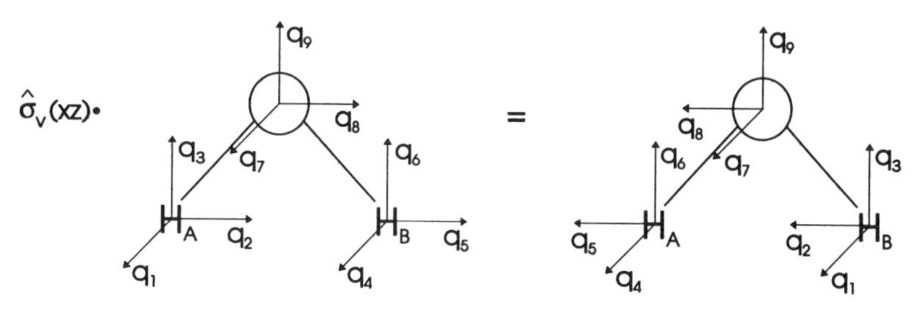

Figure 7.23. The effect of $\hat{\sigma}_v(xz)$ on H_2O.

These characters can be easily generated by inspection without writing down the complete matrices since only the diagonal elements of a matrix are needed to determine the character. For the $\hat{E}$ operation the contribution by each atom to $\chi(\hat{E})$ is 3 so that $\chi^{3N}(\hat{E}) = 3N$. If any displacement vectors are moved from one atom to another by a symmetry operation, then they contribute zero to the total character of that symmetry operation. The total character for any operation is the sum of the contributions from each atom. Any atom for which the displacement vectors are rotated by θ contributes $1 + 2\cos\theta$ to the total character $\chi(\hat{C}_n)$ since the rotation matrix is

$$\mathbf{D}(\hat{C}_\theta) = \begin{pmatrix} \cos\theta & \sin\theta & 0 \\ -\sin\theta & \cos\theta & 0 \\ 0 & 0 & 1 \end{pmatrix} \tag{7.153}$$

and the trace is $1 + 2\cos\theta$. Any $\hat{\sigma}$ operation contributes $+1$ for each atom on the symmetry plane to the total character since the trace of the reflection matrix

$$\mathbf{D}(\hat{\sigma}) = \begin{pmatrix} 1 & 0 & 0 \\ 0 & 1 & 0 \\ 0 & 0 & -1 \end{pmatrix} \tag{7.154}$$

is one. Since $\hat{S}_\theta = \hat{C}_\theta\hat{\sigma}_n$, each atom not displaced contributes $-1 + 2\cos\theta$ to the character for an improper rotation by θ. Finally the application of an inversion operation to an atom inverts the displacement vectors so that $\chi(\hat{\imath}) = -3$ for any atom at the center of symmetry.

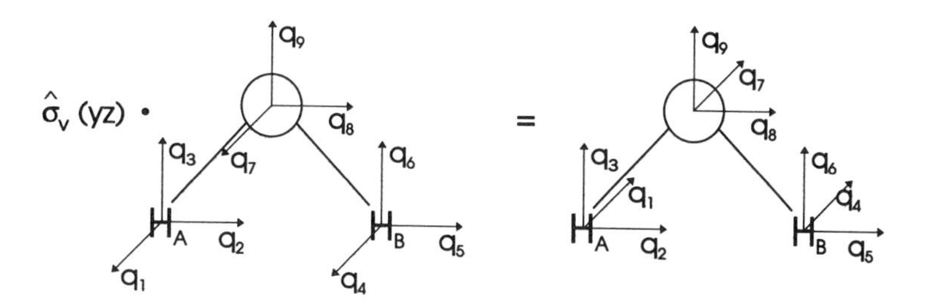

Figure 7.24. The effect of $\hat{\sigma}_v(yz)$ on H_2O.

The 9-dimensional representation can be reduced as

$$\Gamma^{3N} = \oplus \sum a_i \Gamma^i \tag{7.155}$$

using the relationship

$$a_i = \frac{1}{g} \sum_R \chi^{3N}(\hat{R}) \chi^i(\hat{R})^* \tag{7.156}$$

to give

$$\begin{aligned} a_{A_1} &= \tfrac{1}{4}(9 - 1 + 1 + 3) = 3, \\ a_{A_2} &= \tfrac{1}{4}(9 - 1 - 1 - 3) = 1, \\ a_{B_1} &= \tfrac{1}{4}(9 + 1 + 1 - 3) = 2, \\ a_{B_2} &= \tfrac{1}{4}(9 + 1 - 1 + 3) = 3, \end{aligned}$$

so that the reducible representation Γ^{3N} can be written as the direct sum of irreducible representations as

$$\Gamma^{3N} = 3\Gamma^{A_1} \oplus \Gamma^{A_2} \oplus 2\Gamma^{B_1} \oplus 3\Gamma^{B_2} \tag{7.157}$$

for the H_2O molecule. The $3N$ representation, however, still contains three translations and three rotations; these must be removed from the full representation to leave the symmetry representation of the pure vibrational motions.

The symmetry of the translational coordinates can be determined by considering the effect of the symmetry operations on the three Cartesian basis vectors, $\hat{\mathbf{i}}$, $\hat{\mathbf{j}}$, $\hat{\mathbf{k}}$, or on a point $\mathbf{r}$. The reasoning behind this is that the motion of the center of mass is equivalent to the translation of a point through space. For the C_{2v} case translational motion in the x direction behaves like the B_1 line of the character table (Appendix B) with respect to the symmetry operations, translational motion along y behaves like B_2, and along z behaves like A_1, that is,

$$\Gamma^{\text{tr}} = \Gamma^{A_1} \oplus \Gamma^{B_1} \oplus \Gamma^{B_2}. \tag{7.158}$$

The symmetry of the three rotational motions is more difficult to ascertain. Any arbitrary rotation can be expressed in terms of rotations about the x, y, and z-axes, which are designated as R_x, R_y, and R_z. The effect of symmetry operations on an arbitrary rotation can be determined by representing the rotation by a curved arrow to represent the fingers of the right hand, using the right-hand rule (Figure 7.25). The rotation is represented by the counterclockwise movement of the right-hand fingers. The effect of the point group symmetry operations on the movement of the right hand yields the symmetry. The sign of the rotation is given by the right-hand rule, that is, positive for counterclockwise rotation and negative for clockwise rotation. The effect of applying $\hat{C}_2$, $\hat{\sigma}_v(yz)$, and $\hat{\sigma}_v(xz)$ operations on R_z is illustrated in Figures 7.26, 7.27, and 7.28. From these figures, we find $\hat{C}_2 R_z = R_z$, $\hat{\sigma}_v(xz) R_z = -R_z$, $\hat{\sigma}_v(yz) R_z = -R_z$ and, of course, $\hat{E} R_z = R_z$. We find from the C_{2v} character table that R_z behaves like the

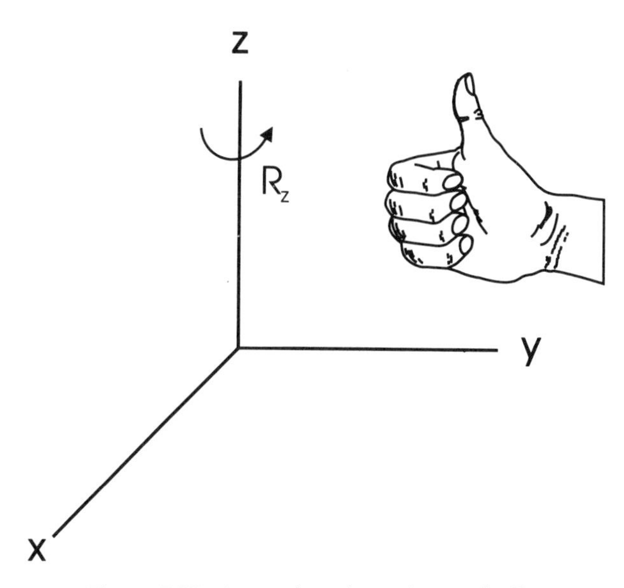

Figure 7.25. A rotation about the z-axis, R_z.

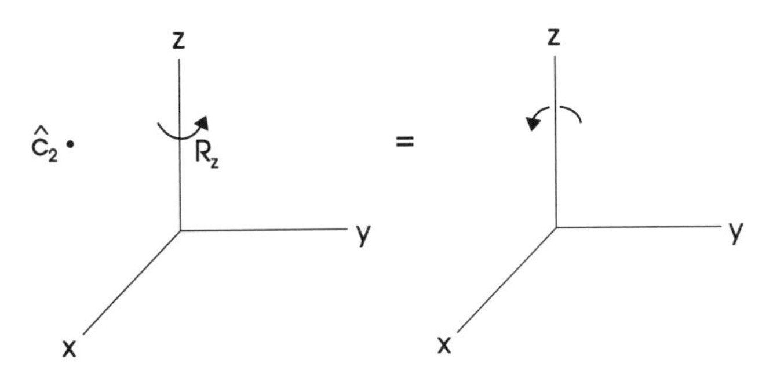

Figure 7.26. The effect of $\hat{C}_2$ on R_z.

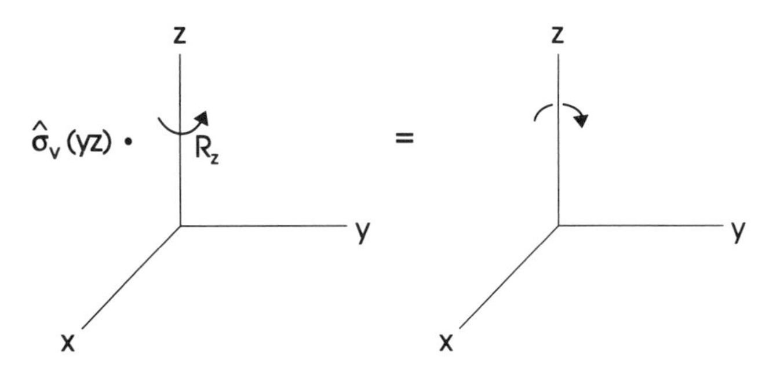

Figure 7.27. The effect of $\hat{\sigma}_v(yz)$ on R_z.

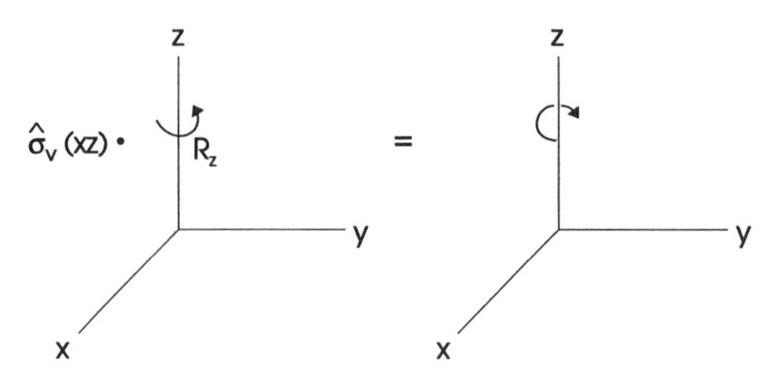

Figure 7.28. The effect of $\hat{\sigma}(xz)$ on R_z.

A_2 line. Similarly, R_x and R_y are found to have B_2 and B_1 symmetry, respectively, so that

$$\Gamma^{\text{rot}} = \Gamma^{A_2} \oplus \Gamma^{B_1} \oplus \Gamma^{B_2}. \tag{7.159}$$

Removing the translational and rotational degrees of freedom from Γ^{3N} leaves

$$\Gamma^{\text{vib}} = 2\Gamma^{A_1} \oplus \Gamma^{B_2}. \tag{7.160}$$

At first sight this procedure might appear to be flawed because the problem was set up using Cartesian displacement coordinates instead of normal coordinates. The mass-weighted Cartesian displacement coordinates q_i are related, however, to the normal coordinates Q_i by an orthogonal transformation

$$\mathbf{Q} = \mathbf{lq} \qquad \text{or} \qquad \mathbf{q} = \mathbf{l}^t\mathbf{Q} \tag{7.161}$$

in which the 6 (or 5) rotations and translations are included in the set of Q_i.

The $3N$-dimensional matrix representation is generated from the equation

$$(\mathbf{q}')^t = \mathbf{q}^t\mathbf{D}^{3N}(\hat{R}) \tag{7.162}$$

whereby substituting equation (7.161) for $\mathbf{q}$ into equation (7.162) gives

$$(\mathbf{l}^t\mathbf{Q}')^t = (\mathbf{l}^t\mathbf{Q})^t\mathbf{D}^{3N}(\hat{R}) \tag{7.163}$$

or

$$(\mathbf{Q}')^t\mathbf{l} = \mathbf{Q}^t\mathbf{l}\mathbf{D}^{3N}(\hat{R}). \tag{7.164}$$

Upon multiplying equation (7.164) from the right by the inverse $\mathbf{l}^t$ we obtain

$$(\mathbf{Q}')^t = \mathbf{Q}^t\mathbf{l}\mathbf{D}^{3N}(\hat{R})\mathbf{l}^t = \mathbf{Q}^t\mathbf{D}^Q(\hat{R}) \tag{7.165}$$

so that

$$\mathbf{D}^Q(\hat{R}) = \mathbf{l}\mathbf{D}^{3N}(\hat{R})\mathbf{l}^t. \tag{7.166}$$

The matrix representation generated by using normal modes is thus related by a similarity transformation to that generated using mass-weighted Cartesian displacement coordinates. This means that the two representations are equivalent and have the same characters. Thus the symmetries of the normal modes of vibration Q_i are correctly generated by using the q_i.

For the H_2O molecule the three modes correspond to the[19]

symmetric stretching mode	$\nu_1(a_1)$	$3657\ \text{cm}^{-1}$,
bending mode	$\nu_2(a_1)$	$1595\ \text{cm}^{-1}$,
and antisymmetric stretching mode	$\nu_3(b_2)$	$3756\ \text{cm}^{-1}$,

where the rules for labeling normal modes of vibration have been used. Normal modes are labeled in numerical order as ν_1, ν_2, ν_3, and so on. The order used for labeling modes follows the order listed in Herzberg's character tables[18] (sometimes called the Herzberg order) that proceeds as follows: A_1 before A_2 before B_1 before B_2, using C_{2v} as an example. For a given symmetry type the frequencies of the modes are arranged in descending order. Finally, lowercase letters are used to describe individual modes similar to the use of lowercase letters for the individual molecular orbitals (Chapters 9 and 10). The use of capital letters for irreducible representations is restricted to the total vibrational or electronic symmetry of a molecule. In the case of H_2O the two a_1 modes precede the b_2 mode and among the two a_1 modes ν_1 is chosen to be the higher frequency symmetric stretching mode. Degenerate modes are given only a single label so that for NH_3, $3N - 6 = 6$, but the modes are $\nu_1(a_1)$, $\nu_2(a_1)$, $\nu_3(e)$, and $\nu_4(e)$. In this case there are two modes for each frequency ν_3 and ν_4.

Selection Rules for Vibrational Transitions

The vibrational wavefunction for the ground state is given by the totally symmetric product

$$\begin{aligned}\psi_0 &= \phi_0(\xi_1)\phi_0(\xi_2)\cdots\phi_0(\xi_{3N-6})\\ &= N_0 e^{-\xi_1^2/2}e^{-\xi_2^2/2}\cdots e^{-\xi_{3N-6}^2/2}\end{aligned} \tag{7.167}$$

in which

$$\xi_i = Q_i\left(\frac{\omega_i}{\hbar}\right)^{1/2},$$

and N_0 is a normalization constant. The ground state wavefunction belongs to the A_1 irreducible representation, since all of the group operations leave ψ_0 unchanged. If the jth vibrational mode is excited by one quantum, then the wavefunction becomes

$$\begin{aligned}\psi_1 &= \phi_0(\xi_1)\cdots\psi_1(\xi_j)\cdots\psi_0(\xi_{3N-6})\\ &= N_1 e^{-\xi_1^2/2}\cdots H_1(\xi_j)e^{-\xi_j^2/2}\cdots e^{-\xi_{3N-6}^2/2},\end{aligned} \tag{7.168}$$

in which the harmonic oscillator wavefunction for $v_j = 1$ has replaced that for $v_j = 0$. The Hermite polynomial part of the wavefunction is $H_1(\xi_j) = 2\xi_j$ for $v_j = 1$ (Table 7.1). The fact that ξ_j is proportional to Q_j means that ψ_1 is of the same symmetry as Q_j.

The intensity of an infrared transition is given by the absolute square of the transition moment integral

$$\mathbf{M} = \int \psi_f^* \boldsymbol{\mu} \psi_i \, d\tau = \int \psi_f^*(Q) \boldsymbol{\mu}(Q) \psi_i(Q) \, dQ, \tag{7.169}$$

in which ψ_f and ψ_i are final and initial vibrational wavefunctions within the same electronic state, $\boldsymbol{\mu}$ is the dipole moment function, and the integral is over all vibrational coordinates ($dQ = dQ_1 \, dQ_2 \cdots dQ_{3N-6}$). The dipole moment depends on the positions of the nuclei and hence on the set of Q_i. By expressing the dipole moment as a Taylor series expansion

$$\boldsymbol{\mu} = \boldsymbol{\mu}_0 + \sum_{k=1}^{3N-6} \left(\frac{\partial \boldsymbol{\mu}}{\partial Q_k} \right)_0 Q_k + \cdots, \tag{7.170}$$

equation (7.169) becomes

$$\mathbf{M} = \boldsymbol{\mu}_0 \int \psi_f^* \psi_i \, dQ + \sum_{k=1}^{3N-6} \left(\frac{\partial \boldsymbol{\mu}}{\partial Q_k} \right)_0 \int \psi_f^* Q_k \psi_i \, dQ + \cdots. \tag{7.171}$$

The first term on the right-hand side of this expression is zero because the vibrational wavefunctions are orthogonal. For a fundamental vibrational transition

$$\psi_i = \phi_0(\xi_1) \cdots \phi_0(\xi_{3N-6}) \tag{7.172}$$

and

$$\psi_f = \phi_0(\xi_1) \cdots \phi_1(\xi_j) \cdots \phi_0(\xi_{3N-6}), \tag{7.173}$$

in which ψ_f differs from ψ_i only in the jth normal mode so that the intensity of a vibrational mode is

$$I_{f \leftarrow i} \propto \left| \left(\frac{\partial \boldsymbol{\mu}}{\partial Q_j} \right)_0 \right|^2 \left| \int \phi_1(\xi_j) Q_j \phi_0(\xi_j) \, dQ_j \right|^2, \tag{7.174}$$

where all of the terms but the jth one in the sum in equation (7.171) vanish due to the orthogonality of the Hermite polynomials and the neglect of higher order terms. In fact $I_{f \leftarrow i} = 0$ unless a single vibrational mode changes its vibrational quantum number by one unit, leading to the selection rule $\Delta v_i = \pm 1$ (arising from the properties of Hermite polynomials). This selection rule directly follows from the use of the harmonic oscillator wavefunctions for ϕ_i and from the truncation of the expansion of the dipole moment (7.170) at terms linear in Q. Consequently, to a first-order approximation, the intensity of an infrared transition is proportional to the square of the dipole moment derivative (7.174).

Table 7.4. The C_{2v} Character Table

C_{2v}	$\hat{E}$	$\hat{C}_2$	$\hat{\sigma}_v(xz)$	$\hat{\sigma}_v(yz)$	
A_1	1	1	1	1	z
A_2	1	1	−1	−1	R_z
B_1	1	−1	1	−1	x, R_y
B_2	1	−1	−1	1	y, R_x

The intensity of electric dipole-allowed vibrational transitions is given by equations (7.169) and (7.174). The integrand $\psi_f^* \boldsymbol{\mu} \psi_i$ must be totally symmetric and $\Gamma(\psi_f^*) \otimes \Gamma(\boldsymbol{\mu}) \otimes \Gamma(\psi_i)$ must therefore contain the A_1 irreducible representation. For fundamental transitions, ψ_i has A_1 symmetry, while ψ_f^* belongs to the irreducible representation of the jth mode, which is excited up to $v_j = 1$. The dipole moment operator is a vector $\boldsymbol{\mu} = \mu_x \hat{\mathbf{e}}_1 + \mu_y \hat{\mathbf{e}}_2 + \mu_z \hat{\mathbf{e}}_3$ that behaves like the point $\mathbf{r} = x\hat{\mathbf{e}}_1 + y\hat{\mathbf{e}}_2 + z\hat{\mathbf{e}}_3$ when the symmetry operations of the group are applied. This implies that

$$\Gamma(\psi_f^*) \otimes \Gamma(\boldsymbol{\mu}) \otimes \Gamma(\psi_i) = \Gamma(\psi_f^*) \otimes \Gamma(\mathbf{r}), \tag{7.175}$$

and ψ_f must have the same symmetry as x or y or z to make the direct product symmetric. For convenience the symmetry of the Cartesian x, y, and z components are listed to the right of the character table (Table 7.4), as are the rotations R_x, R_y, and R_z about the x-, y-, and z-axes. All three normal modes of H_2O are infrared active since $\nu_1(a_1)$, $\nu_2(a_1)$, and $\nu_3(b_2)$ have the same symmetry as z and y.

Considerable information about the vibrational modes of molecules can be predicted with the use of character tables and a table of characteristic vibrational frequencies (Table 7.5). For example, the chloroform molecule $HCCl_3$ molecule of C_{3v} symmetry has nine normal modes that reduce as

$$\Gamma^{\text{vib}} = 3\Gamma^{a_1} \oplus 3\Gamma^{e}.$$

There are six distinct fundamental vibrational frequencies and all modes are infrared (and Raman) active. Since there are three C—Cl bonds and one C—H bond there must be four stretching modes and $9 - 4 = 5$ bending modes. If the three C—Cl stretching modes are represented by three bond stretching coordinates Δr_1, Δr_2, and Δr_3, then the 3-dimensional representation reduces to $a_1 \oplus e$. Consulting the group frequency table (Table 7.5) one therefore predicts a C—H stretching mode of $2960\ \text{cm}^{-1}$ (a_1) and two C—Cl stretching frequencies (a_1 and e) at $650\ \text{cm}^{-1}$.

The symmetry of the bending modes can be predicted by removing two a_1 modes and one e mode from the total of three a_1 and three e modes. The bending modes must have a_1, e, and e symmetry. For the C—Cl bonds, three bond-bending coordinates can be defined ($\Delta\theta_1$, $\Delta\theta_2$, and $\Delta\theta_3$) and they form a reducible 3-dimensional representation. The three C—Cl bending modes therefore reduce to an

Table 7.5 Infrared group wavenumber table

Group	ν/cm^{-1}	Group	ν/cm^{-1}
≡C—H	3300	—O—H	3600
=C<H	3020	>N—H	3350
≡C—H	2960	>P=O	1295
—C≡C—	2050		
		>S=O	1310
>C=C<	1650		
		≡C—H (bend)	700
≡C—C≡	900		
—S— H	2500		
		=C<H, H (bend)	1100
—N=N—	1600		
>C=O	1700		
—C≡N	2100	—C—H, H, H (bend)	1000
≡C—F	1100		
≡C—Cl	650	>C<H, H (bend)	1450
≡C—Br	560		
≡C—I	500	C≡C—C (bend)	300

a_1 and an e mode. This means that the remaining bending mode must be a C—H bend of e symmetry. Predicting the frequency of bending modes is very difficult, but a bending C—H mode in $CHCl_3$ should have a somewhat lower frequency than in H_2O (say ~1000 cm^{-1}), while the C—Cl bends might be near 300 cm^{-1}. Bond bending mode frequencies are typically half of the bond stretching frequencies. The predictions and observations are summarized below in Table 7.6.

Table 7.6. Vibrational Modes of $CHCl_3$

Mode	Symmetry	Type of Mode	Prediction	Observation[20]
ν_1	a_1	C—H stretch	2960 cm^{-1}	3033 cm^{-1}
ν_2	a_1	symmetric C—Cl stretch	650 cm^{-1}	667 cm^{-1}
ν_3	a_1	symmetric C—Cl bend	300 cm^{-1}	364 cm^{-1}
ν_4	e	C—H bend	1000 cm^{-1}	1205 cm^{-1}
ν_5	e	C—Cl stretch	650 cm^{-1}	760 cm^{-1}
ν_6	e	C—Cl bend	300 cm^{-1}	260 cm^{-1}

The numbering of the frequencies follows the order in the character table and within a given symmetry type the modes are numbered in decreasing frequency order.

Vibration–Rotation Transitions of Linear Molecules

In many respects, the vibration–rotation transitions of linear polyatomic molecules closely resemble those of diatomic molecules. The molecular symmetry of linear polyatomic molecules is either $D_{\infty h}$ or $C_{\infty v}$ and there are $3N - 5$ modes of vibration. Let us consider the H—C≡N and H—C≡C—H molecules as examples. The HCN molecule[19] has three fundamental modes of vibration: the C—H stretching mode $\nu_1(\sigma)$ at 3311 cm^{-1}, the bending mode $\nu_2(\pi)$ at 713 cm^{-1}, and the C≡N stretching mode $\nu_3(\sigma)$ at 2097 cm^{-1}. Notice that linear triatomics are an exception to the frequency numbering scheme because $\nu_2(\pi)$ is reserved for the bending mode. Certain functional group such as C—H and C≡N have characteristic vibrational frequencies. Some of these group frequencies are listed in Table 7.5. The bending mode is doubly degenerate because of the possibility of bending in two mutually orthogonal planes. All modes of HCN are infrared active although ν_3 is very weak.

The acetylene molecule H—C≡C—H has $D_{\infty h} = C_{\infty v} \otimes C_i$ symmetry. There are $3N - 5 = 7$ modes with three stretching modes (the number of stretching modes = the number of bonds) and $7 - 3 = 4$ bending modes. The fundamental modes[19] are:

the symmetric C—H stretch $\nu_1(\sigma_g)$ at 3373 cm^{-1},

the C≡C stretch $\nu_2(\sigma_g)$ at 1974 cm^{-1},

the antisymmetric C—H stretch $\nu_3(\sigma_u)$ at 3295 cm^{-1},

the *trans* bend H—C≡C—H (↓ on the left H, ↑ on the right H) $\nu_4(\pi_g)$ at 612 cm^{-1},

and the *cis* bend H—C≡C—H (↓ on both H) $\nu_5(\pi_u)$ at 729 cm^{-1}.

The numbering of the modes is determined by the conventional order of the irreducible representations in the $D_{\infty h}$ character table. Also, notice that the ν_4 and ν_5 bending modes are doubly degenerate with two modes associated with each frequency. The ground vibrational state has σ_g symmetry, and because z belongs to σ_u and x, y belong to π_u, only the σ_u and π_u modes (ν_3 and ν_5) are infrared active.

The number and types of normal modes can be quickly determined for all linear molecules. If there are N atoms, then there will be $N - 1$ stretching frequencies and $[(3N - 5) - (N - 1)]/2 = N - 2$ bending frequencies. In the case of symmetric molecules of $D_{\infty h}$ symmetry, the g or u labels need to be added by symmetrizing the stretching of bonds or the bending of the molecule. For example, for acetylene there is a symmetric C—H stretching mode of σ_g symmetry ($\nu_1 = 3373\ \text{cm}^{-1}$) and an antisymmetric C—H stretching mode of σ_u symmetry ($\nu_3 = 3295\ \text{cm}^{-1}$).

The fundamental vibrational transitions of linear molecules are either of the Σ–Σ (parallel) type for stretching modes or of the Π–Σ (perpendicular) type for bending modes. For symmetric linear molecules, which belong to the $D_{\infty h}$ point group, g and u subscripts are needed. The terms parallel and perpendicular are used because the transition dipole moment is either parallel (μ_z) or perpendicular (μ_x and μ_y) to the molecular z-axis. Allowed parallel transitions arise from the μ_z component of the transition dipole moment with σ_u symmetry,

$$\Gamma^{\psi_f^*} \otimes \Gamma^{\mu_z} \otimes \Gamma^{\psi_i} = \sigma_u^+ \otimes \sigma_u^+ \otimes \sigma_g^+ = \sigma_g^+, \tag{7.176}$$

while allowed perpendicular transitions arise from the μ_x and μ_y components,

$$\Gamma^{\psi_f^*} \otimes \Gamma^{\mu_{x,y}} \otimes \Gamma^{\psi_i} = \pi_u \otimes \pi_u \otimes \sigma_g^+ = \delta_g \oplus \sigma_g^+ \oplus \sigma_g^-. \tag{7.177}$$

The Σ–Σ transitions can only have P and R branches, so that the appearance of the spectrum closely resembles that of the infrared spectrum of a diatomic molecule (Figure 7.29).

The Π–Σ transitions have P, Q, and R branches as shown in Figure 7.30. The rotational energy levels associated with the Π state are doubly degenerate because $l = \pm 1$, where l is the quantum number of vibrational angular momentum (Chapter 6). As the molecule begins to rotate the two components for a given J begin to split

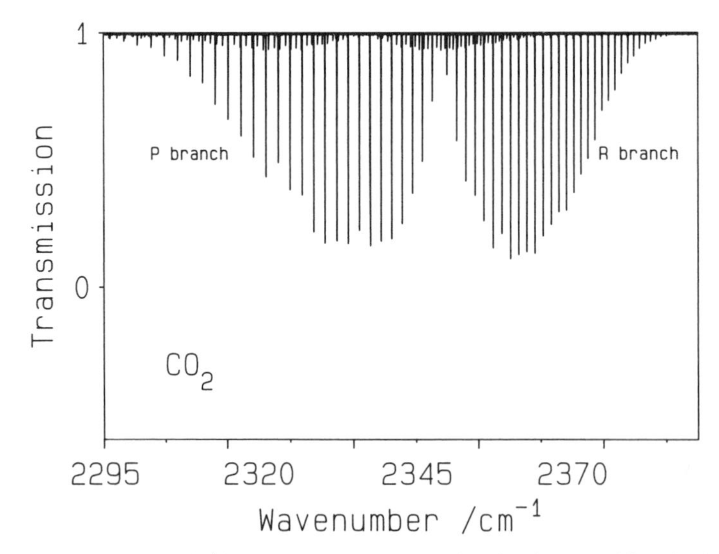

Figure 7.29. The $\nu_3(\sigma_u^+)$ antisymmetric stretching fundamental band of CO_2.

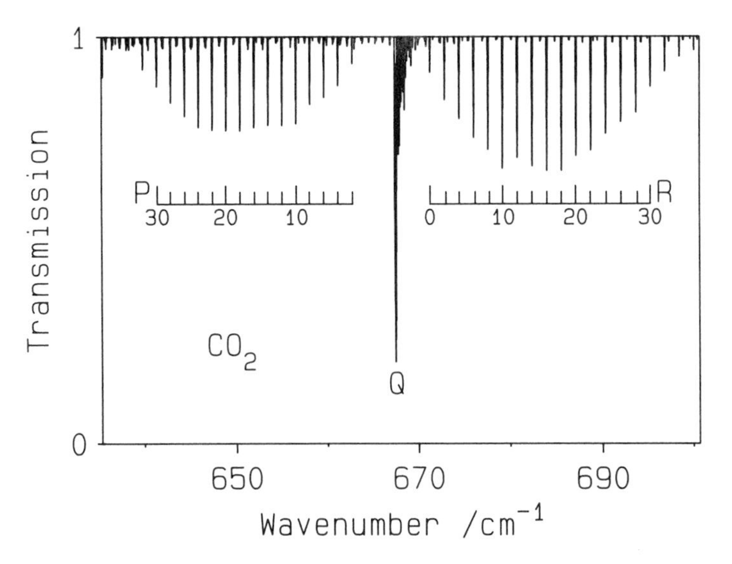

Figure 7.30. The $\nu_2(\pi_u)$ bending fundamental band of CO_2.

slightly because of the interaction of rotational (**J**) with vibrational angular momentum (**l**). The splitting $\Delta\nu$ is proportional to[19]

$$\Delta\nu = qJ(J+1) \tag{7.177}$$

and q is called the l-type doubling constant. It is useful to use parity labels to distinguish the two nearly degenerate levels for each J. There are many different types of parity, but the two most common varieties are total parity and e/f parity. A more detailed description of parity is provided in Chapter 9. Total parity considers the effect of inversion of all coordinates *in the laboratory frame* of the total wavefunction $\psi = \psi_{\text{el}}\psi_{\text{vib}}\psi_{\text{rot}}$. This inversion operation $\hat{E}^*$ inverts the laboratory coordinates of all atoms in a molecule,

$$\hat{E}^*\psi(X_i, Y_i, Z_i) = \psi(-X_i, -Y_i, -Z_i) = (\pm)\psi,$$

and leaves the wavefunction unchanged, except possibly for a change in sign. Total parity can be either positive + (upper sign) or negative − (lower sign). Total parity is commonly used to label the energy levels of atoms as well as the rotational energy levels of diatomic and linear molecules.

This laboratory symmetry operator $\hat{E}^*$ is different from the geometric molecular symmetry operator $\hat{\imath}$ discussed previously. Only $D_{\infty h}$ molecules have $\hat{\imath}$ as a symmetry operator, while all molecules have $\hat{E}^*$ as a symmetry operator. Note that $\hat{E}^*$ is a very peculiar operator because it inverts the entire molecular coordinate system as well as the location of the nuclei. It is therefore a permutation-inversion operator rather than a molecular symmetry operator of the type discussed in Chapters 2 and 3.

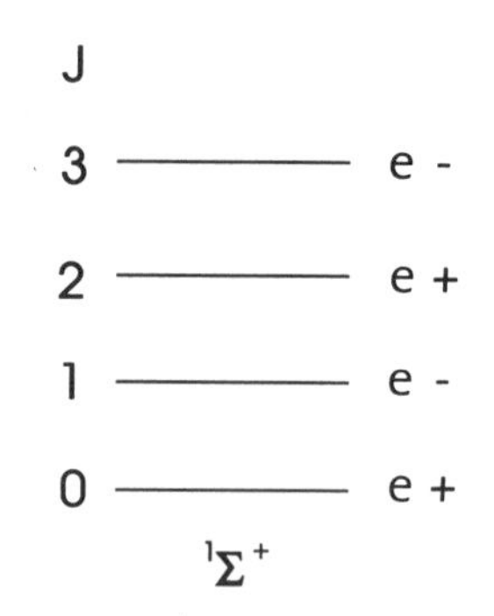

Figure 7.31. Parities of the rotational levels in a $^1\Sigma^+$ state.

The wavefunction can be written as $\psi = \psi_{el}\psi_{vib}\psi_{rot}$ and the effect of $\hat{E}^*$ on each part must be considered. The operation $\hat{E}^*$ leaves the relative positions of the nuclei unchanged, so $\hat{E}^*\psi_{vib} = \psi_{vib}$ for the nondegenerate σ^+ vibrations. The rotational part of the wavefunction, $\psi_{rot} = \psi_{JM}(\theta, \phi)$, changes sign for odd J under the operation $\hat{E}^*$, since

$$\hat{E}^* Y_{JM} = (-1)^J Y_{JM}. \tag{7.178}$$

The effect of $\hat{E}^*$ on ψ_{el} is much more difficult to ascertain because ψ_{el} is a function of internal molecular coordinates. It is possible to show the surprising result[21] that $\hat{E}^*$ in the laboratory frame is equivalent to $\hat{\sigma}_v$ (chosen to be $\hat{\sigma}_v(xz)$, for convenience) in the molecular frame. Thus for the totally symmetric electronic ground state $X^1\Sigma^+$,

$$\hat{E}^*\psi_{el}(x_i, y_i, z_i) = \hat{\sigma}_v\psi_{el}(x_i, y_i, z_i) = +1\psi_{el}. \tag{7.179}$$

The total parity of a linear molecule wavefunction alternates with J as shown in Figure 7.31 for a $^1\Sigma^+$ state. Since this alternation of total parity with J occurs for all electronic states, it is convenient to factor out the J dependence and designate those rotational levels with a total parity of $+(-1)^J$ as e parity and those with a total parity of $-(-1)^J$ as f parity (for half-integer J a total parity of $+(-1)^{J-1/2}$ corresponds to e and $-(-1)^{J-1/2}$ to f).[22] The e/f parity is thus a J independent parity labeling scheme for rovibronic wavefunctions (rovibronic = rotational × vibrational × electronic). All $^1\Sigma^+$ rotational energy levels, therefore, have e parity (Figure 7.31). The e/f parity labels correspond to the residual intrinsic parity of a rotational level after the $(-1)^J$ part has been removed.

The one-photon, electric-dipole selection rule $+ \leftrightarrow -$ is derived by recognizing that the parity of $\boldsymbol{\mu}$ is -1 while the parity of the transition moment integral

$$\int \psi_i^* \boldsymbol{\mu} \psi_f \, d\tau \tag{7.180}$$

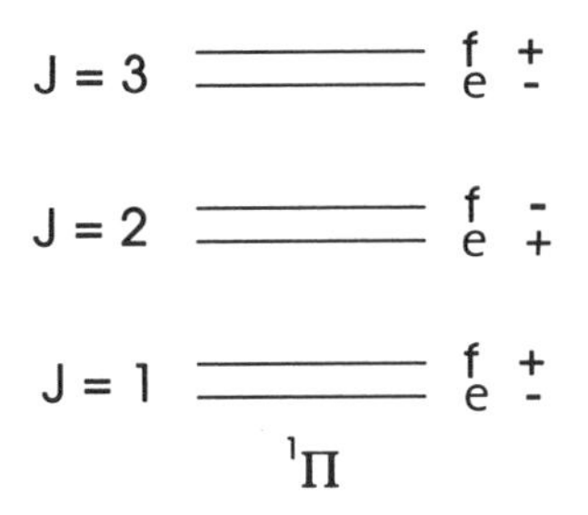

Figure 7.32. Parity labels for the rotational levels of a Π state.

must be +1. This +/− selection rule becomes $e \leftrightarrow e$, $f \leftrightarrow f$ for the P and R branches, and $e \leftrightarrow f$ for the Q branches in the e/f parity labeling scheme.

Parity labeling is essential when nearly degenerate energy levels are present, for example, in Π vibrational states of linear molecules. For a Π state $l = \pm 1$, and it is possible to form linear combinations of the vibrational wavefunctions that are eigenfunctions of $\hat{\sigma}_v$ from the 2-dimensional harmonic oscillator wavefunctions (Chapter 6),

$$\psi_{\text{vib}} = \phi(\rho)(e^{i\phi} \pm e^{-i\phi}). \tag{7.181}$$

Note that $\hat{\sigma}_v \Phi(\phi) = \Phi(\pi - \phi)$ so that the upper sign in equation (7.181) corresponds to f parity, while the lower sign corresponds to e parity. The total parity still changes with J, as shown in Figure 7.32, while the e/f ordering is determined by the sign of q, the l-doubling constant. Therefore, e/f parity labels are convenient for differentiating between the two near-degenerate levels associated with l-type doubling (Figure 7.32).

In a Π–Σ transition (Figure 7.33) the Q branch lines terminate on rotational levels

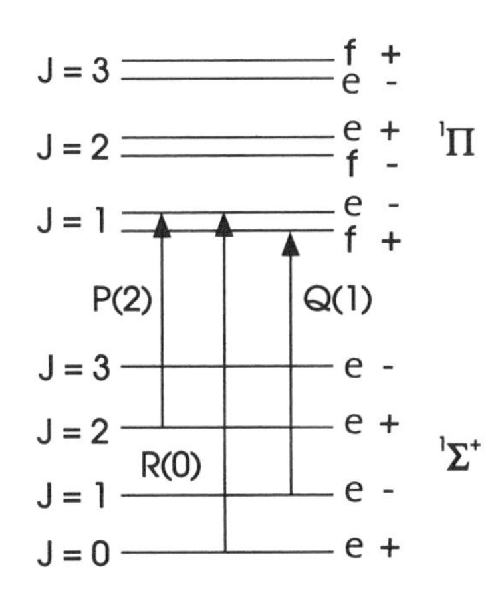

Figure 7.33. Energy-level diagram for a Π–Σ transition.

of the *opposite* parity ($f \leftrightarrow e$), as opposed to P and R branch lines that terminate on rotational levels of the *same* parity ($e \leftrightarrow e$). Thus the usual combination differences involving P and R branches provide rotational constants for the upper and lower levels of e parity only, while analysis of the Q branch yields constants for the upper levels of f parity.

The $+/-$ total parity labels or e/f parity labels are unrelated to the g/u vibrational labels used in $D_{\infty h}$ molecules in which

$$\hat{i}\psi_{\text{vib}} = \pm\psi, \tag{7.182}$$

where $+$ corresponds to g and $-$ corresponds to u. Since $D_{\infty h} = C_{\infty v} \otimes C_i$, the g and u labels are required when the molecule has a center of symmetry.

Nuclear Spin Statistics

An additional symmetry requirement is associated with the constraint placed on molecular wavefunctions by the Pauli exclusion principle. Because identical nuclei are indistinguishable, their exchange can, at most, change the sign of the total wavefunction, which includes nuclear spin or

$$\hat{P}_{12}\psi_{\text{total}} = \pm\psi_{\text{total}}, \tag{7.183}$$

in which the $\hat{P}_{12}$ operator exchanges identical nuclei. For particles with integer nuclear spin ($I = 0, 1, 2, \ldots$), called bosons, the sign in equation (7.183) is found to be positive ($+1$), while for fermions with half-integer nuclear spin ($I = \frac{1}{2}, \frac{3}{2}, \frac{5}{2}, \ldots$) the negative sign ($-1$) applies.

The total wave function can be written as a product of a nuclear spin part and a space part,

$$\psi_{\text{total}} = \psi_{\text{space}}\psi_{\text{spin}} = \psi\psi_{\text{spin}}, \tag{7.184}$$

so that the effect of $\hat{P}_{12}$ on either part can be examined separately. When $\hat{P}_{12}$ operates on the "normal" space part of the total wavefunction of a symmetric linear molecule

$$\hat{P}_{12}\psi = \pm\psi, \tag{7.185}$$

in which $+$ is symmetric or s and $-$ is antisymmetric or a.

Surprisingly, the $\hat{P}_{12}$ permutation operator in the laboratory frame is equivalent to the $\hat{C}_2(y)$ symmetry operator in the molecular frame of a diatomic molecule.[21] This equivalence can be rationalized if one remembers that the $\hat{C}_2(y)$ operation does not, in fact, exchange the nuclei. The $\hat{C}_2$ operator changes the electrons, the nuclear *displacement* vectors, and the rotational variables θ and ϕ, but leaves the positions of the nuclei unaltered. The location of the nuclei define the molecular z-axis, which is not affected by a symmetry operation such as $\hat{C}_2(y)$. Clearly this is physically equivalent to just interchanging the nuclei while leaving the positions of all of the other particles fixed.

The nature of the nuclear spin part of ψ_{total} depends on the particular nuclei under consideration. For example in the F—Be—F or H_2 molecules of $D_{\infty h}$ symmetry, the nuclear spins of F and H are $\frac{1}{2}$. The symmetric and antisymmetric nuclear-spin wavefunctions can be constructed for nuclei A and B with $\alpha = |m_I = +\frac{1}{2}\rangle$, $\beta = |m_I = -\frac{1}{2}\rangle$, as

$$\psi_{\text{spin}} = \begin{cases} \alpha(A)\alpha(B) & \qquad (7.186a) \\ \dfrac{\alpha(A)\beta(B) + \beta(A)\alpha(B)}{\sqrt{2}} \qquad \text{(symmetric)} & \qquad (7.186b) \\ \beta(A)\beta(B) & \qquad (7.186c) \end{cases}$$

and

$$\psi_{\text{spin}} = \frac{\alpha(A)\beta(B) - \beta(A)\alpha(B)}{\sqrt{2}} \qquad \text{(antisymmetric).} \tag{7.187}$$

The total wavefunctions must obey the equation

$$\hat{P}_{12}\psi_{\text{total}} = -\psi_{\text{total}}, \tag{7.188}$$

because H and F nuclei are both fermions. This means that s symmetry spatial wavefunctions must be combined with antisymmetric spin functions, while a symmetry spatial wavefunctions are combined with symmetric spin functions. Since there are three symmetric nuclear spin wavefunctions but only one antisymmetric function, the energy levels with a symmetry have statistical weights three time those of the s levels. This means that, all other things being equal, the transitions from a levels are three times as intense as are those from s levels. Note that s and a labels describe the wavefunction exclusive of nuclear spin.

The $\hat{P}_{12}$ permutation operator in the laboratory frame, or the $\hat{C}_2(y)$ symmetry in the molecular frame, needs to be applied to the total wavefunction

$$\psi = \psi_{\text{el}}\psi_{\text{vib}}\psi_{\text{rot}}. \tag{7.189}$$

Once again one can conclude that for a symmetric vibration and for a symmetric $^1\Sigma_g^+$ electronic state, $\hat{C}_2(y)$ has no effect. The operation of $\hat{C}_2(y)$ on ψ_{rot} replaces θ by $\pi - \theta$ and ϕ by $\pi + \phi$ in $Y_{JM}(\theta, \phi)$ or

$$\hat{C}_2(y)Y_{JM}(\theta, \phi) = Y_{JM}(\pi - \theta, \pi + \phi) = (-1)^J Y_{JM}(\theta, \phi). \tag{7.190}$$

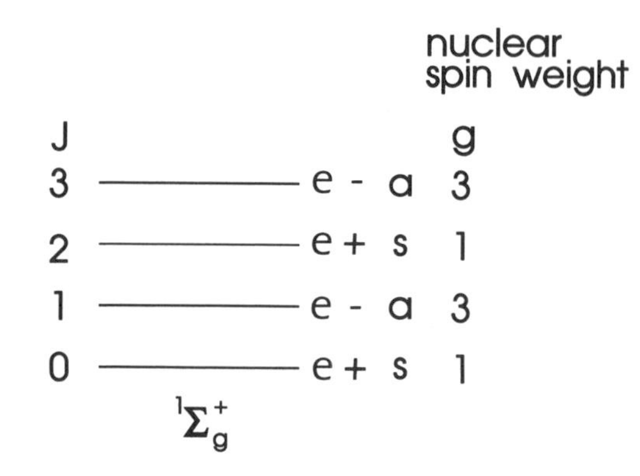

Figure 7.34. Nuclear spin weights and labels for a $^1\Sigma_g^+$ state with two equivalent spin-$\frac{1}{2}$ nuclei.

The energy-level diagram for H_2 or BeF_2 is illustrated in Figure 7.34. The selection rules on *s* and *a* are $s \leftrightarrow s$ and $a \leftrightarrow a$ for transitions. The levels with the larger nuclear spin weighting are designated ortho, while the levels with the smaller weighting are designated para. The effect of nuclear spin statistics can clearly be seen in the spectrum[23] of BeF_2 (Figure 7.35).

The *g* and *u* symmetry labels for an electronic or vibrational state are determined by the $\hat{i}$ symmetry operator acting in the molecular frame or by the $\hat{P}_{12}\hat{E}^*$ product of permutation–inversion operators in the laboratory frame. This is because the $\hat{E}^*$ operator first inverts all coordinates of all particles and then $\hat{P}_{12}$ places the nuclei back into their original positions. The net effect of $\hat{P}_{12}\hat{E}^*$ is thus to invert the electronic coordinates through the origin (i.e., the operator $\hat{i}$). This also means that the *s* or *a* symmetry of a rovibronic wavefunction ($\hat{P}_{12}$ operator) is determined by the *g* or *u*

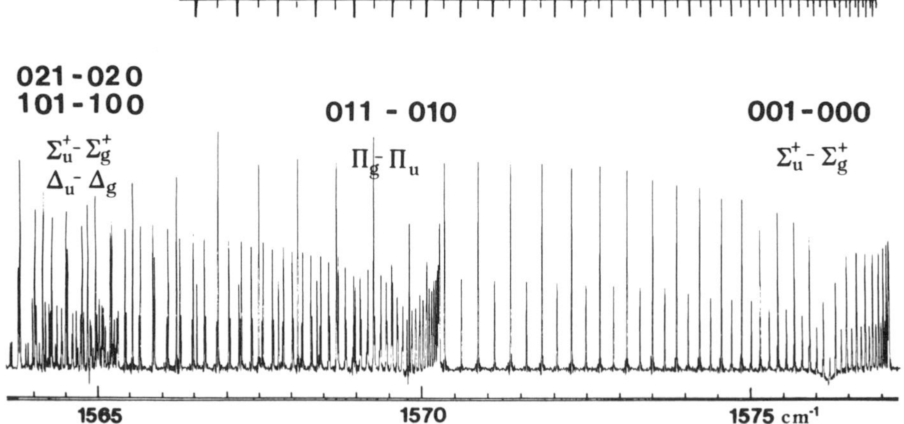

Figure 7.35. Infrared emission spectrum of the $\nu_3(\sigma_u)$ antisymmetric stretching mode of BeF_2.

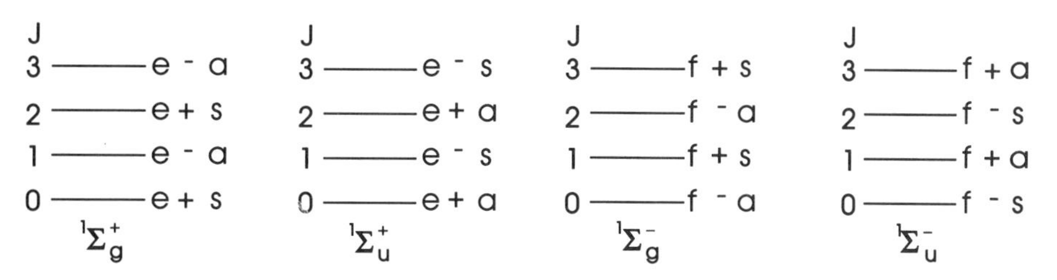

Figure 7.36. Parities for the rotational levels of $^1\Sigma$ states of $D_{\infty h}$ molecules.

symmetry ($\hat{i}$ or $\hat{P}_{12}\hat{E}^*$ operators) of the vibronic part of the wavefunction (vibronic = vibrational × electronic) and by the total parity ($\hat{E}^*$ operator). The various possibilities are illustrated in Figure 7.36.

The exchange of the two equivalent nuclei in a homonuclear diatomic molecule ($\hat{P}_{12}$ operation) can be carried out in other ways equivalent to the operation $\hat{C}_2(y)$. For example, all of the particles are inverted through the origin by the $\hat{E}^*$ operation in the laboratory frame and then the electrons (and the nuclear displacement vectors) are inverted back by the $\hat{i}$ operation in the molecular frame. This leaves the two nuclei exchanged; mathematically $\hat{i}E^* = \hat{P}_{12}\hat{E}^*\hat{E}^* = \hat{P}_{12}$, as required. Thus, the s or a symmetry is determined by the total parity ($\hat{E}^*$ operation) and the g or u symmetry of the electronic state ($\hat{i}$ operation) as shown in Figure 7.36.

Finally, the superscript + or − on the $^1\Sigma^+$ and $^1\Sigma^-$ symbols is necessary in order to distinguish between the effect of $\hat{\sigma}_v$ on the electronic (or vibronic) wavefunctions,

$$\hat{\sigma}_v \psi_{\text{el}} = \pm \psi_{\text{el}}. \tag{7.191}$$

For doubly degenerate vibronic wavefunctions Π, Δ, Φ, and so forth, one component can always be labeled + while the other can be labeled −. However, writing $\Pi^\pm$, $\Delta^\pm$, $\Phi^\pm$ serves no useful purpose, so that only for Σ states is the superscript + or − used.

The effect of nuclear spin statistics is most apparent for molecules such as CO_2 for which the nuclear spins of equivalent nuclei are zero. In this case the equivalent nuclei are bosons so only the s levels are present. The a levels have no antisymmetric nuclear spin functions to combine with and are therefore absent. This means that all of the odd J lines are missing in the infrared spectrum of CO_2 and the spacing between the lines is approximately $4B$. In general the relative nuclear spin weights[5] for two equivalent nuclei are $(I+1)/I$.

Excited Vibrational States of Linear Molecules

The excited vibrational states of linear molecules are obtained by taking direct products of the symmetry species. For example for a doubly excited state ($v_2 = 2$) of Π symmetry one obtains $\Pi \otimes \Pi = \Sigma^+ \oplus [\Sigma^-] \oplus \Delta$. This product, however, can be reduced

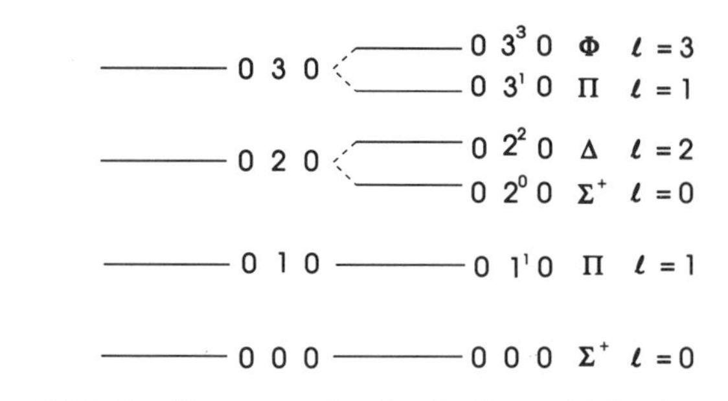

Figure 7.37. Bending energy levels of a linear triatomic molecule.

to a product of a symmetric part $\Sigma^+ \oplus \Delta$ and an antisymmetric part (Σ^-), but only the symmetric part is allowed by symmetry (c.f. the Pauli exclusion principle). In the direct product tables of Appendix C, the antisymmetric part of the product is in square brackets. Notice that $\Pi \otimes \Pi$ is the same as coupling two $l = \pm 1$ vibrational states to make a $l = \pm 2(\Delta)$ and two $l = 0$ states (Σ^+ and Σ^-). However, the $l_1 = +1$, $l_2 = -1$ and $l_1 = -1$, $l_2 = +1$ states are indistinguishable, so that only a Σ^+ results. Note that if the $\Pi \otimes \Pi$ product resulted from a vibronic product where a Π electronic state was coupling to a π bending mode or if each π was from a different vibrational mode, then both Σ^+ and Σ^- states would be present. The stack of energy levels for the bending mode of a linear triatomic molecule, such as HCN, is shown in Figure 7.37. The degenerate energy levels with the same v of the two-dimensional harmonic oscillator are no longer degenerate when anharmonicity is considered.

The energy levels for a collection of $3N - 5$ harmonic oscillators ($3N - 6$ for nonlinear molecules) is

$$G(v_1 v_2 \cdots v_{3N-5}) = \sum_{r=1}^{3N-5} \omega_r \left(v_r + \frac{d_r}{2} \right), \tag{7.192}$$

where d_r is the degeneracy of the rth vibrational mode. For the anharmonic oscillator the modes are no longer independent and cross-terms are present, so that[19]

$$G(v_1 v_2 \cdots v_{3N-5}) = \sum_r \omega_r \left(v_r + \frac{d_r}{2} \right) + \sum_{r \le s} x_{rs} \left(v_r + \frac{d_r}{2} \right)\left(v_s + \frac{d_s}{2} \right) + \sum_{t \le t'} g_{tt'} l_t l_{t'}, \tag{7.193}$$

in which the index t applies to degenerate modes with vibrational angular momentum l_t. As an example, the vibrational energy-level expression for HCN is

$$\begin{aligned} G(v_1 v_2 v_3) = {} & \omega_1(v_1 + \tfrac{1}{2}) + \omega_2(v_2 + 1) + \omega_3(v_3 + \tfrac{1}{2}) \\ & + x_{11}(v_1 + \tfrac{1}{2})^2 + x_{22}(v_2 + 1)^2 + x_{33}(v_3 + \tfrac{1}{2})^2 \\ & + x_{12}(v_1 + \tfrac{1}{2})(v_2 + 1) + x_{13}(v_1 + \tfrac{1}{2})(v_3 + \tfrac{1}{2}) + x_{23}(v_2 + 1)(v_3 + \tfrac{1}{2}) + g l^2. \end{aligned} \tag{7.194}$$

For most large molecules the constants x_{rs} and $g_{tt'}$ are not known but for HCN they are (in cm^{-1}),[24]

$$\begin{aligned}G(v_1 v_2 v_3) = {} & 3441.221(v_1 + \tfrac{1}{2}) + 726.995(v_2 + 1) + 2119.864(v_3 + \tfrac{1}{2}) - 52.490(v_1 + \tfrac{1}{2})^2 \\ & - 2.653(v_2 + 1)^2 - 7.074(v_3 + \tfrac{1}{2})^2 - 19.006(v_1 + \tfrac{1}{2})(v_2 + 1) \\ & - 10.443(v_1 + \tfrac{1}{2})(v_3 + \tfrac{1}{2}) - 2.527(v_2 + 1)(v_3 + \tfrac{1}{2}) + 5.160 l^2, \end{aligned} \tag{7.195}$$

in which higher order terms are dropped.

The energy level diagram is quite complicated even for a triatomic molecule. For example the energy levels of some of the known states of CO_2 are shown in Figure 7.38.

The selection rules for transitions among the excited energy levels are derivable, as usual, from the transition dipole moment integral. The general selection rules can be summarized as $\Delta l = 0, \pm 1$, $g \leftrightarrow u$, and $\Sigma^+ \not\leftrightarrow \Sigma^-$.

The various possibilities are:

1. $\Delta l = 0$ with $l = 0$. This is a parallel transition of the Σ^+–Σ^+ type with P and R branches ($\Delta J = \pm 1$).

2. $\Delta l = \pm 1$. This is a perpendicular transition such as Π–Σ, Δ–Π, and so forth, with P and R branches and a strong Q branch ($\Delta J = 0, \pm 1$).

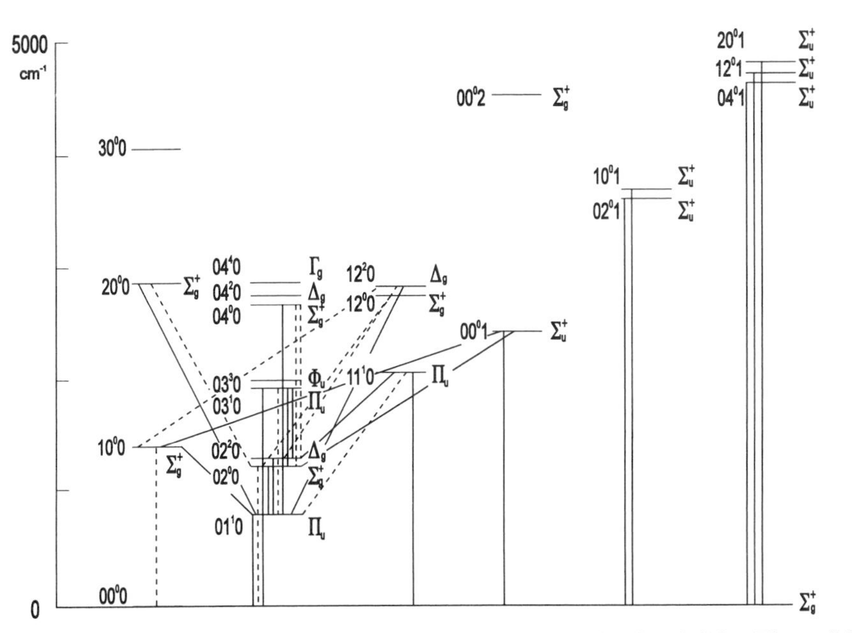

Figure 7.38. Energy-level diagram for some of the vibrational levels of CO_2. The solid lines correspond to infrared transitions, while the dashed transitions are observed by Raman spectroscopy (Chapter 8).

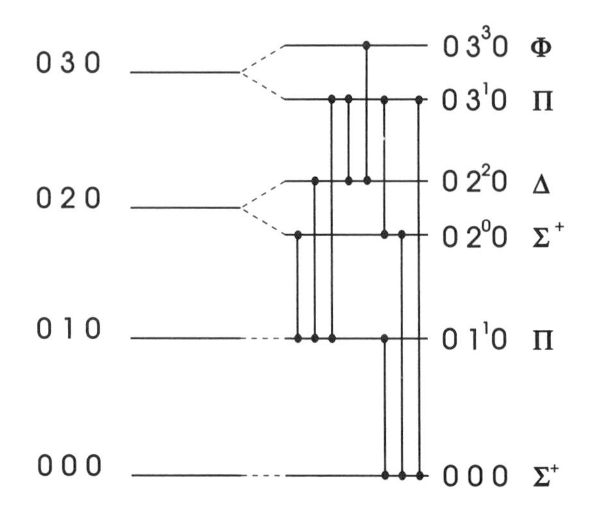

Figure 7.39. All possible allowed transitions among the first four bending energy levels of a triatomic molecule such as HCN.

3. $\Delta l = 0$ with $l \neq 0$. These are transitions of the type Π–Π, Δ–Δ, and so forth with *P* and *R* branches and weak *Q* branches. The *Q* branch lines are rarely observed. The relative intensities of the lines in the various branches are given by the rotational populations and the Hönl–London factors[5] available from many sources (Table 9.5).

Some of the possible transitions are displayed in the energy-level diagram for CO_2 (Figure 7.38). Notice that in addition to fundamentals, overtones, and hot bands, transitions such as $011 \leftarrow 000$ are possible. These transitions, in which the quantum numbers change for two or more modes, are called *combination bands.* For example, all possible allowed transitions among the first four bending energy levels of HCN are illustrated in Figure 7.39.

Vibrational Spectra of Symmetric Tops

Consider a molecule such as CH_3F with $3N - 6 = 9$ modes of vibration. The application of group theory indicates that there are three a_1 modes and three e modes[20] of vibration using the C_{3v} character table. The four bonds in the molecule give rise to four stretching modes: the three C—H stretches and a C—F stretch. The symmetry of the C—H stretching modes are a_1 and e, while the C—F stretch has a_1 symmetry. The remaining five of the nine possible modes must be bending modes. The H—C bending modes can be reduced to a symmetric CH_3 bending mode (umbrella mode) and an antisymmetric CH_3 bending mode of e symmetry. The symmetry of the —C—F bending mode (or CH_3 rock) is e (Figure 7.40).

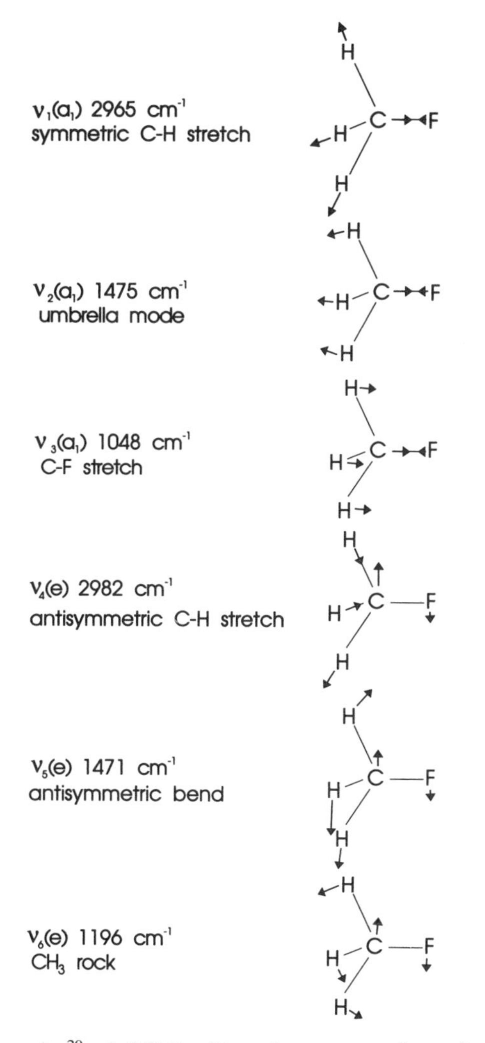

Figure 7.40. Vibrational modes[20] of CH_3F with only one member of each degenerate pair of modes displayed.

Similar to what happens for linear molecules, transitions in symmetric tops can have transition dipole moments parallel to the z-axis (symmetry axis) or perpendicular to the z-axis. Parallel transitions are of the A_1–A_1 type, while perpendicular transitions are of the E–A_1 type. The A_1–A_1 energy-level diagram is given in Figure 7.41. The transitions obey the parallel selection rules $\Delta K = 0$ with $\Delta J = \pm 1$ for the $K' = 0 \leftarrow K'' = 0$ transition and $\Delta J = 0, \pm 1$ transitions for $K \neq 0$. It is useful to note that these are exactly the same selection rules as obtained for the linear molecule, with K playing the role formerly played by l. Indeed the intensity expressions are given by the same Hönl–London factors given in Table 9.4. The transitions associated with each K are called sub-bands. The observed spectrum can be viewed as a superposition of sub-bands as shown in Figure 7.42. At low resolution the band exhibits the

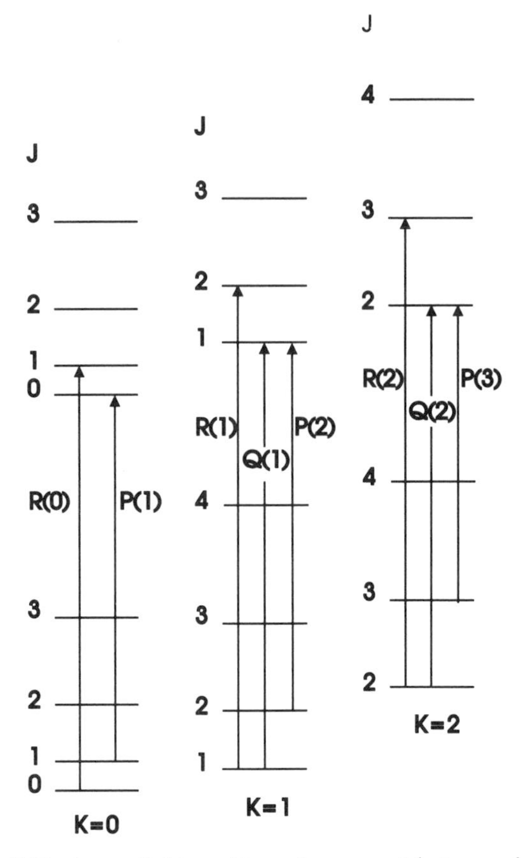

Figure 7.41. A parallel transition of a symmetric top molecule.

characteristic *PQR* pattern, Figure 7.43. When examined at higher resolution, the *K* splittings for each rotational line are resolved (Figure 7.44).

Coriolis Interactions in Molecules

The E–A_1 type transitions of a symmetric top molecule require the addition of Coriolis terms to the vibrational Hamiltonian. Coriolis forces are very important for the doubly degenerate E level.

Consider a molecular reference frame xyz rotating in space relative to the laboratory coordinate system XYZ. This means that the molecular frame of reference is an accelerated coordinate system, which will have "fictitious" centrifugal and Coriolis forces. These forces are not present when the molecule is viewed in the inertial laboratory coordinate system. It is more convenient, however, to work in the molecular frame and to live with the presence of centrifugal and Coriolis forces.

The origin of centrifugal forces is best explained by considering a particle of mass μ

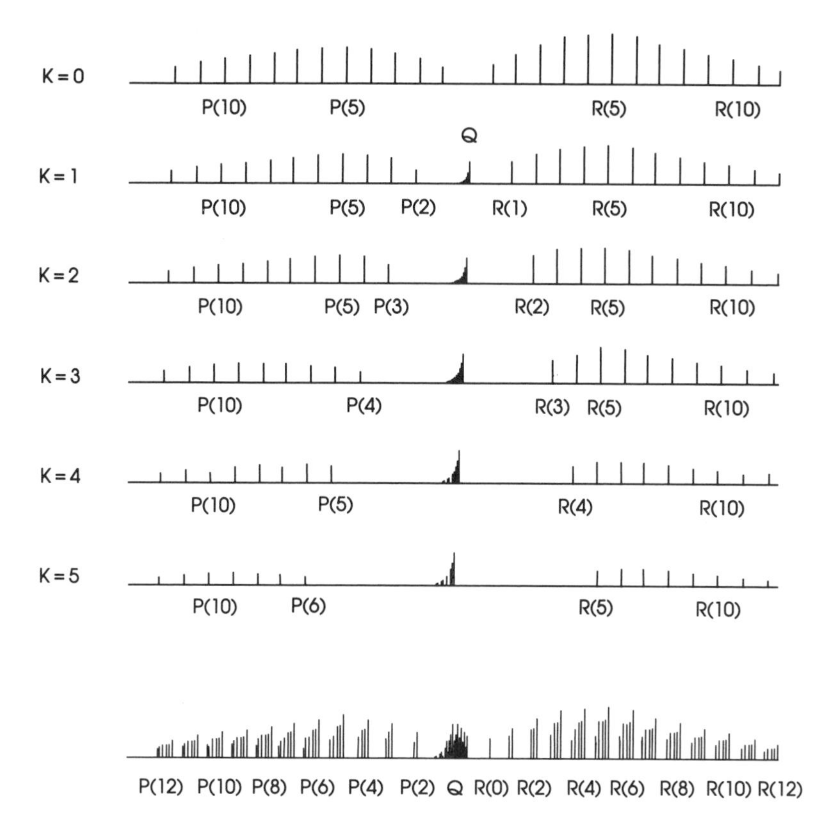

Figure 7.42. Sub-bands of a parallel transition of a symmetric top. In (a) the sub-bands are shown separately and in (b) they are combined to simulate a real spectrum.

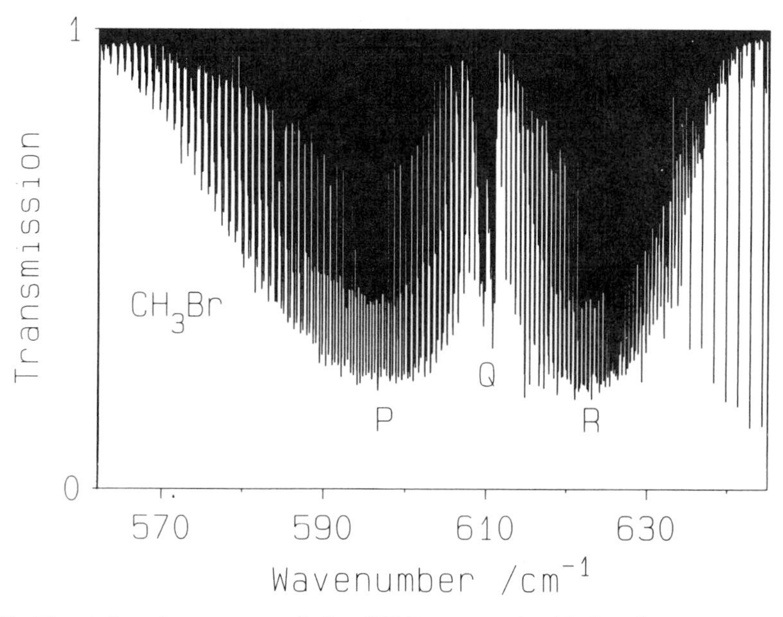

Figure 7.43. The infrared spectrum of the CH_3Br ν_3 mode. Notice the presence of two Q branches, one due to $CH_3{}^{79}Br$ and the other to $CH_3{}^{81}Br$.

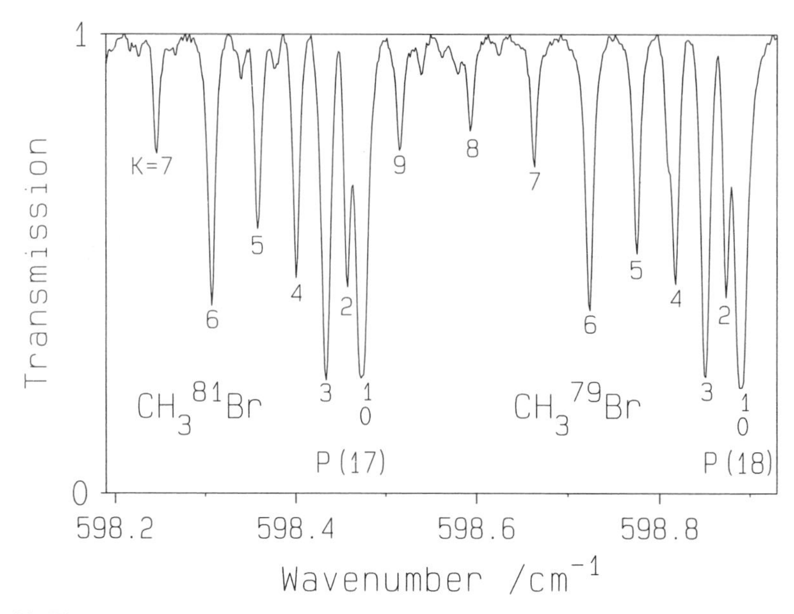

Figure 7.44. The K structure of two rotational lines of the CH_3Br ν_3 mode. The intensity alternation is caused by nuclear spin statistics.

rotating at a constant angular velocity ω (Figure 7.45). This models a rotating diatomic molecule of reduced mass μ as seen from the laboratory frame. Although the magnitude of the velocity of the particle is constant, the direction of the velocity is constantly changing (Figure 7.45). In the laboratory frame the particle is constrained to move in a circle by application of a force of magnitude

$$|\mathbf{T}| = \frac{mv^2}{r} = m\omega^2 r. \tag{7.196}$$

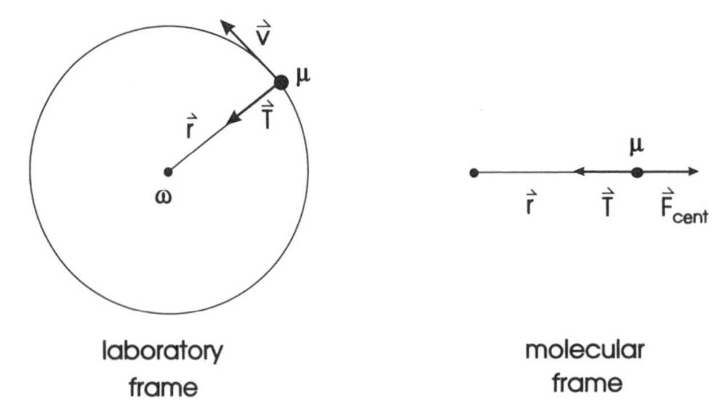

Figure 7.45. A rotating particle of mass μ viewed in the laboratory frame and in the rotating (molecular frame).

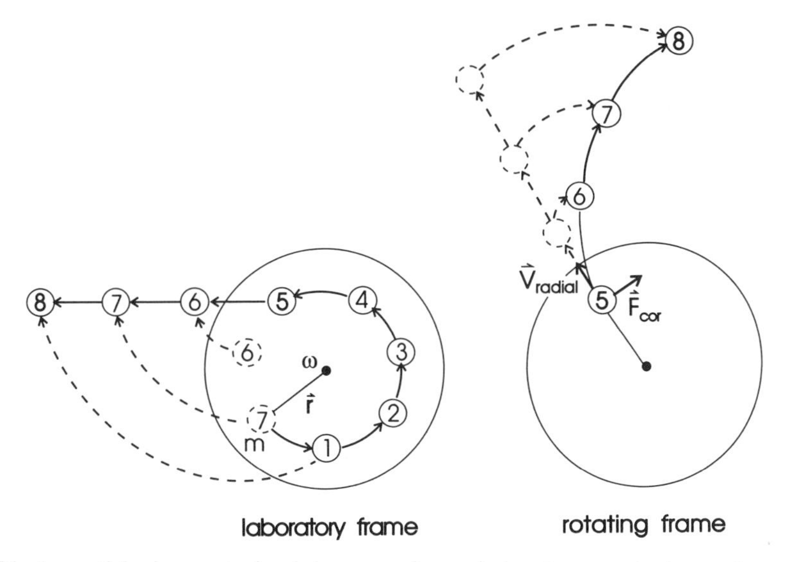

Figure 7.46. A particle is constrained to move in a circle at a constant angular velocity ω (positions 1 to 5). At 5 the particle is released and continues its motion unconstrained by external forces (positions 6 to 8).

The particle is therefore undergoing acceleration, since Newton's second law is

$$\mathbf{T} = \mathbf{F} = \mu \mathbf{a}. \tag{7.197}$$

In the rotating molecular frame (Figure 7.45) the particle is stationary since the angular velocity ω of the particle and of the molecular frame is the same. A new centrifugal force has appeared that exactly balances the force **T**. As anyone who has been in a car that corners sharply can attest, these "fictitious" forces are very real in an accelerated frame of reference.

A Coriolis force is the second type of fictitious force that can appear in an accelerated coordinate system. Consider a particle of mass μ initially moving at a constant angular velocity ω (Figure 7.46). Some time later (position 5 in Figure 7.46) the particle is released and proceeds to move in a straight line at constant velocity because there are no applied forces (Newton's first law). After the particle has been released, the motion in the laboratory frame is simple, but when viewed in the rotating frame the motion appears peculiar. In the frame rotating with an angular velocity $\boldsymbol{\omega}$ the particle moves both radially at a constant velocity $\mathbf{v}_{\text{radial}}$ and veers to the right (Figure 7.46). The particle seems to veer to the right because the rotating frame is rotating out from underneath the particle moving at a constant velocity in the laboratory frame. The motion of the particle to the right is caused by a "fictitious" Coriolis force. The Coriolis force is

$$\mathbf{F}_{\text{cor}} = -2\mu(\boldsymbol{\omega} \times \mathbf{v}') \tag{7.198}$$

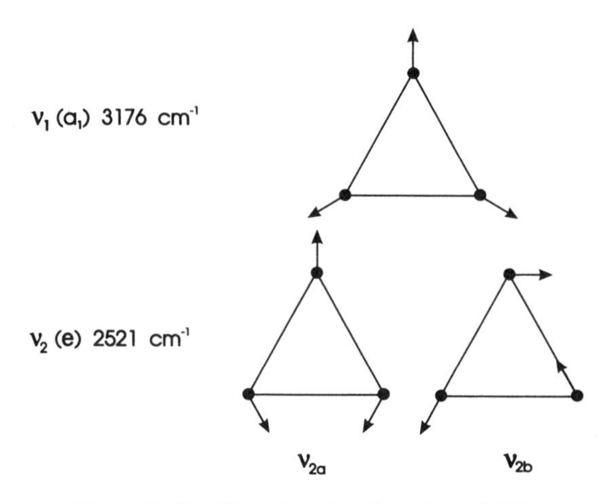

Figure 7.47. The vibrational modes of H_3^+.

where $\mathbf{v}'$ is the velocity in the rotating frame. The magnitude of the Coriolis force is

$$|\mathbf{F}_{cor}| = 2\mu\omega v_{radial}. \tag{7.199}$$

The Coriolis force is responsible for the counterclockwise rotation of tornados and hurricanes in the northern hemisphere. This is because the earth is a rotating reference frame that has a Coriolis force that makes the winds veer to the right in the northern hemisphere.

Coriolis forces are also important in molecules. Consider the H_3^+ molecule that has the structure of an equilateral triangle with D_{3h} symmetry.[25] The $3N - 6 = 3$ modes of vibration are shown in Figure 7.47. The degenerate vibration ν_2 at $2521\ \text{cm}^{-1}$ has two orthogonal modes of vibration ν_{2a} and ν_{2b} that can be chosen as shown in Figure 7.47. If the vibrational mode of the molecule is ν_{2a}, then the Coriolis forces (7.198) act as shown by dotted lines in Figure 7.48. The Coriolis forces acting on ν_{2a} lead to the excitation of ν_{2b}; when the molecule is in ν_{2b}, the Coriolis forces excite ν_{2a}. This is analogous to the strong coupling of two pendula of the same frequency. The molecule will therefore rapidly convert back and forth between ν_{2a} and ν_{2b}. The ν_{2a} and ν_{2b} modes are thus coupled via a first-order Coriolis effect.

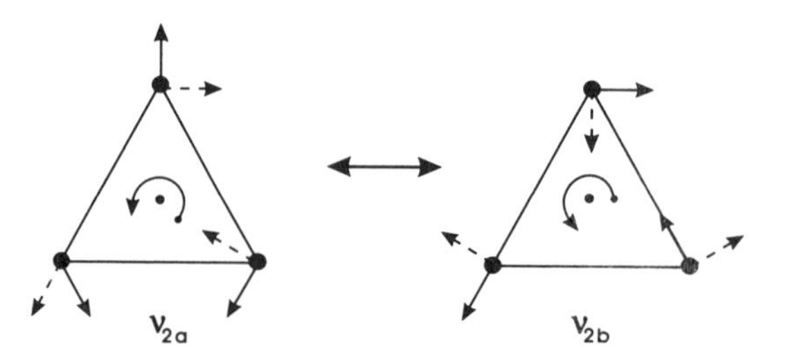

Figure 7.48. Coriolis forces acting on the two degenerate components of the ν_2 mode of H_3^+.

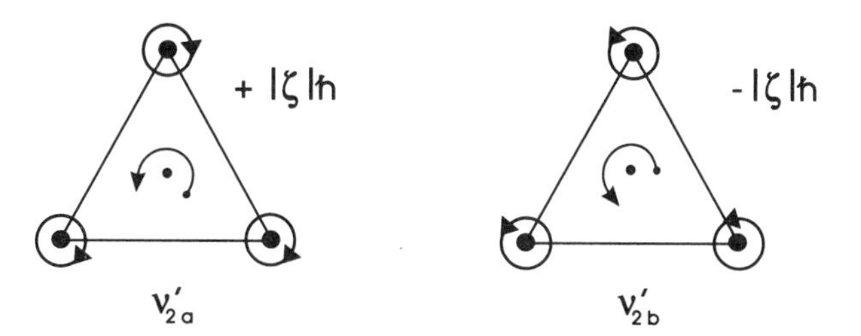

Figure 7.49. The H_3^+ molecule has $\pm|\zeta|\hbar$ units of vibrational angular momentum in the ν_2 mode.

Linear combinations of ν_{2a} and ν_{2b} that have vibrational angular momentum (Figure 7.49) can be formed. In this case the magnitude of the vibrational angular momentum[19] is $\zeta\hbar$ where $-1 \leq \zeta \leq 1$. Notice that unlike linear molecules for which the vibrational angular momentum $l\hbar$ is an integral multiple of $\hbar$, ζ for a symmetric top is not necessarily integral.

Vibrational angular momentum changes the energy-level formula. When vibrational angular momentum is present in a molecule, **J** becomes the vector sum of the rotational angular momentum and the vibrational angular momentum. Note that neither the projection of the vibrational angular momentum ($\hat{\pi}_z$) nor the projection of the rotational angular momentum ($\hat{J}_z - \hat{\pi}_z$) are quantized as integers about the molecular z-axis. The projection of the total angular momentum ($\hat{J}_z$), however, is quantized about the molecular z-axis with a quantum number designated as K. The rotational Hamiltonian operator becomes[19]

$$\hat{H} = \frac{\hat{J}_x^2}{2I_x} + \frac{\hat{J}_y^2}{2I_y} + \frac{(\hat{J} - \hat{\pi}_z)^2}{2I_z}, \tag{7.200}$$

where $\hat{\pi}_z$ is the vibrational angular momentum operator about the symmetric top axis. Expanding, dropping the pure vibrational term, and taking matrix elements give the energy levels as

$$E = BJ(J+1) + (A-B)K^2 \mp 2A\zeta K \qquad \text{(prolate top)}, \tag{7.201a}$$

or

$$E = BJ(J+1) + (C-B)K^2 \mp 2C\zeta K \qquad \text{(oblate top)}, \tag{7.201b}$$

using the vibrational angular momentum of $\zeta\hbar$ units around the top axis. In equations (7.201) the $\mp$ signs are the result of the $\pm|K|$ and $\pm|\zeta|$ possibilities. The rotational and vibrational angular momenta about the top axis can either be in the same direction ($-$ sign) or in the opposite direction ($+$ sign). For historical reasons the $+|\zeta|$ with $+|K|$ (and $-|\zeta|$ with $-|K|$) levels are labeled as $(+l)$ levels and the $+|\zeta|$ with $-|K|$ levels (and

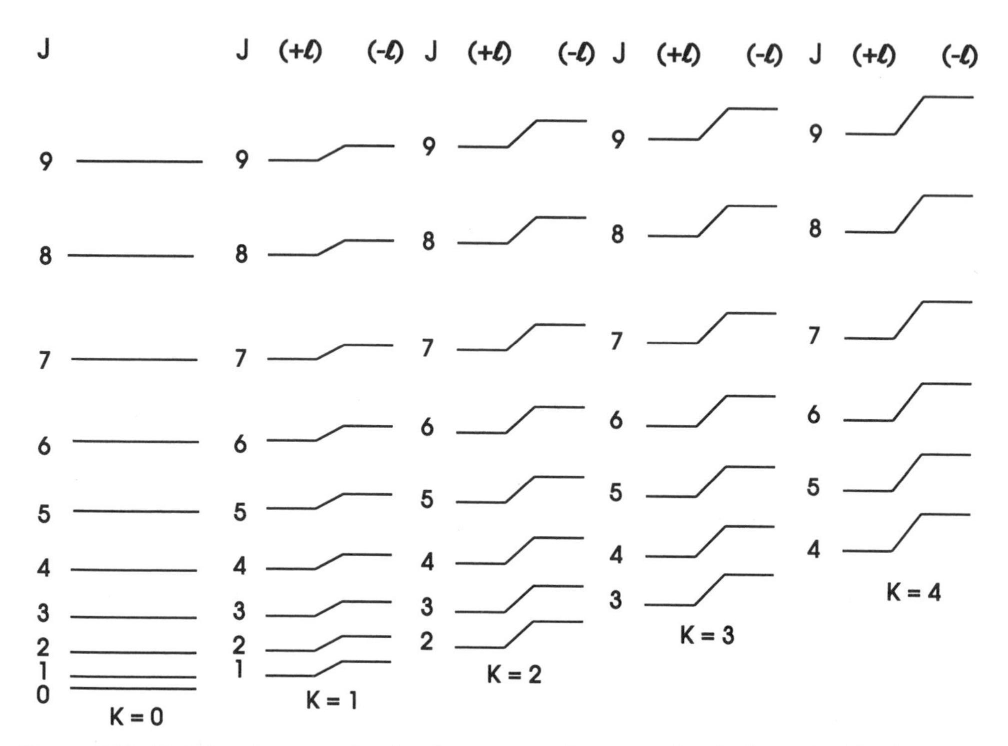

Figure 7.50. Rotational energy levels of a symmetric top molecule in a doubly degenerate vibrational state.

$-|\zeta|$ with $+|K|$) are labeled as $(-l)$ levels. The energy-level diagram for a doubly degenerate vibrational level of a symmetric top molecule is given in Figure 7.50.

The E–A_1 energy-level diagram[19] is given in Figure 7.51. The energy-level structure of an E vibrational state is complicated by the presence of a first-order Coriolis interaction between the two components. The selection rules are $\Delta K = \pm 1$ and $\Delta J = 0, \pm 1$. Note also that for $\Delta K = +1$ the transitions connect to the $(+l)$ stack, while for $\Delta K = -1$ they connect with the $(-l)$ stack. The transition can again be represented by a superposition of sub-bands. Notice how the sub-bands do not line up as they do for a parallel transition, but that they spread out (Figure 7.52). Each sub-band is separated by approximately $2[A(1-\zeta)-B]$. This gives rise to a characteristic pattern of nearly equally spaced Q branches (Figure 7.53).

Infrared Transitions of Spherical Tops

Spherical tops such as CH_4, NH_4^+, SF_6, and C_{60} belong to the point groups T_d, O_h, or I_h. Let us consider the CH_4 molecule.[20] There are $3N-6=9$ modes made up of four stretching modes and five bending modes (Figure 7.54). The four C—H stretching

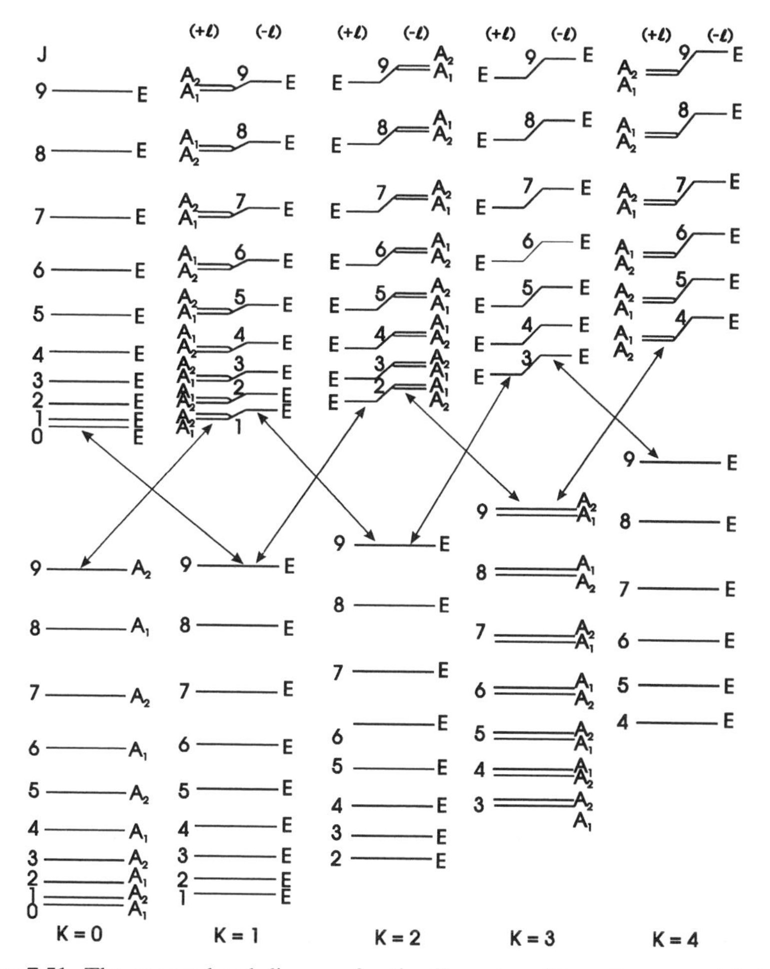

Figure 7.51. The energy-level diagram for the E–A_1 transition of a symmetric top.

coordinates can be reduced to a symmetric a_1 stretch and a triply degenerate antisymmetric C—H stretch of t_2 (or f_2) symmetry. Triply degenerate irreducible representations are labeled as t (or T) by inorganic chemists and electronic spectroscopists, but as f (or F) by many vibrational spectroscopists. The five bends reduce to a pair of e modes and a triply degenerate t_2 bending mode. Only ν_3 and ν_4 are infrared active but all of the modes are Raman active (Chapter 8).

The rotational energy levels of a spherical top are given by $BJ(J+1)$; however, there is both a $(2J+1)$-fold K degeneracy and a $(2J+1)$-fold M degeneracy. The total degeneracy is therefore $(2J+1)^2$ for each rotational level. A more detailed analysis that takes into account the effects of centrifugal distortion and anharmonicity predicts that the K degeneracy is partially lifted. The number of levels into which each J can split can be determined by group theory. These splittings are called cluster splittings and a surprisingly sophisticated theory[26] is required to account for their magnitude. A picture of some cluster splittings of a transition is presented in Figure 7.55.

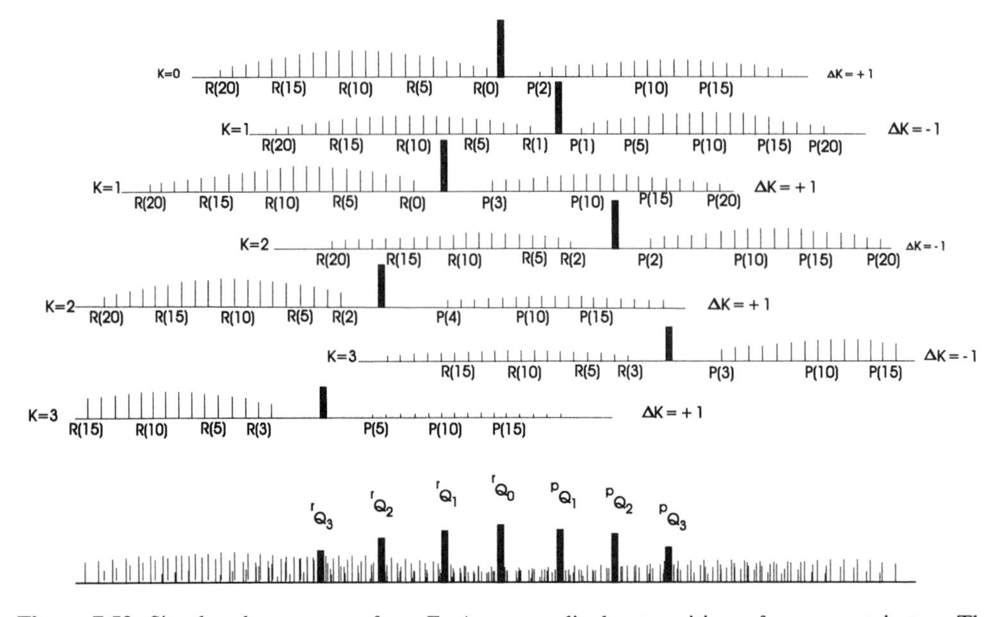

Figure 7.52. Simulated spectrum of an E–A_1 perpendicular transition of a symmetric top. The individual sub-bands are combined to give the total spectrum in the bottom panel. The notation for the Q branches is ${}^{\Delta K}\Delta J_{K''}$.

The E states of a spherical top do not experience first-order Coriolis coupling, so they have the same energy-level pattern as A_1 states. However, T_2 states experience a first-order Coriolis effect and split into three components,[26]

$$F^+(J) = BJ(J+1) + 2B\zeta J, \tag{7.202}$$

$$F^0(J) = BJ(J+1) - 2B\zeta, \tag{7.203}$$

and

$$F^-(J) = BJ(J+1) - 2B\zeta(J+1). \tag{7.204}$$

The energy-level pattern is given in Figure 7.56 for a T_2–A_1 transition. The selection rule is $\Delta J = 0, \pm 1$, but for the transitions

$$\Delta J = +1 \qquad T^-–A_1,$$

$$\Delta J = 0 \qquad T^0–A_1,$$

and

$$\Delta J = -1 \qquad T^+–A_1.$$

The spectrum of a "typical" spherical top is given in Figure 7.57. At low resolution the spectrum exhibits the characteristic PQR contour similar to a symmetric top (Figure 7.43).

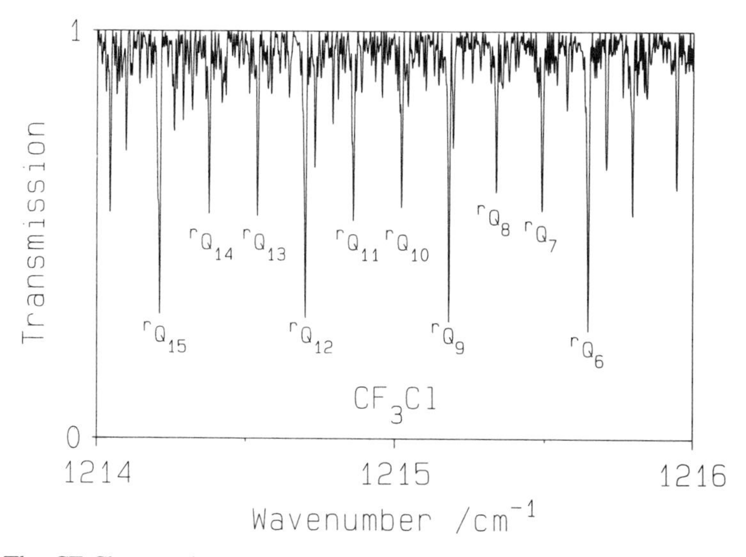

Figure 7.53. The CF_3Cl ν_4 mode exhibiting sub-band Q structure. The intensity variation in the Q branches is due to nuclear spin statistics.

Vibrational Spectra of Asymmetric Tops

The vast majority of polyatomic molecules are asymmetric tops. The H_2O molecule has three vibrational modes (Figure 7.58), with all modes both infrared and Raman active (Chapter 8).

The vibration–rotation transitions of asymmetric tops are classified as a-type, b-type, and c-type, depending on the orientation of the transition dipole moment relative to the principal axes. For example, the oscillating dipole moments of the H_2O ν_1 and ν_2 modes are along the $z(b)$ direction and are classified as b-type transitions. The ν_3 band of H_2O has an oscillating dipole moment along the $y(a)$ direction giving rise to an a-type transition. For molecules of sufficiently low symmetry, such as HOD, hybrid bands can occur, in this case ab-hybrid bands.

The selection rules for a-type, b-type, and c-type transitions are the same as for pure rotational transitions. The selection rules are as follows:

1. a-type bands, with

$$\Delta K_a = 0, \qquad \Delta K_c = \pm 1,$$

and $\Delta J = 0, \pm 1$, except for $K'_a = 0 \leftarrow K''_a = 0$, for which only $\Delta J = \pm 1$ is possible;

2. b-type bands, with

$$\Delta K_a = \pm 1, \qquad \Delta K_c = \pm 1,$$

and $\Delta J = 0, \pm 1$;

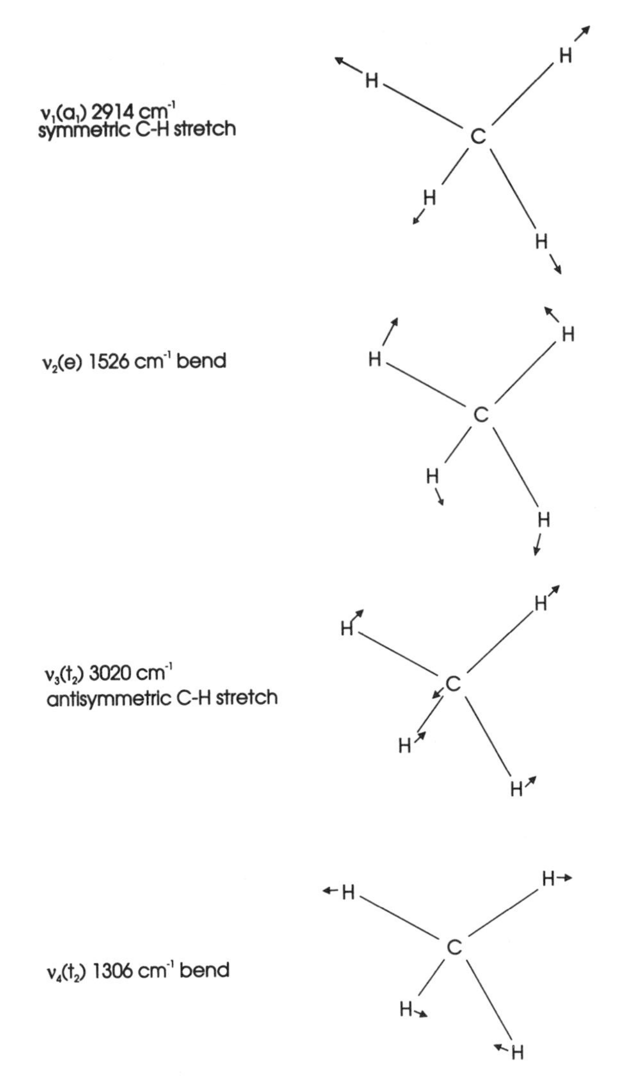

Figure 7.54. The normal modes of vibration of CH_4 with only one member of each degenerate mode shown.

3. c-type bands, with

$$\Delta K_c = 0, \qquad \Delta K_a = \pm 1,$$

and $\Delta J = 0, \pm 1$, but for $K_c' = 0 \leftarrow K_c'' = 0$, $\Delta J = \pm 1$ only.

Since many molecules are near-oblate or near-prolate symmetric tops, the general appearance of asymmetric top bands often resembles either parallel or perpendicular bands of a symmetric top. For example, the a-type transition ν_{11} of C_2H_4 is shown in Figure 7.59. This band is similar in appearance to the parallel transition of a symmetric top.

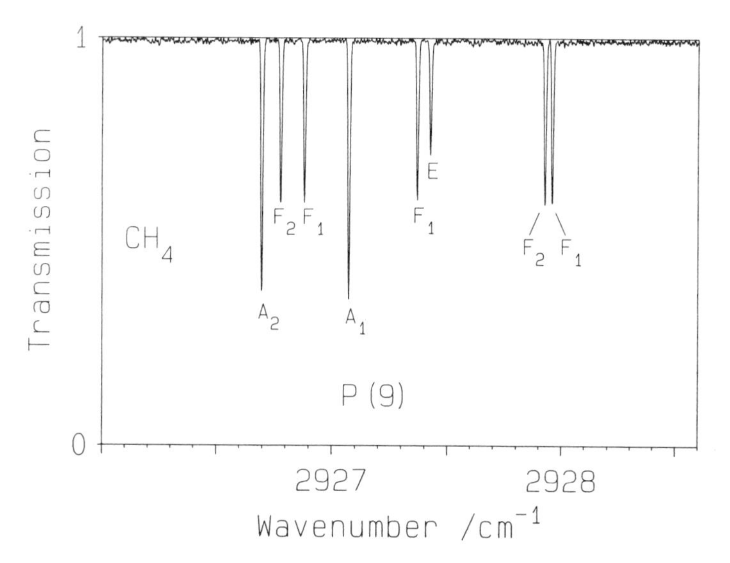

Figure 7.55. The cluster splittings of the $P(9)$ transition of the ν_3 mode of CH_4.

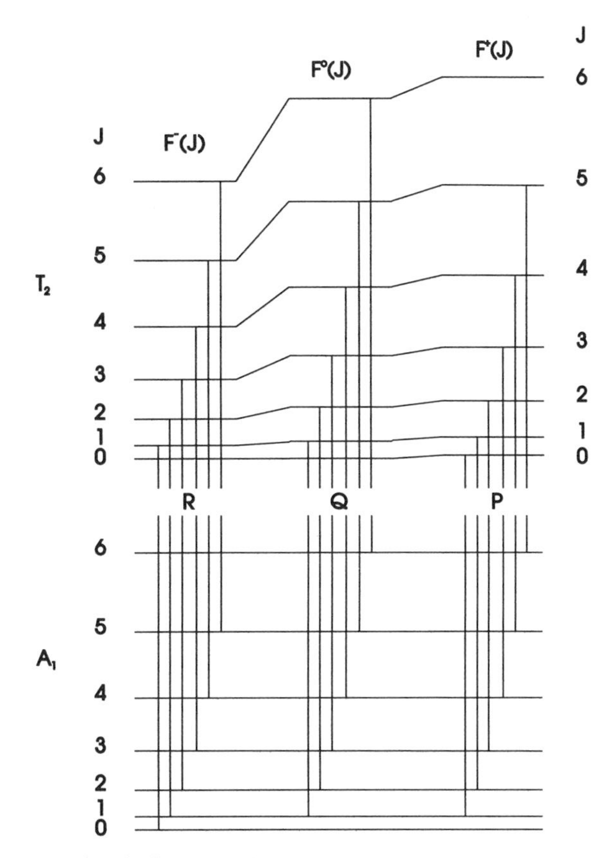

Figure 7.56. The energy-level diagram of a T_2–A_1 vibration–rotation transition of a spherical top.

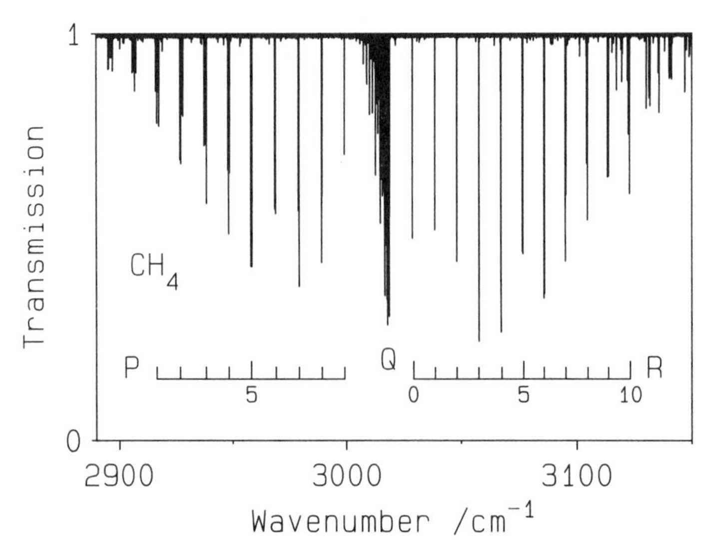

Figure 7.57. The infrared spectrum of the CH_4 ν_3 mode.

Fermi and Coriolis Perturbations

The regular energy-level pattern predicted by the $G(v_i)$ equation (7.192) rarely exists for real molecules. Deviations from a regular pattern are called *perturbations* by spectroscopists. Consider the Raman spectrum (Chapter 8) for CO_2. The ν_1 mode of CO_2 should be strong, while the $2\nu_2$ overtone should be weak. In fact they both have

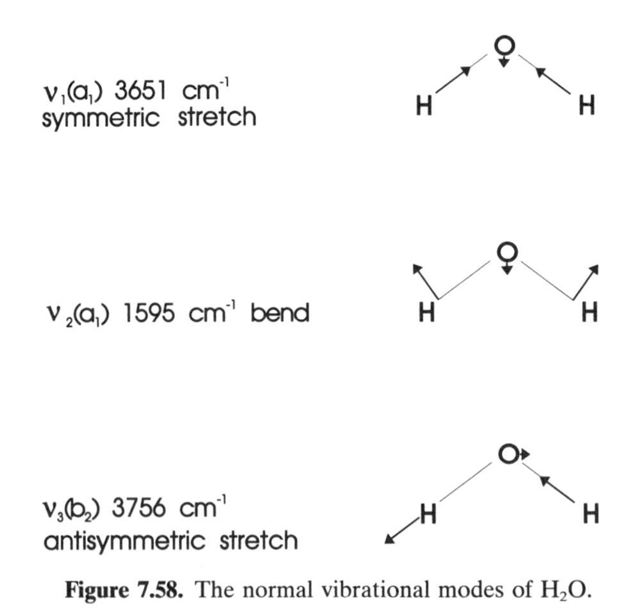

Figure 7.58. The normal vibrational modes of H_2O.

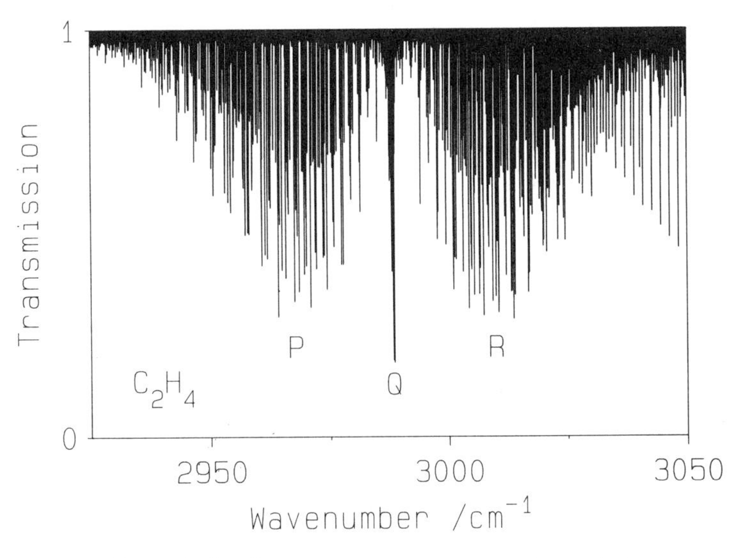

Figure 7.59. The a-type mode ν_{11} of ethylene, C_2H_4.

roughly the same intensity. Moreover, the $2\nu_2$ (σ_g) mode is not present at $2 \times 667\ cm^{-1} = 1334\ cm^{-1}$, but at $1285\ cm^{-1}$ instead (Figure 7.60). The explanation for these discrepancies was provided by Fermi[20]—the ν_1 and $2\nu_2$ vibrational levels have the same symmetry Σ_g^+, so that they can interact. The 02^20 Δ_g state is not affected because it has Δ_g symmetry. This type of interaction between vibrational levels of the same symmetry is known as a *Fermi resonance.*

The unperturbed ν_1 and $2\nu_2$ levels (dashed lines in Figure 7.61) have almost identical energies, but anharmonic terms that were negelected in the simple harmonic oscillator approximation lead to a repulsion between the two levels (Figure 4.1). If the original wavefunctions were ψ_{100} and ψ_{02^00}, then the final mixed wavefunctions are

$$\psi^+ = a\psi_{100} + b\psi_{02^00}$$

$$\psi^- = b\psi_{100} - a\psi_{02^00}$$

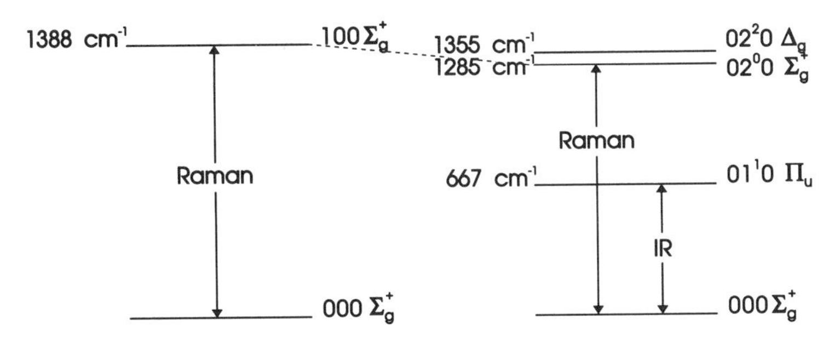

Figure 7.60. Fermi resonance in CO_2.

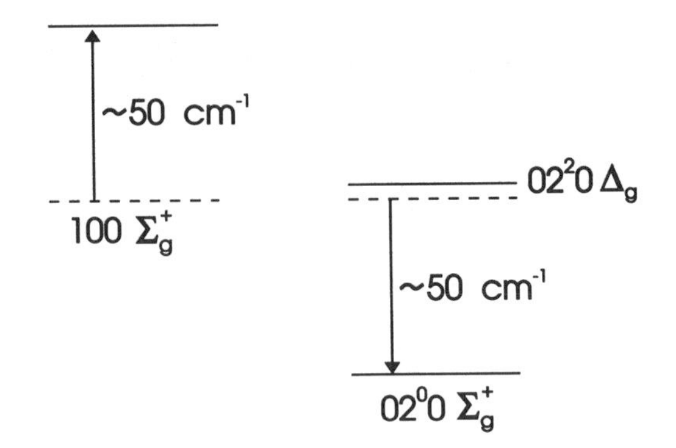

Figure 7.61. Fermi interaction of the 100 Σ_g^+ and 02^{0}0 Σ_g^+ vibrational levels of CO_2.

with $a \approx b \approx 1/\sqrt{2}$ in the case of CO_2. Using second-order perturbation theory, the energy shift is given by

$$\Delta E^{(2)} = \frac{V^2}{E_2^{(0)} - E_1^{(0)}}$$

in which $V = \langle \psi_2 | \hat{V} | \psi_1 \rangle$ is the interaction matrix element and $E_2^{(0)}$ and $E_1^{(0)}$ are the unperturbed energies. Clearly, Fermi resonances occur whenever two states of the same symmetry are in close proximity and are coupled by a nonzero anharmonic interaction term.

Other interactions between levels are possible since the simple harmonic oscillator picture has neglected many types of higher order terms in the vibration–rotation Hamiltonian. In addition to the anharmonic terms responsible for Fermi resonances, Coriolis terms can also perturb the expected regular energy-level pattern. "First order" Coriolis effects have already been discussed for the splittings observed in the E vibrational levels of symmetric tops and the T vibrational levels of spherical tops. These large effects must always be taken into account. In addition "second order" Coriolis effects are possible between states of different symmetry. Since Coriolis interactions involve rotational motion, they occur only in the gaseous state.

If two vibrational levels (ψ_1 and ψ_2) of a molecule are near to each other in energy and they differ in symmetry, such that

$$\Gamma^{\psi_2} \otimes \Gamma^R \otimes \Gamma^{\psi_1}, \qquad R = R_x, R_y, R_z$$

contains a totally symmetric irreducible representation A_1, then a Coriolis resonance is possible. The explanation[27] of this rule is simple: the lowest order Coriolis terms neglected in the total Hamiltonian have the form $p\hat{J}_x$, $q\hat{J}_y$, or $r\hat{J}_z$ (with p, q, and r constants) and they behave like the rotations R_x, R_y, and R_z. Thus if two vibrations differ in symmetry by a rotation about one of the principal axes, then a neglected term in the Hamiltonian can always be found to cause an interaction. Of course, if the states are far apart (hundreds of cm^{-1}), then the effect of the interaction is small since the interaction matrix element is, at most, a few cm^{-1} in size.

An example of a Coriolis resonance exists between the ν_1 and ν_3 vibrational modes[28] of NH_2 (or H_2O) for which

the symmetric stretch $\nu_1(a_1)$ is at 3219 cm^{-1},

the bend $\nu_2(a_1)$ is at 1497 cm^{-1}, and

the antisymmetric stretch $\nu_3(b_2)$ is at 3301 cm^{-1}.

In this case $a_1 \otimes b_2 \otimes b_2 = a_1$ and R_x has b_2 symmetry. The x-axis corresponds to the out-of-plane c axis, so that ν_1 and ν_3 interact via a c-axis Coriolis resonance, which is responsible for some local rotational perturbations in the spectra of ν_1 and ν_3. Notice that $2\nu_2(a_1) \approx 2994\ cm^{-1}$ can also interact with ν_1 via a Fermi resonance. In heavier molecules ν_1, $2\nu_2$, and ν_3 would be too far apart to interact extensively, but in light hydrides such as NH_2 or H_2O the rotational structure covers hundreds of cm^{-1} and there are more possibilities for interaction.

Inversion Doubling and Fluxional Behavior

The rotational energy levels of the ground state of NH_3 are doubled. This was one of the earliest discovered manifestations of fluxional behavior in molecules. The NH_3 molecule can rapidly invert (Figure 7.62) its geometry. The two forms correspond to different enantiomeric forms of the molecule. In fact, for a noninverting molecule such as PH_3 or AsH_3 the two forms (Figure 7.63) of *PHDT* could, in principle, be separated.

For chiral molecules the two forms (enantiomers) have identical energy levels, but a large barrier prevents their interconversion. In NH_3 the barrier for interconversion of these two forms has dropped[29] to 2009 cm^{-1} (Figure 7.62), allowing facile interconversion by tunneling. The energy-level pattern for each form of NH_3 is no longer identical as a result of their mutual interaction. New approximate wavefunctions need to be constructed

$$\psi^+ = \frac{\psi_L + \psi_R}{\sqrt{2}}$$

$$\psi^- = \frac{\psi_L - \psi_R}{\sqrt{2}}$$

by mixing wavefunctions of the left- and right-handed forms. It turns out that the + level (sometimes labeled s for symmetric) lies below the − level (called a for antisymmetric). The + or − are added as superscripts (Figure 7.62) to the vibrational quantum number of the inverting normal mode. This +/− or s/a notation is not

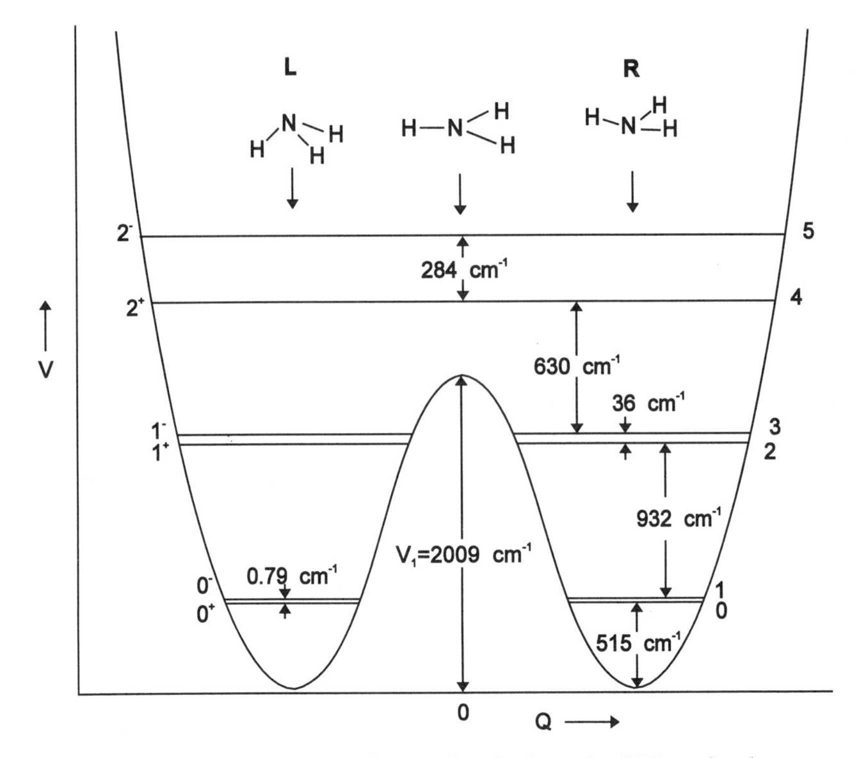

Figure 7.62. Energy-level diagram for the inverting NH_3 molecule.

related to the notation for parity discussed previously. Notice that for the energy levels above the barrier, the inversion splitting becomes a vibrational interval and the numbering on the high right of the diagram is more appropriate.

Fluxional behavior is an effect commonly observed in weakly bound systems such as van der Waals dimers, for example $(H_2O)_2$. Since the inversion of a molecule changes the handedness of the coordinate system, simple geometric symmetry operations are not adequate to describe the molecular motion. Permutations and inversions of the nuclei need to be considered in order to describe these motions. Permutation-inversion group theory[30] has different group operations and group names, but these groups are

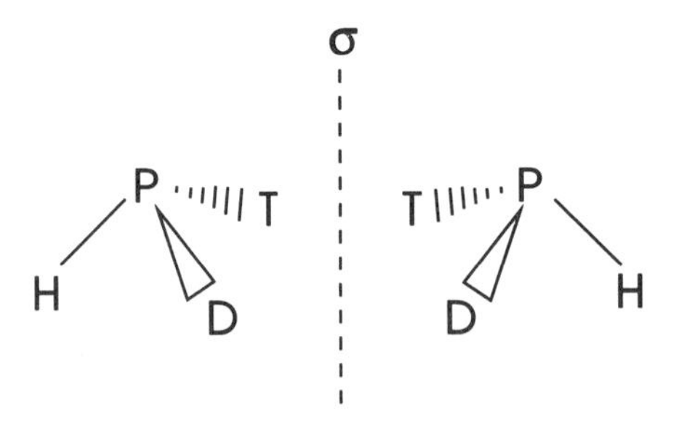

Figure 7.63. The two chiral forms of *PHDT*.

often isomorphic with the more familiar point groups. For example the permutation-inversion point group appropriate for the inverting NH_3 molecule is isomorphic with D_{3h}. The addition of the inversion operation has increased the order of the group from 6 (C_{3v}) to 12 (D_{3h}).

Problems

1. Consider the ethylene molecule of D_{2h} symmetry.
   (a) Determine the number and symmetry of the normal modes of vibration. (Choose x to be out of plane and z along the C–C axis).
   (b) Which modes are infrared active?
   (c) Use projection operators to find the symmetry of the
      (i) C—C stretch
      (ii) C—H stretches
      (iii) In-plane bends
      (iv) Out-of-plane bends
2. Determine the symmetry of the normal modes of vibration for the following molecules:
   (a) H—C≡C—C≡N
   (b) CO_3^{2-}
   (c) $PtCl_4^{2-}$
   (d) *trans*-Glyoxal
3. Fill in the following table:

Molecule	$\tilde{\nu}$(cm^{-1})	Force Constant (N m^{-1})
NH	3133	
NF	1115	
O_2	1555	
N_2	2331	
NO	1876	

4. The molecule 1,1-dichloroethylene ($C_2H_2Cl_2$) is planar with C_{2v} symmetry, the C═C bond coinciding with the C_2 axis. Take this axis as the z-axis and the plane containing the molecule as the (y, z) plane.
   (a) Determine the number and symmetry of the normal modes of vibration.
   (b) How many infrared bands (fundamentals) are there of a-type, b-type, and c-type, respectively? The C_2 axis is the axis of the intermediate moment of inertia I_B.
   (c) Are there any vibrations in this molecule whose fundamental frequency ($\Delta v = 1$) is forbidden in the infrared, but whose first overtone ($\Delta v = 2$) is allowed (given sufficient anharmonicity to permit $\Delta v = 2$)?

5. The positions of the atoms in the molecule N_4S_4 have been determined by x-ray diffraction. In terms of a set of Cartesian coordinates x, y, z placed within the molecule, these are
   N_1: $x = b, y = 0, z = 0$; N_2: $0, b, 0$;
   N_3: $-b, 0, 0$; N_4: $0, -b, 0$;
   S_1: $a, -a, a$; S_2: $-a, a, a$;
   S_3: $a, a, -a$; S_4: $-a, -a, -a$.
   Here the numbers a and b are unrelated parameters of the order of a few angstroms in size.
   (a) To what point group does the molecule belong?
   (b) What are the symmetries of the normal modes of vibration?
   (c) How many different vibrational frequencies does the molecule have?
   (d) How many bands should appear in the infrared absorption spectrum as fundamentals?
   (e) Assume the parameter $a = 3$ Å $= 1.5b$. Compute the moments of inertia as well as A, B, and C. What kind of rotor is the molecule (linear rotor, spherical top, oblate or prolate symmetric top, near oblate or near prolate asymmetric top)?
6. The $2\nu_3$ band of CH_3F is observed near 4.8 μm. The line positions of this parallel band are provided in the table below.
   (a) Determine ν_0, B', B'', D_J', and D_J''.
   (b) Assume a tetrahedral geometry and a "standard" C—H bond length to compute A'' $(= A')$.
   (c) From the B'' constant and the assumption in (b) compute the C—F bond length.

The Line Positions of the $2\nu_3$ Band of CH_3F (in cm^{-1})

J	$R(J)$	$P(J)$	J	$R(J)$	$P(J)$
0	—	—	18	2105.29	2044.05
1	2084.72	—	19	2106.09	—
2	2086.25	—	20	2106.86	2039.01
3	2087.75	2076.13	21	2107.56	2036.42
4	2089.22	2074.27	22	2108.24	2033.81
5	2090.65	2072.41	23	2108.87	2031.15
6	2092.07	2070.49	24	2109.45	2028.44
7	2093.41	2068.55	25	2110.00	2025.69
8	2094.75	2066.55	26	2110.49	2022.95
9	2095.99	—	27	2110.93	2020.10
10	2097.23	2062.43	28	2111.40	—
11	2098.37	2060.31	29	2111.70	2014.33
12	2099.49	2058.10	30	2112.05	2011.38
13	2100.58	2055.88	31	—	2008.39
14	2101.59	2053.60	32	—	2005.37
15	2102.60	2051.27	33	—	2002.26
16	2103.54	2048.92	34	—	—
17	2104.43	2046.50	35	—	1996.00

7.

	1–0 Band			2–0 Band	
J	$P(J)$	$R(J)$	J	$P(J)$	$R(J)$
0	—	2147.0816	0	—	4263.8376
1	—	2150.8565	1	4256.2171	4267.5421
2	2135.5461	2154.5960	2	4252.3023	4271.1774
3	2131.6333	2158.3002	3	4248.3184	4274.7414
4	—	2161.9687	4	4244.2634	4278.2351
5	2123.7001	2165.6015	5	4240.1403	4281.6571
6	2119.6827	2169.1984	6	4235.9477	4285.0096
7	2115.6294	2172.7593	7	4231.6856	4288.2898
8	2111.5434	2176.2840	8	4227.3539	4291.4996
9	2107.4231	2179.7723	9	4222.9549	4294.6379
10	2103.2688	2183.2242	10	4218.4859	4297.7051
11	2099.0815	2186.6395	11	4213.9486	4300.7001
12	2094.8612	2190.0180	12	4209.3431	4303.6240
13	2090.6089	2193.3596	13	4204.6700	4306.4756
14	2086.3215	2196.6642	14	4199.9279	4309.2552
15	2082.0027	2199.9314	15	4195.1186	4311.9619
16	2077.6500	2203.1615	16	4190.2409	4314.5970
17	2073.2647	2206.3539	17	4185.2956	4317.1591
18	2068.8476	2209.5088	18	4180.2830	4319.6487
19	2064.3968	2212.6258	19	4175.2024	4322.0663
20	2059.9148	2215.7040	20	4170.0553	4324.4100
21	2055.4002	2218.7459	21	4164.8411	4326.6807
22	2050.8546	2221.7487	22	4159.5599	4328.8785
23	—	—	23	4154.2115	4331.0029
24	—	2227.6391	24	4148.7969	4333.0537
25	2037.0252	2230.5264			
26	2032.3528	2233.3748			
27	2027.6495	2236.1842			
28	—	2238.9545			

The lines of the fundamental, 1–0, and first overtone, 2–0, vibration–rotation bands of CO are listed in the preceding table, in cm^{-1}.

(a) For each band, determine the five parameters ν_0, B'', D'', B', and D'.

(b) From the band origins, determine ω_e and $\omega_e x_e$.

(c) Determine B_e and α_e.

(d) Compute r_0 and r_e. Why are they different?

(e) Test the Pekeris and Kratzer relationships.

(f) Predict all of the preceding constants for $^{13}C^{16}O$.

8. Even when the rotational structure of a fundamenal band cannot be resolved, it is sometimes possible to extract the rotational constant B from the separation between the maxima in the P and R branches as

$$\Delta\tilde{\nu}_{PR} = 2.3583(BT)^{1/2},$$

where B is the rotational constant (in cm^{-1}) and T is the absolute temperature. Derive this equation.

9. The observed IR bands (in cm^{-1}) of $^{10}BF_3$ and $^{11}BF_3$ are as follows (vs = very

strong; s = strong; m = medium; w = weak):

$^{10}BF_3$	482	718	1370	1505	1838	1985	2243	2385	3008	3263
$^{11}BF_3$	480	691	1370	1454	1838	1932	2243	2336	2903	3214
intensity	s	s	m	vs	w	w	w	w	w	w

The order of increasing vibration frequency of the fundamentals is $\nu_4 < \nu_2 < \nu_1 < \nu_3$. Assign the observed bands. (It might be thought that the 1370 cm^{-1} band is the overtone $2\nu_2$, but this can be ruled out. Why?)

10. A spectroscopist is searching for the LiNNN molecule in the gas phase in the infrared region of the spectrum. By analogy with CaNNN [J. Chem. Phys. **88,** 2112 (1988)] LiNNN should be linear and quite ionic. The N—N bond distance in crystalline azides (M^+NNN^-) is 1.18 Å.
    (a) Estimate a reasonable Li—N bond length (e.g., from ionic radii) and compute a B value from the geometry.
    (b) Determine the number and symmetry of the normal modes of vibration.
    (c) Estimate frequencies for the normal modes by analogy with other azides.
    (d) Describe each IR allowed fundamental transition (i.e., parallel or perpendicular, which branches occur, the spacing of the lines, etc.)
11. For the formaldehyde molecule H_2CO of C_{2v} symmetry:
    (a) Determine the number and symmetries of the normal modes.
    (b) Determine which modes are infrared active.
    (c) Number the normal modes and describe each mode (e.g., symmetric/antisymmetric bend/stretch; in plane/out of plane). For each mode provide an estimated vibrational frequency.
12. (a) To what point group does the dichloroacetylene molecule Cl—C≡C—Cl belong?
    (b) How many fundamental modes will there be for dichloroacetylene?
    (c) Sketch the approximate atomic motion of the normal modes.
    (d) Specify the infrared activity of each.
    (e) Why do the first overtones of the infrared active fundamentals not occur in the IR spectrum?
13. The following table gives the fundamentals and combination bands in the infrared spectrum of acetylene. Fundamentals are very strong (vs), combination lines involving only two fundamentals are of medium (m) intensity, and all others are weak (w). In this slightly idealized version of the spectrum, anharmonicity effects do not occur and $\nu_4 < \nu_5 < \nu_2 < \nu_3 < \nu_1$. Determine the frequencies of the fundamentals and assign the combination bands.

Band position	730	1340	1950	2700	3290	3310	3900	4100	5260	6660	cm^{-1}
Intensity	vs	m	w	m	vs	w	m	m	m	m	

14. Consider the BF_3 molecule of D_{3h} symmetry.
    (a) Determine the number and symmetries of the normal modes.
    (b) Determine the activity of each mode in the infrared.
    (c) Determine the symmetries of each of the following types of internal modes:
        (i) B—F stretching modes;
        (ii) Out-of-plane bending motions;
        (iii) F—B—F angle deformations.
    (d) What are the selection rules for overtone and combination bands in the infrared?
15. Consider the vibrational spectra for the *trans*-difluoroethylene molecule.
    (a) Determine the number and symmetries for the normal modes.
    (b) Determine the activities of each of the modes in the infrared.
    (c) Determine the symmetries of each of the following types of internal modes:
        (i) C—F stretching modes;
        (ii) Out-of-plane bending modes;
        (iii) H—C—F angle deformations;
        (iv) C—C stretching modes.
16. Since the cyanogen molecule C_2N_2 is linear, it has seven fundamental vibrational modes.
    (a) Determine the IR activity of each of these fundamentals.
    (b) Sketch the approximate atomic motions for the vibrational modes. Make sure to symmetrize both the parallel and perpendicular modes.
17. For the *cis*-diimide molecule H—N=N—H of C_{2v} symmetry:
    (a) Determine the number and symmetries of the normal modes of vibration.
    (b) Which modes are IR active?
    (c) Number the normal modes and describe each mode (e.g., symmetric N—H stretch, etc).
    (d) For each mode estimate a vibrational frequency.
18. For the IF molecule, the following spectroscopic constants were recently determined:

$$\omega_e = 610.258\ \text{cm}^{-1}$$

$$\omega_e x_e = 3.141\ \text{cm}^{-1}$$

$$B_e = 0.279711\ \text{cm}^{-1}$$

$$\alpha_e = 0.001874\ \text{cm}^{-1}$$

    (a) Determine the IF bond length (r_e).
    (b) Describe and sketch the fundamental infrared spectrum.
    (c) Calculate the frequency of the $R(2)$ and $P(2)$ transitions for the fundamental band and the first overtone.

19. The infrared spectrum of N_2O has three fundamental bands. Assuming that the structure of N_2O is linear, explain how this spectrum allows you to distinguish between the N—N—O and the N—O—N structures. Sketch the approximate atomic motions of the normal modes.
20. Several of the lines in the $v = 0$ and $v = 1$ transition for HCl have the following frequencies:

$P(J)$ cm^{-1}	J	$R(J)$ cm^{-1}
	0	2906.25
2865.09	1	2925.78
2843.56	2	2944.89
2821.49	3	2963.24
2798.78	4	

   (a) Use these data to determine the band origin ν_o.
   (b) Calculate α_e and B_e.
   (c) Determine the equilibrium internuclear separation r_e to as many significant figures as the data justify.
21. The following bands have been measured in the infrared spectrum of a bent AB_2 molecule:

Frequency (cm^{-1})	Intensity	Frequency (cm^{-1})	Intensity
1200	vs	3600	w
2400	m	3870	m
2670	vs	4700	m
3500	vs	4800	s

   Identify the fundamental, overtone, and combination bands.
22. Prove that Fermi resonance cannot occur between two levels with different values of l.
23. Show that the superposition of two vibrations at right angles to one another gives circular motion if the vibrations have equal amplitudes and differ in phase by 90°.
24. (a) Calculate the **G** matrix for D_2O.
   (b) Using the force constants listed in Table 7.3 calculate the vibrational frequencies of D_2O and compare with the experimental values of $v_1 = 2666$ cm^{-1}, $v_2 = 1179$ cm^{-1} and $v_3 = 2789$ cm^{-1}.
25. (a) Using the vibrational constants (7.183) calculate the symmetry and frequency of all possible vibrational levels of HCN below 3500 cm^{-1}.
   (b) Calculate all possible allowed vibrational transitions between these levels.

References

1. Dunham, J. L. Phys. Rev. **41,** 721 (1932).
2. Townes, C. H. and Schawlow, A. L. *Microwave Spectroscopy,* Dover, New York 1975, pp. 7–9.
3. Schatz, G. C. and Ratner, M. A. *Quantum Mechanics in Chemistry,* Prentice-Hall, Englewood Cliffs, N.J., 1993, pp. 167–175.
4. Townes, C. H. and Schawlow, A. L. *Microwave Spectroscopy,* Dover, New York, 1975, pp. 10–11.
5. Herzberg, G. *Spectra of Diatomic Molecules,* Van Nostrand Reinhold, New York, 1950.
6. Watson, J. K. G. J. Mol. Spectrosc. **45,** 99 (1973) and **80,** 411 (1980).
7. Langhoff, S. R., Bauschlicher, C. W., and Taylor, P. R. J. Chem. Phys. **91,** 5953 (1989).
8. Bragg, S. L., Brault, J. W., and Smith, W. H. Astrophys. J. **263,** 999 (1982).
9. Huber, K. P. and Herzberg, G. *Constants of Diatomic Molecules,* Van Nostrand Reinhold, New York, 1979.
10. Herzberg, G. *Spectra of Diatomic Molecules,* Van Nostrand Reinhold, New York, 1950, pp. 438–441.
11. Dabrowski, I. Can. J. Phys. **62,** 1639 (1984).
12. Le Roy, R. J. and Bernstein, R. B. Chem. Phys. Lett. **5,** 42 (1970).
13. Le Roy, R. J. In *Molecular Spectroscopy: A Specialist Periodical Report,* Vol. I, Chemical Society, London, 1973, p. 113.
14. Brown, W. G. Phys. Rev. **38,** 709 (1931).
15. Tromp, J. W., Le Roy, R. J., Gerstenkorn, S., and Luc, P. J. Mol. Spectrosc. **100,** 82 (1983).
16. Rinsland, C. P., Smith, M. A., Goldman, A., Devi, V. M., and Benner, D. C. J. Mol. Spectrosc. **159,** 274 (1993).
17. Wilson, E. B., Decius, J. C., and Cross, P. C. *Molecular Vibrations,* Dover, New York, 1980.
18. Shimanouchi, T. In *Physical Chemistry: An Advanced Treatise,* vol. IV, H. Eyring, D. Henderson, and W. Jost, Eds., Academic Press, New York, 1970, Chapter 6.
19. Herzberg, G. *Electronic Spectra and Electronic Structure of Polyatomic Molecules,* Van Nostrand Reinhold, New York, 1966.
20. Herzberg G. *Infrared and Raman Spectra of Polyatomic Molecules,* Van Nostrand, Princeton, N.J., 1945.
21. Hougen, J. T. *The Calculation of the Rotational Energy Levels and Rotational Line Intensities in Diatomic Molecules,* NBS Monograph 115, U.S. Government Printing Office, Washington, D.C., 1970; also Hougen, J. T. J. Chem. Phys. **39,** 358 (1963).
22. Brown, J. M. et al. J. Mol. Spectrosc. **55,** 500 (1975).
23. Frum, C. I., Engleman, R., and Bernath, P. F. J. Chem. Phys. **95,** 1435 (1991).
24. Rank, D. H., Skorinko, G., Eastman, D. P., and Wiggins, T. A. J. Opt. Soc. Am. **50,** 421 (1960).
25. Oka, T. Phys. Rev. Lett. **45,** 531 (1980); also Miller, S. and Tennyson, J. J. Mol. Spectrosc. **126,** 183 (1987).
26. Papousek, D. and Aliev, M. R. *Molecular Vibrational-Rotational Spectra,* Elsevier, Amsterdam, 1982, pp. 191–206.
27. Papousek, D. and Aliev, M. R. *Molecular Vibrational-Rotational Spectra,* pp. 169–171.
28. Amano, T., Bernath, P. F., and McKellar, A. R. W. J. Mol. Spectrosc. **94,** 100 (1982).
29. Spirko, V. J. Mol. Spectrosc. **101,** 30 (1983).
30. Bunker, P. R. *Molecular Symmetry and Spectroscopy,* Academic Press, New York, 1979.

General References

Allen, H. C. and Cross, P. C. *Molecular Vib-rotors,* Wiley, N.Y., 1963.

Bellamy, L. J. *The Infrared Spectra of Complex Molecules,* 3rd ed., Chapman & Hall, London, 1976.

Bunker, P. R., *Molecular Symmetry and Spectroscopy,* Academic Press, New York, 1979.

Califano, S., *Vibrational States,* Wiley, New York, 1976.

Colthrup, N. B., Daly, L. H., and Wiberley, S. E. *Introduction to Infrared and Raman Spectroscopy,* 3rd ed., Academic Press, San Diego, 1990.

Cyvin, S. J. *Molecular Vibrations and Mean Square Amplitudes,* Elsevier, Amsterdam, 1968.

Gans, P. *Vibrating Molecules,* Chapman & Hall, London, 1971.

Goldstein, H. *Classical Mechanics,* 2nd ed., Addison-Wesley, Reading, Mass., 1980.

Graybeal, J. D. *Molecular Spectroscopy,* McGraw-Hill, New York, 1988.

Herzberg, G. *Infrared and Raman Spectra of Polyatomic Molecules,* Van Nostrand, New York, 1945.

Herzberg, G. *Spectra of Diatomic Molecules,* Van Nostrand-Reinhold, New York, 1950.

Huber, K. P. and Herzberg, G. *Constants of Diatomic Molecules,* Van Nostrand Reinhold, New York, 1979.

Nakamoto, K. *Infrared and Raman Spectra of Inorganic and Coordination Compounds,* 4th ed., Wiley, New York, 1986.

Jacox, M. Ground-state vibrational frequencies of transient molecules, J. Chem. Ref. Data **13,** 945 (1984).

Jacox, M. Vibrational and electronic energy levels of polyatomic transient molecules, Suppl. 1, **19,** 1387 (1990).

King, G. W., *Spectroscopy and Molecular Structure,* Holt, Reinhart & Winston, New York, 1964.

Papousek, D. and Aliev, M. R. *Molecular Vibrational-Rotational Spectra,* Elsevier, Amsterdam, 1982.

Shimanouchi, T. Tables of molecular vibrational frequencies, Consolidated Vol. I, NSRDS-NBS 39, U.S. Government Printing Office, Washington, D.C., 1972; also Consolidated Vol. II, J. Phys. Chem. Ref. Data **6,** 993 (1977).

Steele, D. *Theory of Vibrational Spectroscopy,* Saunders, Philadelphia, 1971.

Struve, W. S. *Fundamentals of Molecular Spectroscopy,* Wiley, New York, 1989.

Wilson, E. B., Decius, J. G., and Cross, C. *Molecular Vibrations,* Dover, New York, 1980.

Woodward, L. A. *Introduction to the Theory of Molecular Vibrations and Vibrational Spectroscopy,* Oxford University Press, Oxford, 1972.

Chapter 8

The Raman Effect

Background

The Raman effect is a light scattering phenomenon. When light of frequency ν_I or ν_0 (usually from a laser or, in the prelaser era, from a mercury arc lamp) irradiates a sample (Figure 8.1), it can be scattered. The frequency of the scattered light can either be at the original frequency (referred to as *Rayleigh scattering*) or at some shifted frequency $\nu_S = \nu_I \pm \nu_{\text{internal}}$ (referred to as *Raman scattering*). The frequency ν_{internal} is an internal frequency corresponding to rotational, vibrational, or electronic transitions. The vibrational Raman effect is by far the most important, although rotational and electronic Raman effects are also known. For example, the rotational Raman effect provides some of the most accurate bond lengths for homonuclear diatomic molecules.

In discussing the Raman effect some commonly used terms need to be defined (Figure 8.2). Radiation scattering to the lower frequency side (to the "red") of the exciting line is called *Stokes scattering.* The scattered radiation at the same frequency as the incident radiation is called *Rayleigh scattering,* while the light scattered at higher frequencies than the exciting line (to the "blue") is referred to as the *anti-Stokes scattering.* Finally, the magnitude of the shift between the Stokes or the anti-Stokes line and the exciting line is called the *Raman shift,* $\Delta\nu = |\nu_I - \nu_{\text{internal}}|$.

Classical Model

When an electric field is incident on a molecule, the electrons and nuclei respond by moving in opposite directions in accordance with Coulomb's law. The applied electric field therefore induces a dipole moment in the molecule. As long as the applied electric field is not too strong, the induced dipole moment is linearly proportional to the applied electric field, and is given by

$$\boldsymbol{\mu}_{\text{Ind}} = \alpha\mathbf{E}, \tag{8.1}$$

in which the proportionality constant α is called the *polarizability* and is a characteristic of the molecule.

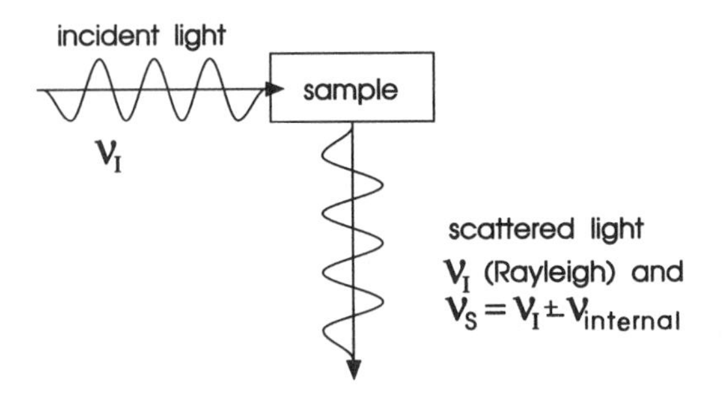

Figure 8.1. Scattering of light by a sample.

The intensity of the scattered light is proportional to the square of the magnitude of the induced oscillating dipole moment. If some internal motion of the molecule (vibrational, rotational, or electronic) modulates this induced oscillating dipole moment, then additional frequencies can appear. Classically, this means that the polarizability has a static term α_0 and a sinusoidal oscillating term with amplitude α_1

$$\alpha = \alpha_0 + \alpha_1 \cos(\omega t) \tag{8.2}$$

with ω $(= \omega_{\text{internal}})$ being some internal angular frequency. For example, a vibrational mode Q_i has

$$\alpha_1 = \left.\frac{\partial \alpha}{\partial Q_i}\right|_{Q_i=0} Q_i \tag{8.3}$$

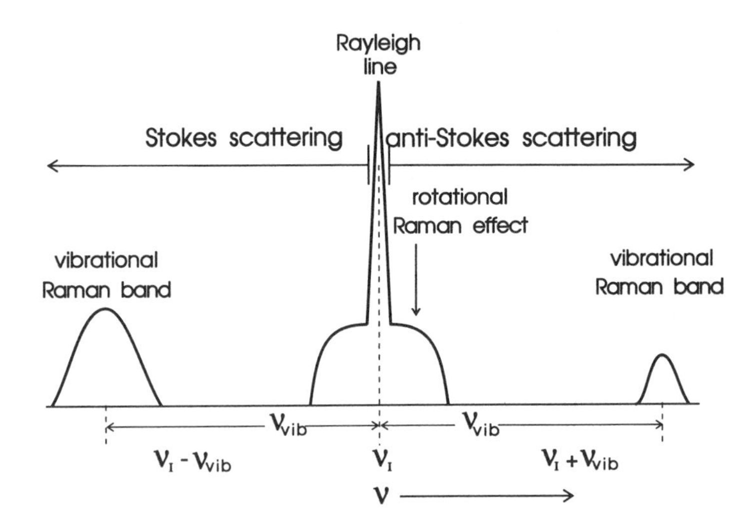

Figure 8.2. Schematic diagram of a Raman spectrum showing vibrational and rotational Raman effects.

so that if the polarizability does not change with vibration, that is, if $(\partial\alpha/\partial Q_i)|_0 = 0$, then there is no vibrational Raman effect. Classically, the oscillating polarizability causes the induced dipole moment to oscillate at frequencies other than the incident ω_I. To see this let us represent **E** as $\mathbf{E}_0 \cos \omega_I t$. Upon substituting equation (8.2) into the magnitude of equation (8.1), we get

$$\mu_{\text{Ind}} = \alpha E = \alpha E_0 \cos \omega_I t \tag{8.4}$$

$$\begin{aligned}\mu_{\text{Ind}} &= (\alpha_0 + \alpha_1 \cos \omega t) E_0 \cos \omega_I t \\ &= \alpha_0 E_0 \cos \omega_I t + \alpha_1 E_0 \cos \omega_I t \cos \omega t \\ &= \alpha_0 E_0 \cos \omega_I t + \frac{\alpha_1 E_0 \cos(\omega_I - \omega)t + \alpha_1 E_0 \cos(\omega_I + \omega)t}{2}.\end{aligned} \tag{8.5}$$

The trigonometric identity

$$\cos \theta \cos \phi = \frac{\cos(\theta - \phi) + \cos(\theta + \phi)}{2} \tag{8.6}$$

has been used in the final step of equation (8.5). The first term is unshifted in frequency and corresponds to Rayleigh scattering (Figure 8.2). The lower frequency term with $\omega_I - \omega$ corresponds to Stokes scattering, while the higher frequency term with $\omega_I + \omega$ corresponds to anti-Stokes scattering (Figure 8.3). This simple classical derivation (8.5) is very deceptive since it predicts that Stokes and anti-Stokes scattering have the same intensity: this is not usually the case.

The energy-level diagram for Stokes and anti-Stokes scattering shows that anti-Stokes scattering will be weaker because the population in the excited vibrational

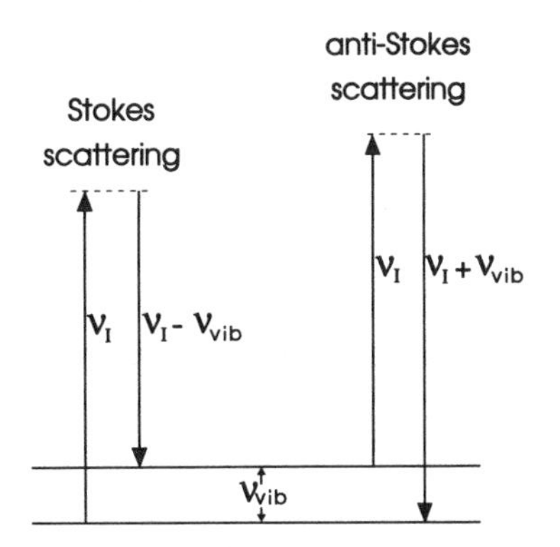

Figure 8.3. Energy-level diagram showing Stokes and anti-Stokes scattering.

level is less than that in the ground state (Figure 8.3). For a classical oscillator the scattering (Rayleigh and Raman) is proportional to the fourth power of the frequency. (The sky is blue because air molecules Rayleigh scatter more blue than red sunlight.) Thus if we introduce the Boltzmann distribution of vibrational populations, the ratio of the intensities of the bands is given by

$$\frac{\text{Anti-Stokes intensity}}{\text{Stokes intensity}} = \frac{(\nu_I + \nu_{\text{vib}})^4 e^{-h\nu_{\text{vib}}/kT}}{(\nu_I - \nu_{\text{vib}})^4} \tag{8.7}$$

for a nondegenerate vibration.

There is one additional complication. For highly symmetric molecules such as CH_4 the induced dipole is always in the same direction as the applied electric field. For less symmetric molecules, however, $\boldsymbol{\mu}_{\text{Ind}}$ and $\mathbf{E}$ can point in different directions. In matrix notation, the equation

$$\boldsymbol{\mu}_{\text{Ind}} = \boldsymbol{\alpha}\mathbf{E} \tag{8.8}$$

becomes

$$\begin{pmatrix} \mu_X \\ \mu_Y \\ \mu_Z \end{pmatrix} = \begin{pmatrix} \alpha_{XX} & \alpha_{XY} & \alpha_{XZ} \\ \alpha_{XY} & \alpha_{YY} & \alpha_{YZ} \\ \alpha_{XZ} & \alpha_{YZ} & \alpha_{ZZ} \end{pmatrix} \begin{pmatrix} E_X \\ E_Y \\ E_Z \end{pmatrix} \tag{8.9}$$

in which $\boldsymbol{\alpha}$ is a 3×3 symmetric matrix. This symmetric matrix is called the *polarizability tensor.*

Quantum Model

In the conversion to quantum mechanics it is found that the intensity of the Raman effect depends, in general, on the square of integrals of the type

$$\int \psi_1^* \alpha_{ij} \psi_0 \, d\tau \qquad i, j = X, Y, \text{ and } Z. \tag{8.10}$$

Specifically, for the vibrational Raman effect the intensity depends on the square of integrals of the type

$$\int \psi_1^* \alpha_{ij} \psi_0 \, dQ \qquad i, j = x, y, z, \tag{8.11}$$

in which ψ_1 and ψ_0 are vibrational wavefunctions, and the integral is evaluated in the molecular coordinate system. The six elements of the polarizability tensor α_{xx}, α_{yy}, α_{zz}, α_{xy}, α_{xz}, and α_{yz} all transform like the binary products of coordinates x^2, y^2, z^2, xy, xz, and yz when the symmetry operations of the point group are applied. The symmetry of these binary products (or properly symmetrized combinations) are listed on the right side of character tables. Thus the direct product

$$\Gamma(\psi_1^*) \otimes \Gamma(\alpha_{ij}) \otimes \Gamma(\psi_0) \tag{8.12}$$

must contain the A_1 irreducible representation in order for the corresponding integral to be nonzero and give a Raman allowed transition.

For example, x^2, y^2, and z^2 for the H_2O molecule have A_1 symmetry, while xy, xz, and yz have A_2, B_1, and B_2 symmetry, respectively. Thus the three normal modes of H_2O $\nu_1(a_1)$, $\nu_2(a_1)$, and $\nu_3(b_2)$ are all Raman active (Figure 8.4).

Notice that if a molecule has a center of symmetry, then both ψ_0 (for fundamentals) and α_{ij} have g symmetry and consequently ψ_1 must also be of g symmetry. Thus all Raman active fundamental transitions have g symmetry if the molecule has a center of symmetry. Correspondingly, all infrared active fundamentals must have u symmetry since $\boldsymbol{\mu}$ has u symmetry. This leads to the rule of mutual exclusion, which states that no fundamental mode of a molecule with a center of symmetry can be both infrared and Raman active. Comparison of infrared and Raman band positions can be a simple but powerful tool in deducing molecular geometry.

For the tetrahedral molecule CCl_4 all four vibrational modes [$\nu_1(a_1)$ 459 cm^{-1}, $\nu_2(e)$ 218 cm^{-1}, $\nu_3(t_2)$ 762 cm^{-1}, $\nu_4(t_2)$ 314 cm^{-1}] (see Figure 8.5) are Raman active.

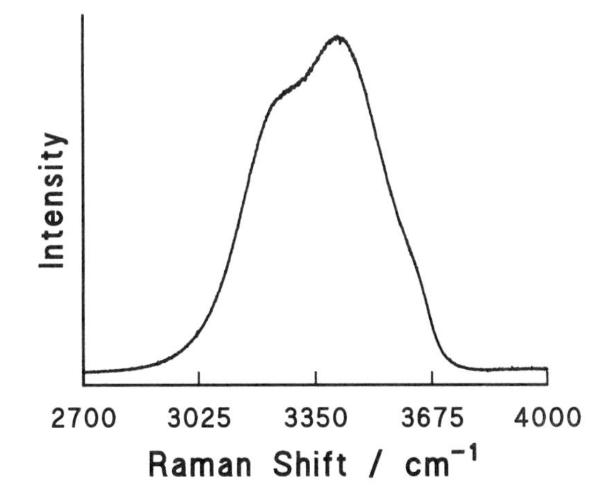

Figure 8.4. Raman spectrum of liquid H_2O in the O—H stretching region. The peak to the right is the ν_1 symmetric stretching mode, while the peak to the left is due to the $2\nu_2$ overtone and the O—H stretching mode of two (or more) hydrogen-bonded H_2O molecules.

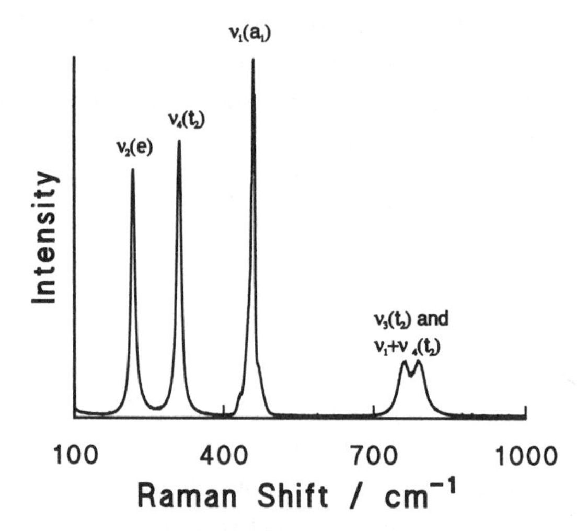

Figure 8.5. Vibrational Raman spectrum of liquid CCl_4.

This is in contrast to the infrared spectrum in which only ν_3 and ν_4 are observed. The partially resolved doublet near 775 cm^{-1} in the Raman spectrum is actually two Fermi resonance transitions (762 cm^{-1}, 790 cm^{-1}) made up of nearly equal mixtures of $\nu_3(t_2)$ and $\nu_1 + \nu_4(t_2)$.

Polarization

The typical perpendicular Raman scattering geometry is shown in Figure 8.6. The intensity of light scattered parallel ($I_{\parallel}$) and perpendicular ($I_{\perp}$) to the incident electric field vector can easily be measured with polarizers. The ratio $\rho = I_{\perp}/I_{\parallel}$, called the *depolarization ratio,* is an important clue in the assignment of a vibrational Raman spectrum because it depends on the symmetry of the vibrational mode.

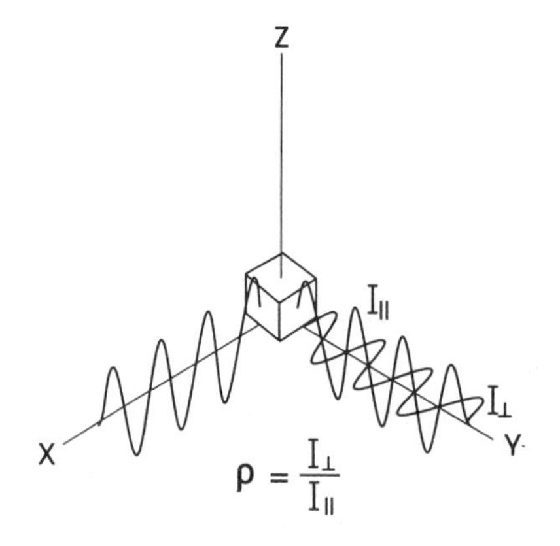

Figure 8.6. Parallel and perpendicular Raman scattering.

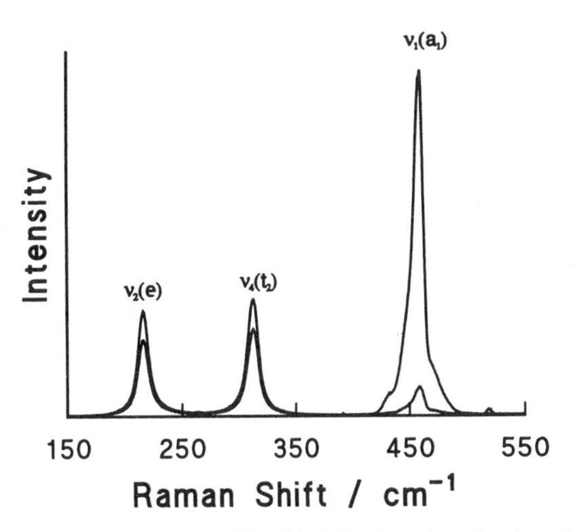

Figure 8.7. Vibrational Raman spectrum of liquid CCl_4 showing the depolarization of the bands. The upper trace corresponds to $I_{\parallel}$ and the lower trace to $I_{\perp}$.

From the theory of the Raman effect, it is known that a symmetric vibration has $0 \leq \rho \leq \frac{3}{4}$ for linearly polarized incident light. For a non-totally symmetric vibration, $\rho = \frac{3}{4}$ for linearly polarized incident light, and the band is said to be depolarized. If unpolarized light is used, for example the mercury arc lamp in the prelaser era, then $\rho = \frac{6}{7}$ for a non-totally symmetric vibration. Thus a measurement of the depolarization ratio will often distinguish between totally symmetric and nonsymmetric vibrations. Totally symmetric vibrations, such as the C—Cl stretching mode ($\nu_1(a_1)$ 459 cm^{-1}) in CCl_4, tend to be strong scatterers with depolarization ratios close to zero (Figure 8.7), whereas this mode is forbidden in the infrared spectrum.

The physical origin of polarized scattering for a symmetric vibration is easy to understand in classical terms. For example, in the case of a symmetric vibration for a spherical top, the induced dipole is always parallel to the incident radiation and the molecule behaves like a tiny sphere (Figure 8.8). The scattered light is also polarized

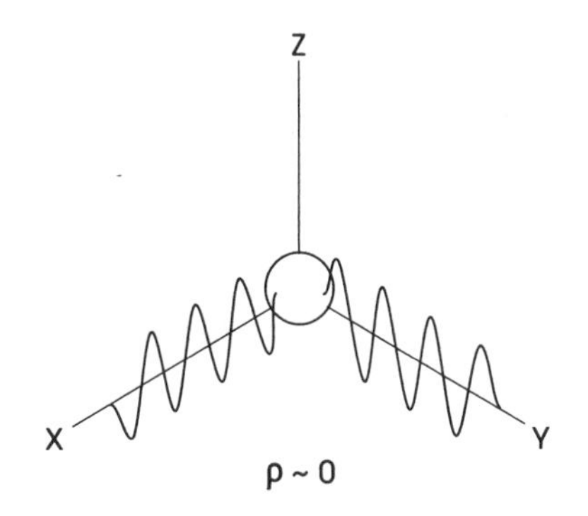

Figure 8.8. Polarized light scattering by a sphere.

parallel to the incident light polarization and $\rho \sim 0$ (Figure 8.8). Molecules with O_h or T_d symmetry behave in this way.

Rotational Raman Effect

The dipole moment induced in a nonrotating molecule when an electric field is applied is given in the laboratory frame by

$$\boldsymbol{\mu} = \boldsymbol{\alpha}\mathbf{E}, \tag{8.13}$$

or, written explicitly, by

$$\begin{pmatrix} \mu_X \\ \mu_Y \\ \mu_Z \end{pmatrix} = \begin{pmatrix} \alpha_{XX} & \alpha_{XY} & \alpha_{XZ} \\ \alpha_{XY} & \alpha_{YY} & \alpha_{YZ} \\ \alpha_{XZ} & \alpha_{YZ} & \alpha_{ZZ} \end{pmatrix} \begin{pmatrix} E_X \\ E_Y \\ E_Z \end{pmatrix}. \tag{8.14}$$

Since the polarizability tensor, like the moment of inertia tensor, is represented by a real symmetric matrix, it is always possible to find an orthogonal transformation that diagonalizes $\boldsymbol{\alpha}$. This new coordinate system is obtained by a rotation of the molecular x, y, and z-axes such that the off-diagonal components of $\boldsymbol{\alpha}$ are eliminated. The transformed coordinates x', y', and z' are called the principal axes of polarizability with

$$\boldsymbol{\alpha}' = \begin{pmatrix} \alpha_{x'x'} & 0 & 0 \\ 0 & \alpha_{y'y'} & 0 \\ 0 & 0 & \alpha_{z'z'} \end{pmatrix}. \tag{8.15}$$

For an asymmetric top molecule such as H_2O, $\alpha_{x'x'} \neq \alpha_{y'y'} \neq \alpha_{z'z'}$, while for a linear or symmetric top molecule $\alpha_{x'x'} = \alpha_{y'y'}$. For a spherical top $\alpha_{x'x'} = \alpha_{y'y'} = \alpha_{z'z'}$, and the molecule behaves like a small sphere when an electric field is applied. It is usually assumed that the coordinate system is the principal coordinate system, and the primes are dropped.

The oscillating electromagnetic field is applied and the scattered light is detected in the laboratory frame of reference. The rotation of the molecule therefore modulates the scattered light for all molecules except spherical top molecules (Figure 8.9).

The rotational Raman effect is less restrictive than is pure rotational spectroscopy because symmetric linear molecules without dipole moments such as Cl_2 and CO_2 have pure rotational Raman spectra. However, spherical tops such as CH_4, SF_6, and C_{60} will not have observable rotational Raman spectra because an anisotropic polarizability tensor is required. In simple terms, an applied electric field can only exert a torque on a molecule if it is more polarizable in one direction than another.

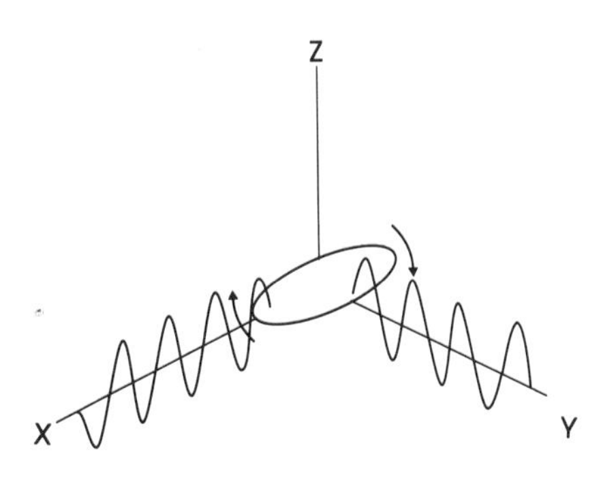

Figure 8.9. Light scattering by a rotating molecule is modulated by the rotational motion.

The rotational selection rules are obtained by evaluating the integrals

$$\int \psi_1^* \alpha_{IJ} \psi_0 \, d\tau = \int \psi_1^* \Big(\sum_{i,j} \Phi_{Ii} \alpha_{ij} \Phi_{jJ} \Big) \psi_0 \, d\tau$$

$$= \sum_{i,j} \alpha_{ij} \int \psi_1^* \Phi_{Ii} \Phi_{jJ} \psi_0 \, d\tau \qquad i, j = x, y, z; \quad I, J = X, Y, Z, \qquad (8.16)$$

in which the Φ_{Ii} are the direction cosines, the ψ_i are rotational wave functions (ψ and Φ are both functions of the Euler angles θ, ϕ, χ (Figure 6.27) and α_{ij} is the polarizability component in the molecular frame. The direction cosines are required (Chapter 6) in order to transform between the laboratory and molecular coordinate systems. Selection rules for rotational Raman spectroscopy are derived from matrix elements of the *products* of the direction cosine matrix elements. As a result, $\Delta J = \pm 2$ transitions are possible. In simple terms, since there are two photons involved in a Raman transition, transitions with $\Delta J = \pm 2$ are possible.

Compare the previous results with pure rotational microwave transitions in which

$$\int \psi_1^* \mu_I \psi_0 \, d\tau = \int \psi_1^* \Big(\sum_i \Phi_{Ii} \mu_i \Big) \psi_0 \, d\tau$$

$$= \sum_i \mu_i \int \psi_1^* \Phi_{Ii} \psi_0 \, d\tau \qquad I = X, Y, Z; \quad i = x, y, z. \qquad (8.17)$$

Again the integration is over the Euler angles, and μ_i is the dipole moment averaged over vibrational and electronic variables. In this case the matrix elements of the direction cosines result in the selection rule, $\Delta J = \pm 1$.

Diatomic Molecules

The selection rules for the rotational Raman effect in linear $^1\Sigma^+$ molecules are $\Delta J = 0, \pm 2$. Only S-branch transitions ($\Delta J = +2$) are observable since the $\Delta J = 0$

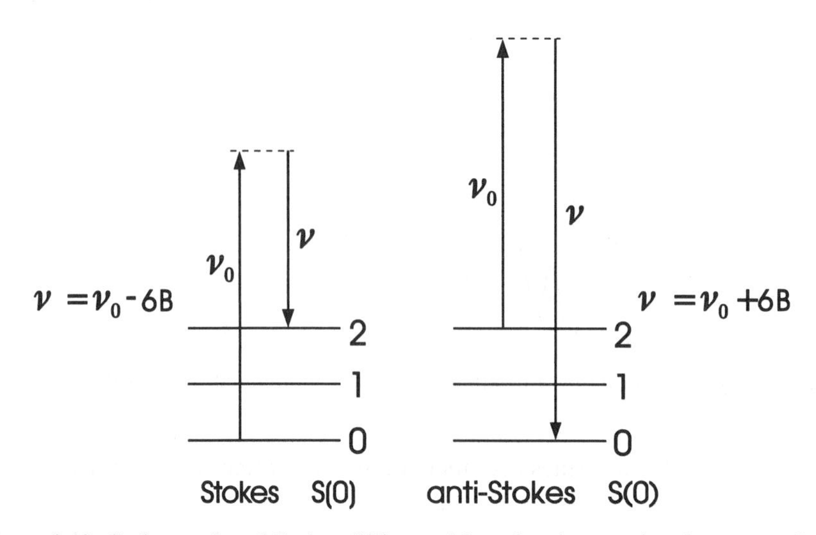

Figure 8.10. Stokes and anti-Stokes $S(0)$ transitions for the rotational Raman effect.

transitions correspond to the unshifted Rayleigh line. The definition of the S branch as $\Delta J = J' - J''$ means that both the Stokes and anti-Stokes transitions are S-branch lines (Figure 8.10), although this seems confusing at first sight.

The transition frequencies are given by

$$\begin{aligned} \nu &= \nu_0 \pm [B(J+2)(J+3) - BJ(J+1)] \\ &= \nu_0 \pm B(4J+6) \end{aligned} \tag{8.18}$$

where $\pm$ corresponds to anti-Stokes and Stokes transitions, respectively. The lines are spaced by about $4B$ from each other. Figure 8.11 shows the rotational Raman spectrum of N_2.

Vibration–Rotation Raman Spectroscopy

Diatomic Molecules

The selection rules for vibration–rotation Raman spectroscopy for $^1\Sigma^+$ diatomic molecules are $\Delta v = \pm 1$ and $\Delta J = 0, \pm 2$. The vibrational transitions with $\Delta v = \pm 2, \pm 3, \ldots$ are allowed for the anharmonic oscillator, similar to infrared vibration–rotation spectroscopy, but are much weaker.

The rotational selection rules $\Delta J = 0, \pm 2$ result in O, Q, and S branches as shown in Figure 8.12. The rotation–vibration Raman spectrum of N_2 is shown in Figure 8.13.

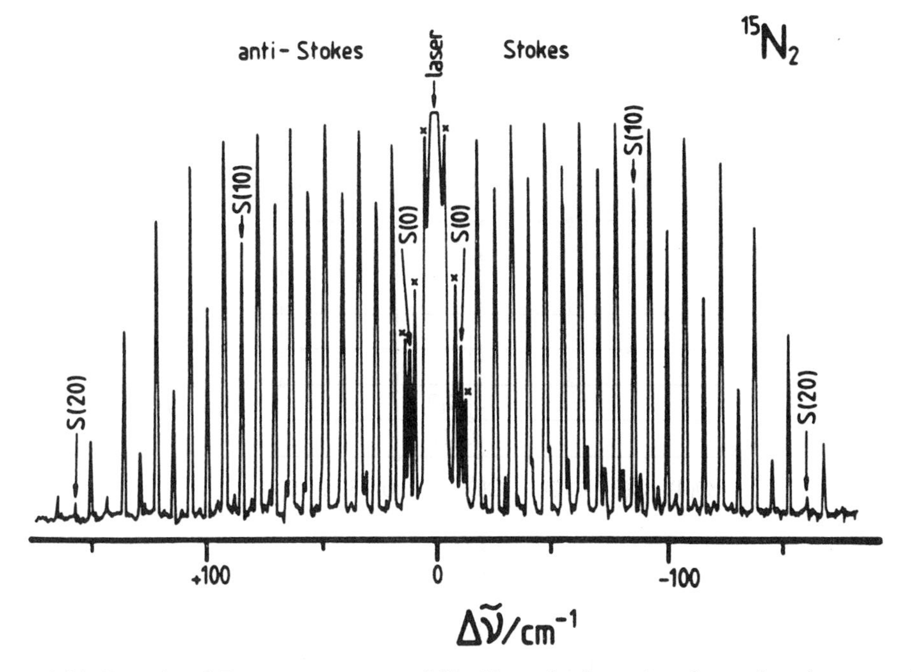

Figure 8.11. Rotational Raman spectrum of N_2. Note the intensity alternation due to nuclear spin statistics and the x's that mark artifacts called grating ghosts.

The equations for the three branches are

$$\nu_S = \nu_0 - [6B' + (5B' - B'')J'' + (B' - B'')(J'')^2] \qquad J'' = 0, 1, 2, \ldots \qquad (8.19)$$

$$\nu_Q = \nu_0 - [(B' - B'')J'' + (B' - B'')(J'')^2] \qquad J'' = 0, 1, 2, \ldots \qquad (8.20)$$

and

$$\nu_O = \nu_0 - [2B' - (3B' + B'')J'' + (B' - B'')(J'')^2] \qquad J'' = 2, 3, \ldots \qquad (8.21)$$

in which $\nu_0 = \nu_l - \Delta G_{1/2}$ for the $1 \leftarrow 0$ Stokes spectrum.

There has been a renaissance in Raman spectroscopy with the availability of lasers, Fourier transform spectrometers, and sensitive array detectors. Although Rayleigh scattering is weak and Raman scattering even weaker (typically 10^{-6} of the incident radiation), Raman spectroscopy has a number of important attributes.

Raman spectroscopy has different selection rules than do direct electronic, vibrational, and rotational spectroscopies, so it provides complementary information. Raman spectroscopy uses visible light to obtain electronic, vibrational, and rotational information about molecules. Since the technology for generating, manipulating, and detecting visible light is often more advanced than the corresponding infrared and millimeter wave technology, this can be an important experimental advantage. The water molecule is a relatively weak Raman scatterer but a strong infrared absorber. Because of this Raman spectroscopy is often the technique of choice for the

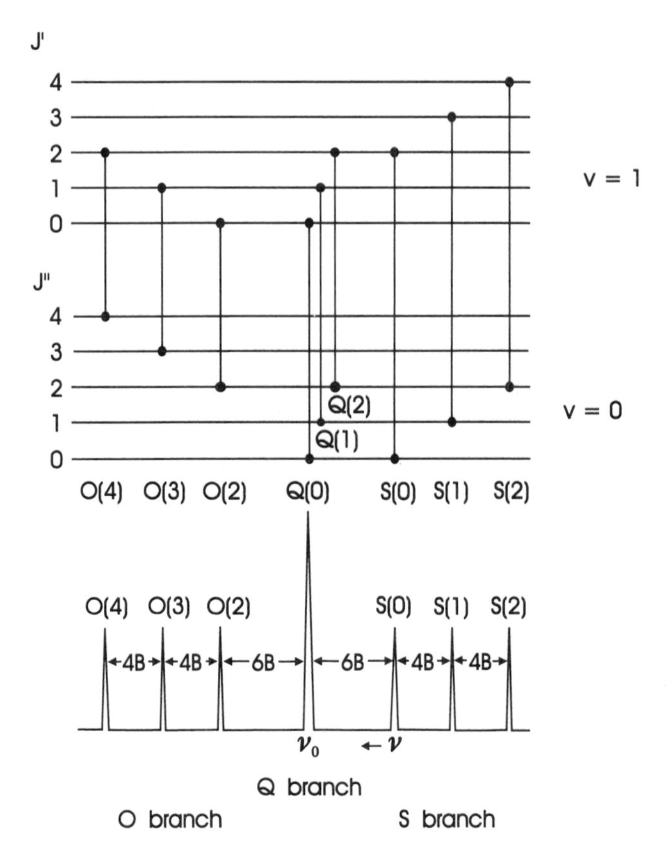

Figure 8.12. Energy-level diagram and spectrum for vibrational Raman scattering of a linear molecule.

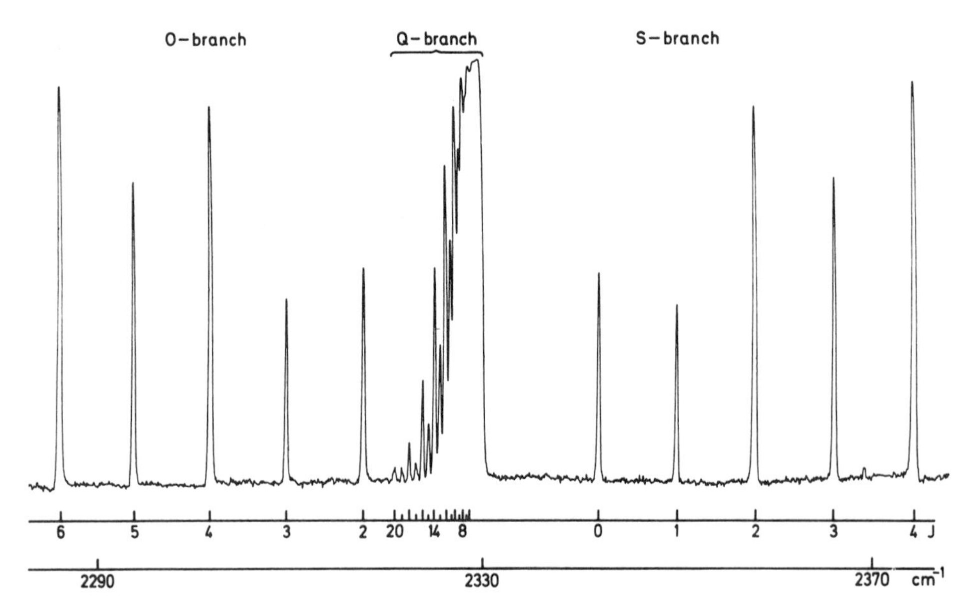

Figure 8.13. Vibration–rotation Raman spectrum of N_2. Note the intensity alternation due to nuclear spin statistics.

vibrational spectroscopy of molecules in aqueous environments. For example, the vibrational spectroscopy of biological samples (which are altered by dehydration) is usually best carried out by Raman scattering.

Problems

1. Which normal modes of ethylene are Raman active? (See Problem 1 of Chapter 7.)
2. For the molecules in Problem 2 of Chapter 7, which modes are infrared active and which are Raman active?
3. Discuss the Raman activity of the normal modes of the molecules in Problems 4, 5, 12, 14, 15, 16, and 17 of Chapter 7.
4. For the ICl molecule the following spectroscopic constants are listed in Huber and Herzberg's book

$$\omega_e = 384.293 \text{ cm}^{-1}$$
$$\omega_e x_e = 1.501 \text{ cm}^{-1}$$
$$B_e = 0.1141587 \text{ cm}^{-1}$$
$$\alpha_e = 0.0005354 \text{ cm}^{-1}$$

   (a) Predict the pure rotational Raman spectrum. What will be the Raman shift of the two lines closest to the exciting laser line?
   (b) Predict the pattern of the Stokes vibration–rotation Raman spectrum. What will be the Raman shifts of the $S(0)$ and $O(2)$ lines from the exciting laser line at 5145 Å?
5. Fill in the following table with a yes (Y) or a no (N) to indicate allowed spectroscopic transitions.

Molecule	Pure Rotational	Vibrational	Rotational Raman	Vibrational Raman
H_2O				
SF_6				
CS_2				
N_2O				
Allene				
Benzene				
Cl_2				

6. The vibrational Raman spectrum of the SO_3^{2-} anion of C_{3v} symmetry exhibits four

bands in aqueous solution: 966 cm^{-1} (strong, p); 933 cm^{-1} (shoulder, dp); 620 cm^{-1} (weak, p); and 473 cm^{-1} (dp) (p = polarized; dp = depolarized). Assign the symmetries of the bands and describe the motion of the normal modes.

General References

Anderson, A. *The Raman Effect,* Dekker, New York, 1973.

Califano, S. *Vibrational States,* Wiley, New York, 1976.

Colthup, N. B., Daly, L. H., and Wilberley, S. E. *Introduction to Infrared and Raman Spectroscopy,* 3rd ed., Academic Press, San Diego, 1990.

Herzberg, G. *Spectra of Diatomic Molecules,* Van Nostrand Reinhold, New York, 1950.

Herzberg, G. *Infrared and Raman Spectra of Polyatomic Molecules,* Van Nostrand Reinhold, New York, 1945.

Huber, K. P. and Herzberg, G. *Constants of Diatomic Molecules,* Van Nostrand Reinhold, New York, 1979.

Koningstein, J. A. *Introduction to the Theory of the Raman Effect,* Reidel, Dordrecht, The Netherlands, 1972.

King, G. W. *Spectroscopy and Molecular Structure,* Holt, Reinhart & Winston, New York, 1964.

Long, D. A. *Raman Spectroscopy,* McGraw-Hill, London, 1977.

Steele, D. *Theory of Vibrational Spectroscopy,* Saunders, Philadelphia, 1971.

Tobin, M. C. *Laser Raman Spectroscopy,* Wiley, New York, 1971.

Wilson, E. B., Decius, J. C., and Cross, P. C. *Molecular Vibrations,* Dover, New York, 1980.

Zare, R. N. *Angular Momentum,* Wiley, New York, 1988.

Chapter 9

Electronic Spectroscopy of Diatomic Molecules

Orbitals and States

Within the Born–Oppenheimer approximation, the electronic Schrödinger equation for a diatomic molecule A—B is

$$\hat{H}_{\text{el}}\psi_{\text{el}} = E_{\text{el}}\psi_{\text{el}}, \tag{9.1}$$

with the Hamiltonian operator given by

$$\hat{H}_{\text{el}} = \frac{-\hbar^2}{2m_e}\sum_i \nabla_i^2 - \sum_i \frac{Z_A e^2}{4\pi\varepsilon_0 r_{Ai}} - \sum_i \frac{Z_B e^2}{4\pi\varepsilon_0 r_{Bi}} + \frac{Z_A Z_B e^2}{4\pi\varepsilon_0 r_{AB}} + \sum_i \sum_{j>i} \frac{e^2}{4\pi\varepsilon_0 r_{ij}}. \tag{9.2}$$

The approximate solution of equation (9.1) is accomplished by assuming that ψ_{el} is made up of molecular orbitals (MOs) and that each MO is a linear combination of atomic orbitals (LCAO). More precisely, an approximate solution is written in the form of a determinant wavefunction

$$\psi_{\text{el}} = |\phi_1(1)\bar{\phi}_1(2)\phi_2(3)\cdots| \tag{9.3}$$

with each ϕ_i being a molecular orbital of the form

$$\phi_i = \sum_j (C_{iA_j}\phi_j^A + C_{iB_j}\phi_j^B), \tag{9.4}$$

in which ϕ_j^A and ϕ_j^B are atomic orbitals localized on atoms A and B, respectively.

The electronic structure of diatomic molecules can therefore be derived from a consideration of the shapes of the molecular orbitals constructed by linear combinations of atomic orbitals. The atomic orbitals of the constituent atoms are in an environment with reduced symmetry ($D_{\infty h}$ or $C_{\infty v}$, rather than K_h) in the diatomic

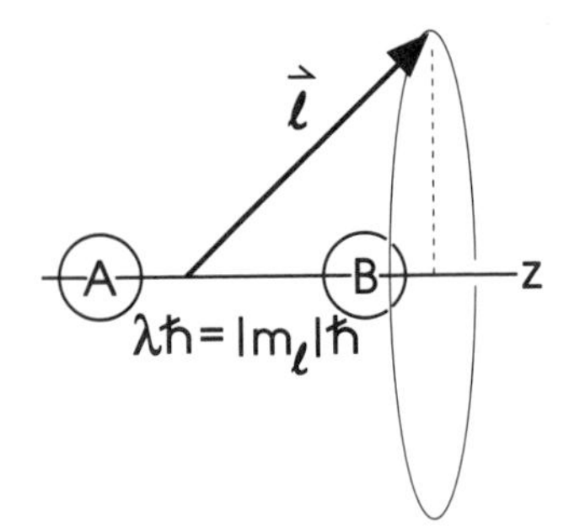

Figure 9.1. The precession of **l** about the internuclear axis.

molecule. As the symmetry is reduced from spherical to axial, each electron with orbital angular momentum l will begin to precess about the internuclear axis (Figure 9.1). This sort of "intramolecular Stark effect" means that although l is no longer a good quantum number, the projection of **l** onto the internuclear axis, m_l, remains useful. The sign of m_l is determined by the left- or right-handed circulation of the electron around the internuclear axis, and the electronic wavefunction is an eigenfunction of the $\hat{l}_z$ operator, that is,

$$\hat{l}_z\psi = -i\hbar\frac{\partial}{\partial\phi}Ae^{im_l\phi} = m_l\hbar\psi = \pm\,|m_l|\,\hbar\psi. \tag{9.5}$$

The direction of the circulation of an electron around the internuclear axis cannot affect the energy, so there is a double degeneracy for $\lambda \equiv |m_l| > 0$. It is useful to label the atomic orbitals in a diatomic molecule by

$$\lambda = |m_l| = 0, 1, 2, \ldots, l. \tag{9.6}$$

The irreducible representations of the point group K_h are s, p, d, f, etc.; similarly σ, π, δ, ϕ, etc., are the irreducible representations of $C_{\infty v}$ (Table 9.1). The labels g and u are appended as subscripts for $D_{\infty h}$ molecules.

Table 9.1. Correlation of Atomic and Molecular Orbitals

Atomic Orbital l	Molecular Orbital λ
s	$s\sigma$
p	$p\sigma$, $p\pi$
d	$d\sigma$, $d\pi$, $d\delta$
f	$f\sigma$, $f\pi$, $f\delta$, $f\phi$

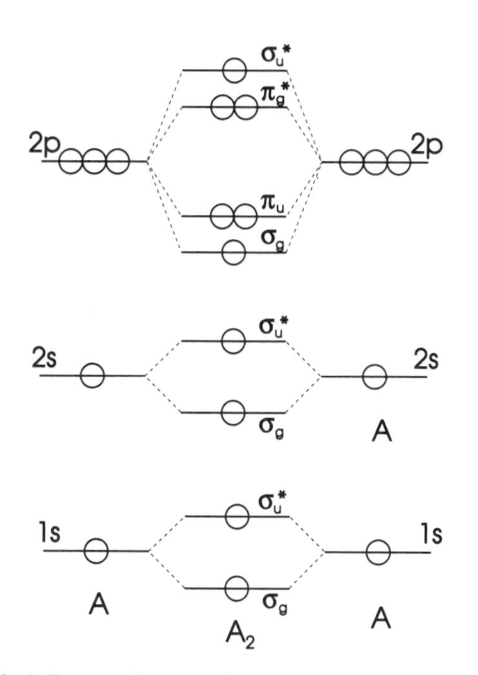

Figure 9.2. Molecular orbital diagram for second-row diatomic molecules. Note that for Li_2 to N_2 the order of the $\pi_u(2p)$ and $\sigma_g(2p)$ orbitals are switched.

The atomic orbitals are combined to give the molecular orbitals as shown in Figure 9.2 for a homonuclear diatomic molecule made of atoms of the second row of the periodic table.

For O_2 the electrons are added to the molecular orbitals to give a

$$(\sigma_g 1s)^2(\sigma_u^* 1s)^2(\sigma_g 2s)^2(\sigma_u^* 2s)^2(\sigma_g 2p)^2(\pi_u 2p)^4(\pi_g^* 2p)^2$$

configuration. The net bond order [(number of bonding electrons − number of antibonding electrons)/2] is 2 for O_2. From a $(\pi_g^*)^2$ configuration the electronic states that result are given by the direct product

$$\Pi_g \otimes \Pi_g = \Sigma_g^+ \oplus [\Sigma_g^-] \oplus \Delta_g \tag{9.7}$$

with square brackets around the antisymmetric part of the product. Since the two π_g^* electrons are identical in O_2, care must be taken not to violate the Pauli exclusion principle. The exchange of the two identical electrons (fermions) requires that the total wavefunction $\psi_{\text{orbital}}\psi_{\text{spin}}$ be antisymmetric. This means that the symmetric part (Σ_g^+ and Δ_g) of the $\Pi_g \otimes \Pi_g$ product combines with the antisymmetric electron spin part $[(\alpha(1)\beta(2) - \alpha(2)\beta(1))/\sqrt{2}]$, while the antisymmetric orbital part (Σ_g^-) combines with the symmetric electron spin part $[\alpha(1)\alpha(2), (\alpha(1)\beta(2) + \alpha(2)\beta(1))/\sqrt{2}, \beta(1)\beta(2)]$.

The $(\pi_g^*)^2$ configuration of oxygen therefore gives rise to the $^1\Sigma_g^+$, $^3\Sigma_g^-$, and $^1\Delta_g$ electronic states. Since Hund's rules apply to molecules as well as to atoms, $^3\Sigma_g^-$ is expected to lie lowest in energy. If the two electrons are in different orbitals $(\pi)^1(\pi')^1$, then twice as many states are possible, namely those associated with the terms $^{1,3}\Sigma^+$, $^{1,3}\Sigma^-$, and $^{1,3}\Delta$. Note that in the direct product tables in Appendix C, the antisymmetric part of the product is enclosed in square brackets.

The notation for the electronic states of diatomic molecules parallels that for atoms, with the symbol $^{2S+1}\Lambda_\Omega$ used in place of $^{2S+1}L_J$, and with $\Lambda = \sum \lambda_i$. Capital Greek letters are used for the multi-electron molecular labels, while lowercase Greek letters are used for one-electron orbital labels. For example, the electronic configuration of O_2 has two unpaired electrons with $\lambda_1 = \pm 1$ and $\lambda_2 = \pm 1$. The possible values of the total orbital angular momentum $\Lambda = \lambda_1 + \lambda_2$ are $\pm 2, 0, 0$, which translates into Δ, Σ^+, and Σ^- electronic states. Notice that since the λ_i are projections of the angular momenta of electrons along the internuclear axis, they add as scalars rather than as vectors.

For a diatomic molecule the total angular momentum (exclusive of nuclear spin) is the vector sum of orbital (**L**), spin (**S**), and rotational (**R**) angular momenta, $\mathbf{J} = \mathbf{L} + \mathbf{S} + \mathbf{R}$ (Figure 9.3). The total angular momentum **J** has a projection of $\Omega\hbar$ units of angular momentum along the molecular axis and (as always) M_J along the space fixed Z-axis (Figure 9.4). The various angular momenta and their projections on the molecular axis are summarized in Table 9.2. Notice that the name given to the projection of **S** along the internuclear axis is Σ. This is completely unrelated to the fact that $\Lambda = 0$ is also called a Σ state! The Ω quantum number is sometimes appended as a subscript to label a particular spin component.

For $\Lambda > 0$ there is a double orbital degeneracy corresponding to the circulation of the electrons in a clockwise or counterclockwise direction. This degeneracy remains for $\Omega > 0$ and it is customary to use $|\Omega|$ to represent both values. For example, a $^3\Sigma_g^-$ state has $^3\Sigma_{1_g}^-$ and $^3\Sigma_{0_g^+}^-$ spin components ($\Lambda = 0, \Sigma = \pm 1, 0$), while a $^2\Pi$ state has $^2\Pi_{3/2}$ and

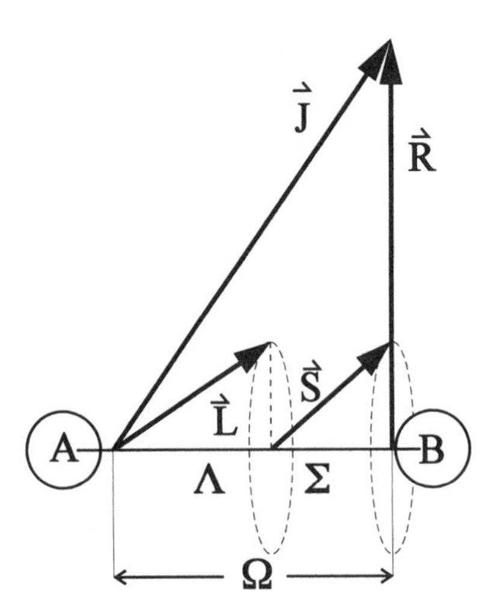

Figure 9.3. Angular momenta in a diatomic molecule.

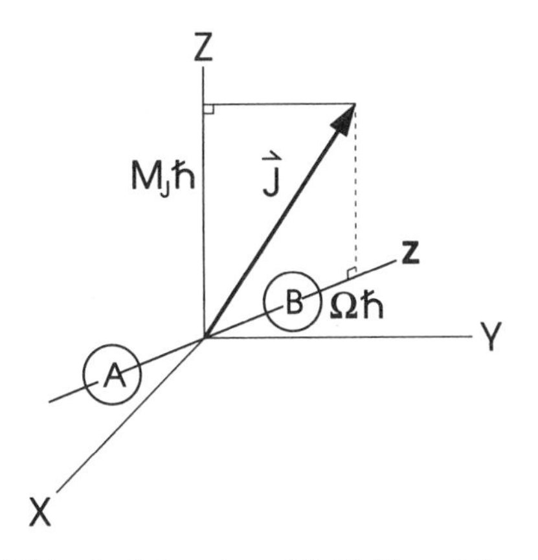

Figure 9.4. Components of **J** in the laboratory (*X, Y, Z*) and the molecular (*x, y, z*) frames.

$^2\Pi_{1/2}$ $(\Lambda = 1, \Sigma = \pm 1/2)$ spin components. Notice that there are always $2S + 1$ spin components, labeled by their $|\Omega|$ values except when $S > |\Lambda| > 0$. In that case there is a notational problem in labeling the $2S + 1$ spin components, so $\Omega = |\Lambda| + \Sigma$ is used instead of $|\Omega|$. For example, for a $^4\Pi$ state $(S = 3/2, \Lambda = 1)$ the spin components are labeled as $^4\Pi_{5/2}$, $^4\Pi_{3/2}$, $^4\Pi_{1/2}$, and $^4\Pi_{-1/2}$. The electronic states of diatomic molecules are also labeled with letters: *X* is reserved for the ground state, while *A, B, C*, and so on, are used for excited states of the same multiplicity $(2S + 1)$ as the ground state, in order of increasing energy. States with a multiplicity different from that of the ground state are labeled with lowercase letters *a, b, c,* and so on, in order of increasing energy. This convention is illustrated by the energy-level diagram of the low-lying electronic states of O_2 in Figure 9.5.

The possible electron transitions among the energy levels are determined by the selection rules:

1. $\Delta\Lambda = 0, \pm 1$. The transitions Σ–Σ, $\Pi - \Sigma$, $\Delta - \Pi$, and so forth, are allowed.

2. $\Delta S = 0$. Transitions that change multiplicity are very weak for molecules formed

Table 9.2. Angular Momenta in Diatomic Molecules

Angular Momentum	Projection on Molecular Axis (units of $\hbar$)
J	$\Omega = (\Lambda + \Sigma)$
L	Λ
S	Σ
R	—
N = **R** + **L**	Λ

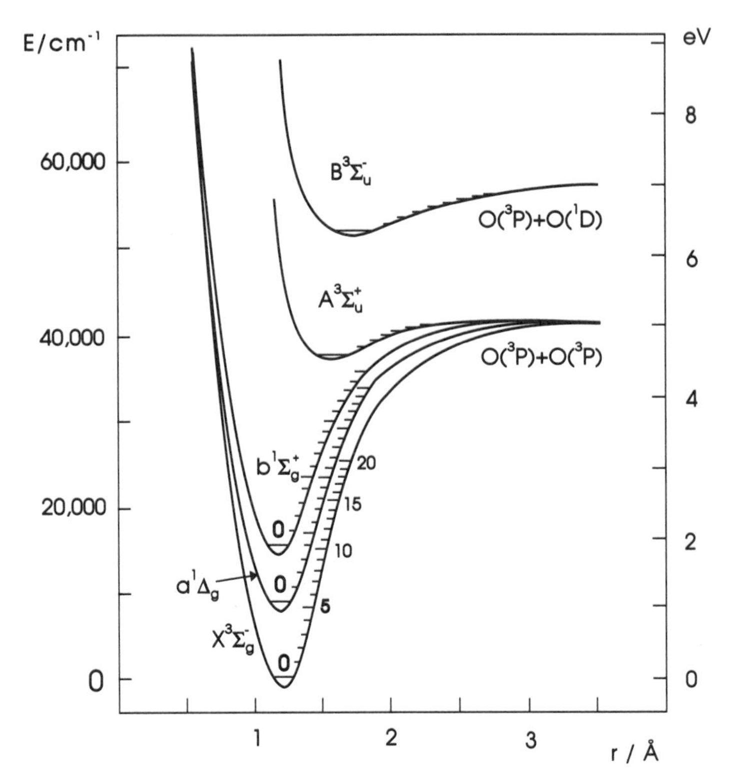

Figure 9.5. The low-lying electronic states of the O_2 molecule.

from light atoms, but as spin-orbit coupling increases in heavy atoms, transitions with $\Delta S \neq 0$ become more strongly allowed.

3. $\Delta\Sigma = 0$ (for Hund's case (a), see below).

4. $\Delta\Omega = 0, \pm 1$.

5. $\Sigma^+ - \Sigma^+$, $\Sigma^- - \Sigma^-$, but not $\Sigma^+ - \Sigma^-$. This selection rule is a consequence of the μ_z transition dipole moment having Σ^+ symmetry. Notice that $\Sigma^+ - \Pi$ and $\Sigma^- - \Pi$ transitions are both allowed.

6. $g \leftrightarrow u$. The transitions ${}^1\Pi_g - {}^1\Pi_u$, ${}^1\Sigma_u^+ - {}^1\Sigma_g^+$, and so forth, are allowed for centrosymmetric molecules.

For example, transitions among the first three electronic states of O_2 ($b^1\Sigma_g^+$, $a^1\Delta_g$, and $X^3\Sigma_g^-$, Figure 9.5) are forbidden, but the $B^3\Sigma_u^- - X^3\Sigma_g^-$ transition is allowed. The B–X transition of O_2, called the Schumann–Runge system, is responsible for the absorption of UV light for wavelengths $\lambda < 200$ nm in the earth's atmosphere. The vacuum UV region begins at 200 nm because of the absorption of radiation by O_2 in air.

Vibrational Structure

An electronic transition is made up of vibrational bands, each of which is in turn made up of rotational lines. The presence of many vibrational bands, labeled as $v'-v''$, explains why electronic transitions are often called band systems. The terms, band system, band, and line, date from the early days of spectroscopy and refer to the appearance of electronic transitions recorded photographically on glass plates with a spectrograph. Although lasers and Fourier transform spectrometers have largely replaced classical techniques, a reproduction of such a photographic plate is presented in Figure 9.6.

The CN free radical occurs prominently in many plasmas that contain carbon and nitrogen impurities.[1] It will serve as an example of an electronic spectrum of a diatomic molecule. The ground state of CN $X^2\Sigma^+$ arises from the configuration $\cdots(\pi 2p)^4(\sigma 2p)^1$, while the first excited state $A^2\Pi_i$ arises from the $(\pi 2p)^3(\sigma 2p)^2$

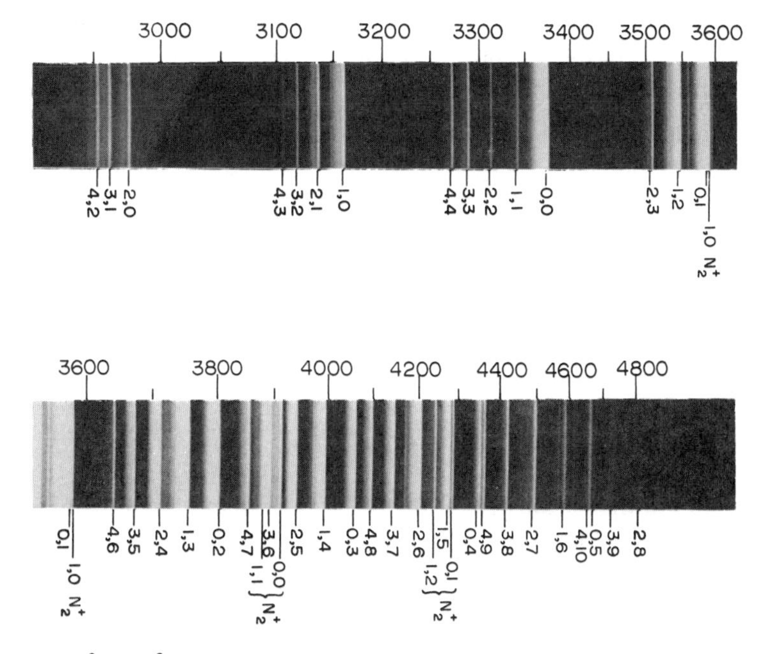

Figure 9.6. The $C^3\Pi_u-B^2\Pi_g$ second positive system of N_2. The pairs of numbers indicate the vibrational bands v', v'', and the wavelength scale is in angstroms.

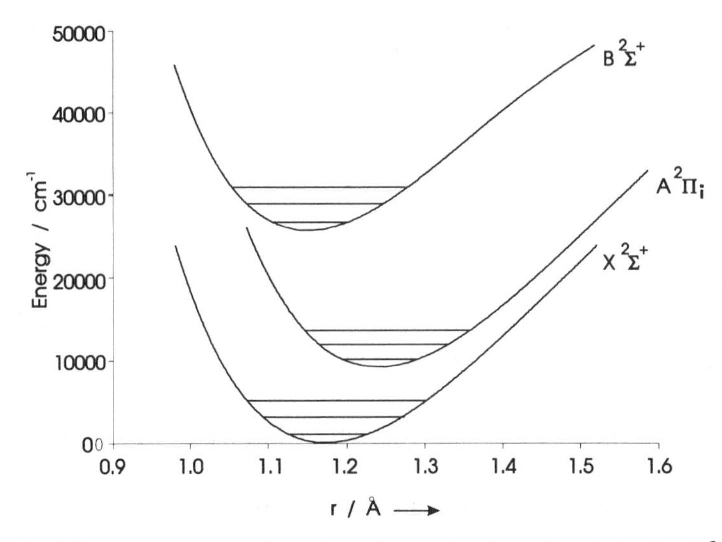

Figure 9.7. Potential energy curves and the lowest few vibrational levels for the $X^2\Sigma^+$, $A^2\Pi_i$, and $B^2\Sigma^+$ states of CN.

configuration (Figure 9.7). The subscript *i* stands for "inverted," which means that the $^2\Pi_{1/2}$ spin component lies above the $^2\Pi_{3/2}$ component. For non-ionic heteronuclear diatomic molecules, such as CN, the electronic configurations can be obtained with the help of Figure 9.2, with the *g* and *u* labels deleted.

The vibrational structure is organized into sequences and progressions. A group of bands with the same Δv is called a *sequence* so that the 0–0, 1–1, 2–2 bands form the $\Delta v = 0$ sequence, while the 0–1, 1–2, and 2–3 bands form the $\Delta v = -1$ sequence (Figure 9.6). When the excited-state and ground-state vibrational constants are similar, bands of the same sequence cluster together. A series of bands all connected to the same vibrational level v, such as 3–1, 2–1, 1–1, and 0–1, is called a *progression.* Upper state progressions connect into the same lower vibrational level, while lower state progressions connect to the same upper vibrational level.

The vibrational band positions of an electronic transition are obtained from the usual vibrational energy-level expression, that is,

$$\nu_{v'v''} = T_e + \omega_e'(v' + \tfrac{1}{2}) - \omega_e'x_e'(v' + \tfrac{1}{2})^2 + \cdots - \omega_e''(v'' + \tfrac{1}{2}) + \omega_e''x_e''(v'' + \tfrac{1}{2})^2 + \cdots \quad (9.8)$$

in which $T_e = E_{\text{upper}} - E_{\text{lower}}$ is the energy separation between the potential minima of the two electronic states.

The intensities of the various vibrational bands are determined by three factors: the intrinsic strength of the electronic transition, the populations of the vibrational levels, and the squared overlap integral of the two vibrational wavefunctions, called the *Franck–Condon factor.* Franck–Condon factors result from the application of a more general rule called the *Franck–Condon principle.* This principle has both classical and quantum mechanical versions.

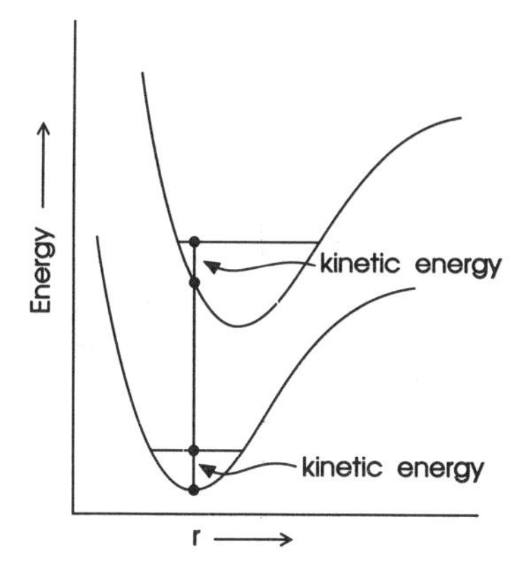

Figure 9.8. Electronic transitions occur vertically on an energy-level diagram and conserve the kinetic energy.

The classical version of the Franck–Condon principle is based on the idea that electronic transitions occur very quickly, in less than 10^{-15} s. In such a short time the nuclei do not have time to move, so vibration, rotation, and translation are "frozen" during an electronic transition. On a potential-energy diagram, therefore, electronic transitions occur vertically at the initial r value. The kinetic energy (but not the potential energy) is the same immediately before and after an electronic transition.

The presence of vibrational levels can be added to the classical picture (Figure 9.8) by quantizing the energy levels (Figure 9.7). Furthermore, vibrating diatomic molecules (except for $v = 0$) spend more time at the classical inner and outer turning points of the vibrational motion than in the middle, so that transitions can be approximated as occurring at the turning points. For example, Figure 9.9 predicts the 0–18 band of the B–X transition of Br_2 to be strong in emission, while Figure 9.7 predicts the 2–0 band of the $A^2\Pi$–$X^2\Sigma^+$ transition of CN to be strong in absorption.

The quantum mechanical version of the Franck–Condon principle is based on the fact that the intensity of a given transition is proportional to the square of the transition moment integral

$$\mathbf{M}_{ev} = \int \psi^*_{e'v'} \boldsymbol{\mu} \psi_{e''v''} \, d\tau, \tag{9.9}$$

and the Born–Oppenheimer approximation $\psi_{ev} = \psi_e \psi_v$. The rotational motion of the diatomic molecule is ignored here, but its inclusion does not change the derivation. The dipole moment operator can be broken into an electronic part and a nuclear part, namely

$$\boldsymbol{\mu} = \sum q_i \mathbf{r}_i = \sum_j q_j \mathbf{r}_j + \sum_\alpha q_\alpha \mathbf{r}_\alpha = \boldsymbol{\mu}_e + \boldsymbol{\mu}_N \tag{9.10}$$

in which the sum over the charged particles in the molecule is broken into a sum over

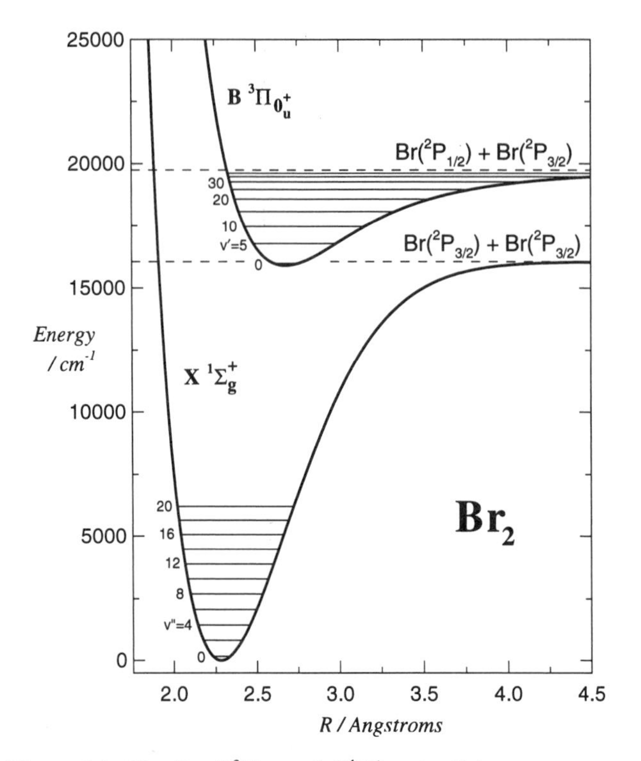

Figure 9.9. The Br_2 $B^3\Pi_{0_u^+}$ and $X^1\Sigma_g^+$ potential energy curves.

electrons (index j) and a sum over nuclei (index α). The transition dipole moment operator then becomes

$$\begin{aligned} \mathbf{M}_{ev} &= \int \psi_{e'}^{*}\psi_{v'}(\boldsymbol{\mu}_e + \boldsymbol{\mu}_N)\psi_{e''}\psi_{v''}\, d\tau \\ &= \int \psi_{e'}^{*}\boldsymbol{\mu}_e\psi_{e''}\, d\tau_{\text{el}} \int \psi_{v'}^{*}\psi_{v''}\, d\tau_N \\ &\quad + \int \psi_{e'}^{*}\psi_{e''}\, d\tau_{\text{el}} \int \psi_{v'}^{*}\boldsymbol{\mu}_N\psi_{v''}\, d\tau_N. \end{aligned} \tag{9.11}$$

The last term on the right-hand side of the $\mathbf{M}_{ev}$ expression (9.11) is zero since the electronic wavefunctions of two different states are orthogonal. Finally the transition dipole moment $\mathbf{M}_{ev}$ is obtained as

$$\mathbf{M}_{ev} = \mathbf{R}_e\langle v' \mid v'' \rangle, \tag{9.12}$$

in which

$$\mathbf{R}_e = \int \psi_{e'}^{*}\boldsymbol{\mu}_e\psi_{e''}\, d\tau_{\text{el}} \tag{9.13}$$

is the electronic transition dipole moment and

$$\langle v' \mid v'' \rangle = \int \psi_{v'}^{*} \psi_{v''} \, dr \tag{9.14}$$

is the vibrational overlap integral. The intensity of a vibronic transition is proportional to the square of the transition moment integral, namely

$$I_{e'v'e''v''} \propto |\mathbf{R}_e|^2 \, q_{v'-v''} \tag{9.15}$$

in which $q_{v'-v''} = |\langle v' \mid v'' \rangle|^2$ is called the *Franck–Condon factor.* Note that although the vibrational wavefunctions are all orthogonal within one electronic state, they are not orthogonal between two different electronic states. The electronic transition dipole moment $|\mathbf{R}_e|$ has a value of about 1 debye for an allowed transition, while q ranges between 0 and 1, depending on the extent of overlap. The value of $|\mathbf{R}_e|^2$ measures the intrinsic strength of an electronic transition, while the Franck–Condon factor determines how the intensity is distributed among the vibrational bands.

An alternative derivation of equation (9.15) follows the method in Chapter 7 (equations (7.48) to (7.50)) and allows the possibility of μ depending on r. This derivation clearly shows the approximate nature of equation (9.12) and has correction terms as given in equation (7.50).

The intensity of the vibrational bands of an electronic transition is determined by the population of the vibrational levels, the intrinsic strength of a transition ($\mathbf{R}_e$) and the Franck–Condon factors. In the case of a common initial vibrational state, the relative intensity of two bands is given by a ratio of Franck–Condon factors. The magnitudes of the vibrational overlap integrals can be estimated from a picture of vibrational wavefunctions (e.g., Figure 7.5). As shown in Figure 9.10, when the two electronic states have similar r_e and ω_e values, then the $\Delta v = 0$ sequence of diagonal

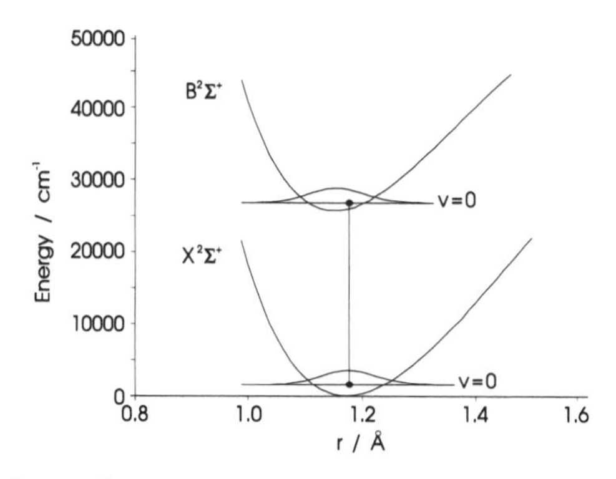

Figure 9.10. For the $B^2\Sigma^+–X^2\Sigma^+$ transition of CN, the 0–0 band will be strong because of favorable vibrational overlap.

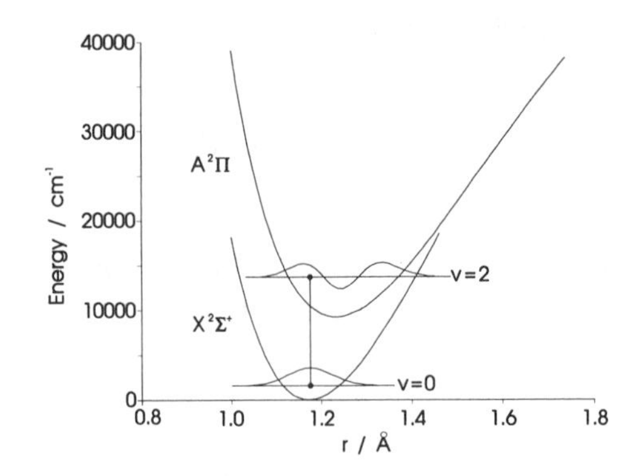

Figure 9.11. The 2–0 band of the $A^2\Pi$-$X^2\Sigma^+$ transition of CN has a large Franck–Condon factor.

bands 0–0, 1–1, 2–2 are strong because of the optimal overlap of the $\Delta v = 0$ vibrational wavefunctions of the upper and lower electronic states. In contrast, when the equilibrium internuclear separation r_e of the two electronic states is significantly different, the off-diagonal bands ($\Delta v \neq 0$) have particularly good vibrational overlaps (Figure 9.11).

The vibrational bands of an electronic band system can be conveniently organized into a Deslandres table. A Deslandres table is a 2-dimensional array of vibrational band energies, constructed as shown in Table 9.3. The terms diagonal and off-diagonal bands refer implicitly to a Deslandres table. The differences between the adjacent rows and columns give estimates for the vibrational intervals of the ground and excited electronic states.

Rotational Structure of Electronic Transitions of Diatomic Molecules

Singlet–Singlet Transitions

The rotational structure of singlet–singlet electronic transitions is identical to that of the vibrational transitions of a linear molecule, as discussed in Chapter 7. In Chapter 7 it was shown for a linear polyatomic molecule that the projection of the total angular momentum along the internuclear axis originates from the vibrational angular momentum **l**. For electronic states the angular momentum projection along the z-axis originates from the orbital motion of electrons (**L**). As shown in Figure 9.3, the projection of **L** along the internuclear axis is denoted Λ and is analogous to the vibrational angular momentum quantum number l. As discussed in Chapter 7, three types of transitions are possible:

Table 9.3. Deslandres Table of the Band Heads (Defined Below) of the N_2 $C^3\Pi_u$–$B^3\Pi_g$ Second Positive System[2]

v' \ v''	0		1		2		3		4		5		6		7
0	29,652.8	*1703.6*	27,949.2	*1674.7*	26,274.5	*1647.2*	24,627.3	*1611.4*	23,015.9	*1596.2*	21,419.7	*1550.4*	19,869.3	*1532.5*	18,336.8
	1990.7		*1992.2*		*1992.4*		*1992.8*		*1987.1*		*1994.6*		*1985.9*		*1996.0*
1	31,643.5	*1702.1*	29,941.4	*1674.5*	28,266.9	*1646.8*	26,620.1	*1617.1*	25,003.0	*1588.7*	23,414.3	*1559.1*	21,855.2	*1522.4*	20,332.8
	1939.8		*1937.1*		*1945.0*		*1939.1*		*1939.2*		*1939.9*		*1944.8*		*1931.7*
2	33,583.4	*1704.8*	31,878.5	*1666.6*	30,211.9	*1652.7*	28,559.2	*1617.0*	26,942.2	*1588.0*	25,354.2	*1554.2*	23,800.0	*1535.5*	22,264.5
	1869.8		*1872.6*		*1864.0*		*1870.7*		*1876.3*		*1871.9*		*1869.3*		*1872.8*
3	35,453.1	*1702.0*	33,751.1	*1675.2*	32,075.9	*1646.0*	30,429.9	*1611.4*	28,818.5	*1592.4*	27,226.1	*1556.8*	25,669.3	*1532.0*	24,137.3
	1752.1		*1771.3*		*1775.8*		*1777.3*		*1771.5*		*1784.8*		*1782.6*		*1776.2*
4	37,205.2	*1682.8*	35,522.4	*1670.7*	33,851.7	*1644.5*	32,207.2	*1617.2*	30,590.0	*1579.1*	29,010.9	*1559.0*	27,451.9	*1538.4*	25,913.5

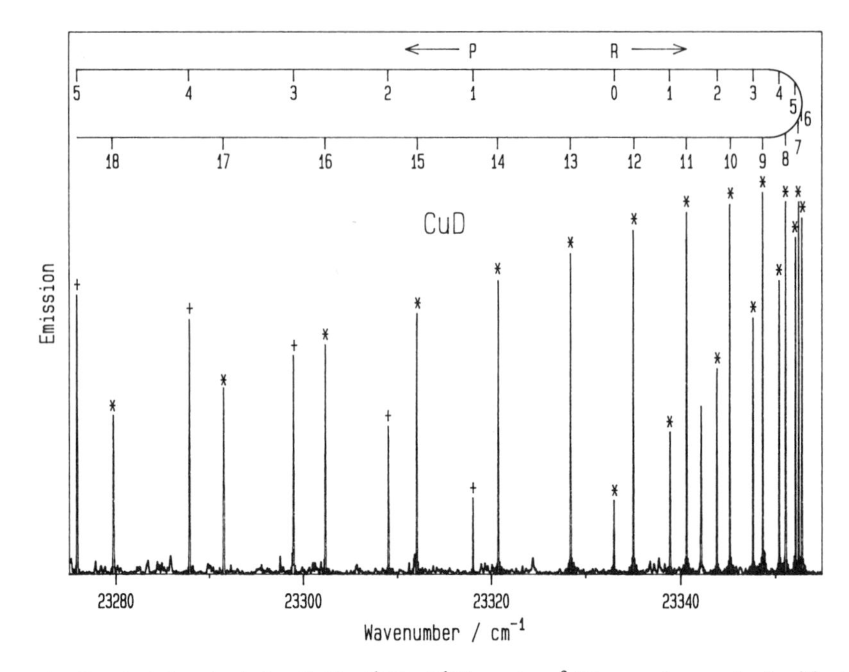

Figure 9.12. The 0–0 band of the CuD $A^1\Sigma^+ - X^1\Sigma^+$ system.[3] The peaks marked with + are P branch transitions, while those marked with * belong to the R branch.

1. $\Delta\Lambda = 0$, $\Lambda'' = \Lambda' = 0$. As shown in Figure 9.12, $^1\Sigma^+ - ^1\Sigma^+$ (or $^1\Sigma^- - ^1\Sigma^-$) transitions have only P and R branches ($\Delta J = \pm 1$). $^1\Sigma - ^1\Sigma$ transitions are parallel transitions, with the transition dipole moment lying along the z-axis.

2. $\Delta\Lambda = \pm 1$. As shown in Figure 9.13, $^1\Pi - ^1\Sigma^+$, $^1\Pi - ^1\Sigma^-$, $^1\Delta - ^1\Pi$, and so on, transitions have strong Q branches as well as P and R branches, with $\Delta J = 0, \pm 1$. These transitions have a transition dipole moment perpendicular to the molecular axis, and hence are designated as perpendicular transitions.

3. $\Delta\Lambda = 0$, $\Lambda' = \Lambda'' \neq 0$. Transitions such as $^1\Pi - ^1\Pi$, $^1\Delta - ^1\Delta$, and so on, are characterized by weak Q branches (for small Λ) and strong P and R branches ($\Delta J = 0, \pm 1$).

The total power $P_{J'J''}$ (in watts/m^3) emitted by an excited rovibronic state $|nv'J'\rangle$ is given by the expression[5]

$$P_{J'J''} = \frac{16\pi^3}{3\varepsilon_0 c^3} \frac{n_{J'}}{(2J'+1)} \nu^4_{J'J''} q_{v'v''} |\mathbf{R}_e|^2 S_{J''} \tag{9.16}$$

in which $n_{J'}$ is the excited-state population in molecules/m^3, $\nu_{J'J''}$ is the transition frequency in Hz, $q_{v'v''}$ is the Franck–Condon factor, $\mathbf{R}_e$ is the electronic transition dipole moment in coulomb-meters and $S_{J''}$ is a rotational line strength term called a Hönl–London factor (Table 9.4). This equation is obtained from the expression for

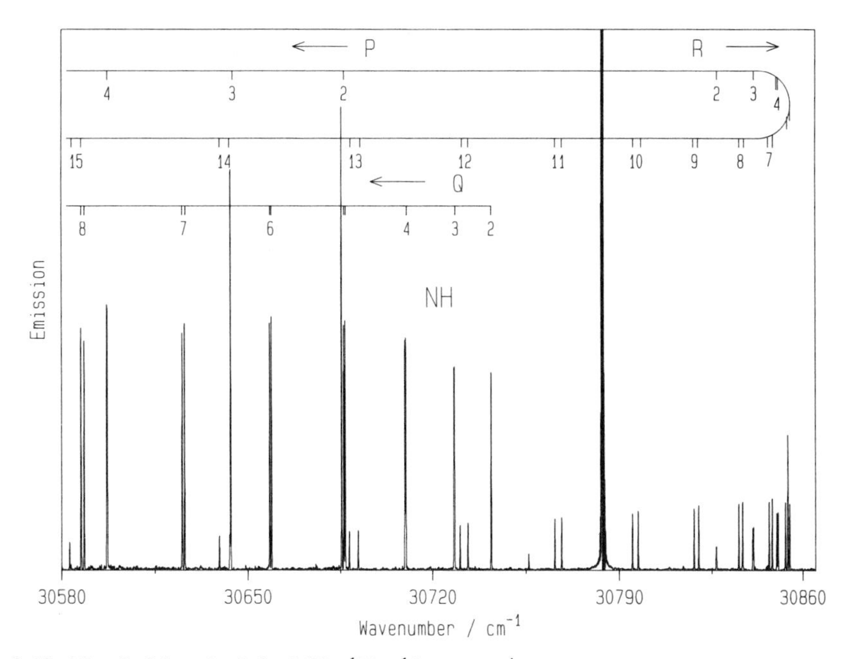

Figure 9.13. The 0–0 band of the NH $c^1\Pi$–$a^1\Delta$ system.[4] The doubling of the lines in the R and Q branches is due to Λ-doubling in the $c^1\Pi$ state. The very intense line near 30790 cm^{-1} is due emission from the He atom.

Table 9.4. Hönl–London Factors

$\Delta\Lambda = 0$

$$S_J^R = \frac{(J''+1+\Lambda'')(J''+1-\Lambda'')}{J''+1}$$

$$S_J^Q = \frac{(2J''+1)\Lambda^2}{J''(J''+1)}$$

$$S_J^P = \frac{(J''+\Lambda'')(J''-\Lambda'')}{J''}$$

$\Delta\Lambda = +1$

$$S_J^R = \frac{(J''+2+\Lambda'')(J''+1+\Lambda'')}{4(J''+1)}$$

$$S_J^Q = \frac{(J''+1+\Lambda'')(J''-\Lambda'')(2J''+1)}{4J''(J''+1)}$$

$$S_J^P = \frac{(J''-1-\Lambda'')(J''-\Lambda'')}{4J''}$$

$\Delta\Lambda = -1$

$$S_J^R = \frac{(J''+2-\Lambda'')(J''+1-\Lambda'')}{4(J''+1)}$$

$$S_J^Q = \frac{(J''+1-\Lambda'')(J''+\Lambda'')(2J''+1)}{4J''(J''+1)}$$

$$S_J^P = \frac{(J''-1+\Lambda'')(J''+\Lambda'')}{4J''}$$

the Einstein A factor (equation (1.52)) by multiplication by $h\nu$ to convert from photons/s to watts, then multiplication by $n_{J'}$ to account for the number density of excited states, and finally the substitution of $q_{v'v''}|\mathbf{R}_e|^2 S_{J''}/(2J'+1)$ for $|\mu_{10}|^2$. The Hönl–London factors are derived from the properties of symmetric top wavefunctions.[6] The relative intensities of the rotational lines in a band of an electronic transition are given by the Hönl–London factors S_J of Table 9.4.

For electronic transitions the rotational constants of the two states can differ significantly. Consider the expressions for *P*, *Q*, and *R* branches,

$$\nu_P = \nu_0 - (B' + B'')J'' + (B' - B'')(J'')^2, \tag{9.17a}$$

$$\nu_R = \nu_0 + 2B' + (3B' - B'')J'' + (B' - B'')(J'')^2, \tag{9.17b}$$

and

$$\nu_Q = \nu_0 + (B' - B'')J''(J'' + 1). \tag{9.18}$$

If $B' < B''$, then the spacing between the lines in the *P* branch will increase as J'' increases (9.17a), while the spacings between the lines of the *R* branch (9.17b) will decrease (Figure 9.12). At some point the lines in the *R* branch will pile up and then turn around. This pile up of lines is called a *band head* and is a characteristic feature of many electronic transitions. Even at low resolution (Figure 9.14) a band head has a characteristic edge structure due to the overlap of many rotational lines. Conversely, if $B' > B''$, then the band head will be in the *P* branch and the band is said to be blue (or violet) degraded (or degraded to shorter wavelengths). If the band head is in the *R* branch, then the band is described as red degraded (or degraded to longer wavelengths), as illustrated in Figure 9.14.

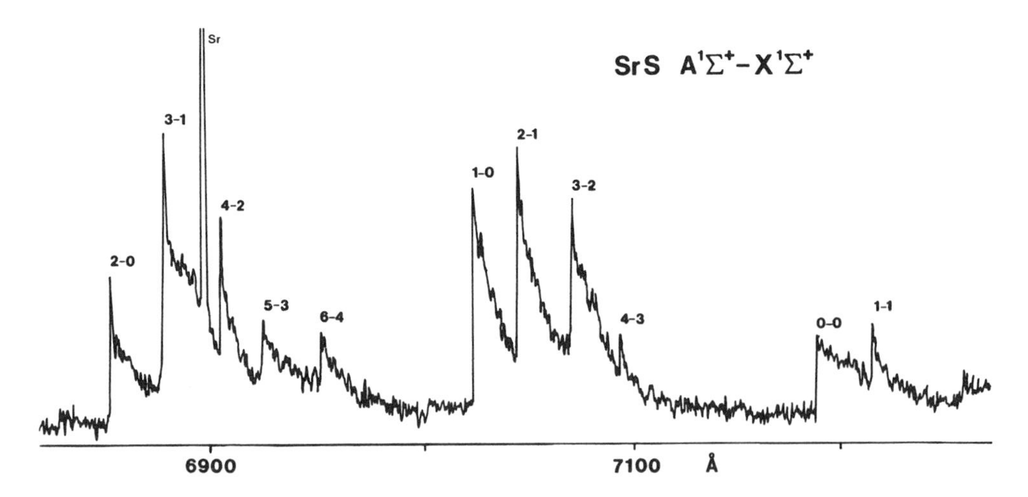

Figure 9.14. The low-resolution laser excitation spectrum of the $A^1\Sigma^+–X^1\Sigma^+$ transition of SrS.[7]

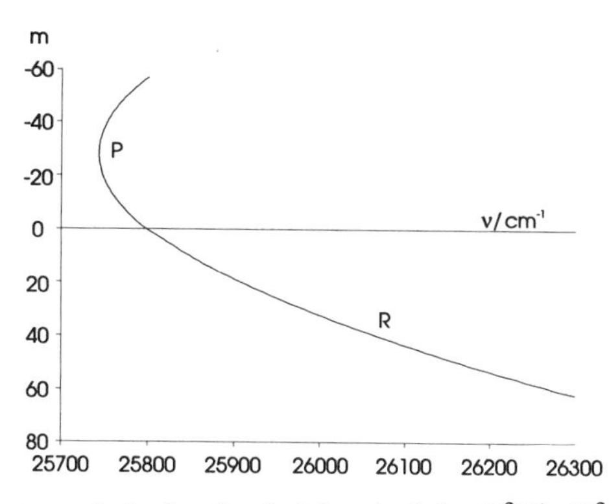

Figure 9.15. Fortrat parabola for the 0–0 band of the $B^2\Sigma^+$–$X^2\Sigma^+$ transition of CN.[1]

The energy expressions for P and R branches (9.17) can be combined into a single expression by defining an index $m = J'' + 1$ for the R branch and $m = -J''$ for the P branch,

$$\nu_{P,R} = \nu_0 + (B' + B'')m + (B' - B'')m^2. \tag{9.19}$$

For the Q branch $m = J''$ and the expression for the line positions is

$$\nu_Q = \nu_0 + (B' - B'')m(m + 1). \tag{9.20}$$

The $\nu_{P,R}$ expression (9.19) can be plotted as a function of m, or more commonly m is plotted as a function of ν, to give what is called a Fortrat parabola (Figure 9.15).

A Fortrat parabola is helpful in visualizing the rotational structure of a vibrational band. The head will occur in the Fortrat parabola when

$$\frac{d\nu}{dm} = 0 = (B' + B'') + 2m(B' - B''), \tag{9.21}$$

or when

$$m_H = -\frac{(B' + B'')}{2(B' - B'')}, \tag{9.22}$$

with the head-origin separation being given by

$$\nu_H - \nu_O = -\frac{(B' + B'')^2}{4(B' - B'')}. \tag{9.23}$$

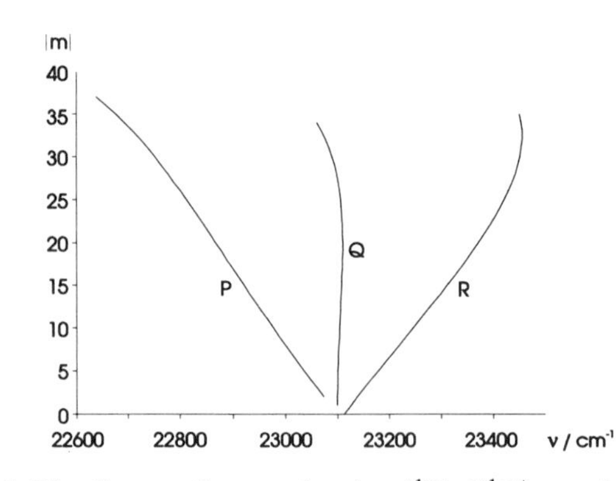

Figure 9.16. The Fortrat diagram for the $A^1\Pi$–$X^1\Sigma^+$ transition of BD.[8]

Sometimes the Fortrat diagram is plotted as a function of $|m|$ so that the P and R branches no longer fall on the same parabola (Figure 9.16). In addition the inclusion of centrifugal distortion terms, which is necessary for high-J lines, distorts the curve from a parabolic shape (Figure 9.16).

For $^1\Pi$ states (and all states with $\Lambda > 0$) there is a degeneracy associated with the two possible values $\pm|\Lambda|$. The effect of this double degeneracy on a Π–Σ transition has already been discussed in Chapter 7. In the case of an electronic transition the small splitting in the $^1\Pi$ state is called Λ-doubling[9] rather than l-type doubling but effects are completely analogous. Only the names and the physical origins of the splittings are different. Thus the energy-level diagram for a Π–Σ transition (Figure 7.33) applies to both vibrational transitions of linear polyatomic molecules and electronic transitions of diatomic (and linear) molecules.

Nonsinglet Transitions

The rotational structure of nonsinglet states is more complex than that of singlet states because of the effect on the rotational structure of both spin and orbital angular momenta. In general, the Hamiltonian operator is separated into electronic, vibrational, and rotational parts, with the effect of spin-orbit coupling within a state accounted for by the addition of the spin-orbit operator,

$$\hat{H}_{so} = A\mathbf{L} \cdot \mathbf{S}, \tag{9.24}$$

so that

$$\hat{H} = \hat{H}_{el} + \hat{H}_{vib} + \hat{H}_{rot} + \hat{H}_{so}. \tag{9.25}$$

The methodology for determining the rotational energies for nonsinglet states is to formulate this Hamiltonian in terms of the various angular momentum operators **J**, **N**, **S**, **R**, and **L** (Table 9.2) and to select suitable basis functions. In the next step a matrix representation of the Hamiltonian can be constructed and diagonalized to obtain the energy eigenvalues and wavefunctions (eigenvectors).

Since the focus of this section is the rotational structure, it is convenient simply to treat the vibronic expectation value,

$$\langle \hat{H}_{\text{el}} + \hat{H}_{\text{vib}} \rangle = E_{ev}, \tag{9.26}$$

as a constant. Explicit forms for $\hat{H}_{\text{rot}}$ and $\hat{H}_{\text{so}}$ are[9]

$$\begin{aligned} \hat{H}_{\text{rot}} &= B(\mathbf{R})^2 = B(\mathbf{J} - \mathbf{L} - \mathbf{S})^2 \\ &= B(\hat{J}^2 - \hat{J}_z^2) + B(\hat{S}^2 - \hat{S}_z^2) + B(L^2 - L_z^2) \\ &\quad - B(\hat{J}^+\hat{L}^- + \hat{J}^-\hat{L}^+) - B(\hat{J}^+\hat{S}^- + \hat{J}^-\hat{S}^+) + B(\hat{L}^+\hat{S}^- + \hat{L}^-\hat{S}^+) \end{aligned} \tag{9.27}$$

and

$$\hat{H}_{\text{so}} = A\mathbf{L} \cdot \mathbf{S} = A\hat{L}_z\hat{S}_z + A\,\frac{\hat{L}^+\hat{S}^- + \hat{L}^-\hat{S}^+}{2} \tag{9.28}$$

in which

$$J^\pm = J_x \pm iJ_y, \qquad L^\pm = L_x \pm iL_y, \qquad S^\pm = S_x \pm iS_y$$

are raising and lowering operators. In equations (9.27) and (9.28) *all* operators are in the molecular frame so that $[\hat{L}_x, \hat{L}_y] = i\hbar\hat{L}_z$, $[\hat{S}_x, \hat{S}_y] = i\hbar\hat{S}_z$, but $[\hat{J}_x, \hat{J}_y] = -i\hbar\hat{J}_z$ (anomalous). Notice that the commutators associated with the components of **L** and **S** are the same in both the laboratory and molecular frames, while those associated with **J** are anomalous in the molecular frame. This means that $\hat{J}^+$ is a lowering operator and $\hat{J}^-$ is a raising operator, that is,

$$\hat{J}^- |\Omega JM\rangle = \hbar[J(J+1) - \Omega(\Omega+1)]^{1/2} |\Omega+1, JM\rangle \tag{9.29}$$

$$\hat{J}^+ |\Omega JM\rangle = \hbar[J(J+1) - \Omega(\Omega-1)]^{1/2} |\Omega-1, JM\rangle. \tag{9.30}$$

Next a suitable basis set needs to be chosen. Since the Hamiltonian is comprised of electronic, vibrational, and rotational terms it is convenient to use a simple product basis set,

$$|\text{el}\rangle\, |\text{vib}\rangle\, |\text{rot}\rangle = |n\Lambda S\Sigma\rangle\, |v\rangle\, |\Omega JM\rangle, \tag{9.31}$$

in which n and v label the electronic state and vibrational level. The basis functions are simultaneous eigenfunctions of the operators $\hat{H}_{\text{el}}$, $\hat{H}_v$, $\hat{L}_z$, $\hat{S}^2$, $\hat{S}_z$, $\hat{J}_z$, $\hat{J}^2$, and $\hat{J}_Z$, that is,

$$\hat{H}_{\text{el}}\,|n\Lambda S\Sigma\rangle = E_{\text{el},n}\,|n\Lambda S\Sigma\rangle \tag{9.32}$$

$$\hat{H}_{\text{vib}}\,|v\rangle = E_{\text{vib}}\,|v\rangle \tag{9.33}$$

$$\hat{L}_z\,|\Lambda S\Sigma\rangle = \Lambda\hbar\,|\Lambda S\Sigma\rangle \tag{9.34}$$

$$\hat{S}^2\,|\Lambda S\Sigma\rangle = S(S+1)\hbar^2\,|\Lambda S\Sigma\rangle \tag{9.35}$$

$$\hat{S}_z\,|\Lambda S\Sigma\rangle = \Sigma\hbar\,|\Lambda S\Sigma\rangle \tag{9.36}$$

$$\hat{J}_z\,|\Omega JM\rangle = \Omega\hbar\,|\Omega JM\rangle \tag{9.37}$$

$$\hat{J}_Z\,|\Omega JM\rangle = M\hbar\,|\Omega JM\rangle \tag{9.38}$$

$$\hat{J}^2\,|\Omega JM\rangle = J(J+1)\hbar^2\,|\Omega JM\rangle. \tag{9.39}$$

The off-diagonal matrix elements that are associated with $\hat{J}^{\pm}$ and $\hat{S}^{\pm}$ are

$$\langle\Omega \mp 1 JM|\,\hat{J}^{\pm}\,|\Omega JM\rangle = \hbar[J(J+1) - \Omega(\Omega \mp 1)]^{1/2} \tag{9.40}$$

$$\langle\Lambda S\Sigma \pm 1|\,\hat{S}^{\pm}\,|\Lambda S\Sigma\rangle = \hbar[S(S+1) - \Sigma(\Sigma \pm 1)]^{1/2}. \tag{9.41}$$

Notice that a label for **L** is not present in the basis set because, in general, the electronic wavefunction is no longer an eigenfunction of $\hat{L}^2$, that is,

$$\hat{L}^2\psi \neq L(L+1)\hbar^2\psi, \tag{9.42}$$

even though the projection of **L** along the internuclear axis is well defined, with

$$\hat{L}_z\psi = \hbar\Lambda\psi. \tag{9.43}$$

The rapid precessional motion of **L** around the internuclear axis (Figure 9.1) prevents the experimental determination of the magnitude of **L**. This means that matrix elements of $\hat{L}^2$ or $\hat{L}^+$ are simply constants whose values can be determined by *ab initio* calculation. Usually the matrix elements of these operators are just absorbed into the electronic band origins.

$^2\Sigma^+$ *States*

For $^2\Sigma^+$ states there are two basis functions,

$$|n, \Lambda = 0, S = \tfrac{1}{2}, \Sigma = \tfrac{1}{2}\rangle\,|v\rangle\,|\tfrac{1}{2}JM\rangle = |^2\Sigma_{1/2}\rangle \tag{9.44}$$

and

$$|n, \Lambda = 0, S = \tfrac{1}{2}, \Sigma = -\tfrac{1}{2}\rangle\, |v\rangle\, |-\tfrac{1}{2}JM\rangle = |^2\Sigma_{-1/2}\rangle \tag{9.45}$$

in which Ω, Λ, and Σ are signed quantum numbers. The diagonal matrix elements of $\hat{H}$, equation (9.27), are in units of cm^{-1}

$$\begin{aligned}\langle^2\Sigma_{1/2}|\, \hat{H}\, |^2\Sigma_{1/2}\rangle &= \langle^2\Sigma_{-1/2}|\, \hat{H}\, |^2\Sigma_{-1/2}\rangle \\ &= B(J(J+1) - \tfrac{1}{4}) + B(\tfrac{3}{4} - \tfrac{1}{4}) \\ &= B(J + \tfrac{1}{2})^2.\end{aligned} \tag{9.46}$$

The off-diagonal matrix elements of $-B(\hat{J}^+\hat{S}^- + \hat{J}^-\hat{S}^+)$ couple the $^2\Sigma_{1/2}$ state to the $^2\Sigma_{-1/2}$ state, namely

$$\begin{aligned}&\langle^2\Sigma_{-1/2}|\, -B(\hat{J}^+\hat{S}^-)\, |^2\Sigma_{1/2}\rangle \\ &\quad = -B(J(J+1) - \Omega(\Omega-1))^{1/2}(S(S+1) - \Sigma(\Sigma-1))^{1/2} = -B(J + \tfrac{1}{2}).\end{aligned} \tag{9.47}$$

Collecting the diagonal and off-diagonal elements of $\mathbf{H}$ (in cm^{-1}) gives

$$\mathbf{H} = \begin{pmatrix} B(J+\tfrac{1}{2})^2 & -B(J+\tfrac{1}{2}) \\ -B(J+\tfrac{1}{2}) & B(J+\tfrac{1}{2})^2 \end{pmatrix}. \tag{9.48}$$

Although this matrix can be diagonalized as it stands, the diagonalization can be avoided by transforming the basis functions as

$$|^2\Sigma^+(e)\rangle = \frac{|^2\Sigma^+_{1/2}\rangle + |^2\Sigma^+_{-1/2}\rangle}{\sqrt{2}} \tag{9.49}$$

$$|^2\Sigma^+(f)\rangle = \frac{|^2\Sigma^+_{1/2}\rangle - |^2\Sigma^+_{-1/2}\rangle}{\sqrt{2}} \tag{9.50}$$

where e and f are parity basis functions (see below in the section on the symmetry of diatomic energy levels, which follows). The Hamiltonian matrix in the transformed basis set becomes

$$\mathbf{H}' = \begin{pmatrix} \overset{e}{B(J+\tfrac{1}{2})^2 - B(J+\tfrac{1}{2})} & \overset{f}{0} \\ 0 & B(J+\tfrac{1}{2})^2 + B(J+\tfrac{1}{2}) \end{pmatrix}. \tag{9.51}$$

If the diagonal elements of $\mathbf{H}'$ are expressed in terms of the integral quantum number

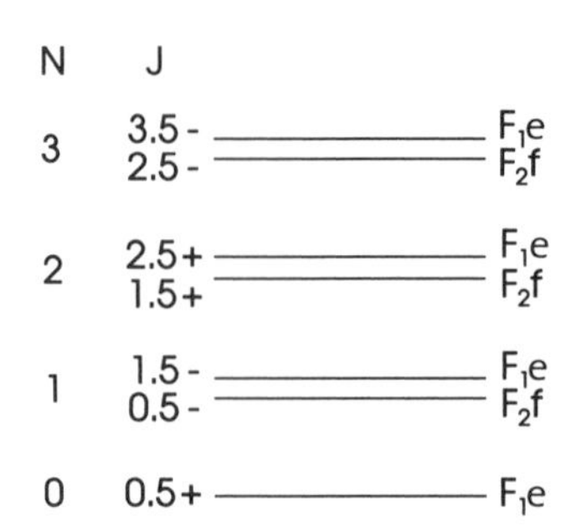

Figure 9.17. Energy-level diagram of a $^2\Sigma^+$ state.

N (i.e., the total angular momentum exclusive of electronic spin, Table 9.2) rather than J, then they simplify to

$$F_1(e) = B(N+1)^2 - B(N+1) = BN(N+1) \tag{9.52a}$$

$$F_2(f) = BN^2 + BN = BN(N+1). \tag{9.52b}$$

In equation (9.52a) N is $J - \frac{1}{2}$ for the e levels, while in equation (9.52b) N is $J + \frac{1}{2}$ for f levels. The energy-level pattern for $^2\Sigma^+$ states is thus identical to that for $^1\Sigma^+$ states if N is used rather than J (Figure 9.17). The subscripts 1 and 2 in equations (9.52) are chosen in accordance with the spectroscopic custom that F_1 has $J = \text{maximum } (N+S)$. Although the energy levels corresponding to the two J values for each N ($F_1(N)$ and $F_2(N)$) are exactly degenerate at this level of theory, inclusion of the spin-rotation interaction[9] term

$$\hat{H}_{SR} = \gamma \mathbf{N} \cdot \mathbf{S}$$

splits the levels by the amount $\gamma(N + \frac{1}{2})$.

$^2\Pi$ *States*

There are four possible basis functions for a $^2\Pi$ state,

$$|^2\Pi_{3/2}\rangle = |n, \Lambda = 1, S = \tfrac{1}{2}, \Sigma = \tfrac{1}{2}\rangle\, |v\rangle\, |\Omega = \tfrac{3}{2} JM\rangle \tag{9.53a}$$

$$|^2\Pi_{1/2}\rangle = |n, \Lambda = 1, S = \tfrac{1}{2}, \Sigma = -\tfrac{1}{2}\rangle\, |v\rangle\, |\Omega = \tfrac{1}{2} JM\rangle \tag{9.53b}$$

$$|^2\Pi_{-1/2}\rangle = |n, \Lambda = -1, S = \tfrac{1}{2}, \Sigma = \tfrac{1}{2}\rangle\, |v\rangle\, |\Omega = -\tfrac{1}{2} JM\rangle \tag{9.53c}$$

$$|^2\Pi_{-3/2}\rangle = |n, \Lambda = -1, S = \tfrac{1}{2}, \Sigma = -\tfrac{1}{2}\rangle\, |v\rangle\, |\Omega = -\tfrac{3}{2} JM\rangle. \tag{9.53d}$$

The derivation of the sixteen possible matrix elements of $\hat{H}$, equations (9.27) and (9.28), is slightly more involved than for $^2\Sigma^+$ states but results in the matrix (in cm^{-1})

$$\mathbf{H} = \begin{pmatrix} \frac{A}{2}+B[(J+\frac{1}{2})^2-2] & -B[(J+\frac{1}{2})^2-1]^{1/2} & 0 & 0 \\ -B[(J+\frac{1}{2})^2-1]^{1/2} & -\frac{A}{2}+B(J+\frac{1}{2})^2 & 0 & 0 \\ 0 & 0 & -\frac{A}{2}+B(J+\frac{1}{2})^2 & -B[(J+\frac{1}{2})^2-1]^{1/2} \\ 0 & 0 & -B[(J+\frac{1}{2})^2-1]^{1/2} & \frac{A}{2}+B[(J+\frac{1}{2})^2-2] \end{pmatrix} \tag{9.54}$$

In this matrix, the vibronic term energy E_{ev} needs to be added along the diagonal. Moreover, the constant matrix element of $\langle B(\hat{L}^2 - \hat{L}_z^2)\rangle$ has been included in E_{ev}. In this case the e/f parity basis functions are written as

$$|^2\Pi_{3/2}e/f\rangle = \frac{|^2\Pi_{3/2}\rangle \pm |^2\Pi_{-3/2}\rangle}{\sqrt{2}} \tag{9.55}$$

$$|^2\Pi_{1/2}e/f\rangle = \frac{|^2\Pi_{1/2}\rangle \pm |^2\Pi_{-1/2}\rangle}{\sqrt{2}} \tag{9.56}$$

in which the upper (lower) sign refers to $e(f)$ parity. The resultant Hamiltonian matrix for e-parity is

$$\begin{matrix} & |^2\Pi_{3/2}e/f\rangle & |^2\Pi_{1/2}e/f\rangle \end{matrix}$$

$$\mathbf{H}' = \begin{pmatrix} \frac{A}{2}+B[(J+\frac{1}{2})^2-2] & -B[(J+\frac{1}{2})^2-1]^{1/2} \\ -B[(J+\frac{1}{2})^2-1)^{1/2} & -\frac{A}{2}+B(J+\frac{1}{2})^2 \end{pmatrix} \tag{9.57}$$

with identical matrix elements for f-parity.

There are two limiting cases for the energy levels of a $^2\Pi$ state depending upon the extent of spin-orbit coupling, as measured by the relative size of the diagonal $A\mathbf{L}\cdot\mathbf{S}$ term in the Hamiltonian. When A is large ($A \gg BJ$), it is referred to as Hund's case (a) coupling, while when A is small ($A \ll BJ$), it is referred to as Hund's case (b) coupling.

The Hund's case (a) $^2\Pi$ Hamiltonian matrix is characterized by well-separated diagonal matrix elements $\Delta E = H'_{11} - H'_{22} \approx A$, and by a relatively small off-diagonal matrix element $H_{12} = -B[(J + \frac{1}{2})^2 - 1]^{1/2} \approx -BJ$. In this case, second-order perturbation theory (Chapter 4) predicts that the upper energy level is shifted up by $V^2/\Delta E = (H_{12})^2/\Delta E$ and the lower energy level down by $-(H_{12})^2/\Delta E$. There are thus two widely separated $^2\Pi_{3/2}$ and $^2\Pi_{1/2}$ spin components, with energies given by

$$E_{3/2} = \frac{A}{2} + B[(J + \tfrac{1}{2})^2 - 2] + \frac{B^2[(J + \frac{1}{2})^2 - 1]}{A}, \tag{9.58}$$

and

$$E_{1/2} = -\frac{A}{2} + B(J + \tfrac{1}{2})^2 - \frac{B^2[(J + \frac{1}{2})^2 - 1]}{A}. \tag{9.59}$$

The energy-level expressions (9.58) and (9.59) can be simplified by defining

$$A_{\text{eff}} = A - 2B, \tag{9.60}$$

$$B_{3/2\text{eff}} = B + \frac{B^2}{A} = B\left(1 + \frac{B}{A}\right), \tag{9.61}$$

and

$$B_{1/2\text{eff}} = B - \frac{B^2}{A} = B\left(1 - \frac{B}{A}\right), \tag{9.62}$$

so that the energy-level expressions become

$$E_{3/2} = \frac{A_{\text{eff}}}{2} + B_{3/2\text{eff}}[(J + \tfrac{1}{2})^2 - 1] \tag{9.63}$$

and

$$E_{1/2} = -\frac{A_{\text{eff}}}{2} + B_{1/2\text{eff}}[(J + \tfrac{1}{2})^2 - 1]. \tag{9.64}$$

Thus, the rotational levels of the $^2\Pi_{1/2}$ and $^2\Pi_{3/2}$ spin components are well separated for each J value and each component has its own effect B value but with the actual rotational constant given by the average

$$B = \frac{B_{1/2} + B_{3/2}}{2}. \tag{9.65}$$

At this stage of development the e and f components for each rotational energy level are exactly degenerate. If Λ-doubling[9] is taken into account, the $^2\Pi_{1/2}$ rotational levels are split by $p(J + \frac{1}{2})$ and the $^2\Pi_{3/2}$ rotational levels are split by $f(p, q)(J + \frac{1}{2})^3$, in which $f(p, q)$ is a parameter that depends on p and q. The two Λ-doubling constants, p and q, account for interactions with distant $^2\Sigma$ states. As illustrated in Figure 9.18 each rotational energy level J is twofold degenerate and this degeneracy is lifted by Λ-doubling interactions in a rotating molecule.

If the splitting between the $^2\Pi_{1/2}$ and $^2\Pi_{3/2}$ spin components is small, then the

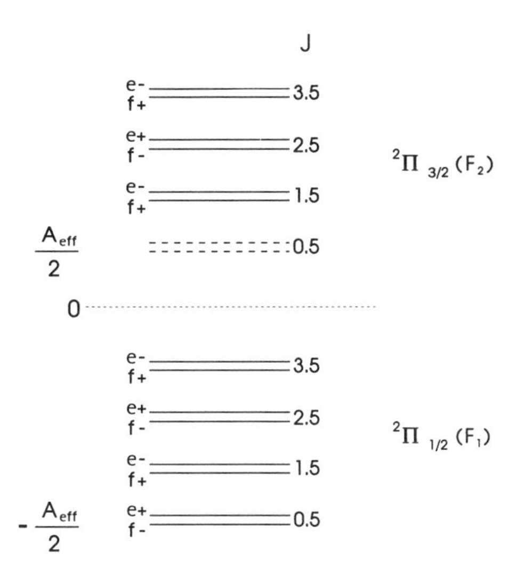

Figure 9.18. The energy-level pattern for a Hund's case (a) $^2\Pi$ state.

Hund's case (b) energy-level pattern applies. In this case, $A - 2B \ll BJ$ (i.e., $(H_{11} - H_{22}) \ll H_{12}$) so that the two spin components are strongly coupled. In this case, the 2×2 Hamiltonian matrix must be diagonalized (Chapter 4) in order to obtain the two energy levels.

For Hund's case (b), the spin-orbit coupling constant is zero ($A = 0$). The two energy levels are given by the expression

$$
\begin{aligned}
E_{\pm} &= \frac{H_{11} + H_{22}}{2} \pm \frac{[(\Delta E)^2 + 4H_{12}^2]^{1/2}}{2} \\
&= B[(J + \tfrac{1}{2})^2 - 1] \pm \frac{[4B^2 + 4B^2((J + \tfrac{1}{2})^2 - 1)]^{1/2}}{2} \\
&= B[(J + \tfrac{1}{2})^2 - 1] \pm B(J + \tfrac{1}{2}). \qquad (9.66)
\end{aligned}
$$

For a Hund's case (b) $^2\Pi$ state it is useful, as in the $^2\Sigma^+$ case, to introduce the integral quantum number $N = J + \frac{1}{2}$ for the E_+ level and $N = J - \frac{1}{2}$ for the E_- level. Thus the F_1 and F_2 energy levels are given by

$$F_2(N) = E_+ = B(N^2 - 1) + BN = B[N(N+1) - 1] \qquad (9.67a)$$

and

$$F_1(N) = E_- = B((N+1)^2 - 1) - B(N+1) = B[N(N+1) - 1]. \qquad (9.67b)$$

The energy levels for a Hund's case (b) $^2\Pi$ state are given by $B[N(N+1) - 1]$, so that

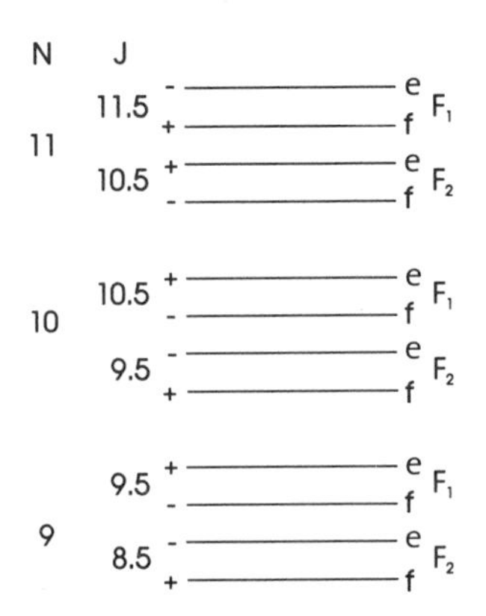

Figure 9.19. Energy-level diagram for a Hund's case (b) $^2\Pi$ state.

for each N there is a fourfold degeneracy, F_{1e}, F_{1f}, F_{2e}, and F_{2f} (Figure 9.19). The F_1 and F_2 labels are defined so that $J = N + \frac{1}{2}$ for F_1 and $J = N - \frac{1}{2}$ for F_2.

In general the rotational energy levels for Hund's case (b) $^{2S+1}\Lambda$ states are given by $BN(N+1)$ and there are $2(2S+1)$ degenerate levels for $\Lambda \neq 0$ and $2S+1$ degenerate levels for $\Lambda = 0$ (Σ^+ or Σ^- states). In contrast (Figure 9.20) for a Hund's case (a) $^{2S+1}\Lambda$ state there are $2S+1$ well-separated spin components in which the rotational energy levels for each spin component are given by $B_{\Omega,\text{eff}}J(J+1)$. For Hund's case (a) each J is doubly degenerate ($\Lambda > 0$), corresponding to $\pm\Lambda$ (or e/f parity states). For Σ states,

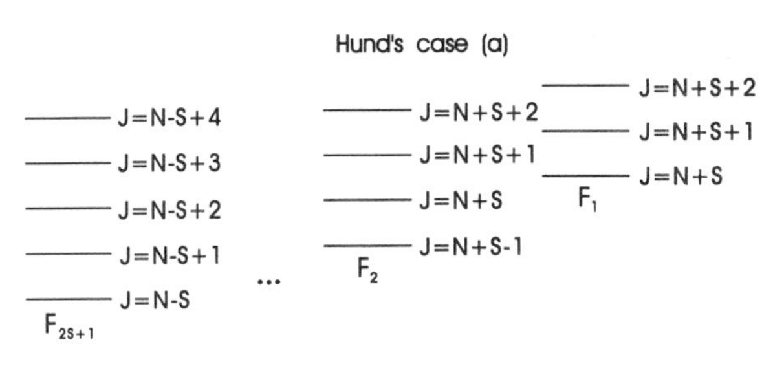

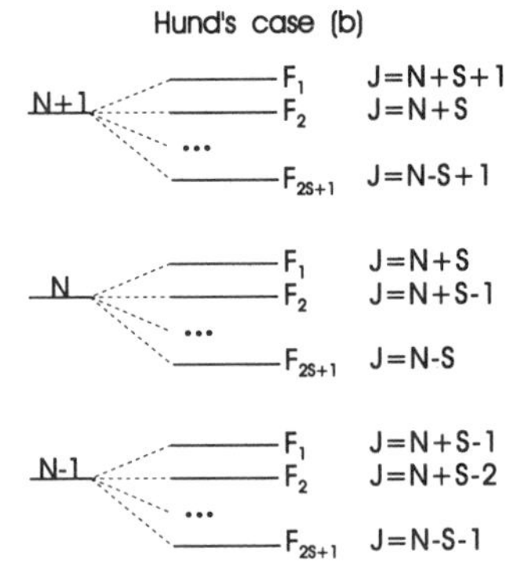

Figure 9.20. Hund's case (a) and Hund's case (b) energy-level patterns for a $^{2S+1}\Lambda$ state.

Hund's case (a) coupling is rare since there is no contribution by the diagonal spin-orbit term to the energy levels. Second-order spin-orbit coupling can, however, split a $^{2S+1}\Sigma$ state into $S+1$ spin components.

The energy-level patterns described by Hund's cases (a) and (b) are extreme limiting cases. The accurate description of molecular energy levels requires the addition of centrifugal distortion, spin-rotation coupling, spin-spin coupling, and Λ-doubling terms, as well as the numerical diagonalization of the Hamiltonian matrix.[9] There are also two additional Hund's coupling cases that are not uncommon: Hund's cases (c) and (d).

In Hund's case (c) the spin-orbit coupling becomes larger than BJ as well as larger than the separation between electronic states, that is, $A \gg BJ$ and $A \gg \Delta E_{\text{states}}$. In this case the various spin components from several $^{2S+1}\Lambda$ terms occur in the same energy region and their wavefunctions become mixed through off-diagonal (i.e., between $^{2S+1}\Lambda$ terms) spin-orbit coupling. The complete spin-orbit Hamiltonian,

$$\hat{H}_{\text{so}} = \Sigma a_i \hat{\mathbf{l}}_i \cdot \hat{\mathbf{s}}_i, \tag{9.68}$$

causes both large spin-orbit splittings and large mixing of the electronic wavefunctions.

The strong coupling of **L** and **S** means that only J and Ω are good quantum numbers for pure case (c) coupling. Since L, S, Λ, and Σ no longer have any meaning, the conventional $^{2S+1}\Lambda_\Omega$ notation is misleading; the Hund's case (c) notation is just Ω. For example, the $B^3\Pi_{0_u^+}$ state of I_2 should be labeled as $B0_u^+$ using Hund's case (c) notation. Since spin-orbit coupling increases rapidly with increasing atomic number Z, electronic states of diatomic molecules containing heavy elements tend toward Hund's case (c). For Hund's case (c) coupling, the $^{2S+1}\Lambda_\Omega$ spin components of a given $^{2S+1}\Lambda$ term are like independent electronic states and are labeled by the good quantum number Ω.

The rotational energy-level pattern for a Hund's case (c) state is similar to the Hund's case (a) pattern (Figure 9.21). The rotational energy-level expression is $B_{\text{eff}}J(J+1)$ for each Ω state. For $\Omega \neq 0$ each level is doubled due to the $\pm\Omega$ degeneracy, which can be lifted by Ω-type doubling interactions. For both Hund's cases (a) and (c) the spin components are labeled by Ω, but only for Hund's case (a) does a spin component originate from a specific $^{2S+1}\Lambda$ term. In Hund's case (c) the Ω state is a mixture of many $^{2S+1}\Lambda_\Omega$ basis functions.

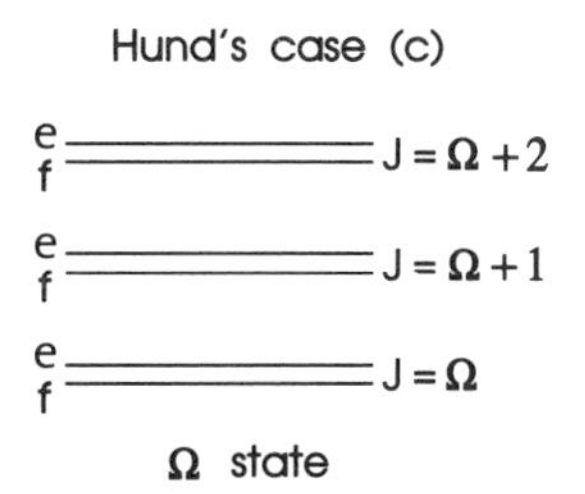

Figure 9.21. Hund's case (c) rotational energy-level pattern.

Table 9.5. Hund's Coupling Cases

Coupling case	Good Quantum Numbers	Rotational Energy Expression	Degeneracy of Each Rotational State
Hund's case (a)	Λ, S, Σ, Ω	$BJ(J+1)$	2 (1 for Σ states)
Hund's case (b)	Λ, S, Σ	$BN(N+1)$	$2(2S+1)$ ($2S+1$ for Σ states)
Hund's case (c)	Ω	$BJ(J+1)$	2 (1 for Σ states)
Hund's case (d)	L, Ω, S, Σ	$BR(R+1)$	$(2L+1)(2S+1)$

Hund's case (d) applies to Rydberg electronic states in which an electron is excited to an orbital with a large principal quantum number. In this case, the Rydberg electron is so distant from the nuclei that **l** and **s** for the Rydberg electron couple only weakly to the internuclear axis. For a pure Hund's case (d) state the rotational energy-level expression is $BR(R+1)$ and each level has a $(2S+1)(2L+1)$ degeneracy. The $2L+1$ and $2S+1$ degeneracy is associated with the M_L and M_S degeneracy of a very atomic-like Rydberg electron. The atomic-like character of the Rydberg molecular orbital allows L to be a good quantum number in this case. The various Hund's coupling cases[10] are summarized in Table 9.5.

The Symmetry of Diatomic Energy Levels: Parity

Nothing causes as much confusion in the study of the spectra of diatomic molecules as does the concept of parity. The difficulty lies in there being several different types of parity, such as g/u, e/f, $+/-$, and s/a. The basic idea is, however, quite simple: if the Hamiltonian operator and a symmetry operator commute, then a set of simultaneous eigenfunctions of the two operators can be found; that is,

$$[\hat{H}, \hat{O}_S] = 0 \tag{9.69}$$

implies that

$$\hat{H}\psi_\pm = E_\pm \psi_\pm \tag{9.70}$$

and that

$$\hat{O}_S \psi_\pm = \pm\psi_\pm. \tag{9.71}$$

If a molecule has a certain symmetry, then the effect of the associated symmetry operator can be used to label wavefunctions and their energy levels. Care must be taken, however, to specify *both* the symmetry operator and the part of the total Hamiltonian under consideration. Failure to do so is the real source of confusion about parity.

Total (+/−) *Parity*

If the symmetry operation is $\hat{E}^*$ (sometimes called $\hat{I}$) and the total Hamiltonian including electronic, vibrational, and rotational parts (but not nuclear spin) is used, then one obtains total parity. The $\hat{E}^*$ operator inverts all of the coordinates of the particles (nuclei and electrons) in the laboratory frame with the origin at the center of mass (Figure 9.4), that is,

$$\begin{aligned}\hat{E}^*\psi(X_i, Y_i, Z_i) &= \psi(-X_i, -Y_i, -Z_i)\\ &= \pm\psi(X_i, Y_i, Z_i).\end{aligned} \tag{9.72}$$

The $\hat{E}^*$ operator is a symmetry operator because all the relative positions of the particles are the same before and after inversion, and consequently the energy levels are unchanged by application of $\hat{E}^*$. Note, however, that the sign of the wavefunction can change under $\hat{E}^*$ since

$$\hat{E}^*(\hat{E}^*)\psi = \psi. \tag{9.73}$$

Total parity is the parity often used in nuclear and atomic physics. Physicists have long noted that the properties of a right-handed system are identical with those of a left-handed system (except for the process of β-decay) since the $\hat{E}^*$ symmetry operator converts one into the other.

The $\hat{E}^*$ symmetry operator is used to divide all rovibronic energy states into two groups by means of the equation

$$\hat{E}^*\psi = \hat{E}^*(\psi_{\text{el}}\psi_{\text{vib}}\psi_{\text{rot}}) = \pm\psi. \tag{9.74}$$

All rovibronic energy levels for which the upper sign applies have positive (+) total parity, while all rovibronic levels for which the lower sign applies have negative (−) total parity. The effects of $\hat{E}^*$ on the electronic, vibrational, and rotational parts of the total wavefunction need to be determined individually.

The effect of $\hat{E}^*$ on the vibrational part of the diatomic wavefunction is easy to ascertain. The inversion operation leaves the vibrational part of the wavefunction unchanged because ψ_{vib} is a function only of the magnitude of the internuclear separation r. Inversion of the coordinates of all particles leaves r unchanged, so that

$$\hat{E}^*\psi_{\text{vib}}(r) = \psi_{\text{vib}}(r). \tag{9.75}$$

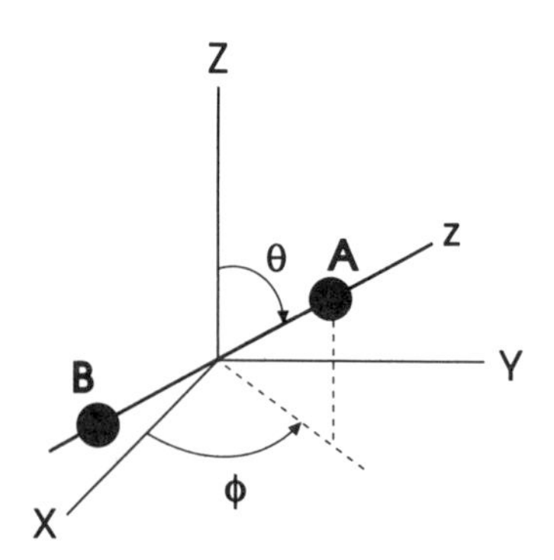

Figure 9.22. Diatomic molecule coordinate system.

The effect of E^* on the rotational wavefunction is more complicated. For instance, Figure 9.22 shows that the $\hat{E}^*$ operation replaces θ by $\pi - \theta$ and ϕ by $\phi + \pi$ in $\psi_{\text{rot}}(\theta, \phi)$. For $^1\Sigma^+$ states $\psi_{\text{rot}} = Y_{JM}(\theta, \phi)$. We have already discussed the effect of inversion in the laboratory frame on spherical harmonics, the result being

$$\hat{E}^* Y_{JM} = (-1)^J Y_{JM}. \tag{9.76}$$

In general, however, $\psi_{\text{rot}} = |\Omega JM\rangle$, and the properties of these rotational wavefunctions need to be considered. The rotational wavefunctions $|\Omega JM\rangle$ are identical to the symmetric top wavefunctions $|KJM\rangle$ discussed in Chapter 6, and it is found that[10]

$$\hat{E}^* |\Omega JM\rangle = (-1)^{J-\Omega} |-\Omega JM\rangle. \tag{9.77}$$

The final part of the problem, the effect of inversion on ψ_{el}, is most difficult because ψ_{el} is known in the molecular frame. The calculation of ψ_{el} is made under the Born–Oppenheimer approximation so that the nuclei are fixed in space, thus $\psi_{\text{el}}(x_i, y_i, z_i)$ is known but $\psi_{\text{el}}(X_i, Y_i, Z_i)$ is not. Since the inversion operation changes the sign of the laboratory coordinates (X_i, Y_i, Z_i) of the particles, the effect of $\hat{E}^*$ on $\psi_{\text{el}}(x_i, y_i, z_i)$ is not obvious. Hougen has considered this problem[10,11] and has shown that $\hat{E}^*$ in the laboratory frame is equivalent to the symmetry operation of reflection $\hat{\sigma}_v$ in the molecular frame. The equivalence of the permutation-inversion operation $\hat{E}^*$ in the laboratory frame and the ordinary point group operation of reflection $\hat{\sigma}_v$ in the symmetry plane of the diatomic molecule is an important (not obvious) result. The effect of $\hat{\sigma}_v$ on the coordinates of an electron is established by replacing (x_i, y_i, z_i) by $(x_i, -y_i, z_i)$, if the reflection is arbitrarily chosen to be in the xz plane of the molecule. Without considering the detailed form of the spin and orbital parts of ψ_{el}, it is not possible to extend the treatment further. It turns out that the effect of $\hat{\sigma}_v$ on the spin and orbital parts is[10]

$$\hat{\sigma}_v |S\Sigma\rangle = (-1)^{S-\Sigma} |S, -\Sigma\rangle, \tag{9.78}$$

and

$$\hat{\sigma}_v\,|\Lambda\rangle = \pm(-1)^\Lambda\,|-\Lambda\rangle. \tag{9.79}$$

The effect of $\hat{\sigma}_v$ on a $|\Lambda = 0\rangle$ orbital part of the electronic wavefunction is particularly interesting since one obtains

$$\hat{\sigma}_v\,|\Lambda = 0\rangle = \pm\,|\Lambda = 0\rangle.$$

The $\pm$ signs are written as superscripts on the term symbols, Σ^+ and Σ^-; they correspond to

$$\hat{\sigma}_v\,|\Sigma^\pm\rangle = \pm\,|\Sigma^\pm\rangle. \tag{9.80}$$

The superscript $\pm$ sign in the term symbol for Σ states indicates the effect of the $\hat{\sigma}_v$ symmetry operator on only the orbital part of the electronic wavefunction. The addition of superscript $\pm$ to the term symbol for $\Lambda > 0$ (Π, Δ, Φ, etc., states) is not necessary since the levels always occur as a $\pm$ pair because of the twofold orbital degeneracy (Figure 9.18).

The effect of the $\hat{\sigma}_v$ operator (equivalent to $\hat{E}^*$) on the total wavefunction is determined by combining equations (9.77), (9.78), and (9.79)

$$\begin{aligned}\hat{\sigma}_v(\psi_{el}\psi_{vib}\psi_{rot}) &= \hat{\sigma}_v(|n\Lambda S\Sigma\rangle\,|v\rangle\,|\Omega JM\rangle) \\ &= (-1)^{J-2\Sigma+S+\sigma}\,|n, -\Lambda, S, -\Sigma\rangle\,|v\rangle\,|-\Omega JM\rangle,\end{aligned} \tag{9.81}$$

in which $\sigma = 0$ for all states except Σ^- states for which $\sigma = 1$. Since the $\hat{\sigma}_v$ operation changes the signs of Λ, Σ, and Ω, the parity eigenfunctions are linear combinations of the basis functions, namely

$$|{}^{2S+1}\Lambda_\Omega \pm\rangle = \frac{|{}^{2S+1}\Lambda_\Omega\rangle \pm (-1)^{J-2\Sigma+S+\sigma}\,|{}^{2S+1}\Lambda_{-\Omega}\rangle}{\sqrt{2}} \tag{9.82}$$

with

$$\hat{\sigma}_v\,|{}^{2S+1}\Lambda_\Omega \pm\rangle = \pm\,|{}^{2S+1}\Lambda_\Omega \pm\rangle. \tag{9.83}$$

The selection rules on total parity are derived, as usual, by requiring a totally symmetric integrand for the transition moment integral,

$$\int \psi_f^* \boldsymbol{\mu} \psi_i \, d\tau. \tag{9.84}$$

The transition moment operator has $(-)$ parity since

$$\hat{E}^*\boldsymbol{\mu} = -\boldsymbol{\mu}, \tag{9.85}$$

so that only $+ \leftrightarrow -$ transitions are allowed for one-photon electric dipole transitions.

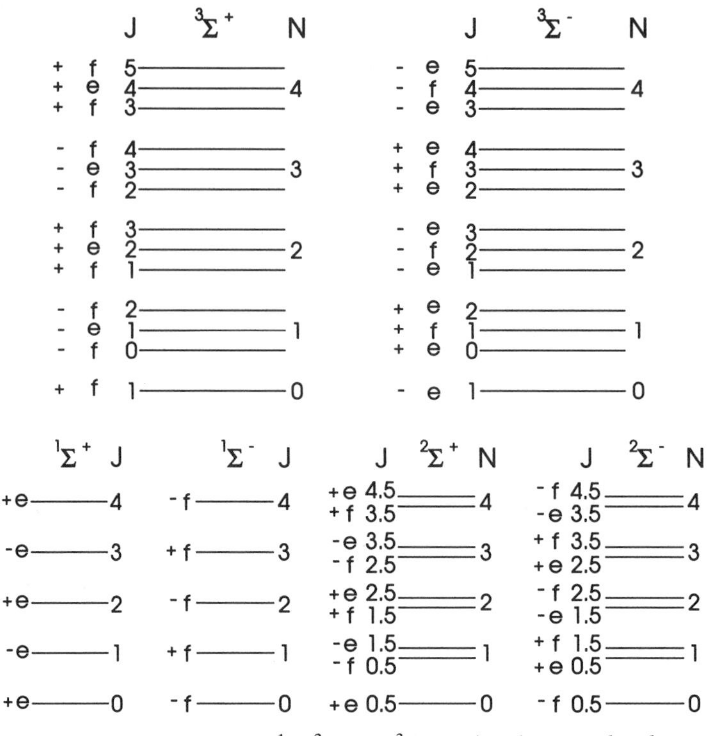

Figure 9.23. Parity of $^1\Sigma$, $^2\Sigma$, and $^3\Sigma$ rotational energy levels.

Rotationless (e/f) *Parity*[12]

Notice that the total parity changes sign with J (Figure 9.23) because of the phase factor $(-1)^J$ in equation (9.77). This alternation of the total parity with J is always present, and hence it is useful to factor it out by defining e and f parity as[12]

$$\hat{E}^*\psi = +(-1)^J\psi \quad \text{for } e, \tag{9.86a}$$

and

$$\hat{E}^*\psi = -(-1)^J\psi \quad \text{for } f \tag{9.86b}$$

for integer J. Similarly, for half-integer J,

$$\hat{E}^*\psi = +(-1)^{J-1/2}\psi \quad \text{for } e, \tag{9.87a}$$

and

$$\hat{E}^*\psi = -(-1)^{J-1/2}\psi \quad \text{for } f, \tag{9.87b}$$

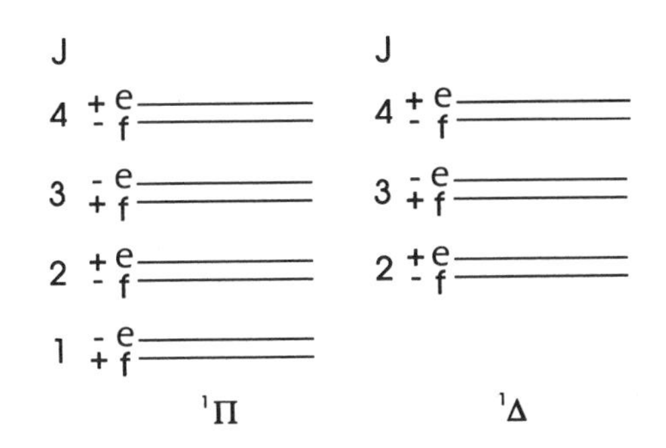

Figure 9.24. Parity of $^1\Pi$ and $^1\Delta$ rotational energy levels.

in which ψ is the total rovibronic wavefunction. In the molecular frame, $\hat{E}^*$ is replaced by $\hat{\sigma}_v$. Notice that e and f parity is a "residual parity" or a rotationless parity describing the total parity with the rotational part removed.

Because the alternation of sign with J has been removed, e/f parity is more convenient to use than total parity. All rotational energy levels of $^1\Sigma^+$ states have e parity, while $^1\Sigma^-$ rotational energy levels have f parity (Figure 9.23). For $^1\Pi$ states all of the rotational energy levels occur as e/f pairs (Figure 9.24). The $+\leftrightarrow-$ selection rule for total parity becomes $e\leftrightarrow e$ and $f\leftrightarrow f$ for P and R branches, while $e\leftrightarrow f$ for Q branches.

The use of e/f parity also suppresses an annoying $(-1)^{J-2\Sigma+S+\sigma}$ factor in the definition of the parity eigenfunctions (9.32) since

$$|^2\Pi_{3/2}e/f\rangle = \frac{|^2\Pi_{3/2}\rangle \pm |^2\Pi_{-3/2}\rangle}{\sqrt{2}}, \tag{9.88}$$

$$|^2\Pi_{1/2}e/f\rangle = \frac{|^2\Pi_{1/2}\rangle \pm |^2\Pi_{-1/2}\rangle}{\sqrt{2}}, \tag{9.89}$$

$$|^2\Sigma^+_{1/2}e/f\rangle = \frac{|^2\Sigma^+_{1/2}\rangle \pm |^2\Sigma^+_{-1/2}\rangle}{\sqrt{2}}, \tag{9.90}$$

and

$$|^2\Sigma^-_{1/2}e/f\rangle = \frac{|^2\Sigma^-_{1/2}\rangle \mp |^2\Sigma^-_{-1/2}\rangle}{\sqrt{2}}. \tag{9.91}$$

Gerade/Ungerade (g/u) *Parity*

For homonuclear diatomic molecules, the point group $D_{\infty h}$ contains an inversion operation $\hat{\imath}$. This inversion operation is applied in the molecular frame, unlike the $\hat{E}^*$

symmetry operation that is applied in the laboratory frame. Moreover, $\hat{E}^*$ (or $\hat{\sigma}_v$) is a symmetry operation for all diatomic molecules, while $\hat{\imath}$ is a symmetry operation only for homonuclear diatomic molecules.

It is useful to classify only the electronic orbital part of the wavefunction with the aid of $\hat{\imath}$. (The operation $\hat{\imath}$ leaves the vibrational, rotational, and electron spin parts of the wavefunction unchanged in any case.) Now the location of the nuclei define the molecular z-axis, so the inversion operation $\hat{\imath}$ means that only the electrons (but not the nuclei or the coordinate system) are inverted through the center of the molecule,

$$\hat{\imath}\psi_{\text{el}}(x_i, y_i, z_i) = \psi_{\text{el}}(-x_i, -y_i, -z_i) = \pm\psi_{\text{el}}(x_i, y_i, z_i) \tag{9.92}$$

or

$$\hat{\imath}\,|\Lambda\rangle = \pm\,|\Lambda\rangle, \tag{9.93}$$

where the + sign corresponds to g (gerade) parity and − to u (ungerade) parity. The g or u parity is appended as a subscript to the term symbol of a diatomic molecule, for example, ${}^2\Sigma_g^+$, ${}^3\Sigma_u^-$, ${}^1\Delta_u$. Thus g and u are used to classify just the electron orbital part of the total wavefunction. The selection rule $g \leftrightarrow u$ applies for electric dipole-allowed transitions since the transition dipole moment $\boldsymbol{\mu}$ is of u parity.

Symmetric/Antisymmetric (s/a) *Parity*

For homonuclear diatomic molecules additional symmetry labels, consisting of s (for symmetric) and a (for antisymmetric), can be used to classify the rotational energy levels. The Pauli exclusion principle demands that the total wavefunction *including nuclear spin* be symmetric or antisymmetric with respect to interchange of the two identical nuclei. This interchange of two identical nuclei is described by the operator $\hat{P}_{12}$ in the laboratory frame. Experimentally it is found that if the identical nuclei are bosons ($I = 0, 1, 2, \ldots$), then the total wavefunction (including nuclear spin) is symmetric with respect to the $\hat{P}_{12}$ operator, while if the identical nuclei are fermions ($I = \frac{1}{2}, \frac{3}{2}, \frac{5}{2}, \ldots$), then the total wavefunction is antisymmetric with respect to $\hat{P}_{12}$, that is,

$$\hat{P}_{12}(\psi\psi_{\text{nuc}}) = +(\psi\psi_{\text{nuc}}) \quad \text{for bosons,} \tag{9.94}$$

and

$$\hat{P}_{12}(\psi\psi_{\text{nuc}}) = -(\psi\psi_{\text{nuc}}) \quad \text{for fermions.} \tag{9.95}$$

The total wavefunction is written as the product of the "normal" wavefunction ψ, which includes electron spin, orbital, vibrational, and rotational parts (as discussed earlier) and a nuclear spin wavefunction ψ_{nuc}.

Consider the molecules H_2 or F_2 in which the nuclei are fermions with $I = \frac{1}{2}$. In this case there are four nuclear spin wavefunctions, three of them symmetric,

$$\psi_{nuc}(\text{sym.}) = \begin{cases} \alpha(1)\alpha(2) \\ \dfrac{[\alpha(1)\beta(2) + \beta(1)\alpha(2)}{\sqrt{2}} \\ \beta(1)\beta(2) \end{cases} \tag{9.96}$$

and one of them antisymmetric,

$$\psi_{nuc}(\text{antisym.}) = \frac{\alpha(1)\beta(2) - \beta(1)\alpha(2)}{\sqrt{2}}. \tag{9.97}$$

The symmetric ψ_{nuc} wavefunctions must be combined with antisymmetric normal wavefunctions (excluding nuclear spin) ψ to give an overall antisymmetric product $\psi\psi_{nuc}$. These rovibronic wavefunctions are therefore antisymmetric with respect to $\hat{P}_{12}$ and are labeled a. Similarly, the single antisymmetric nuclear spin function must be combined with a symmetric normal wavefunction (excluding nuclear spin) to give an overall antisymmetric product $\psi\psi_{nuc}$. These rovibronic energy levels are therefore symmetric (with respect to $\hat{P}_{12}$) and are labeled s. In the case of H_2, the a rotational energy levels (called *ortho levels*) have three times the statistical weight of the s rotational levels (called *para levels*). By convention ortho levels always have the larger statistical weight and para levels the smaller (see Chapter 7).

Which rotational energy levels of a homonuclear diatomic molecule are s and which are a with respect to the $\hat{P}_{12}$ operation? The problem is that $\hat{P}_{12}$ is a permutation-inversion operation that switches (permutes) the two nuclei of a homonuclear diatomic molecule so that $\hat{P}_{12}$ is not an ordinary group symmetry operation.

The $\hat{P}_{12}$ operation can be applied to the wavefunction in two steps. First, all of the electrons and all of the nuclei are inverted through the origin by applying the $\hat{E}^*$ operation in the laboratory frame; then only the electrons are inverted back by applying $\hat{\imath}$ in the molecular frame. Thus all + rotational energy levels have s symmetry for g electronic states or a symmetry for u electronic states. Similary all $-$ rotational energy levels have a symmetry for g electronic states and s symmetry for u electronic states (Figure 9.25).

Now the $\hat{E}^*$ operation is represented by $\hat{\sigma}_v$ in the molecular frame so that

$$\hat{P}_{12} = \hat{\sigma}_v^{xz}\hat{\imath} = \hat{\sigma}_v^{xz}\hat{\sigma}_v^{xz}\hat{C}_2(y) = \hat{C}_2(y). \tag{9.98}$$

This means that the application of $\hat{P}_{12}$ in the laboratory frame is equivalent to $\hat{C}_2(y)$ in the molecular frame.[10] Similarly the $\hat{\imath}$ operation in the molecular frame $[\hat{\imath} = \hat{\sigma}_v^{xz}\hat{C}_2(y)]$

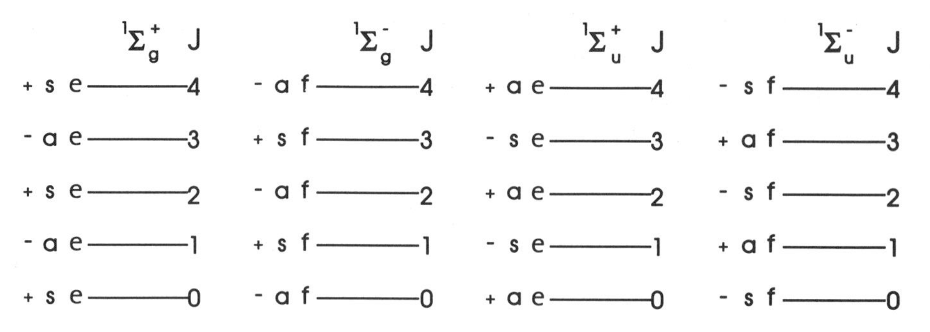

Figure 9.25. Parity for $^1\Sigma_g$ and $^1\Sigma_u$ rotational energy levels.

is equivalent to $\hat{E}^*\hat{P}_{12}$ in the laboratory frame.[10]. This is a very sensible result since the $\hat{E}^*$ operation inverts the electrons and nuclei, while $\hat{P}_{12}$ switches the nuclei back leaving only the electrons inverted.

Consider the ground state of the O_2 molecule of $^3\Sigma_g^-$ symmetry. The even N values have $-$ parity (a) and the odd N values $+$ parity (s) (Figure 9.26). The nuclear spin of ^{16}O is zero, so only symmetric nuclear spin wavefunctions ψ_{nuc} are possible. The symmetric ψ_{nuc} wavefunction must be combined with the s symmetry ψ function because the oxygen nuclei are bosons. This means that the a levels of O_2 (even N) cannot exist and are therefore missing in the spectrum. The s and a symmetry levels are therefore very useful in establishing the relative intensities of rotational lines. In general the relative nuclear spin weight of ortho levels relative to para levels is given

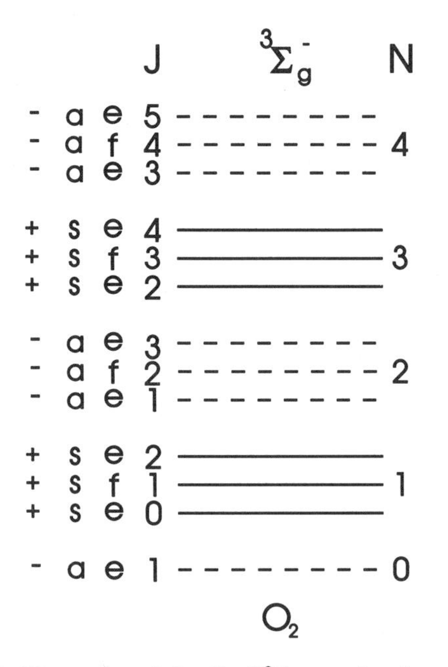

Figure 9.26. The parity of the O_2 $X^3\Sigma_g^-$ rotational energy levels.

by $[(2I+1)(I+1)]/[(2I+1)I] = (I+1)/I$. The electric dipole selection rule for s and a symmetry is $s \leftrightarrow s$ and $a \leftrightarrow a$ since electronic transitions cannot simultaneously flip nuclear spins.

Dissociation and Predissociation

Under certain conditions the dissociation energies of diatomic molecules can be determined by electronic spectroscopy. If the two potential curves involved in an electronic transition have very different r_e values (Figure 9.9), then the Franck–Condon factors favor long progressions in the upper state v for absorption or in the lower state v in emission (Figure 9.27). The resulting progressions can be extrapolated using the Le Roy–Bernstein method (Chapter 7) to determine the dissociation energies. These studies are carried out by electronic spectroscopy. Note that if D_0' is known, D_0'' can be calculated by using the equation

$$T_{00} = x - D_0' + D_0'', \tag{9.99}$$

provided the atomic transition energy x is known (Figure 9.27).

It is quite common for two potential curves to cross (Figure 9.28). Of course, if the two crossing curves have the same symmetry, then an avoided crossing occurs because of the noncrossing rule. (The noncrossing rule states that when two potential energy curves of the same symmetry cross, there will always be a mixing of the two

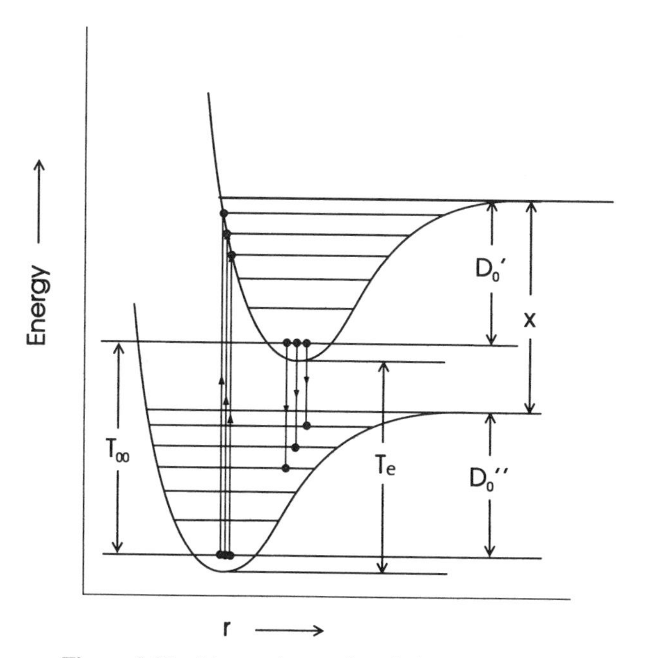

Figure 9.27. Absorption and emission progressions.

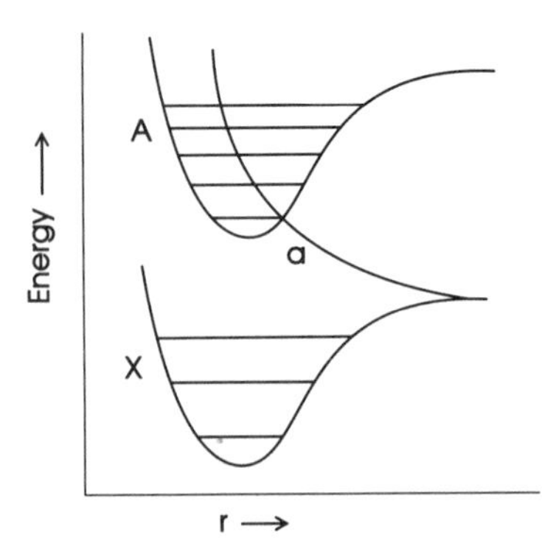

Figure 9.28. Predissociation of a state labeled A by a state labeled a.

wavefunctions to give an avoided crossing.) In any case, in the vicinity of the crossing point an interaction, no matter how small, exists between the diabatic (nonmixed) curves. The interaction acts as a perturbation to mix the wavefunctions of the two states near the crossing point. This means that the "bound" wavefunctions of the A state (Figure 9.28) have some "free" a-state character near the crossing point (and vice versa). The A–X transition, therefore, will display predissociation, where some levels of the A state are able to dissociate. This type of predissociation is present in NH for which the states are $X^3\Sigma^-$, $A^3\Pi$ and the repulsive state is $^5\Sigma^-$.[13] In general, predissociation occurs whenever there is substantial overlap between a bound vibrational wavefunction and a free vibrational wavefunction (approximately a sine wave) from another electronic state. This overlap is a maximum near a curve crossing (Figure 9.28), but a curve crossing is not necessary for predissociation.

Predissociation is surprisingly common in the spectroscopy of diatomic molecules. The presence of predissociation is evident when rotational or vibrational structure abruptly terminates or if the lines become broad (Chapter 1), in accordance with the Heisenberg uncertainty principle ($\Delta E \Delta t > \hbar$). The energy corresponding to the wavelength at which predissociation occurs provides an upper limit to the ground-state dissociation energy.

Problems

1. For the diatomic molecules Na_2, CO, CO^+, SO, and NO:
   (a) What are the lowest energy configurations and the bond order?
   (b) What terms arise from each configuration and which term lies lowest in energy?
   (c) For each term what levels arise and which level lies lowest in energy?
   (d) Predict the lowest energy excited electronic level for each molecule.
2. The CrO molecule has a $\sigma\delta^2\pi$ configuration.
   (a) What electronic states arise from this configuration?
   (b) What is the ground state of CrO and is it regular or inverted?
   (c) If the spin-orbit coupling constant A is 63 cm^{-1} for the ground state, calculate the energy-level pattern of the spin components.

3. For the $^{63}Cu_2$ molecule the following spectroscopic constants were recently determined for the $B^1\Sigma^+ - X^1\Sigma^+$ electronic transition:

$$T_e = 21{,}757.619 \text{ cm}^{-1}$$

$$\omega_e' = 246.317 \text{ cm}^{-1} \qquad \omega_e'' = 266.459 \text{ cm}^{-1}$$

$$\omega_e x_e' = 2.231 \text{ cm}^{-1} \qquad \omega_e x_e'' = 1.035 \text{ cm}^{-1}$$

$$B_e' = 0.098847 \text{ cm}^{-1} \qquad B_e'' = 0.108781 \text{ cm}^{-1}$$

$$\alpha_e' = 0.000488 \text{ cm}^{-1} \qquad \alpha_e'' = 0.000620 \text{ cm}^{-1}$$

   (a) Construct the Deslandres table for $0 \le v' \le 3$ and $0 \le v'' \le 3$.
   (b) Is the 1–0 band degraded to longer or shorter wavelengths?
   (c) At what J and wavenumber will the 1–0 band head occur?

4. The following spectroscopic constants were determined for the SrS $A^1\Sigma^+ - X^1\Sigma^+$ transition:

$$T_e = 13{,}932.707 \text{ cm}^{-1}$$

$$\omega_e' = 339.145 \text{ cm}^{-1} \qquad \omega_e'' = 388.264 \text{ cm}^{-1}$$

$$\omega_e x_e' = 0.5524 \text{ cm}^{-1} \qquad \omega_e x_e'' = 1.280 \text{ cm}^{-1}$$

$$r_e' = 2.51160 \text{ Å} \qquad r_e'' = 2.43968 \text{ Å}$$

$$D_0 = 3.48 \text{ eV} \qquad D_0 \approx 3 \text{ eV}$$

   (a) Calculate the Morse potential parameters for the $X^1\Sigma^+$ and $A^1\Sigma^+$ states of SrS and graph the potential curves.
   (b) Predict the strongest band in the emission spectrum from $v' = 0$ and in the absorption spectrum from $v'' = 0$.

5. The following vibrational bands were observed for the first negative system of N_2 (in cm^{-1}):

17046.5	23992.5	28399.7
17373.2	24144.9	30221.1
19121.4	25542.1	30306.3
19416.6	25739.4	30355.0
19692.7	25913.5	30371.6
21229.1	26065.5	32489.8
21491.1	26183.6	32500.3
21734.5	26261.4	
21952.1	27908.6	
23368.3	28051.2	
23600.6	28169.7	
23807.9	28254.1	

(a) Assign the vibrational quantum numbers to the bands and construct a Deslandres table. (*Hint*: Draw a stick spectrum of the data).
(b) Derive a set of spectroscopic constants that reproduce this data set.

6. Derive the Hamiltonian matrices (9.54) and (9.57).

References

1. Prasad, C. V. V. and Bernath, P. F. J. Mol. Spectrosc. **156,** 327 (1992).
2. Tyte, D. C. and Nicholls, R. W. *Identification Atlas of Molecular Spectra,* Vol. 2, October 1964.
3. Fernando, W. T. M. L., O'Brien, L. C., and Bernath, P. F. J. Mol. Spectrosc. **139,** 461 (1990).
4. Ram, R. S. and Bernath, P. F. J. Opt. Soc. Am. B. **3,** 1170 (1986).
5. Whiting, E. E., Schadee, A., Tatum, J. B., Hougen, J. T., and Nicholls, R. W. J. Mol. Spectrosc. **80,** 249 (1980).
6. Zare, R. N. *Angular Momentum,* Wiley, New York, 1988, pp. 283–286.
7. Pianalto, F. S., Brazier, C. R., O'Brien, L. C., and Bernath, P. F. J. Mol. Spectrosc. **132,** 80 (1988).
8. Fernando, W. T. M. L. and Bernath, P. F. J. Mol. Spectrosc. **145,** 392 (1991).
9. Lefebvre-Brion, H. and Field, R. W. *Perturbations in the Spectra of Diatomic Molecules,* Academic Press, Orlando, Fla., 1986.
10. Hougen, J. T. *The Calculation of Rotational Energy Levels and Rotational Line Intensities in Diatomic Molecules,* U.S. Government Printing Office, NBS Monograph 115, Washington, D.C., 1970.
11. Hougen, J. T. J. Chem. Phys. **37,** 1433 (1962); also, ———. J. Chem. Phys. **39,** 358 (1963).
12. Brown, J. M., et al. J. Mol. Spectrosc. **55,** 500 (1975).
13. Brazier, C. R., Ram, R. S., and Bernath, P. F. J. Mol. Spectrosc. **120,** 381 (1986).

General References

Dunford, H. B. *Elements of Diatomic Molecular Spectra,* Addison-Wesley, Reading, Mass., 1968.

Graybeal, J. D. *Molecular Spectroscopy,* McGraw-Hill, New York, 1988.

Harris, D. C. and Bertolucci, M. D. *Symmetry and Spectroscopy,* Dover, New York, 1989.

Herzberg, G. *Spectra of Diatomic Molecules,* Van Nostrand Reinhold, New York, 1950.

Hollas, J. M. *High Resolution Spectroscopy,* Butterworths, London, 1982.

Hougen, J. T. *The Calculation of Rotational Energy Levels and Rotational Line Intensities in Diatomic Molecules,* U.S. Government Printing Office, NBS Monograph 115, Washington, D.C., 1970.

Huber, K. P. and Herzberg, G. *Constants of Diatomic Molecules,* Van Nostrand Reinhold, New York, 1979.

King, G. W. *Spectroscopy and Molecular Structure,* Holt, Rinehart and Winston, New York, 1964.

Lefebvre-Brion, H. and Field, R. W. *Perturbations in the Spectra of Diatomic Molecules,* Academic Press, Orlando, Fla. 1986.

Pearse, R. W. B. and Gaydon, A. G., *The Identification of Molecular Spectra,* Chapman & Hall, London, 1976.

Rosen, B. *Spectroscopic Data Relative to Diatomic Molecules,* Pergamon Press, Oxford, 1970.

Struve, W. S. *Fundamentals of Molecular Spectroscopy,* Wiley, New York, 1989.

Steinfeld, J. I., *Molecules and Radiation,* 2nd Ed., MIT Press, Cambridge, Mass., 1985.

Zare, R. N. *Angular Momentum,* Wiley, New York, 1988.

Chapter 10
Electronic Spectroscopy of Polyatomic Molecules

Orbitals and States

The spectroscopy of polyatomic molecules has much more variety than the spectroscopy of diatomic molecules. Rather than try to survey this vast field, only a few topics are considered in order to provide the reader with a glimpse of it.

Molecular orbital theory is the key to understanding the electronic structure of polyatomic molecules. The electronic Schrödinger equation

$$\hat{H}_{\text{el}}\psi = E_{\text{el}}\psi \tag{10.1}$$

can be solved approximately by constructing a set of molecular orbitals in which each molecular orbital is a linear combination of atomic orbitals

$$\phi_{\text{MO}} = \sum c_i \phi_i, \tag{10.2}$$

in the same manner as has been discussed for diatomic molecules. The total wavefunction is a Slater determinant of MOs,

$$\psi = |\phi_{\text{MO}}(1)\bar{\phi}_{\text{MO}}(2)\cdots|. \tag{10.3}$$

These electronic wavefunctions are constructed in such a way that they are eigenfunctions of the symmetry operators for a specific molecular point group. Moreover, since the symmetry operators commute with the electronic Hamiltonian, that is,

$$[\hat{H}_{\text{el}}, \hat{O}_R] = 0, \tag{10.4}$$

these wavefunctions are simultaneously eigenfunctions of the electronic Hamiltonian. Thus, the electronic wavefunctions are classified by the irreducible representations of the appropriate molecular point group and the electronic wavefunction belongs to a

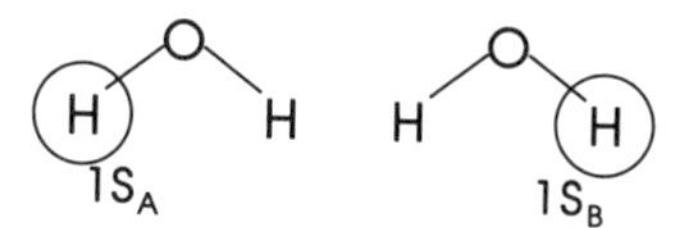

Figure 10.1. The 1*s* orbitals on the H atoms in the H_2O molecule.

particular irreducible representation. For example, because the point group symmetry of water is C_{2v} the electronic wavefunctions of water must have A_1, A_2, B_1, or B_2 symmetry.

Walsh's Rules: Qualitative Molecular Orbital Theory

A great deal of useful insight into the electronic structure of molecules can be gained through the use of qualitative molecular orbital theory. (Of course, even more information is provided by quantitative molecular orbital theory, but this is itself a subfield of chemistry as extensive as molecular spectroscopy.) The application of rudimentary molecular orbital theory in predicting the geometry and electronic structure of molecules was pioneered by A. D. Walsh in a series of influential papers published in the early 1950s. In honor of Walsh's work, particularly in deducing whether the geometry of a triatomic molecule is linear or bent, qualitative molecular orbital predictions can be obtained from a set of simple rules known as "Walsh's rules."

The dihydride triatomics with the structural formula AH_2 can be either linear ($D_{\infty h}$) or bent (C_{2v}). Consider the possible molecular orbitals of the H_2O molecule formed by simple linear combination of the valence 1*s* orbitals of H (Figure 10.1) and the 2*s* and 2*p* orbitals of O. The two 1*s* atomic hydrogen orbitals are not suitable wavefunctions since they do not have the correct C_{2v} symmetry, that is,

$$\hat{C}_2 1s_A = 1s_B. \tag{10.5}$$

The first step is to form symmetry-adapted linear combinations (SALCs) of atomic orbitals either by inspection or by the use of projection operators.

The four irreducible representations of the C_{2v} point group give rise to four projection operators:

$$\hat{P}^{A_1} = \sum \chi_R^{A_1} \hat{O}_R = \hat{O}_E + \hat{O}_{C_2} + \hat{O}_{\sigma_v} + \hat{O}_{\sigma_v'} \tag{10.6}$$

$$\hat{P}^{A_2} = \hat{O}_E + \hat{O}_{C_2} - \hat{O}_{\sigma_v} - \hat{O}_{\sigma_v'} \tag{10.7}$$

$$\hat{P}^{B_1} = \hat{O}_E - \hat{O}_{C_2} + \hat{O}_{\sigma_v} - \hat{O}_{\sigma_v'} \tag{10.8}$$

$$\hat{P}^{B_2} = \hat{O}_E - \hat{O}_{C_2} - \hat{O}_{\sigma_v} + \hat{O}_{\sigma_v'}. \tag{10.9}$$

The application of these four operators to the $1s_A$ function projects out two functions of the appropriate symmetry, namely

$$\phi_{a_1} = \frac{1s_A + 1s_B}{\sqrt{2}}, \tag{10.10}$$

$$\phi_{b_2} = \frac{1s_A - 1s_B}{\sqrt{2}}, \tag{10.11}$$

upon normalization (assuming orthonormal atomic basis functions). With the oxygen atom situated at the origin of the coordinate system, the symmetries of the oxygen valence atomic orbitals s, p_x, p_y, and p_z are a_1, b_1, b_2, and a_1, respectively. The hydrogen and oxygen symmetry orbitals are then combined by forming linear combinations to produce the molecular orbitals shown in Figure 10.2. The O 1*s* core orbital is a nonbonding core orbital with a_1 symmetry, while the O 2*s* orbital is a valence orbital and is the second orbital of a_1 symmetry. Since the O 1*s* a_1 orbital is lower in energy than O 2*s* a_1 it is labeled $1a_1$, while the O 2*s* orbital is labeled $2a_1$. The remaining orbitals are labeled in a similar fashion, in which the numbering starts with the lowest energy orbital for each symmetry type.

The electronic configuration of the ground state of H_2O is

$$(1a_1)^2(2a_1)^2(1b_2)^2(3a_1)^2(1b_1)^2$$

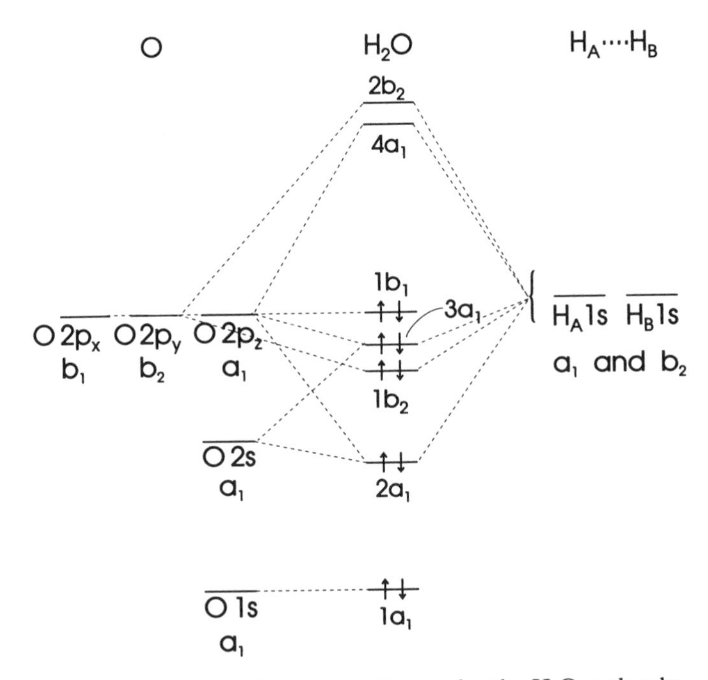

Figure 10.2. Molecular orbital diagram for the H_2O molecule.

and is denoted $\tilde{X}\ {}^1A_1$. The labeling scheme for states, $\tilde{X}, \tilde{A}, \tilde{B}, \tilde{a}, \tilde{b}$, and so on, parallels that of diatomic molecules, but a tilde is added in order to distinguish the $\tilde{A}$ or $\tilde{B}$ states from the A or B labels of the irreducible representations. The orbitals are labeled according to the appropriate irreducible representations using lowercase letters, but the overall symmetry of an electronic state is capitalized. The multiplicity $2S+1$ appears as a left superscript.

The highest occupied molecular orbital (HOMO) in H_2O is the nonbonding, out-of-plane O $2p_x$ orbital. The lowest unoccupied molecular orbital (LUMO) is the strongly antibonding $4a_1$ orbital (Figure 10.2). As is often the case, the first excited configuration of H_2O is obtained by transferring an electron from the HOMO to the LUMO, namely

$$(1a_1)^2(2a_1)^2(1b_2)^2(3a_1)^2(1b_1)^1(4a_1)^1.$$

This configuration gives rise to states with $b_1 \otimes a_1 = b_1$ orbital symmetry and a total electronic spin of $S=0$ or 1. Qualitative molecular orbital theory predicts that there are two states associated with this excited LUMO configuration, $\tilde{A}\ {}^1B_1$ and $\tilde{a}\ {}^3B_1$. Furthermore, Hund's first rule, which states that among all states with the same orbital symmetry arising from the same configuration, the one with the highest multiplicity is lowest in energy, predicts that $\tilde{a}\ {}^3B_1$ is lower in energy than $\tilde{A}\ {}^1B_1$.

The selection rules for electronic transitions are derived through the use of the transition moment integral

$$\int \psi^*_{\mathrm{el},f}\boldsymbol{\mu}\psi_{\mathrm{el},i}\,d\tau. \tag{10.12}$$

In the case of H_2O, the components of $\boldsymbol{\mu}$ have B_1, B_2, and A_1 symmetry, while the initial and final electronic states have A_1 and B_1 symmetry. Thus the $\tilde{A}\ {}^1B_1$–$\tilde{X}\ {}^1A_1$ transition should be a fully allowed, electric-dipole transition in which the electron is transferred out of a nonbonding b_1 orbital into an antibonding a_1 orbital. The H_2O $\tilde{A}\ {}^1B_1$–$\tilde{X}\ {}^1A_1$ transition is present in the 1860- to 1450-Å region of the vacuum ultraviolet.[1] The diffuse nature of this particular electronic transition, typical of electronic transitions observed in the vacuum ultraviolet (VUV) region of the spectrum, is due to molecular photodissociation.

An example of a linear AH_2 molecule is the hypothetical molecule BeH_2. BeH_2 is known to exist as a polymeric solid,[2] however, it is not yet known if BeH_2 can be prepared in the gas phase.

The symmetry-adapted linear combinations of the two hydrogen $1s$ orbitals for linear AH_2 are by inspection

$$\phi_{\sigma_g} = \frac{1s_A + 1s_B}{\sqrt{2}}, \tag{10.13}$$

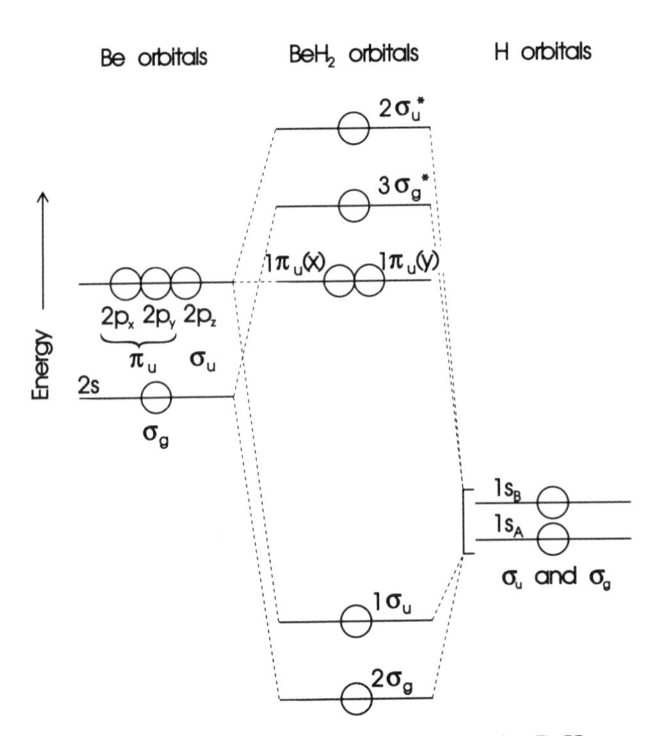

Figure 10.3. Molecular orbital diagram for BeH_2.

and

$$\phi_{\sigma_u} = \frac{1s_A - 1s_B}{\sqrt{2}}. \tag{10.14}$$

The molecular orbital diagram for BeH_2 is given in Figure 10.3. Both the Be 1*s* (not shown) and 2*s* orbitals have σ_g symmetry so that the valence σ_g orbital is labeled as $2\sigma_g$. The ground-state configuration for BeH_2 is $(1\sigma_g)^2(2\sigma_g)^2(1\sigma_u)^2$, which gives an $\tilde{X}\ {}^1\Sigma_g^+$ ground state.

Why is H_2O bent while BeH_2 is predicted to be linear? These predictions can be obtained either from detailed *ab initio* calculations or by constructing a Walsh molecular orbital diagram. A Walsh MO diagram is a correlation diagram based on the change in orbital overlap caused by a change in geometry. The Walsh MO diagram for the AH_2 case is given in Figure 10.4. The $2\sigma_g$ orbital decreases slightly in energy as the molecule bends due to the increased overlap between the hydrogen $1s_A$ and $1s_B$ orbitals, while the bending of the A—H bonds lifts the degeneracy of the nonbonding $1\pi_u$ orbital. The out-of-plane $1b_1$ component remains nonbonding while the in-plane $3a_1$ component becomes strongly bonding as the molecule bends.

The Walsh MO diagram predicts that all AH_2 molecules with four or less valence electrons will be linear in their ground states while all AH_2 molecules with five or more valence electrons will be bent in their ground states. This prediction is confirmed by experimental evidence.

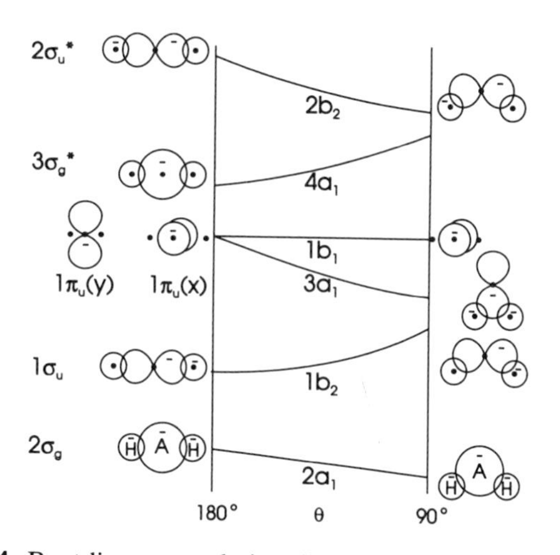

Figure 10.4. Bent-linear correlation diagram for AH_2-type molecules.

The NH_2 molecule with seven valence electrons is another AH_2 example. The ground-state configuration is

$$(1a_1)^2(2a_1)^2(1b_2)^2(3a_1)^2(1b_1)^1,$$

appropriate for a bent molecule. The ground state of NH_2 is predicted and observed to be $\tilde{X}\ ^2B_1$ with the unpaired electron in a nonbonding, out-of-plane p orbital. The first excited configuration is obtained by promoting a $3a_1$ electron to the $1b_1$ orbital, namely

$$(1a_1)^2(2a_1)^2(1b_2)^2(3a_1)^1(1b_1)^2,$$

which gives the $\tilde{A}\ ^2A_1$ state. The $\tilde{A}\ ^2A_1$–$\tilde{X}\ ^2B_1$ transition is electric-dipole allowed and occurs in the visible region of the spectrum. Notice that the NH_2 molecule in the $\tilde{X}\ ^2B_1$ state is predicted to be strongly bent (the observed angle is 103.4°, similar to the bond angle for H_2O), while the first excited state has only one electron in the $3a_1$ orbital so that the geometry in the $\tilde{A}\ ^2A_1$ state is predicted to be considerably less bent. Experimentally the bond angle for the $\tilde{A}\ ^2A_1$ state has been determined to be 144°; this result indicates that NH_2 in the $\tilde{A}\ ^2A_1$ state is closer to being linear than bent at a right angle.[1]

The symmetric nonhydride triatomic BAB is another common type of molecular species. The Walsh MO diagram (bent-linear correlation diagram) is constructed for AB_2 by using the same principles that were used for AH_2. The presence of valence s and p orbitals on all three centers, however, complicates the picture somewhat (see Figure 10.5).

As electrons are added to the AB_2 molecular orbitals, the molecule should be linear as long as the number of valence electrons does not exceed 16. The C_3 molecule with 12 valence electrons and the CO_2 molecule with 16 valence electrons are linear, in

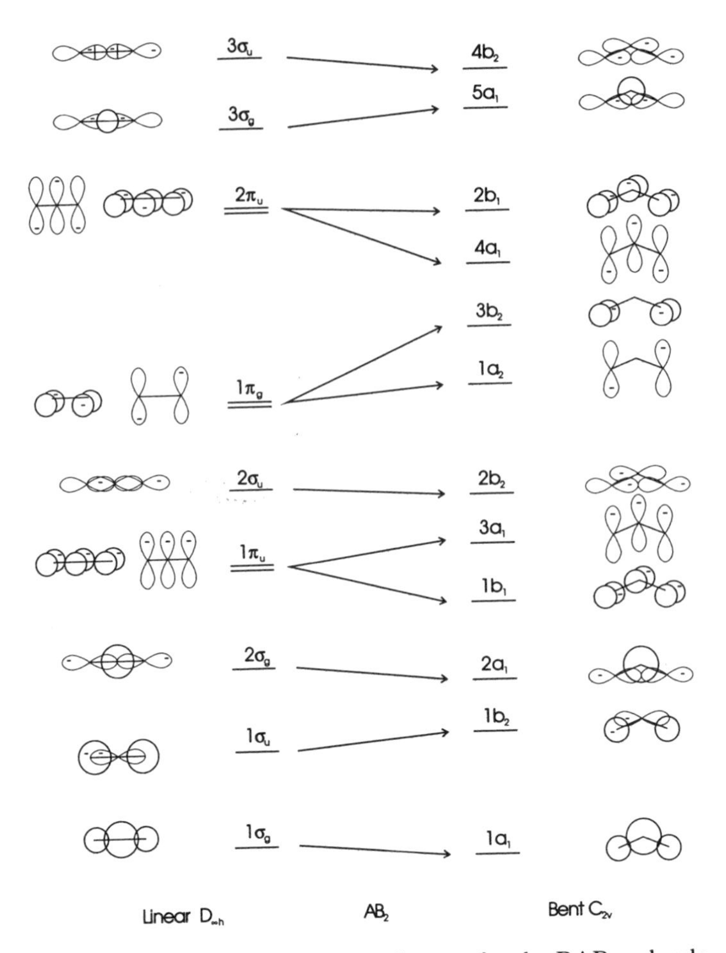

Figure 10.5. Bent-linear correlation diagram for the BAB molecule.

agreement with this prediction. AB_2 molecules with 17 or more electrons are predicted to be bent. Accordingly, NO_2 with 17 valence electrons and O_3 with 18 valence electrons are both bent.[1]

While predictions based on qualitative molecular orbital theory are in most cases reliable, there are some important exceptions. For example, the molecule SiC_2, which is isovalent with linear C_3, is found to be T-shaped.[3] Walsh's rules are only applicable to covalently bonded molecules—not to ionically bonded molecules. The fact that the bonding in SiC_2 is best represented by the ionic species, $Si^+C_2^-$, is an excellent example of why Walsh's rules can fail to predict the correct geometry.

Hückel Molecular Orbital Theory

Aromatic molecules such as benzene and naphthalene can be thought of as containing two types of valence electrons, σ and π. There are localized C—C and C—H σ-bonds, which hold the molecule together (σ-framework), and delocalized π-bonds formed from the out-of-plane carbon p_z orbitals. Hückel MO theory is based on simple principles involving the electronic properties of the π-molecular orbitals. Although Hückel theory is very simple, it manages to capture the essence of π-electronic structure in aromatic molecules.

Hückel MO theory is based on several approximations. The first is that the π-electrons can be treated separately from the σ-electrons in an aromatic molecule. The assumption of σ–π separability is equivalent to assuming that the electronic Hamiltonian can be separated into two parts as

$$\hat{H}_{\text{el}} = \hat{H}_\pi + \hat{H}_\sigma, \tag{10.15}$$

with wavefunctions correspondingly factored as $\psi = \psi_\pi \psi_\sigma$. Similarly, the π-electron Hamiltonian can be broken into a sum of separate one-electron effective Hamiltonians (one for each π-electron), namely

$$\hat{H}_\pi = \sum_{k=1}^{n_\pi} \hat{H}^{\text{eff}}(k). \tag{10.16}$$

The molecular orbitals are expressed as linear combinations of the atomic $p_z(=f_i)$ orbitals, viz.,

$$\phi = \sum_{j=1}^{n_\pi} c_j f_j. \tag{10.17}$$

Substitution of equation (10.17) into the electronic Schrödinger equation then yields the set of homogeneous linear equations

$$\sum_{j=1}^{n_\pi} (H_{ij}^{\text{eff}} - S_{ij}E)c_j = 0. \tag{10.18}$$

The variational principle can then be applied to determine optimal values for the set of coefficients c_j. The variational principle states that the optimum coefficients for an approximate ground-state electronic wavefunction can be obtained by minimizing the electronic energy of the system. In matrix notation, equation (10.18) becomes

$$(\mathbf{H}^{\text{eff}} - \mathbf{S}E)\mathbf{c} = 0 \tag{10.19}$$

in which $\mathbf{S}$ is the overlap matrix and

$$H_{ij}^{\text{eff}} = \langle f_i | \hat{H} | f_j \rangle, \qquad S_{ij} = \langle f_i | f_j \rangle.$$

A nontrivial solution to the secular equation

$$|\mathbf{H} - \mathbf{S}E| = 0 \tag{10.20}$$

consists of n_π eigenvalues E_n with corresponding eigenfunctions

$$\phi_n = \sum_{j=1}^{n_\pi} c_{nj} f_j, \qquad n = 1, \ldots, n_\pi. \tag{10.21}$$

In Hückel theory the integrals H_{ij}^{eff} and S_{ij} are not determined by *ab initio* methods, but instead are determined empirically. Additional approximations are invoked in order to reduce the number of unknown parameters to just two, the integrals α and β, defined as

$$H_{ii}^{\text{eff}} = \int f_i \hat{H}^{\text{eff}} f_i \, d\tau = \alpha, \tag{10.22}$$

and

$$H_{ij}^{\text{eff}} = \int f_i \hat{H}^{\text{eff}} f_j \, d\tau = \beta \quad \text{or} \quad 0. \tag{10.23}$$

If the two carbon atoms are adjacent, then $H_{ij} = \beta$, while if the two carbon atoms are not adjacent, then $H_{ij} = 0$. Furthermore, the overlap matrix is assumed to be the unit matrix, $S_{ij} = \delta_{ij}$. With these assumptions the secular equation takes a simple form in terms of the two parameters α (the Coulomb integral) and β (the resonance or bond integral).

To illustrate Hückel molecular orbital theory, consider the π-electronic structure of the *trans*-butadiene molecule of C_{2h} symmetry (Figure 10.6). The Hückel approximations give rise to the secular equation

$$\begin{vmatrix} \alpha - E & \beta & 0 & 0 \\ \beta & \alpha - E & \beta & 0 \\ 0 & \beta & \alpha - E & \beta \\ 0 & 0 & \beta & \alpha - E \end{vmatrix} = 0 \tag{10.24}$$

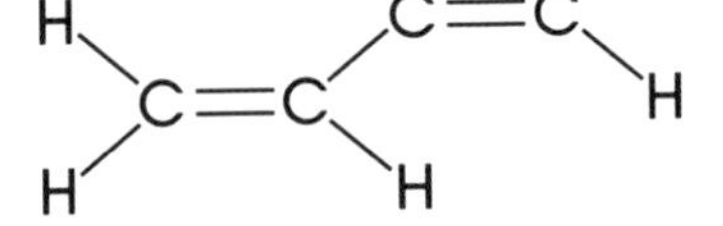

Figure 10.6. The butadiene molecule.

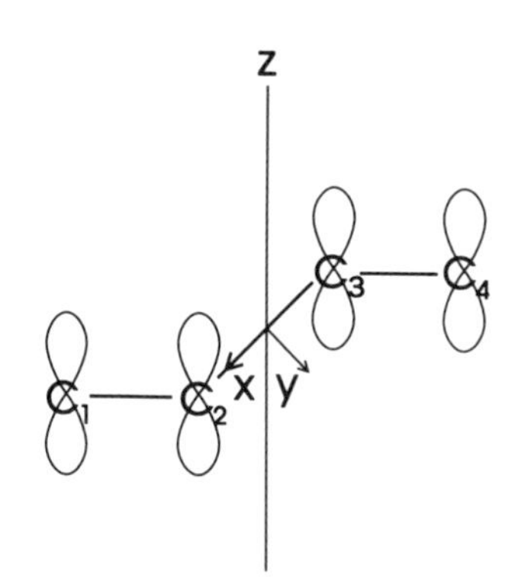

Figure 10.7. Coordinate system for butadiene.

for the four molecular orbitals constructed from the four p_z orbitals, one on each carbon atom (Figure 10.7).

As usual, the application of molecular symmetry simplifies the problem. The four p_z orbitals form a 4-dimensional reducible representation of the C_{2h} point group with characters

	$\hat{E}$	$\hat{C}_2$	$\hat{\imath}$	$\hat{\sigma}_h$
χ^{red}	4	0	0	−4

which can be reduced as $\Gamma^{\text{red}} = 2\Gamma^{a_u} \oplus 2\Gamma^{b_g}$. The symmetry-adapted linear combinations of atomic orbitals can be found by inspection, or by the use of projection operators, to be

$$g_1 = \frac{p_{z1} + p_{z4}}{\sqrt{2}}, \quad (a_u) \tag{10.25a}$$

$$g_2 = \frac{p_{z2} + p_{z3}}{\sqrt{2}}, \quad (a_u) \tag{10.25b}$$

$$g_3 = \frac{p_{z1} - p_{z4}}{\sqrt{2}}, \quad (b_g) \tag{10.25c}$$

$$g_4 = \frac{p_{z2} - p_{z3}}{\sqrt{2}}, \quad (b_g). \tag{10.25d}$$

Using these symmetry-adapted basis functions $\{g_i\}$ rather than $\{p_{zi}\}$ gives matrix elements

$$H'_{11} = \left\langle \frac{p_{z1} + p_{z4}}{\sqrt{2}} \middle| \hat{H} \middle| \frac{p_{z1} + p_{z4}}{\sqrt{2}} \right\rangle \tag{10.26}$$

$$= \alpha$$

$$H'_{22} = \left\langle \frac{p_{z2} + p_{z3}}{\sqrt{2}} \right| \hat{H} \left| \frac{p_{z2} + p_{z3}}{\sqrt{2}} \right\rangle = \alpha + \beta, \tag{10.27}$$

and so on. There is also a new secular determinant,

$$\begin{array}{c} a_u \\ a_u \\ b_g \\ b_g \end{array} \begin{vmatrix} \alpha - E & \beta & 0 & 0 \\ \beta & \alpha + \beta - E & 0 & 0 \\ 0 & 0 & \alpha - E & \beta \\ 0 & 0 & \beta & \alpha - \beta - E \end{vmatrix} = 0. \tag{10.28}$$

Notice that the secular determinant in equation (10.28) has been partitioned into two smaller symmetry blocks because all matrix elements between functions of different symmetry are zero. The four solutions of the secular equation are

$$E_1(1a_u) = \alpha + \frac{(1+\sqrt{5})\beta}{2} = \alpha + 1.618\beta, \tag{10.29a}$$

$$E_2(1b_g) = \alpha + \frac{(-1+\sqrt{5})\beta}{2} = \alpha + 0.618\beta, \tag{10.29b}$$

$$E_3(2a_u) = \alpha + \frac{(1-\sqrt{5})\beta}{2} = \alpha - 0.618\beta, \tag{10.29c}$$

$$E_4(2b_g) = \alpha + \frac{(-1-\sqrt{5})\beta}{2} = \alpha - 1.618\beta, \tag{10.29d}$$

while the associated normalized wavefunctions are

$$\phi_1(1a_u) = 0.3718(p_{z1} + p_{z4}) + 0.6015(p_{z2} + p_{z3}), \tag{10.30a}$$

$$\phi_2(1b_g) = 0.6015(p_{z1} - p_{z4}) + 0.3718(p_{z2} - p_{z3}), \tag{10.30b}$$

$$\phi_3(2a_u) = 0.6015(p_{z1} + p_{z4}) - 0.3718(p_{z2} + p_{z3}), \tag{10.30c}$$

$$\phi_4(2b_g) = 0.3718(p_{z1} - p_{z4}) - 0.6015(p_{z2} - p_{z3}). \tag{10.30d}$$

The atomic orbital composition of the wavefunctions is shown in Figure 10.8.

The lowest energy π-electron configuration of butadiene is $(1a_u)^2(1b_g)^2$ with an $\tilde{X}\ ^1A_g$ ground state. The promotion of an electron from the $1b_g$ HOMO to the $2a_u$ LUMO gives the $(1a_u)^2(1b_g)^1(1a_u)^1$ configuration and the $\tilde{A}\ ^1B_u$ and $\tilde{a}\ ^3B_u$ states. The very diffuse $\tilde{A}\ ^1B_u$–$X\ ^1A_g$ electronic transition has been observed at ~217 nm.[1] Simple Hückel theory predicts the HOMO to LUMO transition to be at $2(0.618)\beta$, giving a value of $-37{,}300\ \text{cm}^{-1}$ for β. Note that both Coulomb integrals (α) and resonance

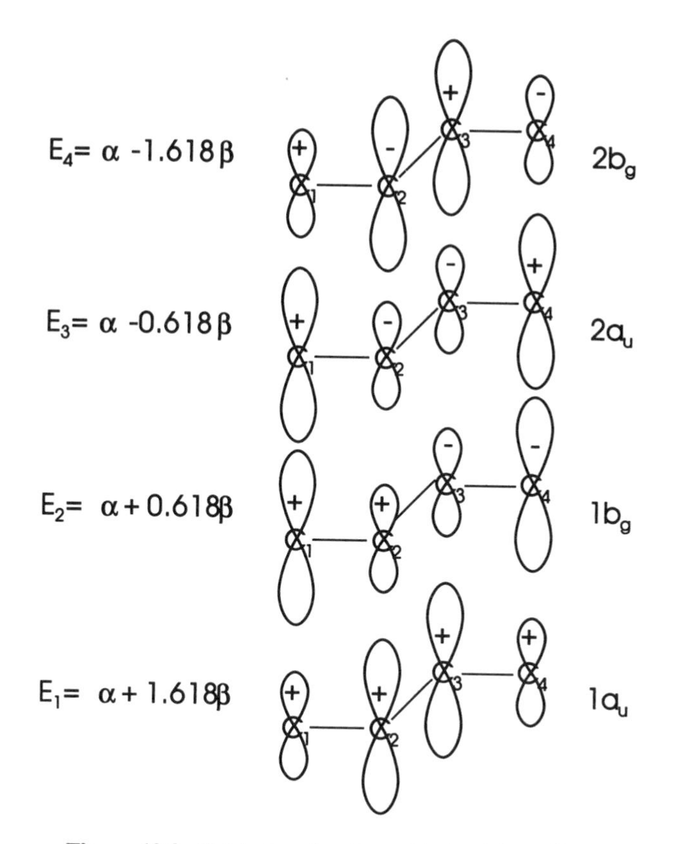

Figure 10.8. Hückel molecular orbitals of butadiene.

integrals (β) are negative numbers due to the choice of the zero of energy (cf. the hydrogen atom, Chapter 5).

In general the electronic Hamiltonian for a linear polyene with n atoms is a symmetric tridiagonal matrix with elements along the diagonal equal to α and sub- or superdiagonal elements equal to β. For this special form of the tridiagonal matrix, the solution to the secular equation is well known and given by[4]

$$E_i = \alpha - 2\beta \cos\left(\frac{\pi i}{n+1}\right), \qquad i = 1, 2, \ldots, n, \tag{10.31}$$

and

$$\phi_i = \left(\frac{2}{n+1}\right)^{1/2} \sum_{j=1}^{n} p_{zj} \sin\left(\frac{i\pi j}{n+1}\right). \tag{10.32}$$

The cyclic π-electron molecules have a different secular equation in Hückel theory.

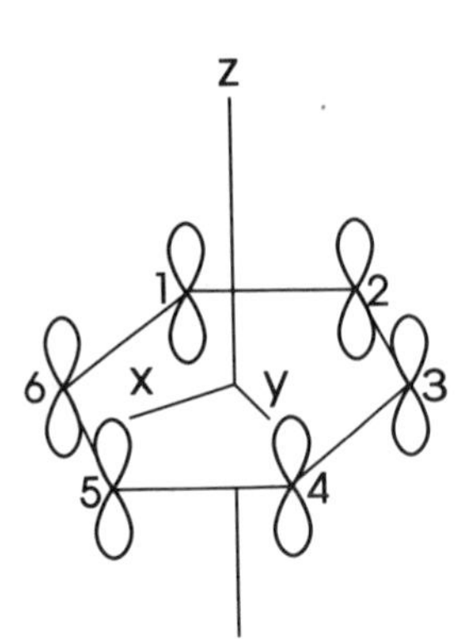

Figure 10.9. Coordinate system for the benzene molecule.

For example, the secular equation for benzene is

$$\begin{vmatrix} \alpha - E & \beta & 0 & 0 & 0 & \beta \\ \beta & \alpha - E & \beta & 0 & 0 & 0 \\ 0 & \beta & \alpha - E & \beta & 0 & 0 \\ 0 & 0 & \beta & \alpha - E & \beta & 0 \\ 0 & 0 & 0 & \beta & \alpha - E & \beta \\ \beta & 0 & 0 & 0 & \beta & \alpha - E \end{vmatrix} = 0. \quad (10.33)$$

The nonzero elements $H_{16} = H_{61} = \beta$ must be added in order to satisfy the cyclic boundary condition that the last carbon atom in the loop around the ring must join to the first carbon atom in the loop.

The application of molecular symmetry is again helpful in simplifying the secular equation in (10.33). The benzene molecule has D_{6h} symmetry (Figure 10.9), but it is simpler to use the rotational subgroup D_6 ($D_{6h} = D_6 \otimes C_i$) for the problem. The appropriate g and u labels for the wavefunctions can be determined either by inspection or by using the symmetry operators of the D_{6h} point group. The six p_z orbitals form a reducible representation with characters

	$\hat{E}$	$\hat{C}_6$	$\hat{C}_3$	$\hat{C}_2$	$\hat{C}_2'$	$\hat{C}_2''$
χ^{red}	6	0	0	0	−2	0

in which the C_6, C_3, and C_2 axes lie along the z-axis, while the C_2' and C_2'' axes lie in the molecular plane. The three C_2' axes bisect carbon atoms, while the three C_2'' axes bisect the carbon–carbon bonds. These characters can be reduced to give

$$\Gamma^{\text{red}} = \Gamma^{a_2} \oplus \Gamma^{b_2} \oplus \Gamma^{e_1} \oplus \Gamma^{e_2}. \quad (10.34)$$

Since the p_z orbitals change sign on application of the $\hat{\sigma}_h$ operator, inspection of the

D_{6h} character table (Appendix B) for the presence of a negative sign for this operation gives

$$\Gamma^{\text{red}} = \Gamma^{a_{2u}} \oplus \Gamma^{b_{2g}} \oplus \Gamma^{e_{1g}} \oplus \Gamma^{e_{2u}}. \tag{10.35}$$

Symmetry-adapted linear combinations of the six p_z orbitals are determined by the use of projection operators, together with the orthogonality condition in the case of the degenerate e_1 and e_2 functions. The six symmeterized functions are

$$f_1 = \frac{p_{z1} + p_{z2} + p_{z3} + p_{z4} + p_{z5} + p_{z6}}{\sqrt{6}} \qquad (a_2) \tag{10.36a}$$

$$f_2 = \frac{p_{z1} - p_{z2} + p_{z3} - p_{z4} + p_{z5} - p_{z6}}{\sqrt{6}} \qquad (b_2) \tag{10.36b}$$

$$f_3 = \frac{p_{z1} - p_{z2} - 2p_{z3} - p_{z4} + p_{z5} + 2p_{z6}}{2\sqrt{3}} \qquad (e_1) \tag{10.36c}$$

$$f_4 = \frac{p_{z1} + p_{z2} - p_{z4} - p_{z5}}{2} \qquad (e_1) \tag{10.36d}$$

$$f_5 = \frac{p_{z1} + p_{z2} - 2p_{z3} + p_{z4} + p_{z5} - 2p_{z6}}{2\sqrt{3}} \qquad (e_2) \tag{10.36e}$$

$$f_6 = \frac{p_{z1} - p_{z2} + p_{z4} - p_{z5}}{2} \qquad (e_2). \tag{10.36f}$$

Constructing the secular determinant in this basis set gives

$$\begin{vmatrix} \alpha + 2\beta - E & 0 & 0 & 0 & 0 & 0 \\ 0 & \alpha - 2\beta - E & 0 & 0 & 0 & 0 \\ 0 & 0 & \alpha + \beta - E & 0 & 0 & 0 \\ 0 & 0 & 0 & \alpha + \beta - E & 0 & 0 \\ 0 & 0 & 0 & 0 & \alpha - \beta - E & 0 \\ 0 & 0 & 0 & 0 & 0 & \alpha - \beta - E \end{vmatrix} = 0, \tag{10.37}$$

which is already diagonal. The energy levels are at $\alpha + 2\beta$, $\alpha - 2\beta$, $\alpha + \beta$ (doubly degenerate), and $\alpha - \beta$ (doubly degenerate) as shown in Figure 10.10.

The π-electron ground-state configuration of benzene is $(a_{2u})^2(e_{1g})^4$ with $\tilde{X}\ {}^1A_{1g}$

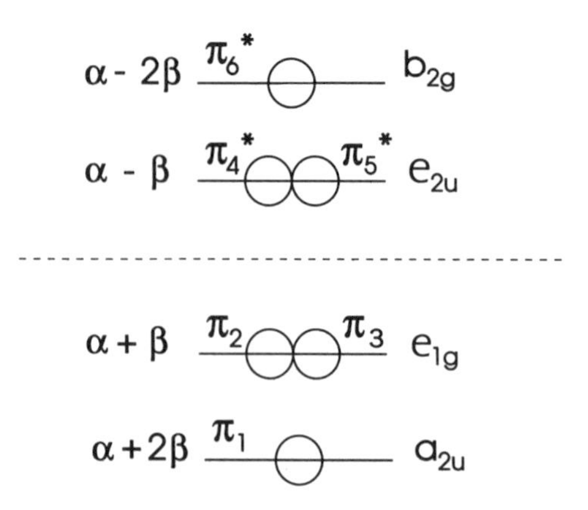

Figure 10.10. Energy-level diagram for the π-electrons of benzene.

being the ground state. The lowest energy excitation promotes one electron out of the e_{1g} orbital into the e_{2u} orbital, which results in the configuration $(a_{2u})^2(e_{1g})^3(e_{2u})^1$. For this configuration the direct product $e_{1g} \otimes e_{2u}$, which reduces to $b_{1u} \oplus b_{2u} \oplus e_{1u}$, gives rise to the states $^1B_{1u}$, $^3B_{1u}$, $^1B_{2u}$, $^3B_{2u}$, $^1E_{1u}$, and $^3E_{1u}$. Since μ_z has A_u symmetry and (μ_x, μ_y) have E_{1u} symmetry, however, only the $^1E_{1u}$–$^1A_{1g}$ transition of benzene is allowed for the D_{6h} point group.

The famous ultraviolet transition of benzene (see Figure 10.12 in the next section) at 260 nm turns out to be the forbidden $\tilde{A}\ ^1B_{1u}$–$\tilde{X}\ ^1A_{1g}$ transition,[1] which becomes allowed as a result of vibronic coupling (see the section in this chapter on vibronic coupling).

Vibrational Structure of Electronic Transitions

Within the Born–Oppenheimer approximation, the separation of vibrational and electronic motion leads to the concept of associating electronic states with potential energy surfaces. For a diatomic molecule, the potential energy function $V(r)$ is a function of a single variable, the internuclear separation r. For a polyatomic molecule the potential energy function $V(Q_i)$ is a function of $3N - 6$ (or 5) internal coordinates, usually expressed in terms of normal modes. The simple one-dimensional diatomic potential energy curve is replaced by a multidimensional potential energy surface for each polyatomic electronic state. Shown in Figure 10.11 is a simple example of a polyatomic potential energy surface for the He(CO) van der Waals molecule.

The solution of the Schrödinger equation for nuclear motion on each potential energy surface of a polyatomic molecule provides the corresponding vibrational frequencies and anharmonicities for each electronic state, namely[1]

$$G(v_i) = \sum_r \omega_r\left(v_r + \frac{d_r}{2}\right) + \sum_{r,s>r} x_{rs}\left(v_r + \frac{d_r}{2}\right)\left(v_s + \frac{d_s}{2}\right) + \sum_{t,t'>t} g_{tt'} l_t l_{t'}. \qquad (10.38)$$

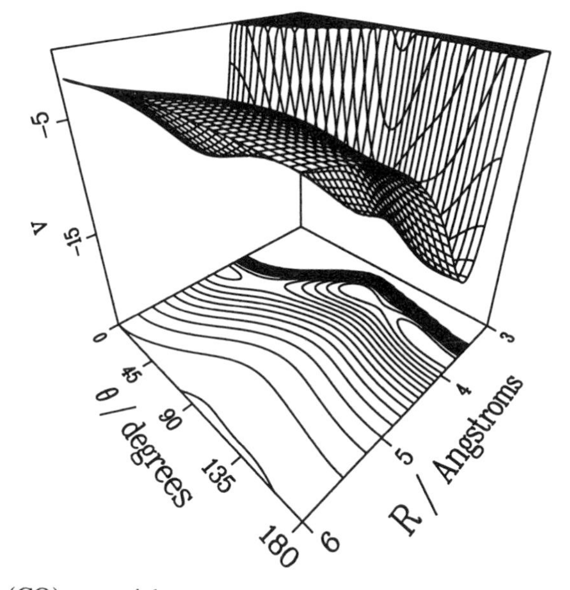

Figure 10.11. The He(CO) potential energy surface as a function of the radial distance R of the He atom from the center of mass of the CO molecule, and the angular position θ of the He atom relative to the CO internuclear axis. The origin at 0° represents a colinear HeCO geometry.

In this equation the ω_r are the harmonic vibrational frequencies each with a corresponding degeneracy d_r, the x_{rs} and $g_{tt'}$ are anharmonic corrections, and the index t refers to degenerate modes with vibrational angular momentum l_t. The polyatomic vibrational equation (10.38) may be compared with the much simpler vibrational energy-level expression for a diatomic molecule, namely

$$G(v) = \omega_e(v + \tfrac{1}{2}) - \omega_e x_e(v + \tfrac{1}{2})^2. \tag{10.39}$$

A vibronic transition frequency is given by the difference between two vibronic term values, namely

$$\nu = T_e + G'(v_1', v_2', \ldots) - G''(v_1'', v_2'', \ldots), \tag{10.40}$$

in which T_e is the minimum potential energy difference between the two states. A vibronic transition can be specified by noting the electronic transition and the vibrational quantum numbers for each of the states, for example, $\tilde{A}$–$\tilde{X}$ $(v_1'v_2'\cdots) - (v_1''v_2''\cdots)$.

A convenient shorthand notation for a vibronic transition of the $\tilde{A}-\tilde{X}$ electronic transition is denoted as

$$\tilde{A}-\tilde{X} \qquad 1^{v_1'}_{v_1''}\, 2^{v_2'}_{v_2''} \cdots .$$

For example, the formaldehyde transition from $\tilde{X}\ {}^1A_1$ ($v_i = 0$) to $\tilde{A}\ {}^1A_2$ ($v_2 = 2$, $v_4 = 1$) can be written either as $\tilde{A}\ {}^1A_2-\tilde{X}\ {}^1A_1$ (020100)–(000000) or as $\tilde{A}\ {}^1A_2-\tilde{X}\ {}^1A_1$, $2^2_0 4^1_0$. Except for triatomic molecules, the second notation is preferable and is now often used for infrared as well as vibronic transitions of polyatomic molecules.

The vibrational selection rules for an allowed electronic transition are determined from the Franck–Condon principle. The intensity of a vibronic transition is proportional to the square of the transition moment integral,

$$\begin{aligned}
\mathbf{M}_{e'v'e''v''} &= \int \psi^*_{e'v'} \boldsymbol{\mu} \psi_{e''v''}\, d\tau_{ev} \\
&= \int \psi^*_{e'} \psi^*_{v'} \boldsymbol{\mu} \psi_{e''} \psi_{v''}\, d\tau_{ev} \\
&= \int \psi^*_{e'} \boldsymbol{\mu} \psi_{e''}\, d\tau_{\text{el}} \int \psi^*_{v'} \psi_{v''}\, d\tau_v \\
&= \mathbf{M}_{e'e''} \int \psi^*_{v_1'} \psi_{v_1''}\, dQ_1 \int \psi^*_{v_2'} \psi_{v_2''}\, dQ_2 \cdots .
\end{aligned} \tag{10.41}$$

Within the realm of the Born–Oppenheimer and normal mode approximations the transition moment integral is comprised of an electronic transition dipole moment $\mathbf{M}_{e'e''}$ and a product of $3N - 6$ (or 5) overlap integrals. For totally symmetric vibrations the selection rule on v is therefore

$$\Delta v_i = 0, \pm 1, \pm 2, \ldots,$$

with the intensity determined by the Franck–Condon factor

$$\left| \int \psi^*_{v_i'} \psi_{v_i''}\, dQ_i \right|^2 \tag{10.42}$$

for that mode. For non-totally symmetric vibrations the Franck–Condon factor vanishes for

$$\Delta v_i = \pm 1, \pm 3, \pm 5, \ldots$$

because the product $\Gamma^{\psi_{v'}} \otimes \Gamma^{\psi_{v''}}$ does not contain the totally symmetric irreducible representation. For an allowed electronic transition, the nonsymmetric vibrational modes obey the selection rule

$$\Delta v_i = \pm 2, \pm 4, \pm 6, \ldots .$$

Vibronic Coupling: The Herzberg–Teller Effect

Often nonsymmetric vibrational transitions occur in an electronic transition with the selection rule

$$\Delta v = \pm 1, \pm 3, \pm 5 \ldots ,$$

although, as discussed earlier, this is forbidden for electric dipole-allowed electronic transitions. The fact that these transitions tend to be relatively weak is indicative of an electronically forbidden character. The $\tilde{A}\ {}^1B_{2u}$–$\tilde{X}\ {}^1A_{1g}$ transition of benzene at 260 nm is a classic example. The $\tilde{A}$–$\tilde{X}$ transition is forbidden (x, y, and z belong to E_{1u} and A_{2u} irreducible representations), but has been observed with moderate intensity in the near UV region (Figure 10.12). The 0–0 origin band is not observed, but $\Delta v_6 = 1$ transitions (v_6 has e_{2g} symmetry) are prominent. In addition a long progression in v_1, the ring breathing mode, is present. (Note that, by convention, the Wilson numbering

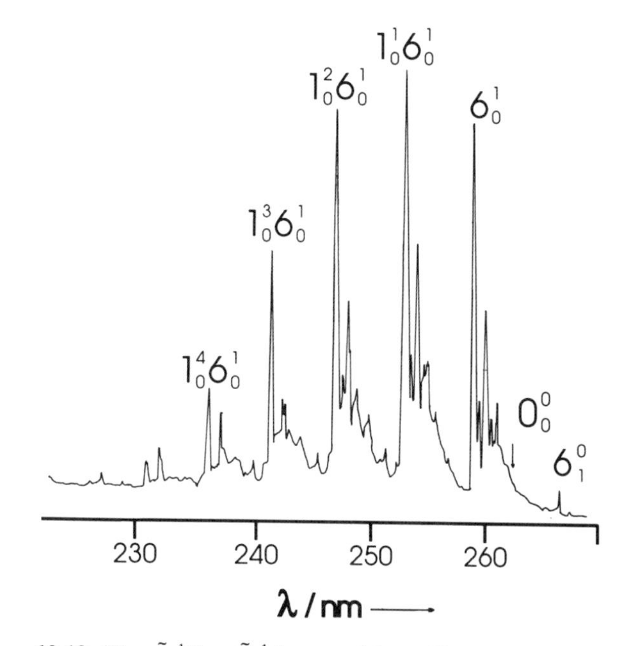

Figure 10.12. The $\tilde{A}\ {}^1B_{2u}$–$\tilde{X}\ {}^1A_{1g}$ transition of benzene near 260 nm.

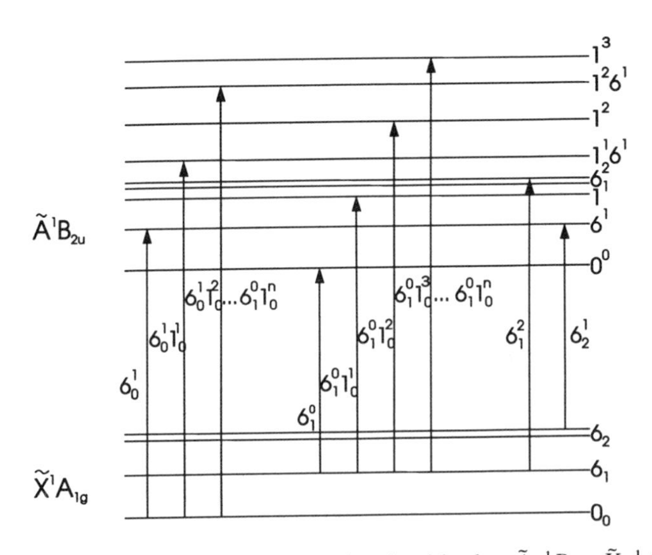

Figure 10.13. Possible vibronic transitions associated with the $\tilde{A}\ ^1B_{2u}$–$\tilde{X}\ ^1A_{1g}$ transition of benzene.

scheme[5,6] for the vibrational modes of benzene is commonly used rather than the Herzberg numbering scheme recommended by Mulliken[7]). The ν_1 mode is totally symmetric (a_{1g}), so there are no symmetry restrictions on Δv_1. The progression in ν_1 (Figure 10.12) means that the benzene ring has D_{6h} symmetry in both ground and $\tilde{A}$ states, but that the size of the ring is different in the two states. A detailed rotational analysis confirms that the C–C bond length increases by +0.0353 Å from 1.397 Å in the $\tilde{X}\ ^1A_{1g}$ ground state.[8] The ν_1 mode, however, always occurs in combination with an odd number of non-totally symmetric vibrations, usually ν_6 (Figures 10.12 and 10.13).

The $\tilde{A}$–$\tilde{X}$ transition is a forbidden electronic transition that becomes allowed by vibronic coupling, as first explained by Herzberg and Teller.[1,9] In this case, the total vibronic symmetry ($\Gamma^{\text{el}} \otimes \Gamma^{\text{vib}} = \Gamma^{\text{vibronic}}$) must be examined. In the case of the benzene $\tilde{A}\ ^1B_{2u}$–$\tilde{X}\ ^1A_{1g}$ 6_0^1 transition, this requires that the transition moment integral

$$\int \psi^*_{\text{vibronic}'} \boldsymbol{\mu} \psi_{\text{vibronic}''}\, d\tau \qquad (10.43)$$

be considered.

The vibronic symmetry of the ground state is A_{1g}, while the vibronic symmetry of the excited state is $B_{2u} \otimes e_{2g} = E_{1u}$. Since μ_x and μ_y have E_{1u} symmetry, the transition moment integral is now nonzero. Provided that the vibrational and electronic degrees of freedom are mixed and cannot be factored, the benzene $\tilde{A}$–$\tilde{X}$ transition becomes vibronically allowed. The inclusion of a single nonsymmetric quantum of vibration changes the overall symmetry of the excited state and permits a transition.

The intensity of a vibronic transition depends on the degree of mixing of the vibrational and electronic wavefunctions. As long as this mixing is not too extensive, it can be estimated by perturbation theory. Consider a set of zeroth-order electronic and harmonic vibrational wavefunctions without the effects of vibronic coupling. The Schrödinger equation for a fixed equilibrium configuration is

$$\hat{H}_e^0 \psi_e^0 = E_e \psi_e^0. \tag{10.44}$$

The electronic Hamiltonian depends parametrically on the value of the vibrational coordinates. Vibronic coupling is derived by expanding the electronic Hamiltonian in the power series

$$\hat{H}_e = \hat{H}_e^0 + \sum_i \left(\frac{\partial H_e}{\partial Q_i}\right)_{Q_i=0} Q_i + \cdots . \tag{10.45}$$

Truncating the expansion to terms linear in Q_i gives the perturbation operator terms

$$\hat{H}' = \sum_i \left(\frac{\partial H_e}{\partial Q_i}\right)_{Q_i=0} Q_i. \tag{10.46}$$

The excited-state wavefunction ψ_f^0 becomes mixed with other zeroth-order electronic states through the perturbation term $\hat{H}'$,

$$\psi_{e'} = \psi_f^0 + \sum_{k \neq f} c_k \psi_k^0 \tag{10.47}$$

with

$$c_k = \frac{\langle \psi_k^0 | \hat{H}' | \psi_f^0 \rangle}{E_f^0 - E_k^0}. \tag{10.48}$$

The degree of mixing is determined by the magnitude of the vibronic coupling matrix element $\langle \psi_k^0 | \hat{H}' | \psi_f^0 \rangle$ and the separation $(E_f^0 - E_k^0)$ between the interacting states. The electronic transition moment integral becomes

$$\begin{aligned} \mathbf{M}_{e'e''} &= \int \psi_{e'}^* \boldsymbol{\mu} \psi_{e''} \, d\tau \\ &= \int (\psi_f^*)^0 \boldsymbol{\mu} \psi_{e''}^0 \, d\tau_e + \sum c_k \int (\psi_k^*)^0 \boldsymbol{\mu} \psi_{e''}^0 \, d\tau_e \end{aligned} \tag{10.49}$$

For the benzene $\tilde{A}\ {}^1B_{2u}$–$\tilde{X}\ {}^1A_{1g}$ transition the first term in equation (10.49) vanishes and the sum is dominated by interaction with the nearby $\tilde{C}\ {}^1E_{1u}$ state. The $\tilde{C}\ {}^1E_{1u}$–$\tilde{X}\ {}^1A_{1g}$ electronic transition is fully allowed. The mode ν_6 of e_{2g} symmetry mixes with the

$\tilde{A}\ ^1B_{2u}$ electronic state to give a vibronic state of E_{1u} symmetry. The $\tilde{A}\ ^1B_{2u}v_6 = 1$ vibronic state mixes with the $\tilde{C}\ ^1E_{1u}$ electronic state of the same symmetry. The $\tilde{A}\ ^1B_{2u}$–$\tilde{X}\ ^1A_{1g}$ electronic transition[1,8] is observed because of intensity "borrowing" or "stealing" from the nearby strong $\tilde{C}\ ^1E_{1u}$–$\tilde{X}\ ^1A_{1g}$ transition through vibronic coupling involving mainly ν_6'.

The interaction matrix element between the two excited electronic states determines the magnitude of the Herzberg–Teller effect. This matrix element can be expressed as

$$\langle\psi_k^0|\,\hat{H}'\,|\psi_f^0\rangle = \langle\psi_k^0|\left(\frac{\partial H}{\partial Q_i}\right)_{Q_i=0} Q_i\,|\psi_f^0\rangle, \tag{10.50}$$

and is nonzero only if the vibronic symmetry $\Gamma^{\psi_f^0}\otimes\Gamma^{Q_i}$ of the $\tilde{A}$ state matches the electronic symmetry of the $\tilde{C}$ state, $\Gamma^{\psi_k^0}$. In this electronic integral $(\partial H/\partial Q_i)_{Q_i=0}$ has the same symmetry as the normal mode Q_i since H itself is totally symmetric. This leads to the selection rule

$$\Delta v_i = \pm 1(\pm 3, \pm 5, \ldots)$$

for the $\tilde{A}$–$\tilde{X}$ vibronically active mode(s).

Jahn–Teller Effect

The Jahn–Teller effect[1,9,10] also violates the selection rule $\Delta v_i = \pm 2, \pm 4, \pm 6, \ldots$ for nonsymmetric vibrations in an electronic transition. This effect in molecules is a consequence of the much celebrated Jahn–Teller theorem. Jahn and Teller[10] proved that any nonlinear molecule in an orbitally degenerate electronic state will always distort in such a way as to lower the symmetry and remove the degeneracy. This is a perfectly general result, but it conveys no information as to the size of the distortion. Both infinitesimal and massive distortions are possible and both are consistent with the Jahn–Teller theorem.

For nonsinglet molecules spin-orbit coupling competes with the Jahn–Teller effect, since spin-orbit splitting also lifts orbital degeneracies independently of the Jahn–Teller effect. The Jahn–Teller theorem is also inapplicable to linear molecules for which undistorted, orbitally degenerate states ($\Pi, \Delta, \Phi, \ldots$) are possible.

The Jahn–Teller effect is a consequence of Born–Oppenheimer breakdown, and it is convenient to use the same approach that was used to describe vibronic coupling. There is a zeroth-order Born–Oppenheimer electronic Hamiltonian and a perturbation operator,

$$\hat{H}' = \sum_i \left(\frac{\partial H_e}{\partial Q_i}\right)_{Q_i=0} Q_i, \tag{10.51}$$

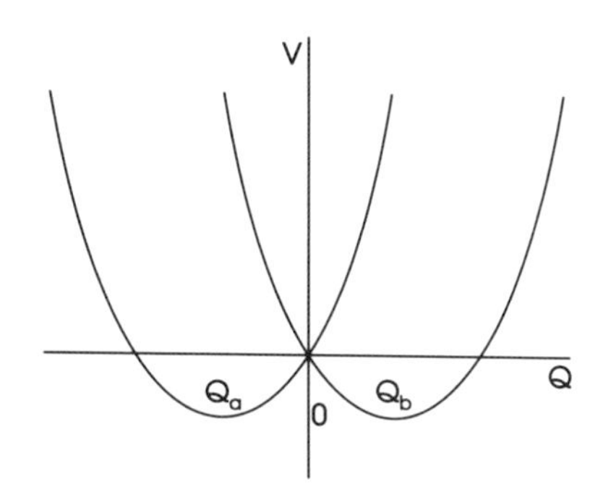

Figure 10.14. A Jahn–Teller distortion along a doubly degenerate vibrational coordinate lifts the degeneracy and stabilizes the molecule.

responsible for the mixing of vibrational and electronic wavefunctions. If ψ_a and ψ_b are linearly independent, orbitally degenerate electronic wavefunctions, then $\hat{H}'$ will lift their degeneracy if there is a nonzero matrix element between them, that is,

$$\int \psi_a^* \hat{H}' \psi_b \, d\tau_e = \sum_i \int \psi_a^* \left(\frac{\partial H_e}{\partial Q_i} \right)_{Q_i=0} Q_i \psi_b \, d\tau \neq 0. \tag{10.52}$$

This electronic integral will be nonzero only if $\Gamma^{Q_i} \otimes (\Gamma^{\psi_a} \otimes \Gamma^{\psi_b})_{\text{sym}}$ contains the totally symmetric irreducible representation because $(\partial H_e / \partial Q_i)_{Q_i=0}$ has the same symmetry as Q_i. Since ψ_a and ψ_b belong to the same irreducible representation, the symmetric product is used to ensure that the Pauli exclusion principle is not violated. Jahn and Teller[10] exhaustively considered all degenerate states of all point groups with respect to all non-totally symmetric vibrational distortions. In all cases, except for linear molecules, a vibrational distortion (Q_i) could be found that broke the symmetry of the molecule. As shown schematically in Figure 10.14, a distortion from the symmetric configuration can be found that lowers the total energy of the molecule. For example, the hexafluorobenzene cation $C_6F_6^+$ has an $\tilde{X}\,^2E_{1g}$ ground state with a small spin-orbit splitting. The $\tilde{X}\,^2E_{1g}$ state of $C_6F_6^+$ distorts, lowering the symmetry from D_{6h} and lifting the degeneracy in the E_{1g} electronic state. The size of the Jahn–Teller distortion can be estimated from the vibronic activity associated with the non-totally symmetric e_{2g} modes in the $\tilde{B}\,^2A_{2u}$–$\tilde{X}\,^2E_{1g}$ electronic transition.[11]

Renner–Teller Effect

Although linear molecules are not subject to the Jahn–Teller effect, they do experience another "name" effect—the Renner–Teller effect.[1,9,12] (Linear molecules are also

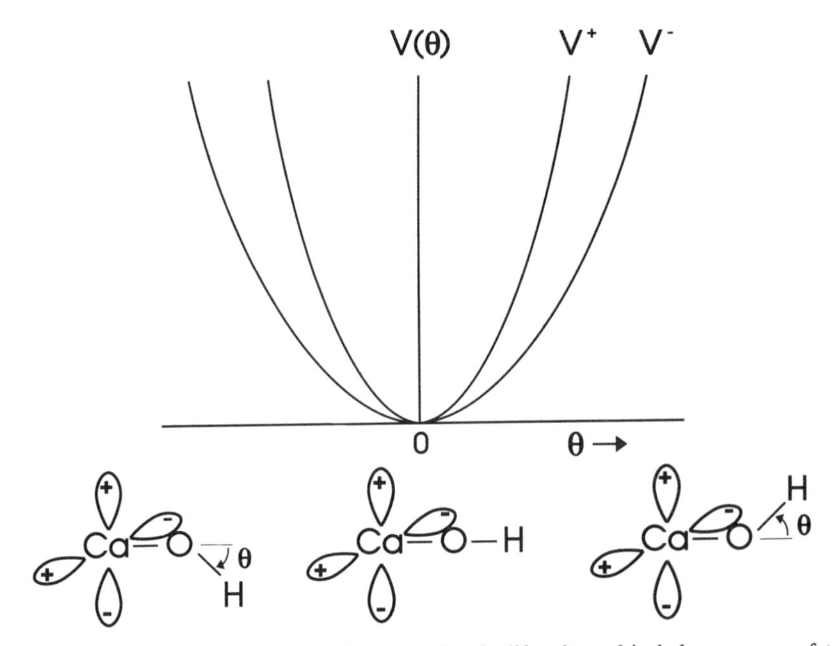

Figure 10.15. The bending motion of a linear molecule lifts the orbital degeneracy of the $\tilde{A}^2\Pi$ state of CaOH.

subject to the Herzberg–Teller effect.) The Renner–Teller effect is the interaction of vibrational and electronic angular momenta in a linear molecule. The levels associated with bending modes are shifted in energy by an interaction that couples vibrational motion to electronic motion for states in which $\Lambda \neq 0$ (i.e., $\Pi, \Delta, \Phi, \ldots$).

The Renner–Teller effect occurs because the double orbital degeneracy is lifted as a linear molecule bends during vibrational motion (Figure 10.15). As the linear molecule bends, the two potential curves V^+ and V^- (corresponding, for example, to the p orbital in the plane of the molecule and the p orbital out of the plane) become distinct. The combined vibrational and electronic motion on these two potential surfaces mixes the zeroth-order vibrational and electronic wavefunctions associated with the linear configuration. The Renner–Teller effect is again a breakdown of the Born–Oppenheimer approximation.

The bending motion of a linear molecule has vibrational angular momentum $l\hbar$ along the z-axis (for a state with v_2 bending quanta $l = v_2, v_2 - 2, \ldots, 0$ or 1). A linear molecule can also have electronic orbital angular momentum with a projection of $\Lambda\hbar$ along the z-axis. For the Renner–Teller effect the coupling of the two requires a new quantum number,

$$K = \Lambda + l. \tag{10.53}$$

Notice that K, Λ, and l are signed numbers, but by convention only the magnitudes are

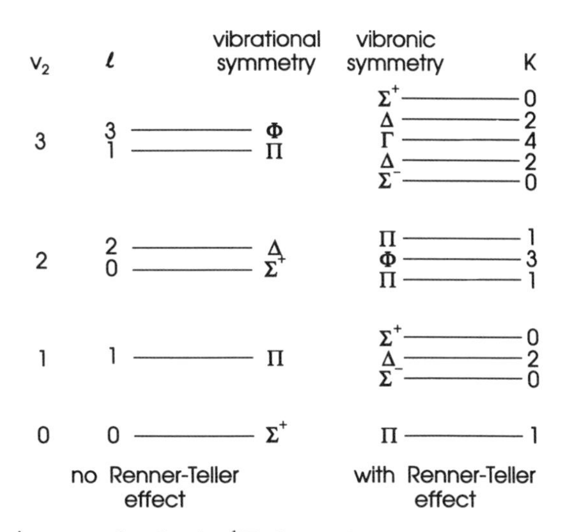

Figure 10.16. Vibronic energy levels of a $^1\Pi$ electronic state of a linear triatomic molecule. The vibrational symmetry is on the left, while the vibronic symmetry (vibrational$\otimes$electronic) is on the right.

usually quoted. The vibrational energy-level pattern appropriate for a $^1\Pi$ electronic state is shown in Figure 10.16. In Figure 10.16 the vibronic symmetries are obtained from the direct product of the vibrational symmetry with the electronic orbital symmetry (Π)

$$\Gamma^{\text{vib}} \otimes \Gamma^{\text{el}} = \Gamma^{\text{vibronic}}. \tag{10.54}$$

Without the Renner–Teller effect the excited bending energy levels in a triatomic molecule are given by the expression gl^2, where g ($=g_2$) has a typical value of a few cm^{-1} because of anharmonicity. With the Renner–Teller effect the vibrational pattern is more complex, with splittings that are typically on the order of tens or hundreds of cm^{-1}, depending upon the magnitude of the electronic–vibrational interaction.

Nonradiative Transitions: Jablonski Diagram

Molecules in liquids, solids, and gases can exchange energy through collisions. Energy deposited in a specific molecule soon dissipates throughout the system because of these *inter*molecular interactions. Interestingly, if energy is deposited in a large, isolated molecule it can also be dissipated through *intra*molecular interactions. For example, electronic excitation can be converted into vibrational excitation in a large molecule without collisions. The various possible processes that can occur for a large organic

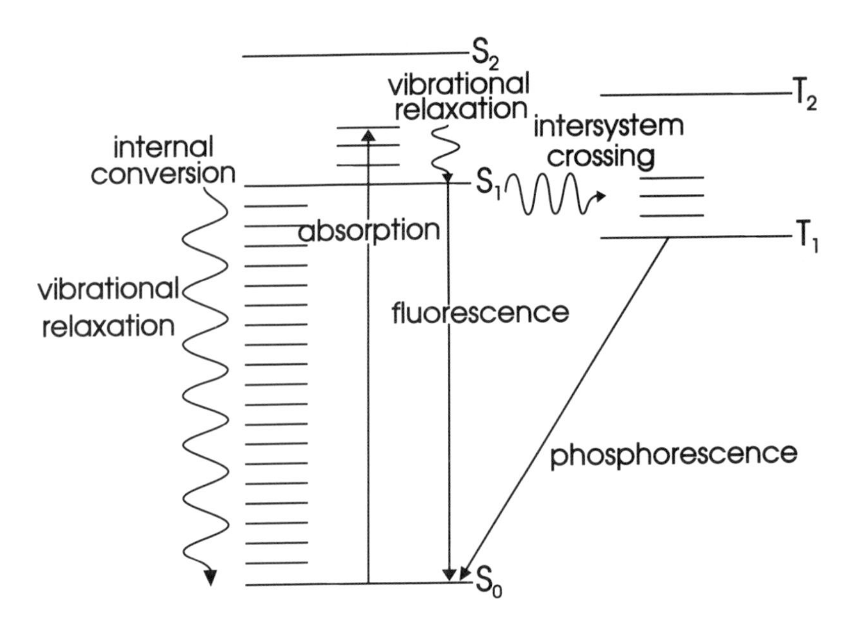

Figure 10.17. Photoprocesses in a large molecule: a Jablonski diagram.

molecule, such as benzene or naphthalene, are summarized in a Jablonski diagram (see Figure 10.17).

In the Jablonski diagram the ground state is a singlet and is labeled S_0 with $S_1, S_2, S_3, \ldots$ used for excited singlet states with increasing energy. For an aromatic molecule such as benzene the $S_1 \leftarrow S_0$ transition is typically a $\pi^* \leftarrow \pi$ excitation. For example, the first excited electronic configuration of benzene $(a_{2u})^2(e_{1g})^3(e_{2u})^1$ gives rise to singlet and triplet states; according to Hund's rules, the lowest energy triplet lies below the lowest energy singlet. The triplet states are labeled as $T_1, T_2, \ldots$ in order of increasing energy.

The fate of an absorbed photon in a large molecule can be described with the aid of the Jablonski diagram. The $S_1 \leftarrow S_0$ absorption is followed by rapid (picosecond) vibrational relaxation to the bottom of the S_1 state in any condensed phase. In the absence of collisions this particular vibrational relaxation process cannot occur without violating the principle of conservation of energy.

There are four possible fates of a large molecule in the electronic state S_1: reaction, fluorescence, internal conversion, or intersystem crossing. From a chemical point of view perhaps the most important possibility is the reaction of the excited molecule with itself or other molecules. This possibility is studied in the vast field of photochemistry and is beyond the scope of this book.

If the molecule in S_1 re-emits a photon, $S_1 \rightarrow S_0$, this process is known as *fluorescence.* If the emitted photon has the same energy as the absorbed photon then the process is known as *resonance fluoresence*; otherwise, if the emitted photon has less energy than the absorbed photon, then the process is known as *relaxed fluorescence.* Small, gas-phase molecules emit resonance fluorescence but large, condensed-phase molecules emit relaxed fluorescence. Fluorescence lifetimes are typically on the order of nanoseconds.

Singlet to triplet conversion for excited states is also possible in which the $S_1 \rightarrow T_1$ process is known as *intersystem crossing*. Emission from the triplet state back down to the ground state is weakly allowed by spin-orbit mixing. The $T_1 \rightarrow S_0$ emission is called *phosphorescence* and typically has a lifetime on the order of milliseconds to seconds for large organic molecules.

A molecule in S_1 can bypass the triplet manifold and directly transfer to high vibrational levels of the ground state. This nonradiative S_1–S_0 process is called *internal conversion*. The processes of internal conversion and intersystem crossing in an isolated gas-phase molecule can only occur because a high density of vibrational levels exists, allowing energy conservation to be satisfied exactly (within the limits of the uncertainty principle based on the lifetime of the excited state).

Photoelectron Spectroscopy

Photoelectron spectroscopy allows the orbital energies of a molecule to be measured directly. In photoelectron spectroscopy the molecule is bombarded with electromagnetic radiation of sufficient energy to ionize the molecule. Typically, vacuum ultraviolet radiation is used to liberate valence electrons, while x-ray radiation is used to dislodge core electrons. Valence photoelectron spectroscopy is of particular interest to chemists since this process provides a means whereby the energies of the bonding orbitals can be measured directly.

The He atom provides a convenient source of vacuum ultraviolet radiation. For instance, one source is the He ^1P–^1S resonance transition at 171,129.148 cm^{-1} (21.217 eV or 584 Å), which involves the promotion of the electron from $1s$ to $2p$. The energy of this transition is more than sufficient to ionize any outer valence electron

$$M + h\nu_0 \rightarrow M^+ + e \tag{10.55}$$

for any molecule. The excess energy that results from the energy difference of the photon minus the binding energy of the electron is distributed as internal energy in M^+ or as kinetic energy of the molecular ion M^+ and the electron. Since an electron is much lighter than a molecule, conservation of momentum requires that the ionized electron will move at a high speed relative to the molecule. In other words, nearly all of the kinetic energy is carried by the electron, so a measurement of the electron energy gives the internal energy of the molecule from the Einstein equation,

$$h\nu_0 = E_{\text{molecule}} + KE_e. \tag{10.56}$$

Ignoring vibrational and rotational energy for the moment, this means that if the photon energy ($h\nu_0$) is known and the electron kinetic energy is measured (KE_e), then the electron binding energy (E_{molecule}) in a particular electronic orbital of the molecule is determined. This means that the photoelectric effect directly measures the binding energy of an electron in a molecule.

For example, the valence photoelectron spectrum of H_2O recorded with He excitation is shown in Figure 10.18. As is customary, the spectrum is not plotted as a function of the measured electron kinetic energy (KE_e) but as a function of $h\nu_0 - KE_e$. In this way the scale directly reads the binding energy of the electron orbital (or the ionization energy of the orbital). The lowest energy ionization at 12.61 eV corresponds to the ionization from the nonbonding out-of-plane orbital of b_1 symmetry (see Figure 10.2). The next ionization (14–16 eV) corresponds to the in-plane H—O bonding orbital of a_1 symmetry. The removal of this bonding electron changes the geometry of the final H_2O^+ ion molecule considerably relative to the geometry of the ground-state molecule. Since the ionization process is very fast, the Franck–Condon principle applies and considerable vibrational structure is observed. The origin band for the ionization of the a_1 electron is at 13.7 eV. Finally, the removal of the in-plane b_2 (Figure 10.2) binding electron with 18–20 eV ionization energy also results in extensive vibrational structure near the origin at 17.2 eV. The final valence orbital at 32 eV, corresponding to the ionization of a $2s$ (a_1) O electron, is not shown in Figure 10.18 because the binding energy is greater than the 21.2 eV of energy that is available from the He *I* source.

The photoelectron spectrum of H_2O gives the binding energies of the four occupied valence molecular orbitals. The y-axis of the qualitative molecular orbital diagram (Figure 10.2) is thus made quantitative. The photoelectron spectrum of H_2O also suggests that molecular orbitals are more than figments of a quantum chemist's imagination. Orbitals exist and their properties can be measured.

Rotational Structure: H_2CO and HCN

The rotational energy levels of linear molecules, symmetric tops, spherical tops, and asymmetric tops have been discussed in previous chapters. The general selection rules have also been considered in the sections on infrared spectroscopy (Chapter 7) and electronic spectroscopy of diatomic molecules (Chapter 9).

Electronic spectra of polyatomic molecules display much more variety than is found in the infrared transitions of polyatomic molecules or electronic transitions of diatomic molecules. The main reason for this is the possibility of an electronic transition inducing large changes in the geometry. It is not uncommon for a molecule, such as HCN, to change point groups or to dissociate into fragments upon electronic excitation. HCN has a linear ground state, while the first excited electronic state has a bent structure so it is necessary to consider a bent-linear rotational level correlation diagram.

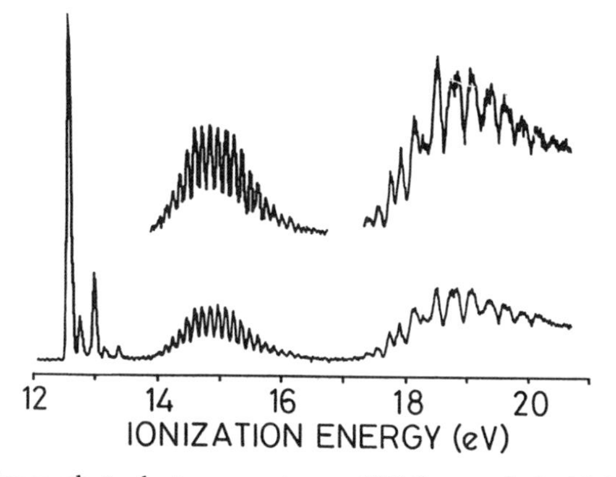

Figure 10.18. Valence photoelectron spectrum of H_2O recorded with He *I* excitation.

The ground state of HCN is $\tilde{X}\ {}^1\Sigma_g^+\ (C_{\infty v})$ with four degrees of vibrational freedom and two degrees of rotational freedom. The $\tilde{A}\ {}^1A''\ (C_s)$ excited electronic state has three degrees of vibrational freedom and three degrees of rotational freedom. It is necessary, therefore, to convert one bending vibration in linear HCN into rotation about the a-axis for bent HCN. A correlation diagram is very helpful in obtaining a qualitative understanding of how energy levels and quantum numbers must change as a function of bending angle. For bent HCN the energy levels are given approximately by (ignoring stretching vibrations)

$$BJ(J+1) + (A-B)K_a^2 + h\nu_b(v_b + \tfrac{1}{2}) \tag{10.57}$$

since the molecule is a near prolate top. In equation (10.57) the K_a quantum number gives the projection of rotational angular momentum about the top axis and $h\nu_b$ is the vibrational quantum of bending. For the linear molecule, the corresponding energy-level expression is

$$BJ(J+1) + gl^2 + h\nu(v+1). \tag{10.58}$$

In equation (10.58) the l quantum number gives the projection of vibrational angular momentum along the z-axis and $h\nu$ is the quantum of bending for the linear molecule. Since J is always a good quantum number, it must remain invariant to any changes in geometry. However, l is transformed into K as the molecule bends (Figure 10.19). The vibrational angular momentum $l\hbar$ becomes $K_a\hbar$ units of rotational angular momentum as the HCN molecule bends.

The $\tilde{A}\ {}^1A''-\tilde{X}\ {}^1\Sigma_g^+$ transition[1,13] of HCN occurs near 1800 Å. This transition shows a long progression in the ν_2 bending mode (and in ν_3, the CN stretch) in both absorption[13] and laser-excited emission. From the Franck–Condon principle, linear–linear transitions will not exhibit a progression in a bending mode. The bands are found to obey the selection rule $K' - l'' = \pm 1$ (in general, $K' - l''$ could be $0, \pm 1$) consistent

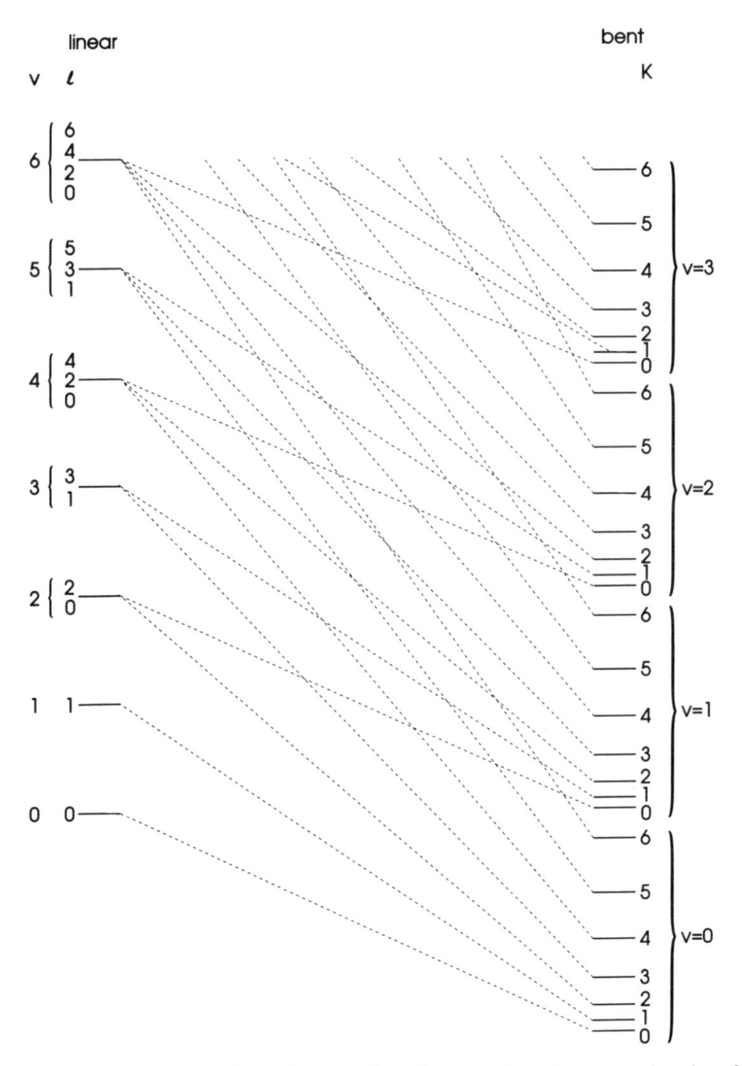

Figure 10.19. Bent-linear correlation diagram for the rotational energy levels of a triatomic molecule.

with the transition dipole moment pointing out of the plane of the bent molecule. The excited $\tilde{A}$ state must therefore have $\tilde{A}\ ^1A''$ symmetry.

The rotational structure of the bands is relatively simple because the excited state is approximately a prolate top with rotational spacings given by $BJ(J+1)$. The selection rule $\Delta J = 0, \pm 1$ appropriate for a perpendicular electronic transition gives a simple P, Q, R structure for each band (Figure 10.20).

Perhaps the most famous electronic spectrum is the $\tilde{A}\ ^1A_2$–$\tilde{X}\ ^1A_1$ transition of formaldehyde,[1,14] observed in the region 3530–2300 Å. The $\tilde{A}$–$\tilde{X}$ transition was the first electronic transition of an asymmetric top molecule in which the rotational structure was understood in great detail.[15] In addition formaldehyde is the simplest molecule

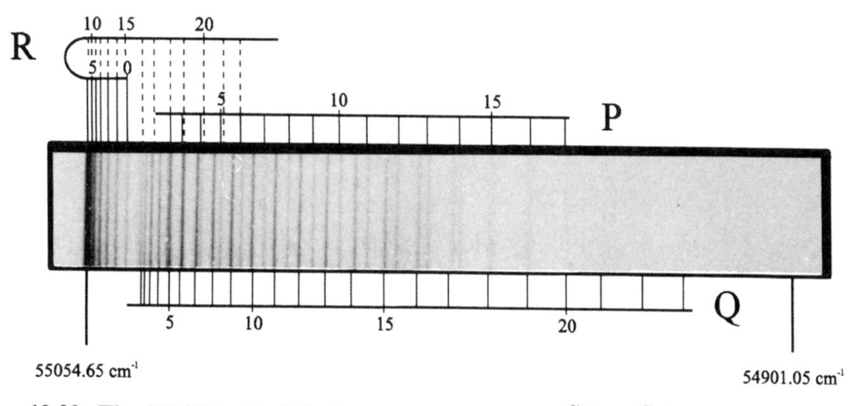

Figure 10.20. The $030(K' = 1)$–000 absorption band of the $\tilde{A}\ {}^1A''$–$\tilde{X}\ {}^1\Sigma^+$ transition of HCN.

with the carbonyl chromophore, and therefore serves as a prototype for more complex aldehydes and ketones.

The electronic structure of formaldehyde can be rationalized with a simple localized molecular orbital picture. The carbonyl chromophore has a set of localized molecular orbitals[15] given in Figure 10.21.

The six valence electrons (two of the four carbon electrons form the C—H bonds) give a $(5a_1)^2(1b_1)^2(2b_2)^2$ ground-state configuration. The HOMO–LUMO transition corresponds to the transfer of a nonbonding, in-plane O_{2p_y} $(2b_2)$ electron to an antibonding C—O π^* $(2b_1)$ orbital. This type of transition is associated with C=O,

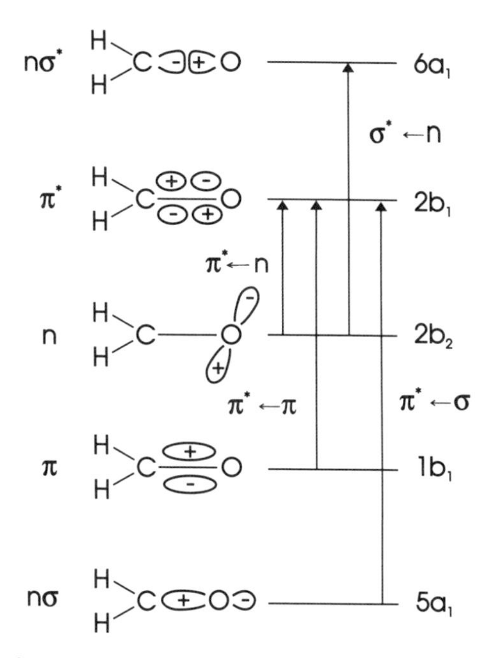

Figure 10.21. Localized molecular orbitals associated with the carbonyl chromophore of formaldehyde.

C=S, —N=O, $—NO_2$, and —O—N=O chromophores and is called an $\pi^* \leftarrow n$ transition. For H_2CO, the first excited-state configuration is $(5a_1)^2(1b_1)^2(2b_2)^1(2b_1)^1$, which gives rise to the $\tilde{A}\ ^1A_2$ and $\tilde{a}\ ^3A_2$ electronic states.

The $\tilde{A}\ ^1A_2$–$\tilde{X}\ ^1A_1$ electronic transition is electric-dipole forbidden because the A_2 irreducible representation is the only one in the C_{2v} point group not associated with one of the components of the dipole moment vector. The $\tilde{A}$–$\tilde{X}$ transition appears weakly through vibronic coupling with ν_4, the out-of-plane bending mode of b_1 symmetry (Figure 10.22). Since the direct product $A_2 \otimes B_1$ gives B_2, the $\tilde{A}\ ^1A_2$–$\tilde{X}\ ^1A_1$ transition borrows intensity from the allowed $\tilde{B}\ ^1B_2$–$\tilde{X}\ ^1A_1$ transition[15] near 1750 Å.

The observed $\tilde{A}\ ^1A_2$–$\tilde{X}\ ^1A_1$ laser excitation spectrum of formaldehyde[16] is shown in Figure 10.23. A tunable laser was used to excite the vibronic levels of the $\tilde{A}$ state of formaldehyde, and the resulting fluorescence was monitored as a function of wavelength.[16] The simple picture of the electronic structure of H_2CO described previously is unable to account for some features of the spectra. The long progression of 1182 cm^{-1} is assigned as the ν_2' carbonyl stretching mode. This long progression appears because of a lengthening of the C—O bond. The vibrational intervals associated with the out-of-plane bending mode in the excited electronic state (ν_4') are, however, very peculiar. In the $\tilde{A}$ state the frequency of the ν_4' mode is 125 cm^{-1}, as compared to the 1167 cm^{-1} value for the ν_4'' mode of the $\tilde{X}$ state.

Surprisingly, the rotational analysis of H_2CO preceded the vibronic analysis. From the moments of inertia, the inertial defect $\Delta(=I_C - I_A - I_B)$ was found to be -0.265 amu Å^2 in the $\tilde{A}$ state, but 0.057 amu Å^2 in the $\tilde{X}$ state.[14] For a perfectly rigid planar molecule, Δ would be exactly zero, but for a real planar molecule (such as the ground state of H_2CO), Δ is a small positive number because of vibrational and electronic motion.

The vibronic and rotational structure of the $\tilde{A}$ state formaldehyde can be understood if the $\tilde{A}$ state is allowed to be slightly nonplanar. This possibility is in accord with the prediction of the Walsh diagram for H_2CO (Figure 10.24). Population of the π^* orbital ($2b_1$) favors a pyramidal geometry. The peculiar vibrational intervals associated with ν_4' occur because of a barrier of 316 cm^{-1} to linearity. The vibrational energy-level pattern is consistent with a potential energy curve for the out-of-plane bending[17] as shown in Figure 10.25.

If the $\tilde{A}$ state of formaldehyde is pyramidal (C_s point group), then at first sight it would seem inappropriate to use the C_{2v} point group to label the electronic state and the vibrational levels. The molecule is only slightly nonplanar, however, with the $v_4' = 1$ vibrational level already above the barrier. Even below the barrier, with the molecule in the ground vibrational level ($v_4' = 0$), the molecule is rapidly inverting like ammonia. In fact, the $v_4' = 0$ to $v_4' = 1$ vibrational levels are pushed closer in energy as the barrier height increases and are analogous to the 0^+–0^- inversion doublet of NH_3 (see the section on inversion doubling in Chapter 7). The "average" structure (but not the equilibrium structure) is planar.

A rigorous theoretical analysis using permutation–inversion operations shows that a group called G_4 of order 4 should be used for the $\tilde{A}$ state.[15,18] However, G_4 is

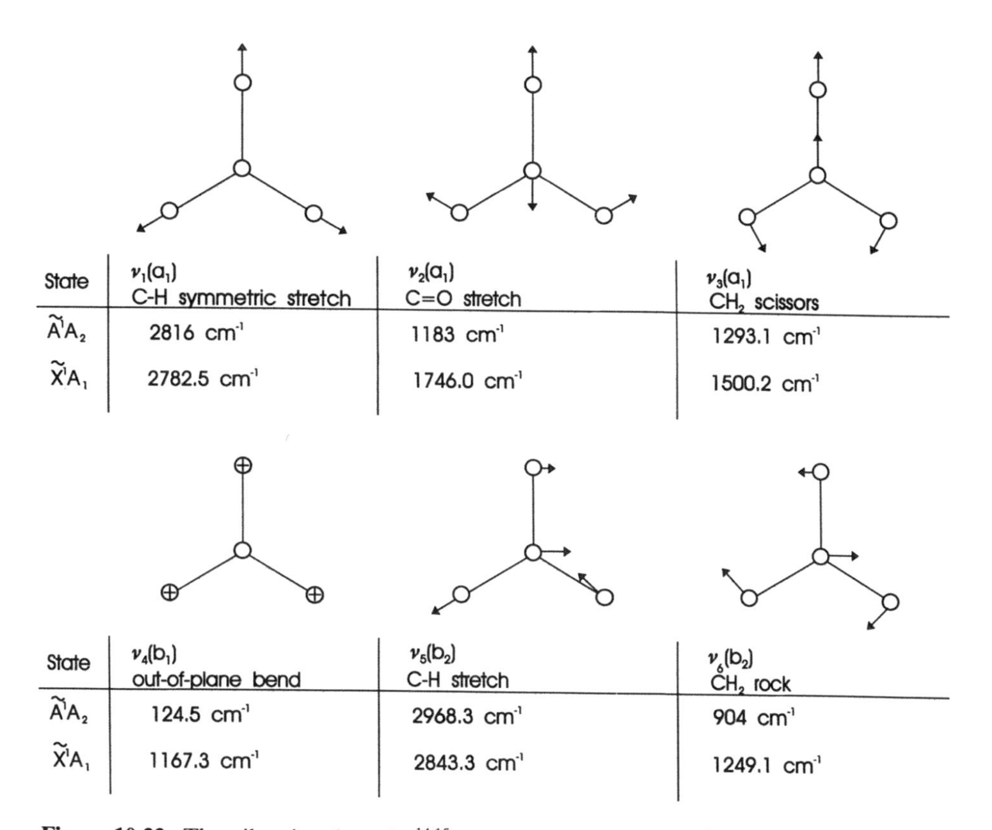

State	$\nu_1(a_1)$ C-H symmetric stretch	$\nu_2(a_1)$ C=O stretch	$\nu_3(a_1)$ CH_2 scissors
$\tilde{A}^1A_2$	2816 cm^{-1}	1183 cm^{-1}	1293.1 cm^{-1}
$\tilde{X}^1A_1$	2782.5 cm^{-1}	1746.0 cm^{-1}	1500.2 cm^{-1}

State	$\nu_4(b_1)$ out-of-plane bend	$\nu_5(b_2)$ C-H stretch	$\nu_6(b_2)$ CH_2 rock
$\tilde{A}^1A_2$	124.5 cm^{-1}	2968.3 cm^{-1}	904 cm^{-1}
$\tilde{X}^1A_1$	1167.3 cm^{-1}	2843.3 cm^{-1}	1249.1 cm^{-1}

Figure 10.22. The vibrational modes[14,15] of formaldehyde in the $\tilde{A}\ ^1A_2$ and $\tilde{X}\ ^1A_1$ states.

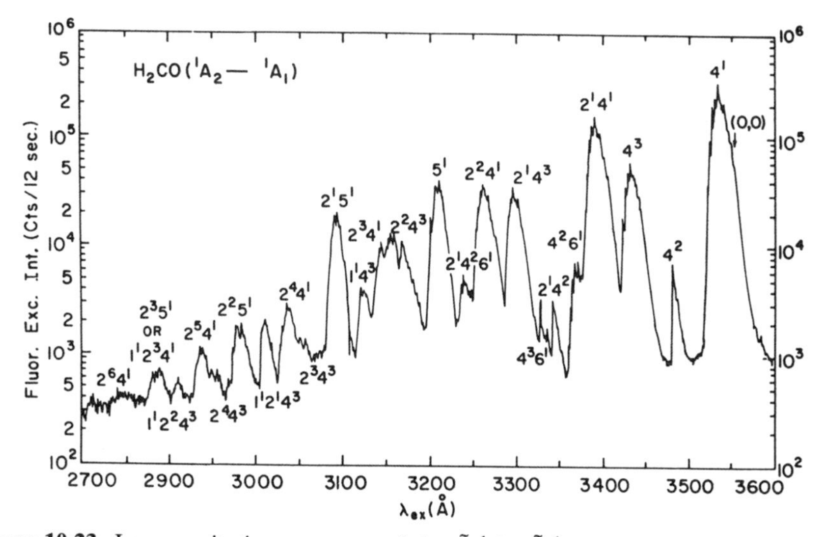

Figure 10.23. Laser excitation spectrum of the $\tilde{A}\ ^1A_2$–$\tilde{X}\ ^1A_1$ transition of formaldehyde.

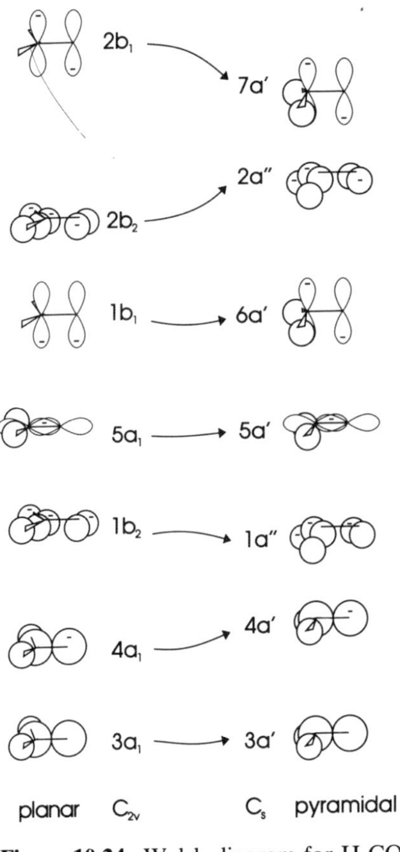

Figure 10.24. Walsh diagram for H_2CO.

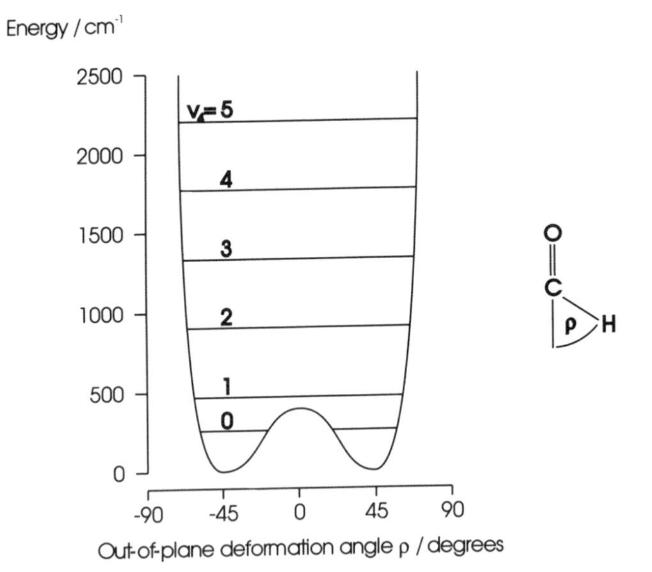

Figure 10.25. The potential energy curve as a function of the out-of-plane bending angle in the $\tilde{A}\ ^1A_2$ state of H_2CO.

isomorphic with C_{2v}, so there is no error in retaining the labels (and results!) derived from the C_{2v} point group. If the $\tilde{A}\ ^1A_2$ state of formaldehyde had a very high barrier to planarity (and no inversion), then the $\tilde{A}\ ^1A''$ label would have to be used.

The rotational analyses of the vibronic bands of the $\tilde{A}\ ^1A_2$–$\tilde{X}\ ^1A_1$ transition also have some peculiarities. H_2CO is a near prolate top with the C=O bond along the a-axis in both electronic states. Many of the strong absorption bands have a vibronic symmetry of B_2 in the $\tilde{A}$ state since the direct product $b_1(v_4) \otimes A_2(\tilde{A}\ ^1A_2)$ gives the B_2 irreducible representation. A B_2–A_1 vibronic transition is associated with the y-component of the transition dipole moment (μ_y). The transition moment therefore lies in the plane of the molecule, but is perpendicular to the a-axis (the C=O bond). This corresponds to the b-axis of the molecule (the c-axis is necessarily out of plane in any planar molecule) so that the strong bands are $\perp$, b-type bands.[14,15] The rotational selection rules are $\Delta K_a = \pm 1$, $\Delta K_c \pm 1$, $\Delta J = 0, \pm 1$ for these bands.

Some perpendicular, c-type bands can also be found in the absorption spectra[14,15] with the selection rules $\Delta K_a = \pm 1$, $\Delta K_c = 0$, $\Delta J = 0, \pm 1$. They must be associated with B_1–A_1 vibronic bands and contain odd quanta of $v_5'(b_2)$ and $v_6'(b_2)$, since the direct product $A_2 \otimes b_2$ gives the B_1 irreducible representation. These bands, in fact, account for as much as one-quarter of the strength of the formaldehyde $\tilde{A}$–$\tilde{X}$ electronic transition.

Finally a few very weak parallel bands (a-type) are found in the $\tilde{A}$–$\tilde{X}$ transition. For these bands the selection rules are $\Delta K_a = 0$, $\Delta K_c = \pm 1$, $\Delta J = 0, \pm 1$. Surprisingly, the origin band (0_0^0) is found in the spectrum with a-type selection rules. The origin band has A_2–A_1 vibronic character and is forbidden by electric dipole selection rules in a vibronically induced electronic transition. Magnetic dipole transitions are possible, however, and the magnetic dipole moment behaves like the rotations R_x, R_y, and R_z in the character table. Callomon and Innes[19] proved that the origin band of the $\tilde{A}\ ^1A_2$–$\tilde{X}\ ^1A_1$ transition has magnetic-dipole character.

Problems

1. Construct a planar-pyramidal correlation diagram (Walsh diagram) for an AH_3-type of molecule. Predict the geometries of the BH_3, CH_3, NH_3, and CH_3^- molecules.
2. With the aid of the Walsh diagram (Figure 10.4) predict the geometry and electronic symmetry of the BH_2, BH_2^+, and BH_2^- molecules. For BH_2 predict the lowest energy electronic transition.
3. Predict the ground electronic states for the following BAB-type molecules: CNC, NCN, BO_2, N_3, CO_2^+, and NO_2^+.
4. The ethylene molecule has a strong diffuse absorption beginning at 2100 Å and extending into the vacuum ultraviolet. A very long progression in the excited-state CH_2 torsional mode is observed. On the basis of simple π-molecular orbital theory

and the Franck–Condon principle, account for these observations and suggest an assignment for the electronic transition.

5. Construct a molecular orbital diagram for the π-electrons of the linear molecule C_4. For the C_4, C_4^+, and C_4^- molecules predict the ground electronic states.
6. The π-electrons of naphthalene ($C_{10}H_8$) can be considered to be confined to a rectangular box of dimension 4 Å by 7 Å ("particle-in-a-box").
   (a) Set up and solve the Schrödinger equation to find the energy levels.
   (b) Add the electrons to the energy-level diagram.
   (c) Which levels correspond to HOMO and LUMO? At what wavelength will the lowest energy transition occur?
7. Consider the π-electron system of the cyclopentadienyl radical, C_5H_5, taken to be of C_{5h} symmetry.
   (a) Determine to which irreducible representation the MOs belong.
   (b) Assuming orthogonality between the $2p_z$ atomic orbitals, determine the normalized MOs.
   (c) Use the Hückel approximation to determine the MO energies.
   (d) What are the ground- and first excited-state configurations for the π-electron system?
   (e) What transitions will be symmetry allowed?
8. Consider the hydrogen peroxide molecule H—O—O—H (assume C_{2h} symmetry).
   (a) For the basis set of ten atomic orbitals including the $1s$ on the two hydrogen atoms, and $2s$, $2p_x$, $2p_y$, and $2p_z$ on the oxygens, determine the symmetries of the ten molecular orbitals.
   (b) Use the projection operator technique to obtain functions that form bases for these irreducible representations.
   (c) Specify the irreducible representations corresponding to each of the molecular orbitals in the following table of intermediate neglect of differential overlap (INDO) eigenvalues and eigenvectors:

Eigenvalues	−1.610	−1.278	−0.772	−0.752	−0.706	−0.577	−0.543	0.165	0.322	0.342 au
	1	2	3	4	5	6	7	8	9	10
1 O_1 $2s$	−0.642	−0.581	0.300	0.040	0.000	−0.218	0.000	−0.260	0.196	−0.069
2 O_1 $2p_x$	0.138	−0.171	0.051	−0.555	0.000	−0.396	0.000	0.550	0.125	−0.406
3 O_1 $2p_y$	−0.064	−0.160	−0.574	−0.282	0.000	0.502	0.000	−0.224	0.406	−0.308
4 O_1 $2p_z$	0.000	0.000	0.000	0.000	−0.707	0.000	−0.707	0.000	0.000	0.000
5 O_2 $2s$	−0.642	0.581	−0.300	0.040	0.000	−0.218	0.000	0.260	0.196	0.069
6 O_2 $2p_x$	−0.138	−0.171	0.051	0.555	0.000	0.396	0.000	0.550	−0.125	−0.406
7 O_2 $2p_y$	0.064	−0.160	−0.574	0.282	0.000	−0.502	0.000	−0.224	−0.406	−0.308
8 O_2 $2p_z$	0.000	0.000	0.000	0.000	−0.707	0.000	0.707	0.000	0.000	0.000
9 H_1 $1s$	−0.254	−0.328	−0.278	−0.332	0.000	0.210	0.000	0.282	−0.530	0.485
10 H_2 $1s$	−0.254	0.328	0.279	−0.332	0.000	0.210	0.000	−0.282	−0.530	−0.485

   (d) Determine the symmetries of the excited configurations formed by promoting an electron from the highest occupied MO to the lowest empty (number 8) and to the next highest (number 9) MO. Will the electronic transitions from the ground (A_g) state to excited states of these symmetries be allowed? Explain your reasoning.

9. For the *s-cis*-butadiene molecule $CH_2{=}CH{-}CH{=}CH_2$ of C_{2v} symmetry:
   (a) What are the symmetries of the molecular orbitals for the π-electrons?
   (b) Using Hückel theory, derive the energies and wavefunctions of the π-molecular orbitals.
   (c) What are the ground-state electronic symmetries of the cation, neutral, and anion of *s-cis*-butadiene?
10. For the allyl free radical CH_2CHCH_2 of C_{2v} symmetry, what are the symmetries of the π-orbitals? For the π-electrons, derive the Hückel molecular orbitals and energies. What is the ground-state π-electron configuration and the electronic symmetry of the allyl free radical? (Pick the x-axis to be out of the plane of the molecule.)

References

1. Herzberg, G. *Electronic Spectra and Electronic Structure of Polyatomic Molecules,* Van Nostrand Reinhold, New York, 1966.
2. Greenwood, N. N., and Earnshaw, A. *Chemistry of the Elements,* Pergamon, Oxford, 1984, pp. 125–126.
3. Michalopoulos, D. L., Geusic, M. E., Langridge-Smith, P. R. R., and Smalley, R. E. J. Chem. Phys. **80,** 3556 (1984).
4. Levine, I. N. *Quantum Chemistry,* 4th ed., Prentice-Hall, Englewood Cliffs, N.J., 1991, pp. 549–572.
5. Wilson, E. B., Decius, J. C., and Cross, P. C. *Molecular Vibrations,* Dover, New York, 1980, pp. 240–272.
6. Miller, F. A. J. Raman Spectrosc. **19,** 219 (1988).
7. Mulliken, R. S. J. Chem. Phys. **23,** 1997 (1955).
8. Riedle, E., Neusser, H. J., and Schlag, E. W. J. Chem. Phys. **75,** 4231 (1981).
9. Fischer, G. *Vibronic Coupling,* Academic Press, London, 1984.
10. Jahn, H. A. and Teller, E. Proc. Roy. Soc. London, Ser. A., **161,** 220 (1937).
11. Miller, T. A. and Bondybey, V. E. In *Molecular Ions: Spectroscopy, Structure and Chemistry,* T. A. Miller and V. E. Bondybey, Eds., North-Holland, Amsterdam, 1983.
12. Jungen, C. and Merer, A. J. In *Molecular Spectroscopy: Modern Research,* Vol. II, K. N. Rao, Ed., Academic Press, New York, 1976, p. 127.
13. Herzberg, G. and Innes, K. K. Can. J. Phys., **35,** 842 (1957).
14. Clouthier, D. J. and Ramsay, D. A. Annu. Rev. Phys. Chem. **34,** 31 (1983).
15. Moule, D. C., and Walsh, A. D. Chem. Rev. **75,** 67 (1975).
16. Miller, R. G. and Lee, E. K. C. Chem. Phys. Lett. **33,** 104 (1975).
17. Jensen, P. and Bunker, P. R. J. Mol. Spectrosc. **94,** 114 (1982).
18. Bunker, P. R. *Molecular Symmetry and Spectroscopy,* Academic Press, New York, 1979.
19. Callomon, J. H., and Innes, K. K., J. Mol. Spectrosc. **10,** 166 (1963).

General References

Bishop, D. M. *Group Theory and Chemistry,* Dover, New York, 1993.

Bunker, P. R. *Molecular Symmetry and Spectroscopy,* Academic Press, New York, 1979.

DeKock, R. L. and Gray, H. B. *Chemical Structure and Bonding,* Benjamin, Menlo Park, Calif., 1980.

Douglas, B. E. and Hollingsworth, C. A. *Symmetry in Bonding and Spectra,* Academic Press, New York, 1985.

Duncan, A. B. F. *Rydberg Series in Atoms and Molecules,* Academic Press, New York, 1971.

Fischer, G. *Vibronic Coupling,* Academic Press, London, 1984.

Gimarc, B. M. *Molecular Structure and Bonding,* Academic Press, New York, 1979.

Graybeal, J. D. *Molecular Spectroscopy,* McGraw-Hill, New York, 1988.

Harris, D. C. and Bertolucci, M. D. *Symmetry and Spectroscopy,* Dover, New York, 1989.

Herzberg, G. *Electronic Spectra and Electronic Structure of Polyatomic Molecules,* Van Nostrand Reinhold, New York, 1966.

Herzberg, G. *The Spectra and Structure of Simple Free Radicals,* Dover, New York, 1988.

Hirota, E. *High Resolution Spectroscopy of Transient Molecules,* Springer-Verlag, Berlin, 1985.

Hollas, J. M. *High Resolution Spectroscopy,* Butterworths, London, 1982.

Hout, R. F., Pietro, W. J., and Hehre, W. J. *A Pictorial Approach to Molecular Structure and Reactivity,* Wiley, New York, 1984.

Jorgensen, W. L. and Salem, L. *The Organic Chemist's Book of Orbitals,* Academic Press, New York, 1973.

King, G. W. *Spectroscopy and Molecular Structure,* Holt, Reinhart & Winston, New York, 1964.

McGlynn, S. P., Vanquickenborne, L. G., Kinoshita, M., and Carroll, D. G. *Introduction to Applied Quantum Chemistry,* Holt, Reinhart & Winston, New York, 1972.

Rabalais, J. W. *Principles of Ultraviolet Photoelectron Spectroscopy,* Wiley, New York, 1977.

Robin, M. B. *Higher Excited States of Polyatomic Molecules,* Academic Press, New York, 1974.

Steinfeld, J. I. *Molecules and Radiation,* 2nd. Ed., MIT Press, Cambridge, Mass., 1985.

Struve, W. S. *Fundamentals of Molecular Spectroscopy,* Wiley, New York, 1989.

Appendix A
Units, Conversions, and Physical Constants

The Cohen and Taylor (1986) recommended values of the fundamental physical constants. The digits in parentheses are the one standard deviation uncertainty in the last digits of the given value.

Quantity	Symbol	Value	Units
Speed of light in vacuum	c	299792458	$\mathrm{m\,s^{-1}}$
Permeability of vacuum	μ_0	$4\pi \times 10^{-7}$	$\mathrm{N\,A^{-2}}$
		$= 12.566370614\ldots$	$10^{-7}\,\mathrm{N\,A^{-2}}$
Permittivity of vacuum	ε_0	$1/\mu_0 c^2$	
		$= 8.854187817\ldots$	$10^{-12}\,\mathrm{F\,m^{-1}}$
Newtonian constant of gravitation	G	6.67259(85)	$10^{-11}\,\mathrm{m^3\,kg^{-1}\,s^{-2}}$
Planck constant	h	6.6260755(40)	$10^{-34}\,\mathrm{J\,s}$
$h/2\pi$	$\hbar$	1.05457266(63)	$10^{-34}\,\mathrm{J\,s}$
Elementary charge	e	1.60217733(49)	$10^{-19}\,\mathrm{C}$
Bohr magneton, $e\hbar/2m_e$	μ_B	9.2740154(31)	$10^{-24}\,\mathrm{J\,T^{-1}}$
Nuclear magneton, $e\hbar/2m_p$	μ_N	5.0507866(17)	$10^{-27}\,\mathrm{J\,T^{-1}}$
Proton mass	m_p	1.6726231(10)	$10^{-27}\,\mathrm{kg}$
		1.007276470(12)	amu
Proton–electron mass ratio	m_p/m_e	1836.152701(37)	
Proton magnetic moment	μ_p	1.41060761(47)	$10^{-26}\,\mathrm{J\,T^{-1}}$
Proton gyromagnetic ratio	γ_p	26,752.2128(81)	$10^{4}\,\mathrm{s^{-1}\,T^{-1}}$
Fine structure constant, $\frac{1}{2}\mu_0 c e^2/h$	α	7.29735308(33)	10^{-3}
Rydberg constant, $\frac{1}{2}m_e c\alpha^2/h$	R_∞	10,973,731.534(13)	$\mathrm{m^{-1}}$
In hertz, $R_\infty c$		3.2898419499(39)	$10^{15}\,\mathrm{Hz}$
In joules, $R_\infty hc$		2.1798741(13)	$10^{-18}\,\mathrm{J}$
In electron volts, $R_\infty hc/\{e\}$		13.6056981(40)	eV
Bohr radius, $\alpha/4\pi R_\infty$	a_0	0.529177249(24)	$10^{-10}\,\mathrm{m}$
Hartree energy, $e^2/4\pi\varepsilon_0 a_0 = 2R_\infty hc$	E_h	4.3597482(26)	$10^{-18}\,\mathrm{J}$
in eV, $E_h/\{e\}$		27.2113961(81)	eV
Electron mass	m_e	9.1093897(54)	$10^{-31}\,\mathrm{kg}$
		5.48579903(13)	10^{-4} amu
Electron magnetic moment	μ_e	928.47701(31)	$10^{-26}\,\mathrm{J\,T^{-1}}$
In Bohr magnetons	μ_e/μ_B	1.001159653193(10)	
In nuclear magnetons	μ_e/μ_N	1838.282000(37)	
Electron g-factor, $2(1 + a_e)$	g_e	2.002319304386(20)	
Avogadro constant	N_A	6.0221367(36)	$10^{23}\,\mathrm{mol^{-1}}$

From E. R. Cohen and B. N. Taylor, The 1986 adjustment of fundamental physical constants, *CODATA Bull.*, No. 63, November 1986 (Pergamon Press, Elmsford, N.Y.).

Quantity	Symbol	Value	Units
Atomic mass constant			
$m_u = m(^{12}C)/12$	u, m_u, amu	1.6605402(10)	10^{-27} kg
Faraday constant	F	96,485.309(29)	C mol^{-1}
Molar gas constant	R	8.314510(70)	J mol^{-1} K^{-1}
Boltzmann constant, R/N_A	k	1.380658(12)	10^{-23} J K^{-1}
In electron volts, $k/\{e\}$		8.617385(73)	10^{-5} eV K^{-1}
In hertz, k/h		2.083674(18)	10^{10} Hz K^{-1}
In wavenumbers, k/hc		69.50387(59)	m^{-1} K^{-1}
Molar volume (ideal gas), RT/p			
$T = 273.15$ K, $p = 101{,}325$ Pa	V_m	22.41410(19)	L/mol
$T = 273.15$ K, $p = 100$ kPa	V_m	22.71108(19)	L/mol
Stefan–Boltzmann			
constant, $(\pi^2/60)k^4/\hbar^3c^2$	σ	5.67051(19)	10^{-8} W m^{-2} K^{-4}
First radiation constant, $2\pi hc^2$	c_1	3.7417748(22)	10^{-16} W m^2
Second radiation constant, hc/k	c_2	0.01438769(12)	m K
Wien displacement law constant,			
$b = \lambda_{max} T = c_2/4.96511423\ldots$	b	2.897756(24)	10^{-3} m K
Electron volt, $(e/C)\ J = \{e\}$ J	eV	1.60217733(49)	10^{-19} J
Standard atmosphere	atm	101,325	Pa
Standard acceleration of gravity	g_n	9.80665	m s^{-2}

Appendix B
Character Tables

C_s	$\hat{E}$	$\hat{\sigma}_h$		
A'	1	1	$x; y; R_z$	$x^2; y^2; z^2; xy$
A''	1	−1	$z; R_x; R_y$	$xz; yz$

C_i	$\hat{E}$	$\hat{i}$		
A_g	1	1	$R_x; R_y; R_z$	$x^2; y^2; z^2; xy; xz; yz$
A_u	1	−1	$x; y; z$	

C_2	$\hat{E}$	$\hat{C}_2$		
A	1	1	$z; R_z$	$x^2; y^2; z^2; xy$
B	1	−1	$x; y; R_x; R_y$	$xz; yz$

C_3	$\hat{E}$	$\hat{C}_3$	$\hat{C}_3^2$		$\varepsilon = \exp(2\pi i/3)$
A	1	1	1	$z; R_z$	$x^2 + y^2; z^2$
E	$\begin{Bmatrix} 1 \\ 1 \end{Bmatrix}$	$\begin{matrix} \varepsilon \\ \varepsilon^* \end{matrix}$	$\begin{matrix} \varepsilon^* \\ \varepsilon \end{matrix}$	$(x, y); (R_x, R_y)$	$(x^2 - y^2, xy); (xz, yz)$

C_4	$\hat{E}$	$\hat{C}_4$	$\hat{C}_2$	$\hat{C}_4^3$		
A	1	1	1	1	$z; R_z$	$x^2 + y^2; z^2$
B	1	−1	1	−1		$x^2 - y^2; xy$
E	$\begin{Bmatrix} 1 \\ 1 \end{Bmatrix}$	$\begin{matrix} i \\ -i \end{matrix}$	$\begin{matrix} -1 \\ -1 \end{matrix}$	$\begin{matrix} -i \\ i \end{matrix}$	$(x, y); (R_x, R_y)$	(xz, yz)

After D. Bishop, *Group Theory and Chemistry*, Dover, New York, 1993.

For groups which can be written as direct products $G = G_1 \otimes C_i$ or $G = G_1 \otimes C_s$, i.e. C_{nh}, D_{nh}, D_{3d}, D_{5d}, S_6, O_h, $D_{\infty h}$ and I_h, the order in which the irreducible representations are listed in the Herzberg character tables (Vol. II, *Infrared and Raman Spectra*; Vol. III, *Electronic Spectra of Polyatomic Molecules*) differs from that given here. By convention it is the order in the Herzberg character tables that is to be used in numbering normal modes of vibration.

C_5	$\hat{E}$	$\hat{C}_5$	$\hat{C}_5^2$	$\hat{C}_5^3$	$\hat{C}_5^4$		$\varepsilon = \exp(2\pi i/5)$
A	1	1	1	1	1	$z; R_z$	$x^2+y^2; z^2$
E_1	1	ε	ε^2	ε^{2*}	ε^*	$(x, y); (R_x, R_y)$	(xz, xy)
	1	ε^*	ε^{2*}	ε^2	ε		
E_2	1	ε^2	ε^*	ε	ε^{2*}		(x^2-y^2, xy)
	1	ε^{2*}	ε	ε^*	ε^2		

C_6	$\hat{E}$	$\hat{C}_6$	$\hat{C}_3$	$\hat{C}_2$	$\hat{C}_3^2$	$\hat{C}_6^5$		$\varepsilon = \exp(2\pi i/6)$
A	1	1	1	1	1	1	$z; R_z$	$x^2+y^2; z^2$
B	1	-1	1	-1	1	-1		
E_1	1	ε	$-\varepsilon^*$	-1	$-\varepsilon$	ε^*	$(x, y); (R_x, R_y)$	(xz, yz)
	1	ε^*	$-\varepsilon$	-1	$-\varepsilon^*$	ε		
E_2	1	$-\varepsilon^*$	$-\varepsilon$	1	$-\varepsilon^*$	$-\varepsilon$		(x^2-y^2, xy)
	1	$-\varepsilon$	$-\varepsilon^*$	1	$-\varepsilon$	$-\varepsilon^*$		

$D_2 = V$	$\hat{E}$	$\hat{C}_2(z)$	$\hat{C}_2(y)$	$\hat{C}_2(x)$		
A	1	1	1	1		$x^2; y^2; z^2$
B_1	1	1	-1	-1	$z; R_z$	xy
B_2	1	-1	1	-1	$y; R_y$	xz
B_3	1	-1	-1	1	$x; R_x$	yz

D_3	$\hat{E}$	$2\hat{C}_3$	$3\hat{C}_2'$		
A_1	1	1	1		$x^2+y^2; z^2$
A_2	1	1	-1	$z; R_z$	
E	2	-1	0	$(x, y); (R_x, R_y)$	$(x^2-y^2, xy); (xz, yz)$

D_4	$\hat{E}$	$2\hat{C}_4$	$\hat{C}_2$	$2\hat{C}_2'$	$2\hat{C}_2''$		
A_1	1	1	1	1	1		$x^2+y^2; z^2$
A_2	1	1	1	-1	-1	$z; R_z$	
B_1	1	-1	1	1	-1		x^2-y^2
B_2	1	-1	1	-1	1		xy
E	2	0	-2	0	0	$(x, y); (R_x, R_y)$	(xz, yz)

D_5	$\hat{E}$	$2\hat{C}_5$	$2\hat{C}_5^2$	$5\hat{C}_2'$		$\alpha = 72°$
A_1	1	1	1	1		$x^2+y^2; z^2$
A_2	1	1	1	-1	$z; R_z$	
E_1	2	$2\cos\alpha$	$2\cos 2\alpha$	0	$(x, y); (R_x, R_y)$	(xz, yz)
E_2	2	$2\cos 2\alpha$	$2\cos\alpha$	0		(x^2-y^2, xy)

D_6	$\hat{E}$	$2\hat{C}_6$	$2\hat{C}_3$	$\hat{C}_2$	$3\hat{C}_2'$	$3\hat{C}_2''$		
A_1	1	1	1	1	1	1		$x^2+y^2; z^2$
A_2	1	1	1	1	−1	−1	$z; R_z$	
B_1	1	−1	1	−1	1	−1		
B_2	1	−1	1	−1	−1	1		
E_1	2	1	−1	−2	0	0	$(x, y); (R_x, R_y)$	(xz, yz)
E_2	2	−1	−1	2	0	0		(x^2-y^2, xy)

C_{2v}	$\hat{E}$	$\hat{C}_2$	$\hat{\sigma}_v(xz)$	$\hat{\sigma}_v(yz)$		
A_1	1	1	1	1	z	$x^2; y^2; z^2$
A_2	1	1	−1	−1	R_z	xy
B_1	1	−1	1	−1	$x; R_y$	xz
B_2	1	−1	−1	1	$y; R_x$	yz

C_{3v}	$\hat{E}$	$2\hat{C}_3$	$3\hat{\sigma}_v$		
A_1	1	1	1	z	$x^2+y^2; z^2$
A_2	1	1	−1	R_z	
E	2	−1	0	$(x, y); (R_x, R_y)$	$(x^2-y^2, xy); (xz, yz)$

C_{4v}	$\hat{E}$	$2\hat{C}_4$	$\hat{C}_2$	$2\hat{\sigma}_v$	$2\hat{\sigma}_d$		
A_1	1	1	1	1	1	z	$x^2+y^2; z^2$
A_2	1	1	1	−1	−1	R_z	
B_1	1	−1	1	1	−1		x^2-y^2
B_2	1	−1	1	−1	1		xy
E	2	0	−2	0	0	$(x, y); (R_x, R_y)$	(xz, yz)

C_{5v}	$\hat{E}$	$2\hat{C}_5$	$2\hat{C}_5^2$	$5\hat{\sigma}_v$		$\alpha = 72°$
A_1	1	1	1	1	z	$x^2+y^2; z^2$
A_2	1	1	1	−1	R_z	
E_1	2	$2\cos\alpha$	$2\cos 2\alpha$	0	$(x, y); (R_x, R_y)$	(xz, yz)
E_2	2	$2\cos 2\alpha$	$2\cos\alpha$	0		(x^2-y^2, xy)

C_{6v}	$\hat{E}$	$2\hat{C}_6$	$2\hat{C}_3$	$\hat{C}_2$	$3\hat{\sigma}_v$	$3\hat{\sigma}_d$		
A_1	1	1	1	1	1	1	z	$x^2+y^2; z^2$
A_2	1	1	1	1	−1	−1	R_z	
B_1	1	−1	1	−1	1	−1		
B_2	1	−1	1	−1	−1	1		
E_1	2	1	−1	−2	0	0	$(x, y); (R_x, R_y)$	(xz, yz)
E_2	2	−1	−1	2	0	0		(x^2-y^2, xy)

C_{2h}	$\hat{E}$	$\hat{C}_2$	$\hat{i}$	$\hat{\sigma}_h$		
A_g	1	1	1	1	R_z	$x^2; y^2; z^2; xy$
B_g	1	−1	1	−1	$R_x; R_y$	$xz; yz$
A_u	1	1	−1	−1	z	
B_u	1	−1	−1	1	$x; y$	

C_{3h}	$\hat{E}$	$\hat{C}_3$	$\hat{C}_3^2$	$\hat{\sigma}_h$	$\hat{S}_3$	$\hat{S}_3^5$		$\varepsilon = \exp(2\pi i/3)$
A'	1	1	1	1	1	1	R_z	$x^2 + y^2; z^2$
E'	1	ε	ε^*	1	ε	ε^*	(x, y)	$(x^2 - y^2, xy)$
	1	ε^*	ε	1	ε^*	ε		
A''	1	1	1	−1	−1	−1	z	
E''	1	ε	ε^*	−1	$-\varepsilon$	$-\varepsilon^*$	(R_x, R_y)	(xz, yz)
	1	ε^*	ε	−1	$-\varepsilon^*$	$-\varepsilon$		

C_{4h}	$\hat{E}$	$\hat{C}_4$	$\hat{C}_2$	$\hat{C}_4^3$	$\hat{i}$	$\hat{S}_4^3$	$\hat{\sigma}_h$	$\hat{S}_4$		
A_g	1	1	1	1	1	1	1	1	R_z	$x^2 + y^2; z^2$
B_g	1	−1	1	−1	1	−1	1	−1		$x^2 - y^2; xy$
E_g	1	i	−1	$-i$	1	i	−1	$-i$	(R_x, R_y)	(xz, yz)
	1	$-i$	−1	i	1	$-i$	−1	i		
A_u	1	1	1	1	−1	−1	−1	−1	z	
B_u	1	−1	1	−1	−1	1	−1	1		
E_u	1	i	−1	$-i$	−1	$-i$	1	i	(x, y)	
	1	$-i$	−1	i	−1	i	1	$-i$		

C_{5h}	$\hat{E}$	$\hat{C}_5$	$\hat{C}_5^2$	$\hat{C}_5^3$	$\hat{C}_5^4$	$\hat{\sigma}_h$	$\hat{S}_5$	$\hat{S}_5^7$	$\hat{S}_5^3$	$\hat{S}_5^9$		$\varepsilon = \exp(2\pi i/5)$
A'	1	1	1	1	1	1	1	1	1	1	R_z	$x^2 + y^2; z^2$
E_1'	1	ε	ε^2	ε^{2*}	ε^*	1	ε	ε^2	ε^{2*}	ε^*	(x, y)	
	1	ε^*	ε^{2*}	ε^2	ε	1	ε^*	ε^{2*}	ε^2	ε		
E_2'	1	ε^2	ε^*	ε	ε^{2*}	1	ε^2	ε^*	ε	ε^{2*}		$(x^2 - y^2; xy)$
	1	ε^{2*}	ε	ε^*	ε^2	1	ε^{2*}	ε	ε^*	ε^2		
A''	1	1	1	1	1	−1	−1	−1	−1	−1	z	
E_1''	1	ε	ε^2	ε^{2*}	ε^*	−1	$-\varepsilon$	$-\varepsilon^2$	$-\varepsilon^{2*}$	$-\varepsilon^*$	(R_x, R_y)	(xz, yz)
	1	ε^*	ε^{2*}	ε^2	ε	−1	$-\varepsilon^*$	$-\varepsilon^{2*}$	$-\varepsilon^2$	$-\varepsilon$		
E_2''	1	ε^2	ε^*	ε	ε^{2*}	−1	$-\varepsilon^2$	$-\varepsilon^*$	$-\varepsilon$	$-\varepsilon^{2*}$		
	1	ε^{2*}	ε	ε^*	ε^2	−1	$-\varepsilon^{2*}$	$-\varepsilon$	$-\varepsilon^*$	$-\varepsilon^2$		

C_{6h}	$\hat{E}$	$\hat{C}_6$	$\hat{C}_3$	$\hat{C}_2$	$\hat{C}_3^2$	$\hat{C}_6^5$	$\hat{i}$	$\hat{S}_3^5$	$\hat{S}_6^5$	$\hat{\sigma}_h$	$\hat{S}_6$	$\hat{S}_3$		$\varepsilon = \exp(2\pi i/6)$
A_g	1	1	1	1	1	1	1	1	1	1	1	1	R_z	$x^2+y^2; z^2$
B_g	1	−1	1	−1	1	−1	1	−1	1	−1	1	−1		
E_{1g}	1	ε	$-\varepsilon^*$	−1	$-\varepsilon$	ε^*	1	ε	$-\varepsilon^*$	−1	$-\varepsilon$	ε^*	(R_x, R_y)	(xz, yz)
	1	ε^*	$-\varepsilon$	−1	$-\varepsilon^*$	ε	1	ε^*	$-\varepsilon$	−1	$-\varepsilon^*$	ε		
E_{2g}	1	$-\varepsilon^*$	$-\varepsilon$	1	$-\varepsilon^*$	$-\varepsilon$	1	$-\varepsilon^*$	$-\varepsilon$	1	$-\varepsilon^*$	$-\varepsilon$		(x^2-y^2, xy)
	1	$-\varepsilon$	$-\varepsilon^*$	1	$-\varepsilon$	$-\varepsilon^*$	1	$-\varepsilon$	$-\varepsilon^*$	1	$-\varepsilon$	$-\varepsilon^*$		
A_u	1	1	1	1	1	1	−1	−1	−1	−1	−1	−1	z	
B_u	1	−1	1	−1	1	−1	−1	1	−1	1	−1	1		
E_{1u}	1	ε	$-\varepsilon^*$	−1	$-\varepsilon$	ε^*	−1	$-\varepsilon$	ε^*	1	ε	$-\varepsilon^*$	(x, y)	
	1	ε^*	$-\varepsilon$	−1	$-\varepsilon^*$	ε	−1	$-\varepsilon^*$	ε	1	ε^*	$-\varepsilon$		
E_{2u}	1	$-\varepsilon^*$	$-\varepsilon$	1	$-\varepsilon^*$	$-\varepsilon$	−1	ε^*	ε	−1	ε^*	ε		
	1	$-\varepsilon$	$-\varepsilon^*$	1	$-\varepsilon$	$-\varepsilon^*$	−1	ε	ε^*	−1	ε	ε^*		

$D_{2h} = V_h$	$\hat{E}$	$\hat{C}_2(z)$	$\hat{C}_2(y)$	$\hat{C}_2(x)$	$\hat{i}$	$\hat{\sigma}(xy)$	$\hat{\sigma}(xz)$	$\hat{\sigma}(yz)$		
A_g	1	1	1	1	1	1	1	1		$x^2; y^2; z^2$
B_{1g}	1	1	−1	−1	1	1	−1	−1	R_z	xy
B_{2g}	1	−1	1	−1	1	−1	1	−1	R_y	xz
B_{3g}	1	−1	−1	1	1	−1	−1	1	R_x	yz
A_u	1	1	1	1	−1	−1	−1	−1		
B_{1u}	1	1	−1	−1	−1	−1	1	1	z	
B_{2u}	1	−1	1	−1	−1	1	−1	1	y	
B_{3u}	1	−1	−1	1	−1	1	1	−1	x	

D_{3h}	$\hat{E}$	$2\hat{C}_3$	$3\hat{C}_2$	$\hat{\sigma}_h$	$2\hat{S}_3$	$3\hat{\sigma}_v$		
A_1'	1	1	1	1	1	1		$x^2+y^2; z^2$
A_2'	1	1	−1	1	1	−1	R_z	
E'	2	−1	0	2	−1	0	(x, y)	(x^2-y^2, xy)
A_1''	1	1	1	−1	−1	−1		
A_2''	1	1	−1	−1	−1	1	z	
E''	2	−1	0	−2	1	0	(R_x, R_y)	(xz, yz)

D_{4h}	$\hat{E}$	$2\hat{C}_4$	$\hat{C}_2$	$2\hat{C}_2'$	$2\hat{C}_2''$	$\hat{i}$	$2\hat{S}_4$	$\hat{\sigma}_h$	$2\hat{\sigma}_v$	$2\hat{\sigma}_d$		
A_{1g}	1	1	1	1	1	1	1	1	1	1		$x^2+y^2; z^2$
A_{2g}	1	1	1	−1	−1	1	1	1	−1	−1	R_z	
B_{1g}	1	−1	1	1	−1	1	−1	1	1	−1		x^2-y^2
B_{2g}	1	−1	1	−1	1	1	−1	1	−1	1		xy
E_g	2	0	−2	0	0	2	0	−2	0	0	(R_x, R_y)	(xz, yz)
A_{1u}	1	1	1	1	1	−1	−1	−1	−1	−1		
A_{2u}	1	1	1	−1	−1	−1	−1	−1	1	1	z	
B_{1u}	1	−1	1	1	−1	−1	1	−1	−1	1		
B_{2u}	1	−1	1	−1	1	−1	1	−1	1	−1		
E_u	2	0	−2	0	0	−2	0	2	0	0	(x, y)	

D_{5h}	$\hat{E}$	$2\hat{C}_5$	$2\hat{C}_5^2$	$5\hat{C}_2$	$\hat{\sigma}_h$	$2\hat{S}_5$	$2\hat{S}_5^3$	$5\hat{\sigma}_v$		$\alpha = 72°$
A_1'	1	1	1	1	1	1	1	1		$x^2+y^2; z^2$
A_2'	1	1	1	−1	1	1	1	−1	R_z	
E_1'	2	$2\cos\alpha$	$2\cos 2\alpha$	0	2	$2\cos\alpha$	$2\cos 2\alpha$	0	(x, y)	
E_2'	2	$2\cos 2\alpha$	$2\cos\alpha$	0	2	$2\cos 2\alpha$	$2\cos\alpha$	0		$(x^2-y^2,, xy)$
A_1''	1	1	1	1	−1	−1	−1	−1		
A_2''	1	1	1	−1	−1	−1	−1	1	z	
E_1''	2	$2\cos\alpha$	$2\cos 2\alpha$	0	−2	$-2\cos\alpha$	$-2\cos 2\alpha$	0	(R_x, R_y)	(xz, yz)
E_2''	2	$2\cos 2\alpha$	$2\cos\alpha$	0	−2	$-2\cos 2\alpha$	$-2\cos\alpha$	0		

D_{6h}	$\hat{E}$	$2\hat{C}_6$	$2\hat{C}_3$	$\hat{C}_2$	$3\hat{C}_2'$	$3\hat{C}_2''$	$\hat{i}$	$2\hat{S}_3$	$2\hat{S}_6$	$\hat{\sigma}_h$	$3\hat{\sigma}_d$	$3\hat{\sigma}_v$		
A_{1g}	1	1	1	1	1	1	1	1	1	1	1	1		$x^2+y^2; z^2$
A_{2g}	1	1	1	1	−1	−1	1	1	1	1	−1	−1	R_z	
B_{1g}	1	−1	1	−1	1	−1	1	−1	1	−1	1	−1		
B_{2g}	1	−1	1	−1	−1	1	1	−1	1	−1	−1	1		
E_{1g}	2	1	−1	−2	0	0	2	1	−1	−2	0	0	(R_x, R_y)	(xz, yz)
E_{2g}	2	−1	−1	2	0	0	2	−1	−1	2	0	0		(x^2-y^2, xy)
A_{1u}	1	1	1	1	1	1	−1	−1	−1	−1	−1	−1		
A_{2u}	1	1	1	1	−1	−1	−1	−1	−1	−1	1	1	z	
B_{1u}	1	−1	1	−1	1	−1	−1	1	−1	1	−1	1		
B_{2u}	1	−1	1	−1	−1	1	−1	1	−1	1	1	−1		
E_{1u}	2	1	−1	−2	0	0	−2	−1	1	2	0	0	(x, y)	
E_{2u}	2	−1	−1	2	0	0	−2	1	1	−2	0	0		

$D_{2d} = V_d$	$\hat{E}$	$2\hat{S}_4$	$\hat{C}_2$	$2\hat{C}_2'$	$2\hat{\sigma}_d$		
A_1	1	1	1	1	1		$x^2+y^2; z^2$
A_2	1	1	1	−1	−1	R_z	
B_1	1	−1	1	1	−1		x^2-y^2
B_2	1	−1	1	−1	1	z	xy
E	2	0	−2	0	0	$(x, y); (R_x, R_y)$	(xz, yz)

D_{3d}	$\hat{E}$	$2\hat{C}_3$	$3\hat{C}_2'$	$\hat{i}$	$2\hat{S}_6$	$3\hat{\sigma}_d$		
A_{1g}	1	1	1	1	1	1		$x^2+y^2; z^2$
A_{2g}	1	1	−1	1	1	−1	R_z	
E_g	2	−1	0	2	−1	0	(R_x, R_y)	$(x^2-y^2, xy); (xz, yz)$
A_{1u}	1	1	1	−1	−1	−1		
A_{2u}	1	1	−1	−1	−1	1	z	
E_u	2	−1	0	−2	1	0	(x, y)	

D_{4d}	$\hat{E}$	$2\hat{S}_8$	$2\hat{C}_4$	$2\hat{S}_8^3$	$\hat{C}_2$	$4\hat{C}_2'$	$4\hat{\sigma}_d$		
A_1	1	1	1	1	1	1	1		$x^2+y^2; z^2$
A_2	1	1	1	1	1	−1	−1	R_z	
B_1	1	−1	1	−1	1	1	−1		
B_2	1	−1	1	−1	1	−1	1	z	
E_1	2	$\sqrt{2}$	0	$-\sqrt{2}$	−2	0	0	(x, y)	
E_2	2	0	−2	0	2	0	0		(x^2-y^2, xy)
E_3	2	$-\sqrt{2}$	0	$\sqrt{2}$	−2	0	0	(R_x, R_y)	(xz, yz)

D_{5d}	$\hat{E}$	$2\hat{C}_5$	$2\hat{C}_5^2$	$5\hat{C}_2'$	$\hat{i}$	$2\hat{S}_{10}^3$	$2\hat{S}_{10}$	$5\hat{\sigma}_d$		$\alpha = 72°$
A_{1g}	1	1	1	1	1	1	1	1		$x^2+y^2; z^2$
A_{2g}	1	1	1	−1	1	1	1	−1	R_z	
E_{1g}	2	$2\cos\alpha$	$2\cos 2\alpha$	0	2	$2\cos\alpha$	$2\cos 2\alpha$	0	(R_x, R_y)	(xz, yz)
E_{2g}	2	$2\cos 2\alpha$	$2\cos\alpha$	0	2	$2\cos 2\alpha$	$2\cos\alpha$	0		(x^2-y^2, xy)
A_{1u}	1	1	1	1	−1	−1	−1	−1		
A_{2u}	1	1	1	−1	−1	−1	−1	1	z	
E_{1u}	2	$2\cos\alpha$	$2\cos 2\alpha$	0	−2	$-2\cos\alpha$	$-2\cos 2\alpha$	0	(x, y)	
E_{2u}	2	$2\cos 2\alpha$	$2\cos\alpha$	0	−2	$-2\cos 2\alpha$	$-2\cos\alpha$	0		

D_{6d}	$\hat{E}$	$2\hat{S}_{12}$	$2\hat{C}_6$	$2\hat{S}_4$	$2\hat{C}_3$	$2\hat{S}_{12}^5$	$\hat{C}_2$	$6\hat{C}_2'$	$6\hat{\sigma}_d$		
A_1	1	1	1	1	1	1	1	1	1		$x^2+y^2; z^2$
A_2	1	1	1	1	1	1	1	−1	−1	R_z	
B_1	1	−1	1	−1	1	−1	1	1	−1		
B_2	1	−1	1	−1	1	−1	1	−1	1	z	
E_1	2	$\sqrt{3}$	1	0	−1	$-\sqrt{3}$	−2	0	0	(x, y)	
E_2	2	1	−1	−2	−1	1	2	0	0		(x^2-y^2, xy)
E_3	2	0	−2	0	2	0	−2	0	0		
E_4	2	−1	−1	2	−1	−1	2	0	0		
E_5	2	$-\sqrt{3}$	1	0	−1	$\sqrt{3}$	−2	0	0	(R_x, R_y)	(xz, yz)

S_4	$\hat{E}$	$\hat{S}_4$	$\hat{C}_2$	$\hat{S}_4^3$		
A	1	1	1	1	R_z	$x^2+y^2; z^2$
B	1	−1	1	−1	z	$x^2-y^2; xy$
E	$\{1$	i	−1	$-i\}$	$(x, y); (R_x, R_y)$	(xz, yz)
	$\{1$	$-i$	−1	$i\}$		

S_6	$\hat{E}$	$\hat{C}_3$	$\hat{C}_3^2$	$\hat{i}$	$\hat{S}_6^5$	$\hat{S}_6$		$\varepsilon = \exp(2\pi i/3)$
A_g	1	1	1	1	1	1	R_z	$x^2+y^2; z^2$
E_g	$\{1$	ε	ε^*	1	ε	$\varepsilon^*\}$	(R_x, R_y)	$(x^2-y^2, xy); (xz, yz)$
	$\{1$	ε^*	ε	1	ε^*	$\varepsilon\}$		
B_u	1	1	1	−1	−1	−1	z	
E_u	$\{1$	ε	ε^*	−1	$-\varepsilon$	$-\varepsilon^*\}$	(x, y)	
	$\{1$	ε^*	ε	−1	$-\varepsilon^*$	$-\varepsilon\}$		

S_8	$\hat{E}$	$\hat{S}_8$	$\hat{C}_4$	$\hat{S}_8^3$	$\hat{C}_2$	$\hat{S}_8^5$	$\hat{C}_4^3$	$\hat{S}_8^7$		$\varepsilon = \exp(2\pi i/8)$
A	1	1	1	1	1	1	1	1	R_z	$x^2+y^2; z^2$
B	1	−1	1	−1	1	−1	1	−1	z	
E_1	$\{1$	ε	i	$-\varepsilon^*$	−1	$-\varepsilon$	$-i$	$\varepsilon^*\}$	$(x, y); (R_x, R_y)$	
	$\{1$	ε^*	$-i$	$-\varepsilon$	−1	$-\varepsilon^*$	i	$\varepsilon\}$		
E_2	$\{1$	i	−1	$-i$	1	i	−1	$-i\}$		(x^2-y^2, xy)
	$\{1$	$-i$	−1	i	1	$-i$	−1	$i\}$		
E_3	$\{1$	$-\varepsilon^*$	$-i$	ε	−1	ε^*	i	$-\varepsilon\}$		(xz, yz)
	$\{1$	$-\varepsilon$	i	ε^*	−1	ε	$-i$	$-\varepsilon^*\}$		

T_d	$\hat{E}$	$8\hat{C}_3$	$3\hat{C}_2$	$6\hat{S}_4$	$6\hat{\sigma}_d$		
A_1	1	1	1	1	1		$x^2 + y^2 + z^2$
A_2	1	1	1	-1	-1		
E	2	-1	2	0	0		$(2z^2 - x^2 - y^2, x^2 - y^2)$
$T_1(F_1)$	3	0	-1	1	-1	(R_x, R_y, R_z)	
$T_2(F_2)$	3	0	-1	-1	1	(x, y, z)	(xy, xz, yz)

O	$\hat{E}$	$8\hat{C}_3$	$3\hat{C}_2$	$6\hat{C}_4$	$6\hat{C}_2'$		
A_1	1	1	1	1	1		$x^2 + y^2 + z^2$
A_2	1	1	1	-1	-1		
E	2	-1	2	0	0		$(2z^2 - x^2 - y^2, x^2 - y^2)$
$T_1(F_1)$	3	0	-1	1	-1	(R_x, R_y, R_z); (x, y, z)	
$T_2(F_2)$	3	0	-1	-1	1		(xy, xz, yz)

O_h	$\hat{E}$	$8\hat{C}_3$	$3\hat{C}_2$	$6\hat{C}_4$	$6\hat{C}_2'$	$\hat{i}$	$8\hat{S}_6$	$3\hat{\sigma}_h$	$6\hat{S}_4$	$6\hat{\sigma}_d$		
A_{1g}	1	1	1	1	1	1	1	1	1	1		$x^2 + y^2 + z^2$
A_{2g}	1	1	1	-1	-1	1	1	1	-1	-1		
E_g	2	-1	2	0	0	2	-1	2	0	0		$(2z^2 - x^2 - y^2, x^2 - y^2)$
$T_{1g}(F_{1g})$	3	0	-1	1	-1	3	0	-1	1	-1	(R_x, R_y, R_z)	
$T_{2g}(F_{2g})$	3	0	-1	-1	1	3	0	-1	-1	1		(xy, xz, yz)
A_{1u}	1	1	1	1	1	-1	-1	-1	-1	-1		
A_{2u}	1	1	1	-1	-1	-1	-1	-1	1	1		
E_u	2	-1	2	0	0	-2	1	-2	0	0		
$T_{1u}(F_{1u})$	3	0	-1	1	-1	-3	0	1	-1	1	(x, y, z)	
$T_{2u}(F_{2u})$	3	0	-1	-1	1	-3	0	1	1	-1		

$C_{\infty v}$	$\hat{E}$	$2\hat{C}(\phi)$	...	$\infty\hat{\sigma}_v$		
$\Sigma^+(A_1)$	1	1	...	1	z	$x^2 + y^2$; z^2
$\Sigma^-(A_2)$	1	1	...	-1	R_z	
$\Pi(E_1)$	2	$2\cos\phi$	...	0	(x, y); (R_x, R_y)	(xz, yz)
$\Delta(E_2)$	2	$2\cos 2\phi$	...	0		$(x^2 - y^2, xy)$
$\Phi(E_3)$	2	$2\cos 3\phi$	...	0		
...	...	...	...	...		

$D_{\infty h}$	$\hat{E}$	$2\hat{C}(\phi)$	...	$\infty\hat{\sigma}_v$	$\hat{i}$	$2\hat{S}(\phi)$	...	$\infty\hat{C}_2'$		
Σ_g^+	1	1	...	1	1	1	...	1		$x^2 + y^2$; z^2
Σ_g^-	1	1	...	-1	1	1	...	-1	R_z	
Π_g	2	$2\cos\phi$	...	0	2	$-2\cos\phi$	...	0	(R_x, R_y)	(xz, yz)
Δ_g	2	$2\cos 2\phi$	...	0	2	$2\cos 2\phi$	...	0		$(x^2 - y^2, xy)$
...	...	...	...	...	...	...	...	...		
Σ_u^+	1	1	...	1	-1	-1	...	-1	z	
Σ_u^-	1	1	...	-1	-1	-1	...	1		
Π_u	2	$2\cos\phi$	...	0	-2	$2\cos\phi$	...	0	(x, y)	
Δ_u	2	$2\cos 2\phi$	...	0	-2	$-2\cos 2\phi$	...	0		
...	...	...	...	...	...	...	...	...		

I_h	$\hat{E}$	$12\hat{C}_5$	$12\hat{C}_5^2$	$20\hat{C}_3$	$15\hat{C}_2$	$\hat{i}$	$12\hat{S}_{10}$	$12\hat{S}_{10}^3$	$20\hat{S}_6$	$15\hat{\sigma}$		
A_g	1	1	1	1	1	1	1	1	1	1		$x^2 + y^2 + z^2$
T_{1g}	3	$2\cos\frac{\pi}{5}$	$2\cos\frac{3\pi}{5}$	0	−1	3	$2\cos\frac{3\pi}{5}$	$2\cos\frac{\pi}{5}$	0	−1	(R_x, R_y, R_z)	
T_{2g}	3	$2\cos\frac{3\pi}{5}$	$2\cos\frac{\pi}{5}$	0	−1	3	$2\cos\frac{\pi}{5}$	$2\cos\frac{3\pi}{5}$	0	−1		
G_g	4	−1	−1	1	0	4	−1	−1	1	0		
H_g	5	0	0	−1	1	5	0	0	−1	1		$(2z^2 - x^2 - y^2, x^2 - y^2, xy, yz, xz)$
A_u	1	1	1	1	1	−1	−1	−1	−1	−1		
T_{1u}	3	$2\cos\frac{\pi}{5}$	$2\cos\frac{3\pi}{5}$	0	−1	−3	$-2\cos\frac{3\pi}{5}$	$-2\cos\frac{\pi}{5}$	0	1	(x, y, z)	
T_{2u}	3	$2\cos\frac{3\pi}{5}$	$2\cos\frac{\pi}{5}$	0	−1	−3	$-2\cos\frac{\pi}{5}$	$-2\cos\frac{3\pi}{5}$	0	1		
G_u	4	−1	−1	1	0	−4	1	1	−1	0		
H_u	5	0	0	−1	1	−5	0	0	1	−1		

Appendix C
Direct Product Tables

The antisymmetric product is in brackets. Since the tables are symmetric about the principal diagonal, the part below the diagonal is omitted.

C_s	A'	A''
A'	A'	A''
A''		A'

C_i	A_g	A_u
A_g	A_g	A_u
A_u		A_g

C_2, C_{2h}[1]	A	B
A	A	B
B		A

C_{2v}	A_1	A_2	B_1	B_2
A_1	A_1	A_2	B_1	B_2
A_2		A_1	B_2	B_1
B_1			A_1	A_2
B_2				A_1

D_2, D_{2h}[1]	A	B_1	B_2	B_3
A	A	B_1	B_2	B_3
B_1		A	B_3	B_2
B_2			A	B_1
B_3				A

C_3, C_{3h},[2] S_6[1]	A	E
A	A	E
E		$[A]+A+E$

After Harris, D. C. and Bertolucci, M. D. *Symmetry and Spectroscopy,* Dover, New York, 1989.

C_{3v}, D_3, D_{3d},[1] D_{3h}[2]	A_1	A_2	E
A_1	A_1	A_2	E
A_2		A_1	E
E			$A_1 + [A_2] + E$

C_4, C_{4h},[1] S_4	A	B	E
A	A	B	E
B		A	E
E			$[A] + A + 2B$

$C_{4v}, D_4, D_{2d}, D_{4h}$[1]	A_1	A_2	B_1	B_2	E
A_1	A_1	A_2	B_1	B_2	E
A_2		A_1	B_2	B_1	E
B_1			A_1	A_2	E
B_2				A_1	E
E					$A_1 + [A_2] + B_1 + B_2$

C_5, C_{5h}[2]	A	E_1	E_2
A	A	E_1	E_2
E_1		$[A] + A + E_2$	$E_1 + E_2$
E_2			$[A] + A + E_1$

C_{5v}, D_5, D_{5d},[1] D_{5h}[2]	A_1	A_2	E_1	E_2
A_1	A_1	A_2	E_1	E_2
A_2		A_1	E_1	E_2
E_1			$A_1 + [A_2] + E_2$	$E_1 + E_2$
E_2				$A_1 + [A_2] + E_1$

C_6, C_{6h}[1]	A	B	E_1	E_2
A	A	B	E_1	E_2
B		A	E_2	E_1
E_1			$[A] + A + E_2$	$2B + E_1$
E_2				$[A] + A + E_2$

C_{6v}, D_6, D_{6h}[1]	A_1	A_2	B_1	B_2	E_1	E_2
A_1	A_1	A_2	B_1	B_2	E_1	E_2
A_2		A_1	B_2	B_1	E_1	E_2
B_1			A_1	A_2	E_2	E_1
B_2				A_1	E_2	E_1
E_1					$A_1 + [A_2] + E_2$	$B_1 + B_2 + E_1$
E_2						$A_1 + [A_2] + E_2$

D_{6d}	A_1	A_2	B_1	B_2	E_1	E_2	E_3	E_4	E_5
A_1	A_1	A_2	B_1	B_2	E_1	E_2	E_3	E_4	E_5
A_2		A_1	B_2	B_1	E_1	E_2	E_3	E_4	E_5
B_1			A_1	A_2	E_5	E_4	E_3	E_2	E_1
B_2				A_1	E_5	E_4	E_3	E_2	E_1
E_1					$A_1+[A_2]+E_2$	E_1+E_3	E_2+E_4	E_3+E_5	$B_1+B_2+E_4$
E_2						$A_1+[A_2]+E_4$	E_1+E_5	$B_1+B_2+E_2$	E_3+E_5
E_3							$A_1+[A_2]+B_1+B_2$	E_1+E_5	E_2+E_4
E_4								$A_1+[A_2]+E_4$	E_1+E_3
E_5									$A_1+[A_2]+E_2$

O, O_h,[1] T_d	A_1	A_2	E	T_1	T_2
A_1	A_1	A_2	E	T_1	T_2
A_2		A_1	E	T_2	T_1
E			$A_1+[A_2]+E$	T_1+T_2	T_1+T_2
T_1				$A_1+E+[T_1]+T_2$	$A_2+E+T_1+T_2$
T_2					$A_1+E+[T_1]+T_2$

$C_{\infty v}$, $D_{\infty h}$[1]	Σ^+	Σ^-	Π	Δ	Φ	Γ	. . .
Σ^+	Σ^+	Σ^-	Π	Δ	Φ	Γ	
Σ^-		Σ^+	Π	Δ	Φ	Γ	
Π			$\Sigma^++[\Sigma^-]+\Delta$	$\Pi+\Phi$	$\Delta+\Gamma$	$\Phi+H$	
Δ				$\Sigma^++[\Sigma^-]+\Gamma$	$\Pi+H$	$\Delta+I$	
Φ					$\Sigma^++[\Sigma^-]+I$	$\Pi+\Theta$	
Γ						$\Sigma^++[\Sigma^-]+K$	
. . .							

I_h[1]	A	T_1	T_2	G	H
A	A	T_1	T_2	G	H
T_1		$A+[T_1]+H$	$G+H$	T_2+G+H	T_1+T_2+G+H
T_2			$A+[T_2]+H$	T_1+G+H	T_1+T_2+G+H
G				$A+[T_1]+[T_2]+G+H$	T_1+T_2+G+2H
H					$A+[T_1]+[T_2]+[G]+G+2H$

K, K_h[1]	S	P	D	F	. . .
S	S	P	D	F	
P		$S+[P]+D$	$P+D+F$	$D+F+G$	
D			$S+[P]+D+[F]+G$	$P+D+F+G+H$	
F				$S+[P]+D+[F]+G+[H]+I$	
. . .					

[1] Add the g-u selection rules, viz., $g \times g = g$; $g \times u = u$; $u \times u = g$.

[2] Add the prime-double prime selection rules, viz., $'x' = '$; $'x'' = ''$; $''x'' = '$.

Appendix D
Introductory Textbooks Covering All of Spectroscopy

In addition to more specialized monographs, there are a number of textbooks that cover the entire field of molecular spectroscopy at an introductory level. Unfortunately many of them are out of print and copies are difficult to obtain.

Atkins, P. W. *Molecular Quantum Mechanics,* 2nd ed., Oxford University Press, Oxford, 1983.

Barrow, G. M. *Introduction to Spectroscopy,* McGraw-Hill, Singapore, 1962.

Banwell, C. N. *Fundamentals of Molecular Spectroscopy,* McGraw-Hill, London, 1983.

Bransden, B. H. and Joachain, C. J. *Physics of Atoms and Molecules,* Longmans, London, 1983.

Chang, R. *Basic Principles of Spectroscopy,* Krieger, Malabar, Fla., 1978.

Dixon, R. N. *Spectroscopy and Structure,* Methuen, London, 1965.

Flygare, W. H. *Molecular Structure and Dynamics,* Prentice-Hall, Englewood Cliffs, N.J., 1978.

Graybeal, J. D. *Molecular Spectroscopy,* McGraw-Hill, New York, 1988.

Harmony, M. D. *Introduction to Molecular Energies and Spectra,* Holt, Reinhart & Winston, New York, 1972.

Harris, D. C. and Bertolucci, M. D. *Symmetry and Spectroscopy,* Dover, New York, 1989.

Hollas, J. M. *Modern Spectroscopy,* 2nd ed., Wiley, Chichester, UK, 1992.

Hollas, J. M. *High Resolution Spectroscopy,* Butterworths, London, 1982.

Karplus, M. and Porter, R. N. *Atoms and Molecules,* Benjamin, New York, 1970.

King, G. W. *Spectroscopy and Molecular Structure,* Holt, Reinhart & Winston, New York, 1964.

Levine, I. N. *Molecular Spectroscopy,* Wiley, New York, 1975.

Richards, W. G. and Scott, P. R. *Structure and Spectra of Molecules,* Wiley, Chichester, UK, 1985.

Steinfeld, J. I. *Molecules and Radiation,* 2nd ed. MIT Press, Cambridge, Mass., 1985.

Struve, W. S. *Fundamentals of Molecular Spectroscopy,* Wiley, New York, 1989.

Thorne, A. P. *Spectrophysics,* 2nd ed., Chapman & Hall, London, 1988.

Walker, S. and Straw, H. *Spectroscopy,* Vols. I, II, and III, Chapman & Hall, London, 1976.

Weissbluth, M. *Atoms and Molecules,* Academic Press, New York, 1978.

Whiffen, D. H. *Spectroscopy,* 2nd ed., Longmans, London, 1972.

Figure Acknowledgments

Figure 1.2. After Figure 0-1 of Harris and Bertolucci.
Figure 1.4. Thanks to S. C. Liao and Dr. J. F. Ogilvie, Academia Sinica.
Figure 1.9. After Figure 2.1 of Letokhov and Chebotayev.
Figure 1.10. After Figure 2.30 of Demtröder.
Figure 1.20. After Figure 2.18 of Svelto.
Figure 2.11. After Table 3–7.1 of Bishop.
Figure 3.5. After Figure 5–2.3 of Bishop.
Figure 5.1. Thanks to G. Herzberg and I. Dabrowski.
Figure 5.13. After Figure 28 of Herzberg, *Atoms.*
Figure 5.14. After Figure 27 of Herzberg, *Atoms.*
Figure 5.15. After Figure 32 of Herzberg, *Atoms.*
Figure 6.15. Reproduced, with permission, from Fleming, J. W., and Chamberlain, J., *Infrared Phys.* **14,** 277 (1974).
Figure 6.19. Thanks to Professor H. W. Kroto, University of Sussex.
Figure 6.22. After Figure 27(a) of Herzberg's *IR and Raman Spectra.*
Figure 6.27. After Figure 3.1 of Zare.
Figure 6.32. Thanks to Professor H. W. Kroto, University of Sussex.
Figure 7.5. Thanks to Professor R. J. Le Roy, University of Waterloo.
Figure 7.10. Reproduced, with permission, from Le Roy, R. J., in *Molecular Spectroscopy: A Specialist Periodical Report,* vol. 1, Barrow, R. F., Long, D. A., and Millen, D. J., eds., Chemical Society, London, 1973, Figure 1.
Figure 7.11. Reproduced, with permission, from Le Roy, R. J., in *Molecular Spectroscopy: A Specialist Periodical Report,* vol. 1, Barrow, R. F., Long, D. A., and Millen, D. J., eds., Chemical Society, London, 1973, Figure 8.
Figure 7.35. Reproduced, with permission, from Frum, C. I., Engleman, R., and Bernath, P. F., *J. Chem. Phys.* **95,** 1435 (1991).
Figure 7.38. After Figure 84 of Herzberg's *IR and Raman Spectra.*
Figure 7.40. After Figure 91 of Herzberg's *IR and Raman Spectra.*
Figure 7.42. After Figure 122 of Herzberg's *IR and Raman Spectra.*
Figure 7.47. After Figure 32 of Herzberg's *IR and Raman Spectra.*
Figure 7.48 and 7.49. After Figure 116 of Herzberg's *IR and Raman Spectra.*
Figure 7.50. After Figure 117 of Herzberg's *IR and Raman Spectra.*

Figure 7.51. After Figure 118 of Herzberg's *IR and Raman Spectra.*
Figure 7.52. After Figure 99 of Herzberg's *Polyatomics.*
Figure 7.54. After Figure 41 of Herzberg's *IR and Raman Spectra.*
Figure 7.56. After Figure 137 of Herzberg's *IR and Raman Spectra.*
Figure 7.58. After Figure 60 of Herzberg's *IR and Raman Spectra.*
Figure 7.62. After Figure 5.64 of Hollas.
Figures 8.4, 8.5, and 8.7. Thanks to R. Bartholomew and Prof. D. Irish, University of Waterloo.
Figure 8.11. Reproduced, with permission, from Hollas, Figure 4.33(a).
Figure 8.13. Reproduced, with permission, from Bendtsen, J., *J. Raman Spectrosc.* **2,** 133 (1974).
Figure 9.2. After Figure 4.27 of DeKock and Gray.
Figure 9.5. After Figure 195 of Herzberg's *Diatomics.*
Figure 9.6. Reproduced, with permission, from Tyte, D. C., and Nicholls, R. W., *Identification Atlas of Molecular Spectra,* vol. 2, 1964.
Figure 9.9. Thanks to Professor R. J. Le Roy, University of Waterloo.
Figure 9.14. Reproduced, with permission, from Pianalto, F. S., Brazier, C. R., O'Brien, L. C., and Bernath, P. F., *J. Mol. Spectrosc.* **132,** 80 (1988).
Figure 10.2. After Figure 5–7 of DeKock and Gray.
Figure 10.3. After Figure 5–4 of DeKock and Gray.
Figure 10.4. After Figure 6.63 of Hollas.
Figure 10.5. After Figure 3 in Chapter 7 of Gimarc.
Figure 10.10. After Figure 5–37 in DeKock and Gray.
Figure 10.11. Thanks to C. Chuaqui and Prof. R. J. Le Roy, University of Waterloo.
Figure 10.12. After Figure 6.92 of Hollas.
Figure 10.13. After Figure 7.13 of Struve.
Figure 10.14. After Figure 11 of Herzberg's *Polyatomics.*
Figure 10.17. After Figure 5–19 of Harris and Bertolucci.
Figure 10.18. Thanks to Professor D. Klapstein, St. Francis Xavier University.
Figure 10.19. After Figure 45 in Herzberg's *Polyatomics.*
Figure 10.20. Thanks to G. Herzberg and I. Dabrowski.
Figure 10.21. After Figure 10.21 of King.
Figure 10.22. After Figure 10.22 of King.
Figure 10.23. Reproduced, with permission, from Miller, R. G., and Lee, E. K. C., *Chem. Phys. Lett.* **33,** 104 (1975).
Figure 10.24. After Figure 3 of Chapter 5 of Gimarc.
Figure 10.25. After Figure 2 of Jensen, P., and Bunker, P. R., *J. Mol. Spectrosc.* **94,** 114 (1982).

References for Figure Acknowledgments

Bishop, D. M., *Group Theory and Chemistry,* Dover, New York, 1993.
DeKock, R. L. and Gray, H. B., *Chemical Structure and Bonding,* Benjamin, Menlo Park, Calif., 1980.

Demtröder, W., *Laser Spectroscopy,* Springer-Verlag, Berlin, 1982.

Herzberg, G., *Atomic Spectra and Atomic Structure,* Dover, New York, 1945.

Herzberg, G., *Spectra of Diatomic Molecules,* Van Nostrand–Reinhold, New York, 1950.

Herzberg, G., *Infrared and Raman Spectra,* Van Nostrand–Reinhold, New York, 1945.

Herzberg, G., *Electronic Spectra of Polyatomic Molecules,* Van Nostrand–Reinhold, New York, 1966.

Gimarc, B. M., *Molecular Structure and Bonding,* Academic Press, New York, 1979.

Harris, D. C. and Bertolucci, M. D., *Symmetry and Spectroscopy,* Dover, New York, 1989.

Hollas, J. M., *High Resolution Spectroscopy,* Butterworths, London, 1982.

King, G. W., *Spectroscopy and Molecular Structure,* Holt, Reinhart & Winston, New York, 1964.

Letokhov, V. S. and Chebotayev, V. P., *Nonlinear Laser Spectroscopy,* Springer-Verlag, Berlin, 1977.

Struve, W. S., *Fundamentals of Molecular Spectroscopy,* Wiley, New York, 1989.

Svelto, O., *Principles of Lasers,* 3rd ed., Plenum, New York, 1989.

Index